Introduction to **Research Methods in Psychology**

Visit the *Introduction to Research Methods in Psychology, third edition* Companion Website at **www.pearsoned.co.uk/howitt** to find valuable **student** learning material including:

- Overview: A short introduction to each chapter gives students a feel for the topics covered
- Multiple choice questions: A set of MCQ's for every chapter allow students to check knowledge and understanding
- Essay questions: Between 6–8 essay questions for every chapter help students to plan for coursework and exams
- Ethical dilemmas: 12 cases, each with different scenarios and questions, encourage students consider the wider implications of a research project
- Guide to statistical computations: A short guide to statistical tools and techniques for easy reference when online
- Roadmaps: A set of visual guides to help students find the right test to use to analyse a set of data

Introduction to Research Methods in Psychology

Third Edition

Dennis Howitt Loughborough University
Duncan Cramer Loughborough University

Prentice Hall
is an imprint of

Harlow, England • London • New York • Boston • San Francisco • Toronto • Sydney • Singapore • Hong Kong
Tokyo • Seoul • Taipei • New Delhi • Cape Town • Madrid • Mexico City • Amsterdam • Munich • Paris • Milan

Pearson Education Limited
Edinburgh Gate
Harlow
Essex CM20 2JE
England

and Associated Companies throughout the world

Visit us on the World Wide Web at:
www.pearsoned.co.uk

First published 2005
Second edition 2008
Third edition published 2011

ISBN 978-0-273-73499-4

British Library Cataloguing-in-Publication Data
A catalogue record for this book is available from the British Library

Library of Congress Cataloging-in-Publication Data
Howitt, Dennis.
Introduction to research methods in psychology / Dennis Howitt, Duncan Cramer. --
3rd ed.
p. cm.
Includes bibliographical references and index.
ISBN 978-0-273-72607-4
1. Psychology--Research--Methodology. I. Cramer, Duncan. II. Title.
BF76.5.H695 2011
150.72--dc22

2010036374

10 9 8 7 6 5 4 3 2 1
14 13 12 11 10

Typeset in 9.5/12pt Sabon by 35
Printed by Ashford Colour Press Ltd, Gosport

Brief contents

Contents

Supporting resources

Visit **www.pearsoned.co.uk/howitt** to find valuable online resources

Companion Website for students

- Overview: A short introduction to each chapter gives students a feel for the topics covered
- Multiple choice questions: A set of MCQ's for every chapter allow students to check knowledge and understanding
- Essay questions: Between 6–8 essay questions for every chapter help students to plan for coursework and exams
- Ethical dilemmas: 12 cases, each with different scenarios and questions, encourage students consider the wider implications of a research project
- Guide to statistical computations: A short guide to statistical tools and techniques for easy reference when online
- Roadmaps: A set of visual guides to help students find the right test to use to analyse a set of data

Also: The Companion Website provides the following features:

- Search tool to help locate specific items of content
- E-mail results and profile tools to send results of quizzes to instructors
- Online help and support to assist with website usage and troubleshooting

For more information please contact your local Pearson Education sales representative or visit **www.pearsoned.co.uk/howitt**

Guided tour

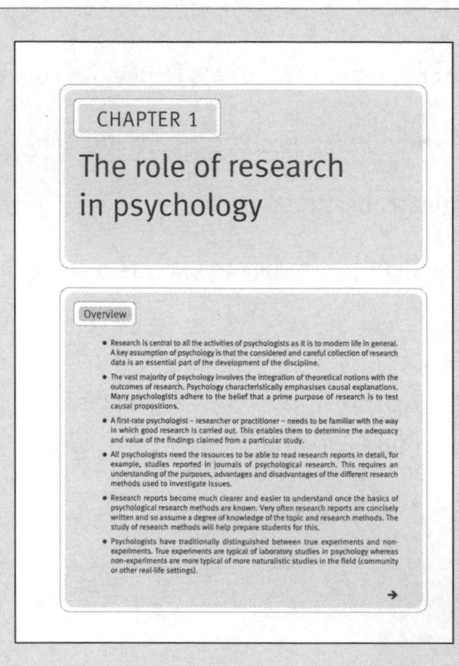

CHAPTER 1

The role of research in psychology

Overview

- Research is central to all the activities of psychologists as it is to modern life in general. A key assumption of psychology is that the considered and careful collection of research data is an essential part of the development of the discipline.
- The vast majority of psychology involves the integration of theoretical notions with the outcomes of research. Psychology characteristically emphasises causal explanations. Many psychologists adhere to the belief that a prime purpose of research is to test causal propositions.
- A first-rate psychologist – researcher or practitioner – needs to be familiar with the way in which good research is carried out. This enables them to determine the adequacy and value of the findings claimed from a particular study.
- All psychologists need the resources to be able to read research reports in detail, for example, studies reported in journals of psychological research. This requires an understanding of the purposes, advantages and disadvantages of the different research methods used to investigate issues.
- Research reports become much clearer and easier to understand once the basics of psychological research methods are known. Very often research reports are concisely written and so assume a degree of knowledge of the topic and research methods. The study of research methods will help prepare students for this.
- Psychologists have traditionally distinguished between true experiments and non-experiments. True experiments are typical of laboratory studies in psychology whereas non-experiments are more typical of more naturalistic studies in the field (community or other real-life settings).

Clear Overview

Introduces the chapter to give students a feel for the topics covered

CHAPTER 9 THE BASIC LABORATORY EXPERIMENT 169

belief that they had been given alcohol was the key causal variable. Since the effects of alcohol are well known, participants believing that they have taken alcohol may behave accordingly. By giving both groups alcohol, both groups will believe that they have taken alcohol. The only thing that varies is the key variable of the amount of alcohol taken. In good experimental research the effectiveness of the experimental manipulation is often evaluated. This is discussed in Box 9.1. In this case, participants in the experiment might be asked about whether or not they believed that they had taken alcohol in a debriefing interview at the end of the study.

The condition having the lower quantity of alcohol is referred to as the *control condition*. The condition having the higher quantity of alcohol may be called the *experimental condition*. The purpose of the control condition is to see how participants behave when they receive less of the variable that is being manipulated.

Box 9.1 Key Ideas

Checks on the experimental manipulation

It can be a grave mistake to assume that simply because an experimental manipulation has been introduced by the researcher that the independent *variable* has actually been effectively manipulated. It might be argued that if the researcher finds a difference between the experimental and control conditions on the dependent variable that the manipulation must have been effective. Things are not that simple.

Assume that we are investigating the effects of anger on memory. In order to manipulate anger, the researcher deliberately says certain *pre-scripted* offensive comments to the participants in the experimental group whereas nice things are said to the participants in the control group. It is very presumptuous to assume that this procedure will work effectively without subjecting it to some test.

For example, the participants might well in some circumstances regard the offensive comments as a joke rather than an insult so the manipulation may make them happier rather than angrier. Alternatively, the control may find the nice comments of the experimenter to be patronising and become somewhat annoyed or angry as a consequence. So there is a degree of uncertainty whether or not the experimental manipulation has actually worked. One relatively simple thing to do in this case would be to get participants to complete a questionnaire about their mood containing a variety of emotions, such as angry, happy and sad, which the participant rates in terms of their own feelings. In this way it would be possible to assess whether the experimental group was indeed angrier than the control group following the anger manipulation.

Alternatively, at the debriefing session following participation in the experiment, the participants could be asked about how they felt after the experimenter said the offensive or nice things. This check would also demonstrate that the manipulation had had a measurable effect on the participants' anger levels.

Sometimes it is appropriate, as part of pilot work trying out one's procedures prior to the study proper, to establish the effectiveness of the experimental manipulation as a distinct step in its own right. Researchers need to be careful not to assume that simply because they obtain statistically significant differences between the experimental and control conditions this is evidence of the effectiveness of their experimental manipulation. If the experimental manipulation has had an effect on the participants but not the one intended, it is vital that the researcher knows this. Otherwise, the conceptual basis for their analysis may be inappropriate. For example, they may be discussing the effects of anger when they should be discussing the effects of happiness.

In our experience, checks on the experimental manipulation are relatively rare in published research and are, probably, even rarer in student research. Yet such checks would seem to be essential. As we have seen, the debriefing session can be an ideal opportunity to interview participants about this aspect of the study along with its other features. The most thorough researchers may also consider a more objective demonstration of the effectiveness of the manipulation as above when the participants' mood was assessed.

Key Ideas

Outlines the important concepts in more depth to give you a fuller understanding

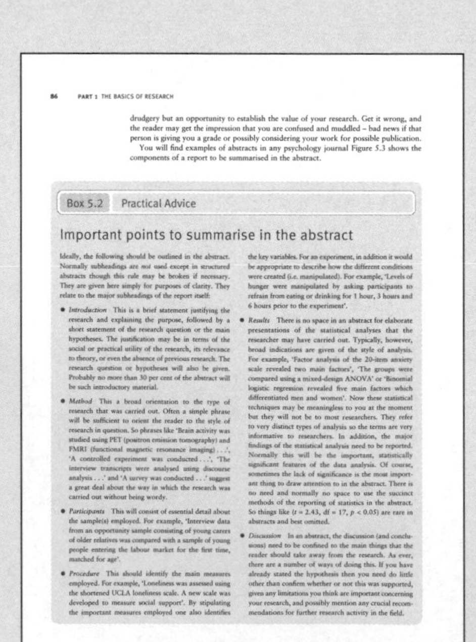

86 PART 1 THE BASICS OF RESEARCH

drudgery but an opportunity to establish the value of your research. Get it wrong, and the reader may get the impression that you are confused and muddled – bad news if that person is giving you a grade or possibly considering your work for possible publication.

You will find examples of abstracts in any psychology journal Figure 5.3 shows the components of a report to be summarised in the abstract.

Box 5.2 Practical Advice

Important points to summarise in the abstract

Ideally, the following should be outlined in the abstract. Normally subheadings are *not* used except in structured abstracts though this *rule* may be broken if necessary. They are given here simply for purposes of clarity. They relate to the major subheadings of the report itself:

- *Introduction* This is a brief statement justifying the research and explaining the purpose, followed by a short statement of the research question or the main hypotheses. The justification may be in terms of the social or practical utility of the research, its relevance to theory, or even the absence of previous research. The research question or hypotheses will also be given. Probably no more than 30 per cent of the abstract will be such introductory material.
- *Method* This a broad orientation to the type of research that was carried out. Often a simple phrase will be sufficient to orient the reader to the style of research in question. So phrases like 'Brain activity was studied using PET (positron emission tomography) and FMRI (functional magnetic resonance imaging) . . .', 'A controlled experiment was conducted . . .', 'The interview transcripts were analysed using discourse analysis . . .' and 'A survey was conducted . . .' suggest a great deal about the way in which the research was carried out without being wordy.
- *Participants* This will consist of essential detail about the sample(s) employed. For example, 'Interview data from an opportunity sample consisting of young carers of older relatives was compared with a sample of young people entering the labour market for the first time, matched for age'.
- *Procedure* This should identify the main measures employed. For example, 'Loneliness was assessed using the shortened UCLA loneliness scale. A new scale was developed to measure social support'. By stipulating the important measures employed one also identifies the key variables. For an experiment, in addition it would be appropriate to describe how the different conditions were created (i.e. manipulated). For example, 'Levels of hunger were manipulated by asking participants to refrain from eating or drinking for 1 hour, 3 hours and 6 hours prior to the experiment'.
- *Results* There is no space in an abstract for elaborate presentations of the statistical analyses that the researcher may have carried out. Typically, however, broad indications are given of the style of analysis. For example, 'Factor analysis of the 20-item anxiety scale revealed two main factors', 'The groups were compared using a mixed-design ANOVA' or 'Binomial logistic regression revealed five main factors which differentiated men and women'. Now these statistical techniques may be meaningless to you at the moment but they will not be to most researchers. They refer to very distinct types of analysis so the terms are very informative to researchers. In addition, the major findings of the statistical analysis need to be reported. Normally this will be the important, statistically significant features of the data analysis. Of course, sometimes the lack of significance is the most important thing to draw attention to in the abstract. There is no need and normally no space to use the succinct methods of the reporting of statistics in the abstract. So things like (t = 2.43, df = 17, p < 0.05) are rare in abstracts and best omitted.
- *Discussion* In an abstract, the discussion (and conclusions) need to be confined to the main things that the reader should take away from the research. As ever, there are a number of ways of doing this. If you have already stated the hypothesis then you need do little other than confirm whether or not this was supported, given any limitations you think are important concerning your research, and possibly mention any crucial recommendations for further research activity in the field.

Practical Advice

Gives you handy hints and tips on how to carry out research in practice

368 PART 4 QUALITATIVE RESEARCH METHODS

Box 22.2 Research Example

Discourse analysis

In research on menstruation, Lovering (1995) talked with 11- and 12-year-old boys and girls in discussion groups. Among a range of topics included on her guide for conducting the discussions were issues to do with menstruation. These included questions such as: 'Have you heard of menstruation?'; 'What have you been told about it?'; 'What do you think happens when a woman menstruates?'; 'Why does it happen?'; and 'Who has told you?' (Lovering, 1995, p. 17). In this way relatively systematic material could be gathered in ways closer to ordinary conversation than would be generated by one-on-one interviews. She took detailed notes of her experiences as soon as possible after the discussion groups using a number of headings (p. 17):

- How she (Lovering) felt
- General emotional tone and reactions
- Non-verbal behaviour
- Content recalled
- Implications and thoughts.

This is a form of diary writing of the sort discussed already in relation to grounded theory. The difference perhaps is that she applied it to the data collection phase rather than the transcription phase. Lovering transcribed the tape-recording herself – eventually using the Jefferson system described in Chapter 19. She also employed a computer-based analysis program (of the sort that NVivo is the modern equivalent). Such a program does not do the analysis for you; it allows you to store and work with a lot of text, highlight or mark particular parts of the text, sort the text and print it out. All of these things can be achieved just using pencil and paper, but a computer is more convenient.

The next stage was to sort the text into a number of categories – initially, she had more than 50. She developed an analysis of the transcribed material partly based on her awareness of a debate about the ways in which male and female bodies are socially construed quite differently. Boys' physical development is regarded as a gradual and unproblematic process, whereas in girls the process is much more problematic. The following excerpts from a transcript illustrate this:

A: They [school teachers] don't talk about the boys very much only the girls = yes = yes.

A: It doesn't seem fair. They are laughing at us. Not much seems to happen to boys.

A: Girl all go funny shapes = yes = like that = yes.

A: Because the boys, they don't really . . . change very much. They just get a little bit bigger.

A: It feels like the girls go through all the changes because we are not taught anything about the boys REALLY.

(Lovering, 1995, pp. 23–4)

Menstruation was learnt about from other people – predominantly female teachers or mothers. Embarrassment dominated, and the impression created was that menstruation was not to be discussed or even mentioned as a consequence. Talk of female bodies and bodily functions by the youngsters features a great deal of sniggering. In contrast, when discussing male bodies things become more ordinary and more matter of fact. Furthermore, boys are also likely to use menstruation as a psychological weapon against girls. That is, menstruation is used to make jokes about and ridicule girls. In Lovering's analysis, this is part of male oppression of females: even in sex education lessons learning about menstruation is associated in girls' minds as being 'laughing at girls'.

Of course, many more findings emerged in this study. Perhaps what is important is the complexity of the process by which the analysis proceeds. It is not possible to say that if the researcher does this and then does that, a good analysis will follow. Nevertheless, it is easy to see how the researcher's ideas relate to key aspects of discourse analytic thinking. For example, the idea that menstruation is used as a weapon of oppression of females clearly has its roots in feminist sexual politics which suggests that males attempt to control females in many ways from domestic violence through rape to, in this example, sex education lessons. One could equally construe this as part of Edwards and Potter's (1993) discursive action model. This suggests, among other things, that in talk, conversation or text, one can see social action unfolding before one's eyes. One does not have to regard talk, text or conversation as the external

Research Example

Explores a real example of research being carried out, giving you an insight into the process

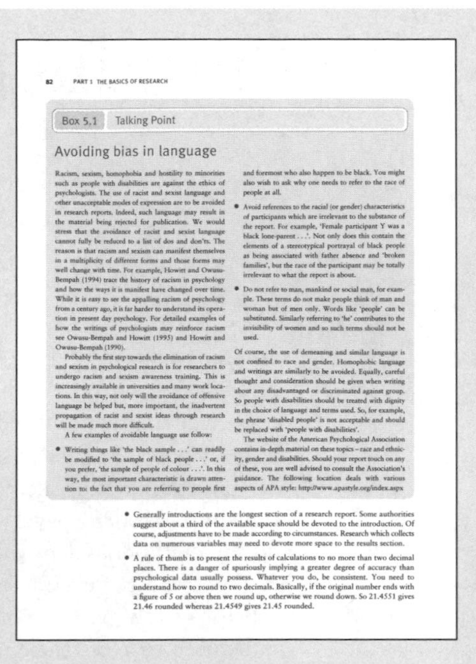

82 PART 1 THE BASICS OF RESEARCH

Box 5.1 Talking Point

Avoiding bias in language

Racism, sexism, homophobia and hostility to minorities such as people with disabilities are against the ethics of psychologists. The use of racist and sexist language and other unacceptable modes of expression are to be avoided in research reports. Indeed, such language may result in the material being rejected for publication. We would stress that the avoidance of racist and sexist language cannot fully be reduced to a list of dos and don'ts. The reason is that racism and sexism can manifest themselves in a multiplicity of different forms and those forms may well change with time. For example, Howitt and Owusu-Bempah (1994) trace the history of racism in psychology and how the ways it is manifest have changed over time. While it is easy to see the appalling racism of psychology from a century ago, it is far harder to understand its operation in present day psychology. For detailed examples of how the writings of psychologists may reinforce racism see Owusu-Bempah and Howitt (1995) and Howitt and Owusu-Bempah (1990).

Probably the first step towards the elimination of racism and sexism in psychological research is for researchers to undergo racism and sexism awareness training. This is increasingly available in universities and many work locations. In this way, not only will the avoidance of offensive language be helped but, more important, the inadvertent propagation of racist and sexist ideas through research will be made much more difficult.

A few examples of avoidable language use follow:

- Writing things like 'the black sample . . .' can readily be modified to 'the sample of black people . . .' or, if you prefer, 'the sample of people of colour . . .'. In this way, the most important characteristic is drawn attention to: the fact that you are referring to people first and foremost who also happen to be black. You might also wish to ask why one needs to refer to the race of people at all.
- Avoid references to the racial (or gender) characteristics of participants which are irrelevant to the substance of the report. For example, 'Female participant Y was a black lone-parent . . .'. Not only does this contain the elements of a stereotypical portrayal of black people as being associated with father absence and 'broken families', but the race of the participant may be totally irrelevant to what the report is about.
- Do not refer to man, mankind or social man, for example. These terms do not make people think of man and woman but of men only. Words like 'people' can be substituted. Similarly referring to 'he' contributes to the invisibility of women and so such terms should not be used.

Of course, the use of demeaning and similar language is not confined to race and gender. Homophobic language and writings are similarly to be avoided. Equally, careful thought and consideration should be given when writing about any disadvantaged or discriminated against group. So people with disabilities should be treated with dignity in the choice of language and terms used. So, for example, the phrase 'disabled people' is not acceptable and should be replaced with 'people with disabilities'.

The website of the American Psychological Association contains in-depth material on these topics – race and ethnicity, gender and disabilities. Should your report touch on any of these, you are well advised to consult the Association's guidance. The following location deals with various aspects of APA style: http://www.apastyle.org/index.aspx

- Generally introductions are the longest section of a research report. Some authorities suggest about a third of the available space should be devoted to the introduction. Of course, adjustments have to be made according to circumstances. Research which collects data on numerous variables may need to devote more space to the results section.
- A rule of thumb is to present the results of calculations to no more than two decimal places. There is a danger of spuriously implying a greater degree of accuracy than psychological data usually possess. Whatever you do, be consistent. You need to understand how to round to two decimals. Basically, if the original number ends with a figure of 5 or above then we round up, otherwise we round down. So 21.4551 gives 21.46 rounded whereas 21.4549 gives 21.45 rounded.

Talking Point

Investigates an important debate or issue in research

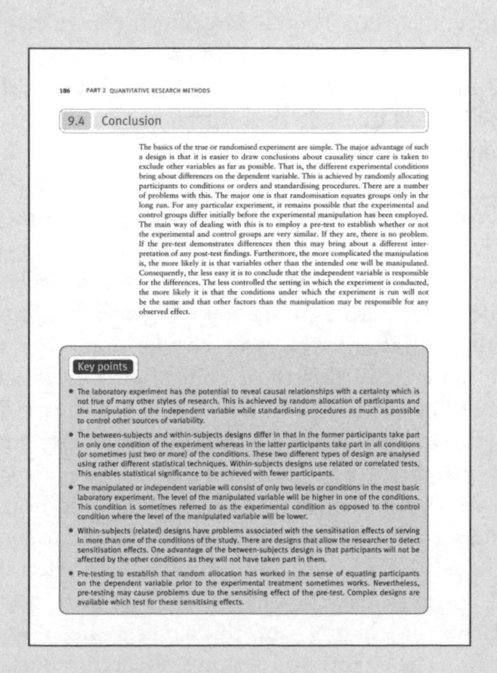

186 PART 2 QUANTITATIVE RESEARCH METHODS

9.4 Conclusion

The basics of the true or randomised experiment are simple. The major advantage of such a design is that it is easier to draw conclusions about causality since care is taken to exclude other variables as far as possible. That is, the different experimental conditions bring about differences on the dependent variable. This is achieved by randomly allocating participants to conditions or orders and standardising procedures. There are a number of problems with this. The major one is that randomisation equates groups only in the long run. For any particular experiment, it remains possible that the experimental and control groups differ initially before the experimental manipulation has been employed. The main way of dealing with this is to employ a pre-test to establish whether or not the experimental and control groups are very similar. If they are, there is no problem. If the pre-test demonstrates differences then this may bring about a different interpretation of any post-test findings. Furthermore, the more complicated the manipulation is, the more likely it is that variables other than the intended one will be manipulated. Consequently, the less easy it is to conclude that the independent variable is responsible for the differences. The less controlled the setting in which the experiment is conducted, the more likely it is that the conditions under which the experiment is run will not be the same and that other factors than the manipulation may be responsible for any observed effect.

Key points

- The laboratory experiment has the potential to reveal causal relationships with a certainty which is not true of many other styles of research. This is achieved by random allocation of participants and the manipulation of the independent variable while standardising procedures as much as possible to control other sources of variability.
- The between-subjects and within-subjects designs differ in that in the former participants take part in only one condition of the experiment whereas in the latter participants take part in all conditions (or sometimes just two or more) of the conditions. These two different types of design are analysed using rather different statistical techniques. Within-subjects designs use related or correlated tests. This enables statistical significance to be achieved with fewer participants.
- The manipulated or independent variable will consist of only two levels or conditions in the most basic laboratory experiment. The level of the manipulated variable will be higher in one of the conditions. This condition is sometimes referred to as the experimental condition as opposed to the control condition where the level of the manipulated variable will be lower.
- Within-subjects (related) designs have problems associated with the sensitisation effects of serving in more than one of the conditions of the study. There are designs that allow the researcher to detect sensitisation effects. One advantage of the between-subjects design is that participants will not be affected by the other conditions as they will not have taken part in them.
- Pre-testing to establish that random allocation has worked in the sense of equating participants on the dependent variable prior to the experimental treatment sometimes works. Nevertheless, pre-testing may cause problems due to the sensitising effect of the pre-test. Complex designs are available which test for these sensitising effects.

Conclusion/Key Points

Each chapter has a conclusion and set of key points to help summarise chapter coverage when you're revising a topic

24 PART 1 THE BASICS OF RESEARCH

ACTIVITIES

1. Choose a recent study that has been referred to either in a textbook you are reading or in a lecture that you have attended. Obtain the original publication. Were the study and its findings correctly reported in the textbook? Do you think that there were important aspects of the study that were not mentioned in the text or the lecture that should have been? If you do think there were important omissions, what are these? Why do you think they were not cited? Did the study test a causal proposition? If so, what was this proposition? If not, what was the main aim of this study? In terms of the designs outlined in this chapter what kind of design did the study use?
2. Either choose a chapter from a textbook or go to the library and obtain a copy of a single issue of a journal. Work through the material and for every study you find, classify it as one of the following:
   - correlational or cross-sectional study
   - longitudinal study
   - experiment – or study with randomised assignment.

   What percentage of each did you find?

Activities

Each chapter concludes with activities to help you test your knowledge and explore the issues further

Introduction

The third edition of *Introduction to Research Methods in Psychology* is one of three books designed to cover the major approaches to psychological research and analysis as they are currently practised. We do not believe that with this intention in mind, research methods and data analysis can be covered satisfactorily in a single volume, though you will find examples of successful textbooks based on this formula in almost any university bookshop. Modern psychology is extremely varied in the styles of research it employs and the methodological and statistical sophistication that it currently enjoys would have been undreamt of even just a few years ago. It does students a disservice to provide them with those few basics which once would have been sufficient but now are hopelessly inadequate to meet their needs. To our minds, the incredible progress of modern psychology means that teaching resources must struggle to keep up to date and to cope with the variety of different educational experiences provided by different universities. At heart, each volume in our trilogy *Introduction to Research Methods in Psychology*, *Introduction to Statistics in Psychology* and *Introduction to SPSS Statistics in Psychology* is modularly constructed. That is, we do not expect that all their contents will be covered by lecturers and other instructors. Instead, there is a menu of largely self-contained chapters from which appropriate selections can be made.

This is illustrated by the coverage of *Introduction to Research Methods in Psychology*. This is unusual in that both quantitative and qualitative research are covered in depth. These are commonly but, in our opinion, wrongly seen as alternative and incompatible approaches to psychological research. For some researchers, there may be an intellectual incompatibility between the two. From our perspective, it is vitally important that students understand the intellectual roots of the two traditions, how research is carried out in these traditions, and what each tradition is capable of achieving. We believe that the student who is so informed will be better placed to make intelligent and appropriate choices about the style of research appropriate for the research questions they wish to address. On its own, the qualitative material in this third edition effectively supports much of the qualitative research likely to be carried out today. There is as much detailed practical advice and theory as is available in most books on qualitative research methods. (If more is required, Dennis Howitt's *Introduction to Qualitative Research in Psychology* [Howitt, 2010] will probably meet your requirements.) But this is in addition to the quantitative coverage, which easily outstrips any competition in terms of variety, depth and authority. We have tried to provide students with resources to help them in ways largely ignored by most other texts. For example, the chapter on literature searches is comprehensive and practical. Similarly, the chapter on ethics meets the most recent standards and deals with them in depth. The chapter on writing research reports places report writing at the centre of the research process rather than as an add-on at the end. We would argue that a student requires an understanding of the nature of research in psychology to be able to write a satisfactory research report. However, we have included a chapter which illustrates many of the problems that are found in research reports in response to requests for such material.

As far as is possible, we have tried to provide students with practical skills as well as the necessary conceptual overview of research methods in modern psychology. Nevertheless, there is a limit to this. The bottom line is that anyone wishing to understand research needs to read research, not merely plan, execute, analyse and write-up research. Hence, almost from the start we emphasise that reading is not merely unavoidable but crucial. Without such additional reading, the point of this book is missed. It is not intended as a jumble of technical stuff too boring to be part of any module other than one on research methods. The material in the book is intended to expand students' understanding of psychology by explaining just how researchers go about creating psychology. At times this can be quite exciting as well as frustrating and demanding.

This is the fifth book the authors have written together. It is also the one that came close to spoiling a long friendship. What became very clear while writing this book is how emotive the topic of research methods can be. We found out, perhaps for the first time, how different two people's thinking can be, even when dealing with seemingly dry topics. As a consequence, rather than smooth over the cracks, making joins when this was not possible, you will find that we have incorporated the differences of opinion. This is no different from the disparity of positions to be found within the discipline itself – probably less so.

The main features of this book are:

- in-depth coverage of both quantitative and qualitative methods;
- a range of pedagogic features including summaries, exercises, boxes and step-by-step instructions where appropriate;
- analysis strategies provided for the research designs discussed;
- detailed information about the structure, purpose and contents of research reports;
- the use of databases and other resources;
- suggestions about how to develop research ideas for projects and similar studies;
- ethics as an integral feature of the work of all psychologists.

Introduction to Research Methods in Psychology is part of the trilogy of books which also includes *Introduction to Statistics in Psychology* and *Introduction to SPSS Statistics in Psychology*. In *Introduction to Research Methods in Psychology* we have tried to make the presentation both clear in terms of the text but with additional visual learning aids throughout the book. The main new additions to the other two more statistically oriented books, apart from colour, are in terms of the statistical techniques of power analysis and moderator effects. These reflect our determination to provide resources to students which are both user-friendly and professionally oriented. Increasingly research is part of many of the different sorts of careers which psychology students enter – we simply hope that our books speed the user towards a considered, mature approach to research.

Introduction to Statistics in Psychology, we feel, remains the best introduction to statistical concepts for students at all levels. The intention is to provide an introduction to statistics for the beginner which will take them through to the professional level unproblematically. *Introduction to SPSS Statistics in Psychology* is a quicker approach to learning and carrying out statistical procedures than *Introduction to Statistics in Psychology*. Instead of detailed explanations of theory together with practical details, *Introduction to SPSS Statistics in Psychology* provides a short conceptual introduction to each statistical routine together with step-by-step screenshots and instructions about their calculation using SPSS Statistics.

Education is a cooperative effort. So should you find errors then please let us know. These can be difficult to spot but easy to correct – some can be made when a book is reprinted. Ideas and comments of any sort would be most welcome.

Acknowledgements

Authors' acknowledgements

The authors would like to express their gratitude to the following for the immense contribution that they have made to this book:

- Janey Webb – our editor at Pearson Education who was indefatigable in planning and progressing this third edition. As ever, Janey was remarkably supportive at every stage.
- Catherine Morrissey – Janey's assistant at Pearson Education kept things flowing but has now moved on to a new job. Good luck to Catherine in that.
- Mary Lince and Georgina Clark-Mazo – they took charge of production of the book and made sure that everything happened when it should have happened and that the impossible was routine.
- Kevin Ancient – was responsible for the text design without which the book would be much harder to read and navigate around.
- Nicola Woowat – provided the cover design which continues the tradition of striking covers for this book and the others in the set.
- Ros Woodward – she copy-edited our manuscript into the lively structure of the final book and generally made the manuscript work as a book.
- Rose James – proof-read the book and spotted the things which the authors never would.
- Annette Musker – was responsible for the index which makes navigation through the book so much easier.
- Louise Newman – was responsible for the accompanying website which does so much to make the book even more useful.

Authors are highly dependent on academic reviewers for new ideas and general feedback. The following were tremendously helpful in planning this third edition:

- Deborah Fantini, University of Essex
- Claire Fox, Keele University
- Koen Lamberts, University of Warwick
- Jane Walsh, National University of Ireland, Galway
- Dr Paul Seager, University of Central Lancashire

Dennis Howitt and Duncan Cramer

Publisher's acknowledgements

We are grateful to the following for permission to reproduce copyright material:

Screenshots

Screenshots 7.3, 7.4, 7.5, 7.6, 7.7 from Thomson Reuters, Thomson Reuters; Screenshot 7.8 from Loughborough University Ex Libris Ltd, Ex Libris Ltd; Screenshots 7.9, 7.10 from reproduced by permission of SAGE Publications, London, Los Angeles, New Delhi and Singapore, from SAGE journals online; Screenshot 7.11 from The PsycINFO® Database, reproduced with permission of the American Psychological Association, publisher of the PsycINFO database, all rights reserved. No further reproduction or distribution is permitted without written permission from the American Psychological Association. Images produced by ProQuest. Inquiries may be made to: ProQuest, P.O. Box 1346, 789 E. Eisenhower Parkweay, Ann Arbor, MI 48106-1346 USA. Telephone (734) 761-7400; E-mail: info@proquest.com; Web-page: www.proquest.com; Screenshot 7.12 from The PsycINFO® Database, reproduced with permission of the American Psychological Association, publisher of the PsycINFO database, all rights reserved. No further reproduction or distribution is permitted without written permission from the American Psychological Association. Images produced by ProQuest. Inquiries may be made to: ProQuest, P.O. Box 1346, 789 E. Eisenhower Parkweay, Ann Arbor, MI 48106-1346 USA. Telephone (734) 761-7400; E-mail: info@proquest.com; Web-page: www.proquest.com, PsycINFO is a registered trademark of the American Psychological Association (APA). The PsycINFO Database content is reproduced with permission of the APA. The CSA Illumina internet platform is the property of ProQuest LLC. and Image published with permission of ProQuest. Further reproduction is prohibited without permission.; Screenshot 7.13 from The PsycINFO® Database, reproduced with permission of the American Psychological Association, publisher of the PsycINFO database, all rights reserved. No further reproduction or distribution is permitted without written permission from the American Psychological Association. Images produced by ProQuest. Inquiries may be made to: ProQuest, P. O. Box 1346, 789 E. Eisenhower Parkweay, Ann Arbor, MI 48106-1346 USA. Telephone (734) 761-7400; E-mail: info@proquest.com; Web-page: www.proquest.com; Screenshot 7.14 from The PsycINFO® Database, reproduced with permission of the American Psychological Association, publisher of the PsycINFO database, all rights reserved. No further reproduction or distribution is permitted without written permission from the American Psychological Association. Images produced by ProQuest. Inquiries may be made to: ProQuest, P.O. Box 1346, 789 E. Eisenhower Parkweay, Ann Arbor, MI 48106-1346 USA. Telephone (734) 761-7400; E-mail: info@proquest.com; Web-page: www.proquest.com; Screenshot 7.15 from The PsycINFO® Database, reproduced with permission of the American Psychological Association, publisher of the PsycINFO database, all rights reserved. No further reproduction or distribution is permitted without written permission from the American Psychological Association. Images produced by ProQuest. Inquiries may be made to: ProQuest, P.O. Box 1346, 789 E. Eisenhower Parkweay, Ann Arbor, MI 48106-1346 USA. Telephone (734) 761-7400; E-mail: info@proquest.com; Web-page: www.proquest.com; Screenshot 20.1 from QSR International Pty Ltd, Courtesy of QSR International Pty Ltd.

Tables

Table on page 241 from http://www.britsocat.com, British Social Attitudes Survey (2007), National Centre for Social Research; Table 21.1 after reproduced by permission of SAGE Publications, London, Los Angeles, New Delhi and Singapore, from J. A. Smith, R. Harre and L. V. Langenhove, Rethinking Methods in Psychology, 'Grounded theory', p. 39, Charmaz, K. (© SAGE, 1995); Table 26.1 adapted from The 100 most eminent psychologists of the 20th century, *Review of General Psychology*, 6, 139–52 (Haggbloom, S. J., Warnick, R., Warnick, J. E., Jones, V. K., Yarbrough, G. L., Russell, T. M. *et al.* 2002), American Psychological Association.

In some instances we have been unable to trace the owners of copyright material, and we would appreciate any information that would enable us to do so.

PART 1

The basics of research

PART 1
The basics of research

CHAPTER 1

The role of research in psychology

Overview

- Research is central to all the activities of psychologists as it is to modern life in general. A key assumption of psychology is that the considered and careful collection of research data is an essential part of the development of the discipline.
- The vast majority of psychology involves the integration of theoretical notions with the outcomes of research. Psychology characteristically emphasises causal explanations. Many psychologists adhere to the belief that a prime purpose of research is to test causal propositions.
- A first-rate psychologist – researcher or practitioner – needs to be familiar with the way in which good research is carried out. This enables them to determine the adequacy and value of the findings claimed from a particular study.
- All psychologists need the resources to be able to read research reports in detail, for example, studies reported in journals of psychological research. This requires an understanding of the purposes, advantages and disadvantages of the different research methods used to investigate issues.
- Research reports become much clearer and easier to understand once the basics of psychological research methods are known. Very often research reports are concisely written and so assume a degree of knowledge of the topic and research methods. The study of research methods will help prepare students for this.
- Psychologists have traditionally distinguished between true experiments and non-experiments. True experiments are typical of laboratory studies in psychology whereas non-experiments are more typical of more naturalistic studies in the field (community or other real-life settings).

- Many psychologists believe that true experiments (laboratory studies) in general provide a more convincing test of causal propositions. Others would dispute this on the grounds that such true experiments often achieve precision at the expense of realism.
- Conducting one's own research is a fast route to understanding research methods. Increasingly, research is seen as an integral part of the training and work of all psychologists irrespective of whether they are practitioners or academics.

1.1 Introduction

Research is exciting – the lifeblood of psychology. To be sure, the subject matter of psychology is fascinating, but this is not enough. Modern psychology cannot be fully appreciated without some understanding of the research methods that make psychology what it is. Although initially psychology provides many intriguing ideas about the nature of people and society, as one matures intellectually the challenges and complexities of the research procedure that helped generate these ideas are increasingly appreciated. Psychological issues are intriguing: for example, why are we attracted to some people and not to others? Why do we dream? What causes depression and what can we do to alleviate it? Can we improve our memory and, if so, how? What makes us aggressive and can we do anything to make us less aggressive? What are the rules which govern everyday conversation? The diversity of psychology means that our individual interests are well catered for. It also means that research methods must be equally diverse if we are to address such a wide range of issues. Psychology comes in many forms and so does good psychological research.

Students often see research methods as a dull, dry and difficult topic which is tolerated rather than enjoyed. They much prefer their other lecture courses on exciting topics such as interpersonal attraction, mental illness, forensic investigation, brain structure and thought. What they overlook is that these exciting ideas are created by active and committed researchers. For these psychologists, psychology and research methods are intertwined – psychology and the means of developing psychological ideas through research cannot be differentiated. For instance, it is stimulating to learn that we are attracted to people who have the same or similar attitudes to us. It is also of some interest to be given examples of the kinds of research which support this idea. But is this all that there is to it? Are there not many more questions that spring to mind? For example, why should we be attracted to people who have similar attitudes to our own? Do opposites never attract? When does similarity lead to attraction and when does dissimilarity lead to attraction? The answer may have already been found to such questions. If not the need for research is obvious. Research makes us think hard – which is the purpose of any academic discipline. The more thinking that we do about research, the better we become at it.

Box 1.1 contains definitions of various concepts such as 'variable' and 'correlation' to which you may need to refer to if you are unfamiliar with these terms.

Box 1.1 Key Ideas

Some essential concepts in research

Cause Something which results in an effect, action or condition.

Data The information from which inferences are drawn and conclusions reached. A lot of data are collected in numerical form but it is equally viable to use data in the form of text for an analysis.

Randomised experiment This refers to a type of research in which participants in research are allocated at random (by chance) to an experimental or control condition. Simple methods of random assignment include flipping a coin and drawing slips of paper from a hat. The basic idea is that each participant has an equal chance of being allocated to the experimental or control conditions. The experimental and control conditions involve differences in procedure related to the hypothesis under examination. So by randomisation, the researcher tries to avoid any systematic differences between the experimental and control conditions prior to the experimental manipulation. Random selection is covered in detail in Chapter 13, pp. 233–236.

Reference In psychology, this refers to the details of the book or article that is the source of the ideas or data being discussed. The reference includes such information as the author, the title and the publisher of the book or the journal in which the article appears.

Variable A variable is any concept that varies and can be measured or assessed in some way. Intelligence, height and social status are simple examples.

1.2 Reading

The best way of understanding psychological research methods is to read in detail about the studies which have been done and build on this. Few psychological textbooks give research in sufficient detail to substitute effectively for this. Developing a better understanding of how research is carried out in a particular area is greatly helped when one reads some of the research work in its original form that lecturers and textbook writers refer to. Admittedly, some psychologists use too much jargon in their writing but ignore these in favour of the many good communicators among them wherever possible. University students spend only a small part of a working week being taught – they are expected to spend much of their time on independent study, which includes reading a great deal as well as independently working on assignments. Glance through any textbook or lecture course reading list and you will see the work of researchers cited. For example, the lecturer or author may cite the work of Byrne (1961) on attraction and similarity of attitude. Normally a list of the 'references' cited is provided. The citation provides information on the kind of work it is (for example, what the study is about) and where it has been presented or published. The details are shown in the following way:

> Byrne, D. (1961). Interpersonal attraction and attitude similarity. *Journal of Abnormal and Social Psychology*, 62, 713–15.

The format is standard for a particular type of publication. Details differ according to what sort of publication it is – a book is referenced differently from a journal article and an Internet source is referenced differently still. For a journal article, the last name of the author is given first, followed by the year in which the reference was published. After this comes the title of the work. Like most research in psychology, Byrne's study was published in a journal. The title of the journal is given next together with the number of the volume in which the article appeared together with the numbers of the first and last pages of the article. These references are generally listed alphabetically according to the last name of the first author in a reference list at the end of the journal article or book. Where there is more than one reference by the same author or authors, they will be listed according to the year the work was presented. This is known as the Harvard system or author–date system. This is described in much more detail in Chapters 5 and 6 which are about writing a research report. We will cite references in this way in this book. However, we will cite very few references compared with psychology texts on other subjects as many of the ideas we are presenting have been previously summarised by other authors (although usually not in the same way) and have been generally accepted for many years.

Many of the references cited in lectures or textbooks are to reports of research that has been carried out to examine a particular question or small set of questions. Research studies have to be selective and restricted in their scope. As already indicated, the prime location for the publication of research is journals. Journals consist of volumes which are usually published every year. Each volume typically comprises a number of issues or parts that come out say every three months but this is variable. The papers or articles that make up an issue are probably no more than 4000 or 5000 words in length though it is not uncommon to find some of them 10 000 words long. Their shortness necessitates their being written concisely. As a consequence, they are not always easy to read and often require careful study in order to master them. An important aim of this book is to provide you with the basic knowledge which is required to read these papers – and even to write them. Often there appear to be obstacles in the way of doing the necessary reading. For example, there are many different psychology journals – too many for individual libraries to stock, so they subscribe to a limited number of them. If the reference that you are interested in is important and is not available locally, then you may be able to obtain it from another library or it is worthwhile trying to obtain a copy (usually called offprints) from the author. Nowadays many papers are readily available in electronic files (usually in Portable Digital Format, PDF) which can be easily accessed or e-mailed as attachments. Chapter 7 on searching the literature suggests how you can access publications which are not held in your own library. Fortunately, it is becoming increasingly common that university libraries subscribe to digital versions of journals. That means that often you can download to your computer articles which, otherwise, would not be available at your university. The convenience of this is significant and there are no overdue fines.

One of the positive things about psychology is that you may have questions about a topic that have not been addressed in lectures or textbooks. For example, you may wonder whether attraction to someone depends on the nature of the particular attitudes that are shared. Are some attitudes more important than others and, if so, what are these? If you begin to ask questions like these while you are reading something then this is excellent. It is the sort of intellectual curiosity required to become a good researcher. Furthermore, as you develop through your studies, you probably will want to know what the latest thinking and research are on the topic. If you are interested in a topic, then wanting to know what other people are thinking about it is only natural. Your lecturers will certainly be pleased if you do. There is a great deal to be learnt about finding out what is happening in any academic discipline. Being able to discover what is happening and what has happened in a field of research is a vitally important skill. Chapter 7 discusses how we go about searching for the current publications on a topic.

1.3 Evaluating the evidence

Psychology is not simply about learning what conclusions have been reached on a particular topic. It is perhaps more important to find out and carefully evaluate the evidence which has led to these conclusions. Why? Well, what if you have always subscribed to the old adage 'opposites attract'? Would you suddenly change your mind simply because you read in a textbook that people with similar attitudes are attracted to each other? Most likely you would want to know a lot more about the evidence. For example, what if you checked and found that the research in support of this idea was obtained simply by asking a sample of 100 people whether they believed that opposites attract? In this case, all the researchers had really established was that people generally thought it was true that people are attracted to other people with similar attitudes. After all, simply because people once believed the world was flat did not make the world flat. It may be interesting to know what people believe, but wouldn't one want different evidence in order to be convinced that attraction actually is a consequence of similarity of attitudes? You might also wonder if it is really true that people once believed the world to be flat. Frequently, in the newspapers and on television, one comes across startling findings from psychological research. Is it wise simply to accept what the newspaper or television report claims or would it be better to check the original research in order to evaluate what the research actually meant?

We probably would be more convinced of the importance of attitude similarity in attraction if a researcher measured how attracted couples were to each other and then showed that those with the most similar attitudes tended to be the most attracted to one another. Even then we might still harbour some doubts. For example, just what do we mean by attraction? If we mean wanting to have a drink with the other person at a pub then we might prefer the person with whom we might have a lively discussion, that is, someone who does not share our views. On the other hand, if willingness to share a flat with a person were the measure of attraction then perhaps a housemate with a similar outlook to our own would be preferred. So we are beginning to see that the way in which we choose to measure a concept (or variable) such as attraction may be vital in terms of the answers we get to our research questions.

It is possibly even more difficult to get a satisfactory measure of attitudes than it is to measure attraction. This is partly because there are many different topics that we can express attitudes about. So, for example, would we expect attraction to be affected in the same way if two people share the view that there is life on Mars than if two people share the same religious views? Would it matter that two people had different tastes in music than if they had different views about openness in relationships? That is, maybe some attitudes are more important than others in determining attraction – perhaps similarity on some attitudes is irrelevant to the attraction two people have for each other. One could study this by asking people about their attitudes to a variety of different topics and then how important each of these attitudes is to them. (Sometimes this is called salience.) Alternatively, if we thought that some attitudes were likely to be more important than others, we could focus on those particular attitudes in some depth. So it should be clear from all of this that the process of evaluating the research in a particular field is not a narrow, nit-picking exercise. Instead it is a process by which new ideas are generated as well as stimulating research to test these new propositions.

These various propositions that we have discussed about the relationship between attraction and similarity are all examples of *hypotheses*. A hypothesis is merely a supposition or proposition which serves as the basis of further investigation, either through the collection of research data or through reasoning. The word hypothesis comes from the Greek word for foundation – perhaps confirming that hypotheses are the foundation on which psychology develops. Precision is an important characteristic of good hypotheses. So, our hypothesis that similarity of attitudes is related to attraction might benefit from refinement. It looks as if we might have to say something more about the attitudes that people have (and what we mean by attraction for that matter) if we are going to pursue our questions any further. If we think that the attitudes have to be important, then the hypothesis should be reformulated to read that *people are more attracted to those with similar attitudes on personally important topics*. If we thought attraction was based on having a similar attitude towards spending money, we should restate the hypothesis to say that *people are more attracted to those with similar attitudes towards spending money*.

The evaluation of research evidence involves examining the general assertion that the researcher is making about an issue and the information or data that are relevant to this assertion. We need to check whether the evidence or data support the assertion or whether the assertion goes beyond what could be confidently concluded. Sometimes, in extreme cases, researchers draw conclusions which seem not to be justified by their data. Any statement that goes beyond the data is speculation or conjecture and needs to be recognised as such. There is nothing wrong with speculation as such since hypotheses, for example, are themselves often speculative in nature. Speculation is necessary in order to go beyond what we already know. However, it needs to be distinguished from what can legitimately be inferred from the data.

1.4 Inferring causality

The concept of causality has been important throughout most of the history of psychology. Other disciplines might consider it almost an obsession of psychology. The meaning of the term is embodied in the phrase 'cause and effect'. The idea is that things that happen in the world may have an effect on other things. So when we speak of a causal relationship between attitude similarity and attraction we mean that attitude similarity is the cause of attraction to another person. Not all data allow one to infer causality with confidence. Sometimes researchers suggest that their research demonstrates a causal relationship when others would claim that it demonstrates no such thing – that there may be a relationship but that one thing did not cause the other. In strictly logical terms, some claims of a causal relationship can be regarded as an error since they are based on research methods which by their nature are incapable of establishing causality with certainty. Frequently research findings may be consistent with a causal relationship but they are, equally, consistent with other explanations.

Pairs of friends:
are their attitudes
similar?
Pairs of strangers:
are their attitudes
dissimilar?

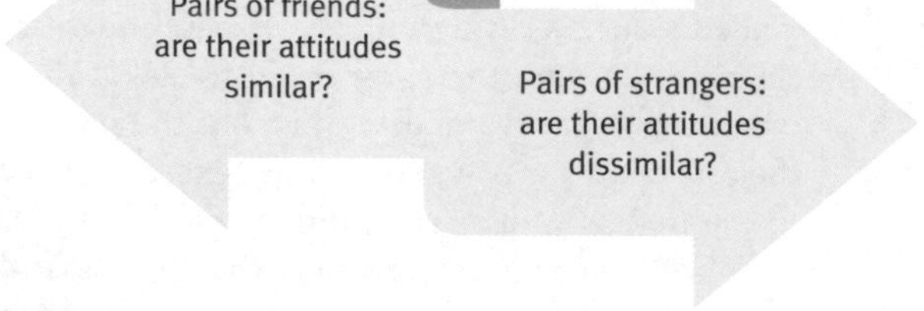

FIGURE 1.1 Looking for causal relationships

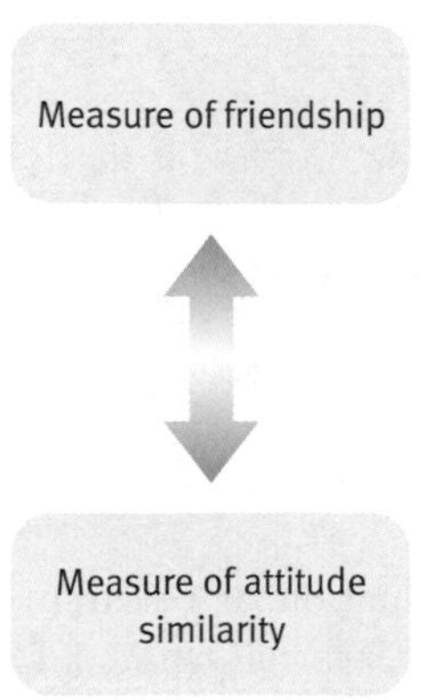

FIGURE 1.2 Cross-sectional study: measures taken at the same point in time

A great deal of psychology has as its focus causes of things even though the word 'cause' is not used directly. Questions such as why we are attracted to one person rather than another, why people become depressed and why some people commit violent crimes are typical examples of this. The sorts of explanation that are given might be, for example, some people commit violent crimes because they were physically abused as children. In other words, physical abuse as a child is a *cause* of adult violent crime. There may be a relationship between physical abuse and violent crime, but does this establish that physical abuse is a cause? To return to our main example, suppose a study found that people who were attracted to each other had similar attitudes. Pairs of friends were compared with pairs of strangers in terms of how similar their attitudes were (see Figure 1.1). It emerged that the friends had more similar attitudes than pairs of strangers. Could we conclude from this finding that this study showed that similar attitudes cause people to be attracted towards one another? If we can conclude this, on what grounds can we do so? If not, then why not?

There are at least three main reasons why we cannot conclude definitively from this study that similar attitudes lead to people liking each other:

- Attraction, measured in terms of friendship, and similarity of attitudes are assessed once and at precisely the same time (see Figure 1.2). As a consequence we do not know which of these two came first. Did similarity of attitudes precede friendship as it would have to if similar attitudes led to people liking each other? Without knowing the temporal sequence, definitive statements about cause and effect are not possible (see Figure 1.3).

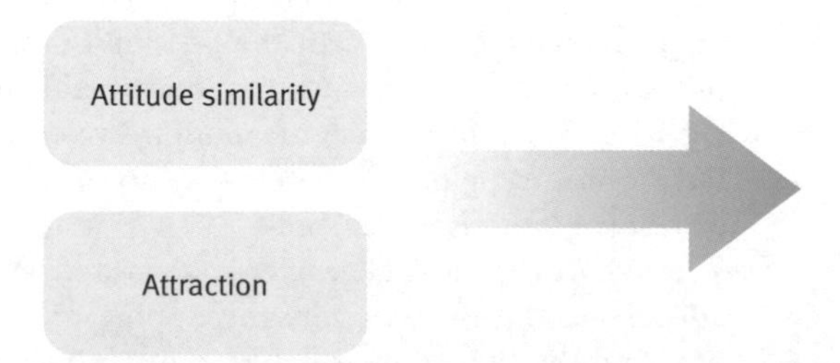

FIGURE 1.3 No time lag between the measurement of attitude similarity and attraction: no evidence of causality

- Friendship may have preceded similarity of attitudes. In other words, friends develop similar attitudes because they happen to like one another for other reasons. Once again the basic problem is that of the temporal sequence. Because this study measures both friendship and similarity of attitudes at the same time we cannot tell which came first. In other words we cannot determine which caused which (see Figure 1.4).

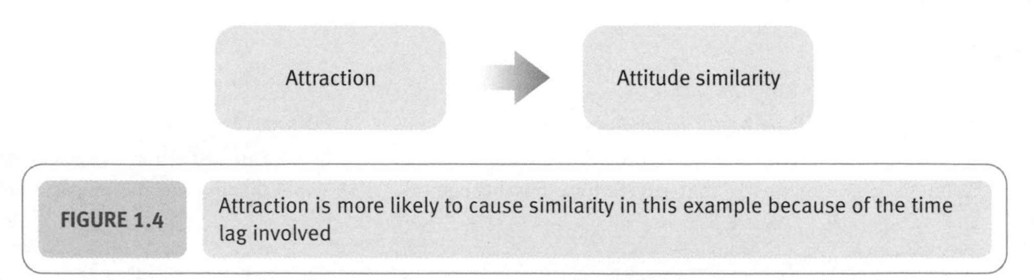

FIGURE 1.4 Attraction is more likely to cause similarity in this example because of the time lag involved

- The development of attraction and similarity may be the result of the influence of a third factor. For example, if one moves to university one begins to be attracted to new people and, because of the general influence of the campus environment, attitudes begin to change. In these circumstances, the relationship between attraction and similarity is not causal (in either direction) but the result of a third factor, which is the effect of the move to campus (see Figure 1.5).

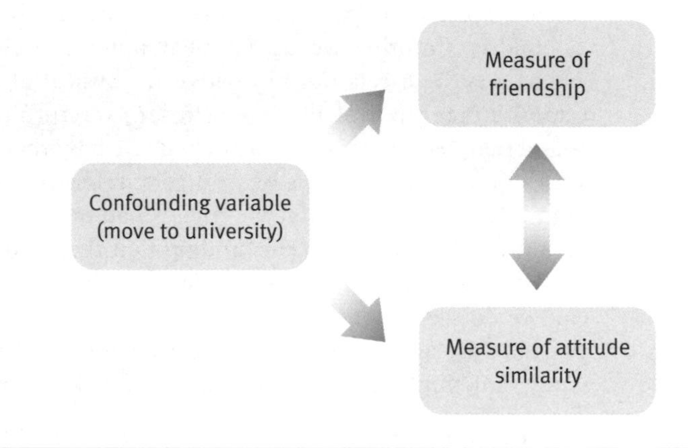

FIGURE 1.5 Confounding variables in research: all measures taken at the same point in time

Care needs to be taken here. It is not being suggested that the research in question is worthless simply because it cannot definitively establish that there is a causal relationship. The findings of the research are clearly compatible with a causal hypothesis and one might be inclined to accept the possibility that it is a causal relationship. Nevertheless, one cannot be certain and may find it difficult to argue against someone who rejects the idea. Such divergence of opinion sometimes becomes a controversy in psychology. Divergence of opinion in research is a positive thing as it leads to new research designed to resolve that disagreement.

Some of the most characteristic research methods of psychology are geared towards addressing the issue of causality. Some of these will be outlined in due course. Most importantly, the contrast between randomised experiments so familiar to psychologists in the form of laboratory experiments and research outside the laboratory has this issue at its root.

The role of causality in psychology is a controversial topic. (See Box 1.2 for a discussion of this.)

Box 1.2 Talking Point

Causal explanations: psychology's weakest link?

It is well worth taking the time to study the history of psychology. This will help you to identify the characteristics of the discipline (e.g. Hergenhahn, 2001; Leahy, 2004). What is very obvious is that psychology has been much more concerned about causality than many of its closely related disciplines – sociology is a good example. There are a number of reasons why this should be the case:

- Psychology was much more influenced by the philosophies of positivism and logical positivism than these other disciplines. Broadly speaking, positivism is a description of the methods of the natural sciences such as physics and chemistry. It basically holds that knowledge is obtained through observation. So the more focused and precise the empirical observation the better. Hence, a precisely defined cause and an equally precisely defined effect would be regarded as appropriate. Positivism is a concept originating in the work of the French sociologist Auguste Comte (1798–1857). It refers to the historical period when knowledge was based on science rather than, say, religious authority. Logical positivism is discussed in some detail in Chapter 17.

- Psychology has traditionally defined itself as much as a biological science as a social science. Consequently, methods employed in the natural sciences have found a substantial place in psychology. In the natural sciences, laboratory studies (experiments) in which small numbers of variables at a time are controlled and studied are common, as they have been in psychology. The success of disciplines such as physics in the nineteenth century and later encouraged psychologists to emulate this approach.

- By emulating the natural sciences approach, psychologists have tended to seek general principles of human behaviour just as natural scientists believe that their laws apply throughout the physical universe. Translated into psychological terms, the implication is that findings from the laboratory are applicable to situations outside the psychological laboratory. Randomised laboratory experiments tend to provide the most convincing evidence of causality – that is what they are designed to do.

Modern psychology is much more varied in scope than it ever was in the past. The issue of causality is not as crucial as it once was. There is a great deal of research that makes a positive contribution to psychology which eschews issues of causality. For example, the forensic psychologist who wishes to predict suicide risk in prisoners does not have to know the causes of suicide among prisoners. So if research shows that being in prison for the first time is the strongest predictor of suicide then this is a possible predictor. It is irrelevant whether the predictor is in itself the direct cause of suicide. There are a multitude of research questions which are not about causality.

Many modern psychologists regard the search for causal relationships as somewhat counterproductive. It may be a good ideal in theory, but in practice it may have a negative influence on the progress of psychology. One reason for this is that the procedures which can help establish causality can actually result in highly artificial and contrived situations, with the researcher focusing on fine detail rather than obtaining a broad view, and the findings of such research are often not all that useful in practice. One does not have to study psychology for long before it becomes more than apparent that there is a diversity of opinion on many matters.

1.5 Types of research and the assessment of causality

In this section we will describe a number of different types of study in order to achieve a broad overview of research methods in psychology. There is no intention to prioritise them in terms of importance or sophistication. They are:

- correlational or cross-sectional studies;
- longitudinal studies;
- experiments – or studies with randomised assignment.

As this section deals largely with these in relation to the issue of causality, all of the types of research discussed below involve a minimum of two variables examined in relation to each other. Types of study which primarily aim to describe the characteristics of things are dealt with elsewhere. Surveys, for example, are discussed in Chapter 13, and qualitative methods are covered in depth in Chapters 17 to 25.

Correlational or cross-sectional studies

Correlational (or cross-sectional) studies are a very common type of research. Basically what happens is that a number of different variables (see Box 1.1) are measured more or less simultaneously for a sample of individuals (see Figure 1.6). Generally in psychology, the strategy is to examine the extent to which these variables measured at a single point in time are associated (that is correlated) with one another. A correlation coefficient is a statistical index or test which describes the degree and direction of the relationship between two characteristics or variables. To say that there is a correlation between two characteristics merely means that there is a relationship between them.

The correlation coefficient is not the only way of testing for a relationship. There are many other statistical techniques which can be used to describe and assess the relationship between two variables. For example, although we could correlate the extent to which people are friends or strangers with how similar are their attitudes using the correlation coefficient there are other possibilities. An equivalent way of doing this is to examine differences. This is what is normally done in this kind of study. One would look at whether there is a difference in the extent to which friends are similar in their attitudes compared with how similar random pairs of strangers are. If there is a difference between the two in terms of degrees of attitude similarity, it means that there is a relationship between the variable friends/strangers and the variable similarity of attitudes. So a test of differences (e.g. the *t*-test) is usually applied rather than a correlation coefficient. A more accurate term for describing these studies is cross-sectional in that they measure variables at one point in time or across a slice or section of time. This alternative term leaves open how we analyse the data statistically since it implies neither a test of correlation nor a test of differences in itself. Issues related to this general topic are discussed in depth in Chapter 4.

FIGURE 1.6 Structure of a cross-sectional study: all measures taken at the same point in time

Correlational or cross-sectional studies are often carried out in psychology's sub-disciplines of social, personality, developmental, educational, and abnormal or clinical psychology. In these areas such research designs have the advantage of enabling the researcher to measure a number of different variables at the same time. Any of these variables might possibly explain why something occurs. It is likely that anything that we are interested in explaining will have a number of different causes rather than a single cause. By measuring a number of different variables at the same time, it becomes possible to see which of the variables is most strongly related to what it is we are seeking to explain.

Confounding variables

A major reason why we cannot infer causality is the problem of the possible influence of unconsidered variables. Sometimes this is referred to as the third variable problem. For example, it could be that both friendship and similarity of attitudes are determined by the area, or kind of area, in which you live (such as a campus as mentioned earlier). You are more likely to make friends with people you meet, who are more likely to live in the same area as you. People living in the same area also may be more likely to have the same or similar attitudes. For example, they may be more likely to share the same religious attitudes or eat the same food. When a researcher asks people such as students to take part in a study they are likely to come from different areas. It could be that it is the area, or kind of area, that people come from that determines both who their friends are and their attitudes. Variables which either wholly or partially account for the relationship between two other variables are known as *confounding* variables. Area, or type of area, could be a confounding variable which we may need to check (see Figure 1.7).

One could try to hold constant in several ways the area from which people come. For example, one could select only people from the same area. In this way the influence of different areas is eliminated. If you did this, then there may still be other factors which account for the fact that people are attracted to others who have similar attitudes to them. It is not always obvious or easy to think what these other factors might be. Because we have to study in this research a number of friendships, it is likely that the people making up these friendships will differ in various ways. It would be very difficult to hold all of these different factors constant. One such additional factor might be age. Pairs of friends are likely to differ in age. Some pairs of friends will be older than other pairs of friends. It could be that any association or relationship between being friends and having similar attitudes is due to age. People are more likely to be friends with people who are similar in age to them. People of a similar age may have similar attitudes, such as the kind of music they like or the kinds of clothes they wear. So age may determine

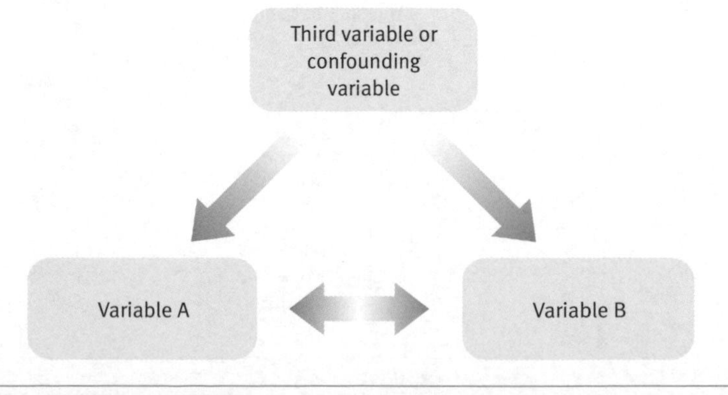

FIGURE 1.7 Stylised diagram of the confounding (third variable) problem

both who becomes friends with whom and what their attitudes are. The easiest way to control for confounding variables is to try to measure them and to control for them statistically. There is another way, which we will mention shortly.

Longitudinal studies

Suppose we measured both friendship and similarity of attitudes at two (or more) different points of time, could we then determine whether friendship led to having similar attitudes? This kind of study is known as a longitudinal study as opposed to a cross-sectional one. It is more difficult to organise this kind of study, but it could and has been done. We would have to take a group of people who did not know each other initially but who would have sufficient opportunity subsequently to get to know each other. Some of these people would probably strike up friendships. One possible group of participants would be first-year psychology students who were meeting together for the first time at the start of their degree. It would probably be best if any participants who knew each other before going to university or came from the same area were dropped from the analysis. We would also need to measure their attitudes towards various issues. Then after a suitable period of time had elapsed, say three or more months, we would find out what friendships had developed and what their attitudes were (see Figure 1.8).

Suppose it were found that students who subsequently became friends started off as having the same or similar attitudes and also had the same or similar attitudes subsequently, could we then conclude that similar attitudes lead to friendship? In addition, those who did not become friends started off dissimilar in attitudes and were still dissimilar three months later. This analysis is illustrated in Figure 1.9 in the left-hand column. Well, it is certainly stronger evidence that similarity of attitudes may result in friendship than we obtained from the cross-sectional study. Nevertheless, as it stands, it is still possible for the sceptic to suggest that there may be other confounding variables which explain both the friendships and the similarity in attitudes despite the longitudinal nature of our new study. It might be that this association between friendship and attitude similarity can be explained in terms of confounding variables (see Figure 1.10). For example, the idea was discussed earlier that people who come from similar kinds of areas may be the most likely to become friends as they find out they are familiar with the same customs or have shared similar experiences. They may also have similar attitudes because they come from similar areas. Thus similarity of area rather than similarity of attitudes could lead to friendships, as illustrated in Figure 1.10.

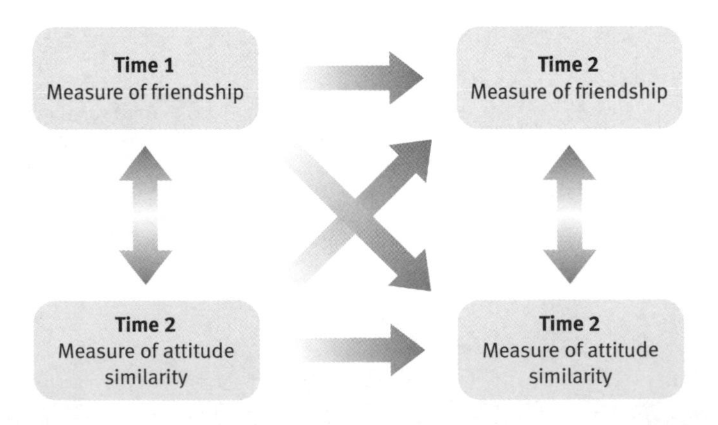

FIGURE 1.8 Longitudinal study of friendship and attitude similarity with variables measured twice

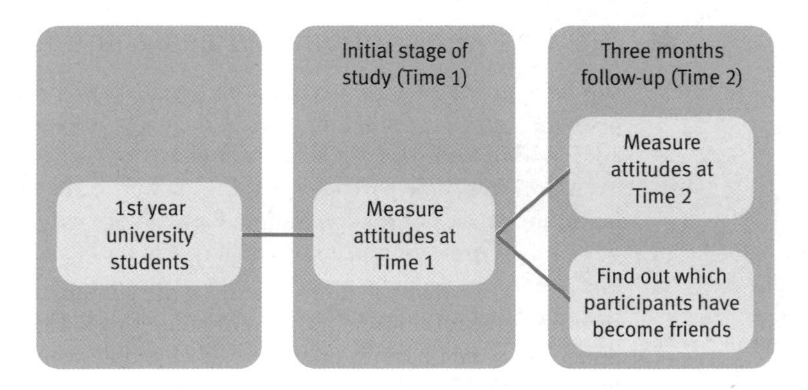

FIGURE 1.9 Study design to assess attitude similarity and development of friendship over time

As with cross-sectional studies, there are statistical methods of controlling these confounding variables in longitudinal studies. Longitudinal studies provide more information than cross-sectional ones. In our example, they will tell us how stable attitudes are. If attitudes are found not to be very stable and if similarity of attitudes determined friendships, then we would not expect friendships to be very stable. Because these studies are more complex, the analyses of their results will be more complicated and will take more effort to understand. As with cross-sectional studies, the major problem is that we fail to take into account all of the confounding factors that may have brought about the results. If we could guarantee to deal with all the confounding variables in this sort of research, it could claim to be an ideal type of research method. Unfortunately, there can be no such guarantee.

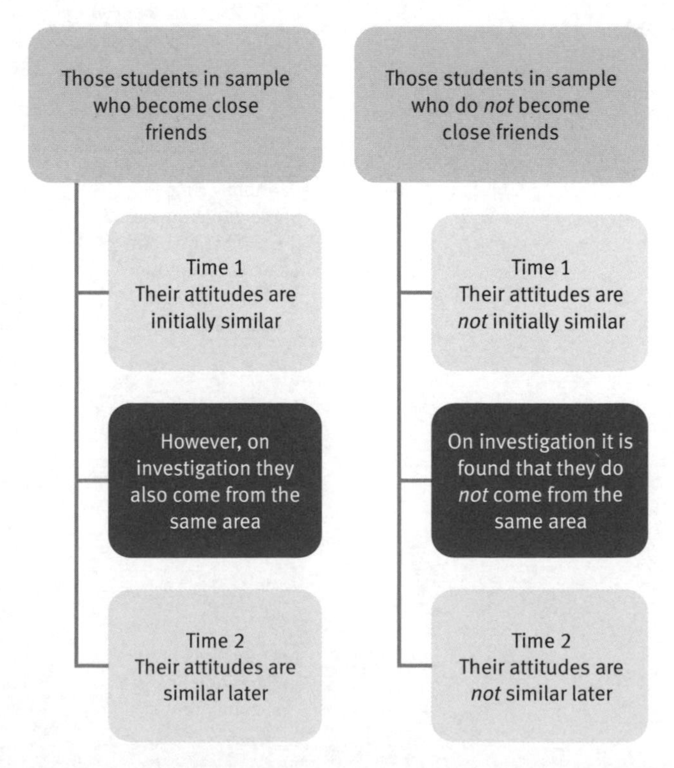

FIGURE 1.10 How a third variable may affect the relationship between friendship and attitudes

Studies with randomised assignment – experiments

We have now identified a basic problem. Researchers simply do not and cannot know just what other variables may affect their key measures. Is there any way in which all confounding variables can be taken into account when we do not know what those variables are? For some psychologists, the answer to the major problems of research design lies in the process of randomisation. Basically we would form two groups of participants who are given the opportunity to interact in pairs and get to know each other better. In one condition, the pairs are formed by choosing one member of the pair at random and then the other member selected at random from the participants who had similar attitudes to the first member of the pair. In the other condition, participants are selected at random but paired with another person dissimilar in attitude to them, again selected at random. By allocating participants to similarity and dissimilarity conditions by chance, any differences between the conditions cannot be accounted for by these confounding variables. By randomising in this way, similarities and dissimilarities in the areas from which the participants come, for example, would be expected to be equalised between groups. This particular example is illustrated in Figure 1.11 and the more general principles of experimental design in Figure 1.12.

The simplest way of randomisation in this example is to allocate participants to the different conditions by tossing a coin. We would have to specify before we tossed the coin whether a coin landing heads facing upwards would mean that the person was paired with a person with the same attitude as them or with a different attitude from them. If we tossed a coin a fixed number of times, say 20 times, then it should come up heads 10 times

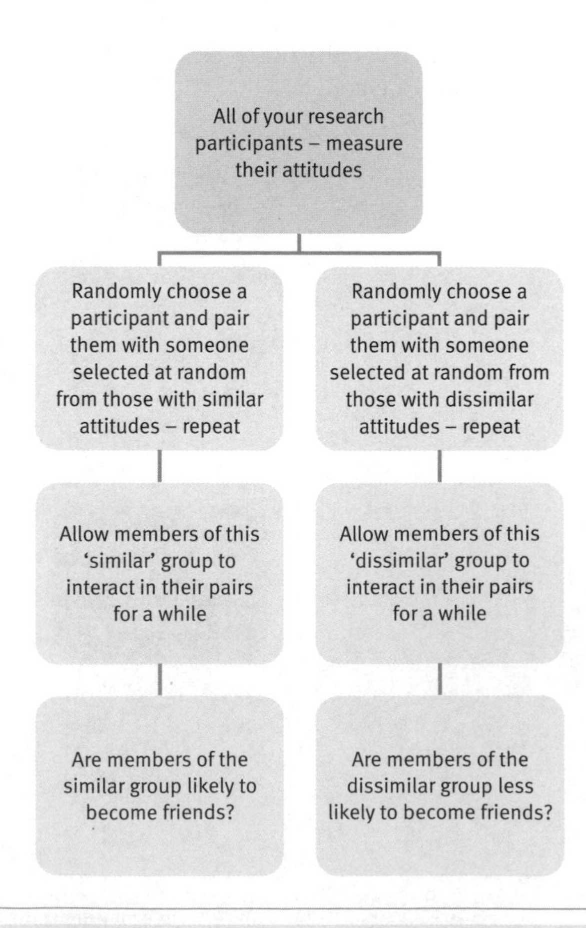

FIGURE 1.11 The experimental design to investigate attitude similarity and friendship

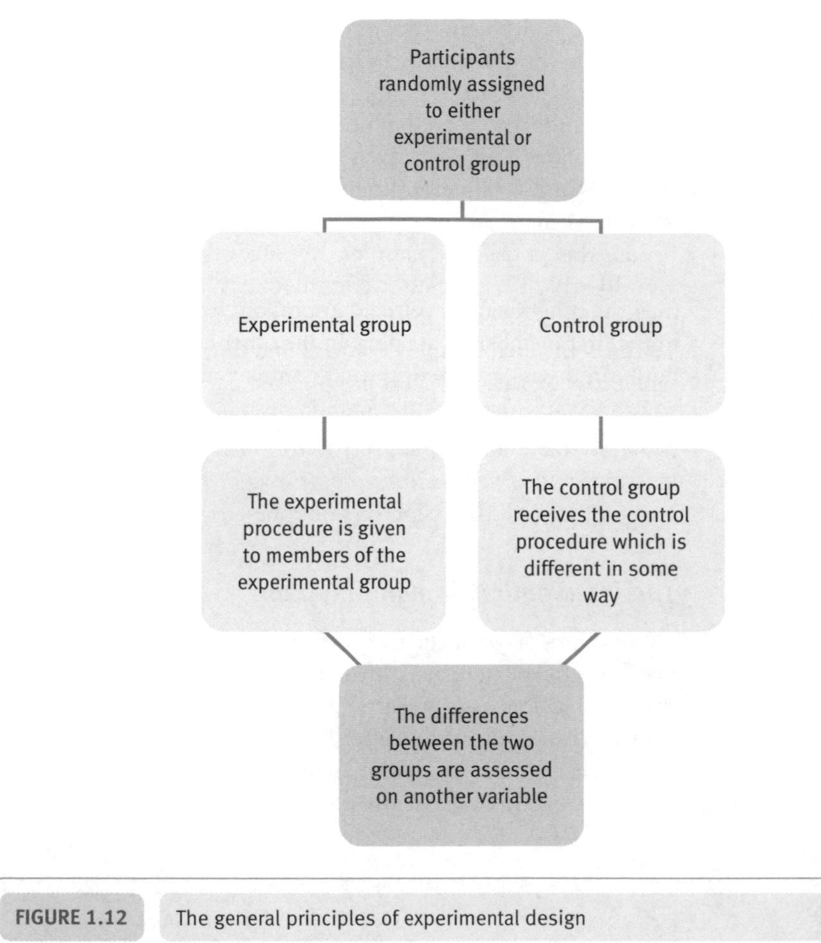

FIGURE 1.12 The general principles of experimental design

and tails 10 times on average. If we had decided that a head means meeting someone with the same attitude, approximately 10 people will have been chosen to meet someone with the same attitude as them and approximately 10 someone with a different attitude from them. This kind of procedure is known as random assignment. People are randomly assigned to different situations which are usually called conditions, groups or treatments. (Actually we have not solved all of the difficulties as we will see later.)

If half the people in our study came from, say, Bristol and half from Birmingham, then about half of the people who were randomly assigned to meeting a person with the same attitude as them would be from Bristol and the remaining half would be from Birmingham, approximately. The same would be true of the people who were randomly assigned to meeting a person with a different attitude from them. About half would be from Bristol and the remaining half would be from Birmingham, approximately. In other words, random assignment should control for the area that people come from by ensuring that there are roughly equal numbers of people from those areas in the two groups. This will hold true for any factor such as the age of the person or their gender. In other words, random assignment ensures that all confounding factors are held constant – without the researcher needing to know what those confounding factors are.

Sampling error

The randomised study is not foolproof. Sampling error will always be a problem. If a coin is tossed any number of times, it will not always come up heads half the time and tails half the time. It could vary from one extreme of no heads to the other extreme of all heads, with the most common number of heads being half or close to half. In other words, the proportion of heads will differ from the number expected by chance. This variability is known as sampling error and is a feature of any study. A sample is the number of people (or units) that are being studied. The smaller that the sample is, the greater the sampling error will be. A sample of 10 people will have a greater sampling error than one of 20 people. Although you may find doing this a little tedious, you could check this for yourself in the following way. Toss a coin 10 times and count the number of heads (this is the first sample). Repeat this process, say 30 times in total, which gives you 30 separate samples of coin tossing. Note the number of times heads comes up for each sample. Now do this again but toss the coin 20 times on each occasion rather than 10 times for each sample. You will find that the number of heads is usually closer to half when tossing the coin 20 times on each occasion rather than 10 times (see Figure 1.13). Many studies will have as few as 20 people in each group or condition because it is thought that the sampling error for such numbers is acceptable. See our companion statistics text, *Introduction to Statistics in Psychology* (Howitt and Cramer, 2011a) for a more detailed discussion of sampling error.

The intervention or manipulation

So, in many ways, if the purpose of one's research is to establish whether two variables are causally related, it is attractive to consider controlling for confounding variables through random assignment of participants to different conditions. To determine whether similar attitudes lead to friendship, we could randomly assign people to meet strangers with either similar or dissimilar attitudes to themselves as we have already described. Remember that we have also raised the possibility that people's attitudes are related to

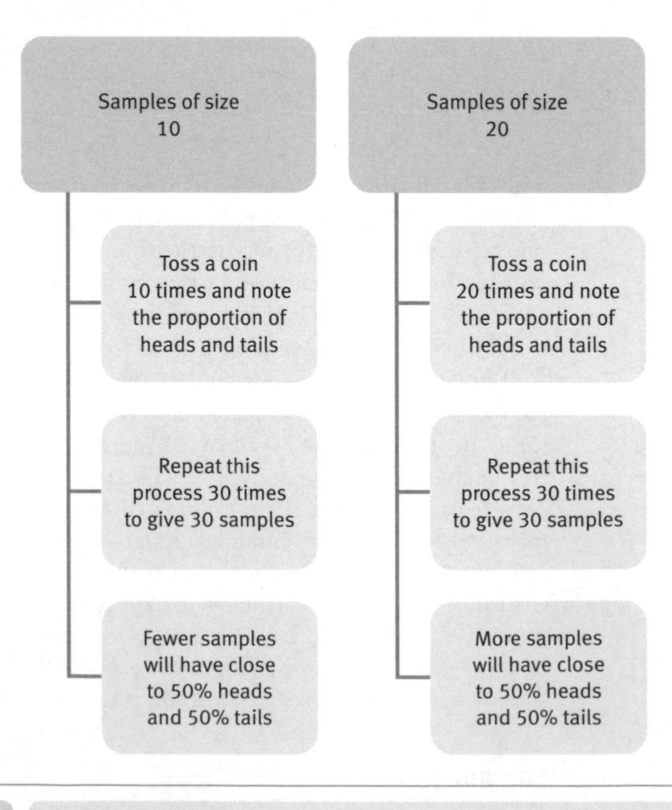
Samples of size
10
Samples of size
20
Toss a coin
10 times and note
the proportion of
heads and tails
Toss a coin
20 times and note
the proportion of
heads and tails
Repeat this
process 30 times
to give 30 samples
Repeat this
process 30 times
to give 30 samples
Fewer samples
will have close
to 50% heads
and 50% tails
More samples
will have close
to 50% heads
and 50% tails

FIGURE 1.13 A sampling 'experiment'

Table 1.1 Manipulating similarity of attitude

Condition	Participant	Stranger
Same attitude	Same as stranger Different from stranger	No acting Act as if the same
Different attitude	Same as stranger Different from stranger	Act as if different No acting

other factors such as the area or kind of area they come from. Assuming that this is the case, participants meeting strangers with the same attitudes as themselves might be meeting people who come from the same area or kind of area as themselves. On the other hand, participants meeting strangers with different attitudes from them may well be meeting people who come from a different area or kind of area to themselves. In other words, we still cannot separate out the effects of having the attitude similarity from the possible confounding effects of area similarity. It is clear that we need to disentangle these two different but interrelated factors. It is not possible to do this using real strangers because we cannot separate the stranger from the place they come from.

Let's consider possible approaches to this difficulty. We need to ensure that the stranger expresses similar attitudes to the participant in the same attitudes condition. That is, if they did not share attitudes with a particular participant, they would nevertheless pretend that they did. In the different attitudes condition, then, we could ensure that the stranger always expresses different attitudes from those of the participant. That is, the stranger pretends to have different attitudes from the participant. See Table 1.1 for an overview of this. In effect, the stranger is now the accomplice, confederate, stooge or co-worker of the researcher with this research design.

The number of times the stranger does not have to act as if they have a different attitude from the one they have is likely to be the same or similar in the two conditions – that is, if participants have been randomly allocated to them. This will also be true for the number of times the stranger has to act as if their attitude is different from the one they have.

Unfortunately, all that has been achieved by this is an increase in complexity of the research design for no other certain gain. We simply have not solved the basic problem of separating similarity of attitude from area. This is because in the same attitude condition some of the strangers who share the same attitude as the participant may well be attractive to the participant actually because they come from the same area as the participant – for example, they may speak with similar accents. Similarly, some of the participants in the different attitudes condition will not be so attracted to the stranger because the stranger comes from a different area. Quite how this will affect the outcome of the research cannot be known. However, the fact that we do not know means that we cannot assess the causal influence of attitude similarity on attraction with absolute certainty.

We need to try to remove any potential influence of place entirely or include it as a variable in the study. Probably the only way to remove the influence of place entirely is by not using a real person as the stranger. One could present information about a stranger's attitude and ask the participant how likely they are to like someone like that. This kind of situation might appear rather contrived or artificial. We could try to make it less so by using some sort of cover story such as saying that we are interested in finding out how people make judgements or form impressions about other people. Obviously the participants would not be told the proposition that we are testing in case their behaviour is affected by being told. For example, they may simply act in accordance with their beliefs about whether or not people are attracted to others with similar attitudes. Not telling them, however, does not mean that the participants do not come to their own conclusions about what the idea behind the study is likely to be and, perhaps, act accordingly.

What we are interested in testing may not be so apparent to the participants because they take part in only one of the two conditions of the study. Consequently they are not so likely to realise what was happening (unless they talked to other people who had already participated in the other condition of the study). We could further disguise the purpose of our study by providing a lot more information about the stranger over and above their attitudes. This additional information would be the same for the stranger in both conditions – the only difference is in terms of the information concerning attitude similarity. In one condition attitudes would be the same as those of the participant while in the other condition they would be different.

If (a) the only difference between the two conditions is whether the stranger's attitudes are similar or dissimilar to those of the participant and (b) we find that participants are more attracted to strangers with similar than with dissimilar attitudes then this difference in attraction must be due to the only difference between the two conditions, that is, the influence of the difference in attitudes. Even then there are problems in terms of how to interpret the evidence. One possibility is that the difference in attraction is not directly due to differences in attitudes themselves but to factors which participants associate with differences in attitudes. For example, participants may believe that people with the same attitudes as themselves may be more likely to come from the same kind of area or be of the same age. Thus it would be these beliefs which are responsible for the differences in attraction. In other words when we manipulate a variable in a study we may, in fact, inadvertently manipulate other variables without realising it. We could try to hold these other factors constant by making sure that the stranger was similar to the participant in these respects, or we could test for the effects of these other factors by manipulating them as well as similarity of attitude.

This kind of study where:

- the presumed cause of an effect is manipulated,
- participants are randomly assigned to conditions, and
- all other factors are held constant

was called a *true experiment* by Campbell and Stanley (1963). In the latest revision of their book, the term 'true' has been replaced by 'randomised' (Shadish, Cook and Campbell, 2002, p. 12). If any of the above three requirements do not hold then the study may be described as a *non-experiment* or *quasi-experiment*. These terms will be used in this book. True or randomised experiments are more common in the sub-disciplines of perception, learning, memory and biological psychology where it is easier to manipulate the variables of interest. The main attraction of true experiments is that they can provide logically more convincing evidence of the causal impact of one variable on another. There are disadvantages which may be very apparent in some fields of psychology. For example, the manipulation of variables may result in very contrived and implausible situations as was the case in our example. Furthermore, exactly what the nature of the manipulation of variables has achieved may not always be clear. Studies are often conducted to try to rule out or to put forward plausible alternative interpretations or explanations of a particular finding. These are generally beneficial to the development of knowledge in that field of research. We will have more confidence in a research finding if it has been confirmed or replicated a number of times, by different people, using different methods and adopting a critical approach.

It should be clear by now that the legitimacy of assertions about causal effects depends on the research design that has been used to study them. If we read claims that a causal effect has been established, then we might be more convinced if we find that the studies in question which showed this effect were true experiments rather than quasi-experiments.

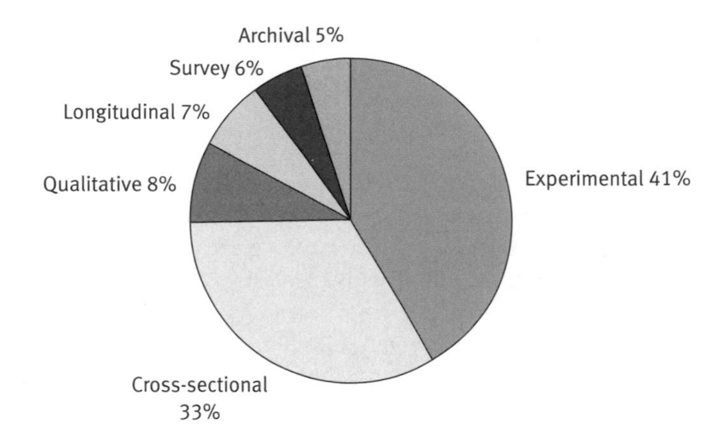

FIGURE 1.14 Different types of design in 200 PsycINFO articles

Furthermore, how effectively the causal variable was manipulated also needs to be considered. Is it possible, as we have seen, that other variables were inadvertently varied at the same time? The nature of the design and of any manipulations that have been carried out are described in journal articles in the section entitled 'Method'.

These and other designs are discussed in more detail in subsequent chapters. Few areas of research have a single dominant method. However, certain methods are more characteristic of certain fields of psychology than others. The results of a survey of a random sample of 200 studies published in the electronic bibliographic database PsycINFO in 1999 (Bodner, 2006) revealed that a variety of research designs are common but dominated by experimental studies. The findings are summarised in Figure 1.14.

Studies investigating the content of psychology journals are not frequent and this is the most recent one. Knowing about the strengths and weaknesses of research designs should help you to be in a better position to critically evaluate their findings. There is more on design considerations in later chapters. A comparison of the main research designs is given in Figure 1.15.

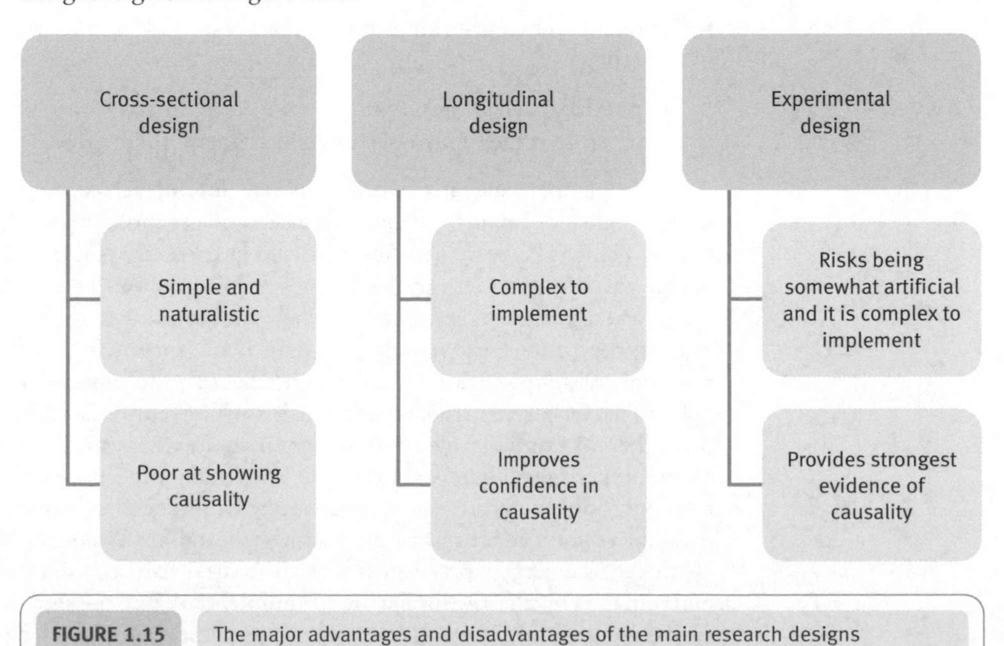

FIGURE 1.15 The major advantages and disadvantages of the main research designs

1.6 Practice

Psychologists believe in the importance of the empirical testing of research ideas. Consequently, doing research is a requirement of most degrees in psychology. For example, to be recognised by the British Psychological Society as a practising psychologist you need to show that you have a basic understanding of research methodology and the skills to carry it out. This is the case even if you do not intend to carry out research in your profession. Training in research is an important part of the training of most practitioners such as educational and clinical psychologists. Practising psychologists simply cannot rely on academic psychologists to research all of the topics from which psychological practice might benefit. The concept of practitioner–researcher has developed in recent years. This is the idea that practitioners such as occupational psychologists and forensic psychologists have a responsibility to carry out research to advance practice in their field of work. To be brutally frank, a student who is not prepared to develop their research skills is doing themselves and the discipline of psychology no favours at all.

1.7 Conclusion

Most psychological ideas develop in relation to empirical data. Propositions are made, tested and emerge through the process of collecting and analysing data. The crucial activity of psychologists is the dissemination of ideas and findings which emerge largely through empirical work in the many fields of psychology. The prime location to find such developments and ideas is in the academic and practitioner journals which describe the outcomes of psychological research. Other important contexts for this are academic and practitioner conferences geared to the presentation of ongoing research developments in psychology and, to a lesser degree, academic books. These various forms of publication and presentation serve a dual purpose:

- To keep psychologists abreast with the latest thinking and developments in their fields of activity.
- To provide psychologists with detailed accounts of developing research ideas and theory so that they may question and evaluate their value.

Although the issue of causality has had less of a role in psychological research in recent years, it remains a defining concern of psychology – and is less typical of some related fields. The basic question involved in causality is the question of whether a particular variable or set of variables causes or brings about a particular effect. Many would argue, though this is controversial, that the best and most appropriate way of testing causal propositions is by conducting 'true' experiments in which participants have been randomly assigned to conditions which reflect the manipulation of possible causal variables. The archetypal true experiment is the conventional laboratory experiment. Even then, there is considerable room for doubt as to what variable has been manipulated in a true experiment. It is important to check out the possibility that the experimental manipulation has not created effects quite different from the ones that were intended. Alternative interpretations of the findings should always be a concern of psychologists. However, the biggest problem is that there are many variables which simply cannot be manipulated by the researcher: for example, it is not possible to manipulate variables such as schizophrenia, gender, social economic status or intelligence for the convenience of testing ideas using true experiments. However, the variety and stimulation of using the more naturalistic or realistic research methods which are often the only rational choice in field settings is a challenge which many psychologists find rewarding.

Often these are described as non-experimental designs which, from some points of view, might be regarded as a somewhat pejorative term. It is a bit like describing women as non-men. It implies that the randomised experiment is the right and proper way of doing psychology. The truth is that there is no right and proper way of intellectual progress. The development of psychology is *not* dependent on any single study but the collective activity of a great many researchers and practitioners. Until there is widespread acceptance and adoption of an idea, it is not possible to judge its value.

Key points

- As a research-based discipline, psychology requires a high degree of sophistication about research and research methods, even as part of the training of psychologists. A vital link in this process is the research article or paper published in academic journals. All psychologists should be able to critically, but constructively, evaluate and benefit from reading such publications.
- Research articles take time to read as they will refer to other research as well as principles of research methodology with which one at first may not be familiar. As one becomes more familiar with the research in an area and with the principles of doing research, the importance of the contents of research papers becomes much quicker and easier to appreciate.
- One major feature of a research study is the design that it uses. There are various designs. A very basic distinction is between what has been called a true or randomised experiment and everything else which can be referred to as a non-experiment.
- True experiments involve the deliberate manipulation of what is presumed to be the causal variable, the random assignment of participants to the conditions reflecting that manipulation and the attempt to hold all other factors constant. Whether or not true experiments should hold a hallowed place in psychology is a matter of controversy. Many researchers and practitioners never have recourse to their use or even to the use of their findings.
- Even if their value is accepted, the use of true experiments is not straightforward. The manipulation of the presumed causal variable and holding all other factors constant is often very difficult. Consequently, a study is never definitive in itself since it requires further research to rule out alternative interpretations of the manipulation by allowing for particular factors which were not held constant in the original study.
- Psychologists generally favour the true experiment because it appears to be the most appropriate way for determining causal effects. If you have manipulated only one variable, held all else constant and found an effect, then that effect is likely to be due to the manipulation. At the same time, however, this is also a potential fatal flaw of true experiments. In real life, variables do not operate independently and one at a time so why should research assume that they do?
- Furthermore, it is not always possible to manipulate the variable presumed to be a cause or to manipulate it in a way which is not contrived. Anyway, not all psychologists are interested in testing causal propositions. Hence the trend for psychologists to increasingly use a wide variety of non-experimental research designs.
- Because of the centrality of research to all aspects of psychology, psychology students are generally required and taught to carry out and write up research. This experience should help them understand what research involves. It also gives them an opportunity to make a contribution to a topic that interests them.

ACTIVITIES

1. Choose a recent study that has been referred to either in a textbook you are reading or in a lecture that you have attended. Obtain the original publication. Were the study and its findings correctly reported in the textbook? Do you think that there were important aspects of the study that were not mentioned in the text or the lecture that should have been? If you do think there were important omissions, what are these? Why do you think they were not cited? Did the study test a causal proposition? If so, what was this proposition? If not, what was the main aim of this study? In terms of the designs outlined in this chapter what kind of design did the study use?

2. Either choose a chapter from a textbook or go to the library and obtain a copy of a single issue of a journal. Work through the material and for every study you find, classify it as one of the following:

   - correlational or cross-sectional study
   - longitudinal study
   - experiment – or study with randomised assignment.

   What percentage of each did you find?

CHAPTER 2

Aims and hypotheses in research

Overview

- Different research methods are effective at doing different things. There are methods which are particularly good at describing a phenomenon in some detail, estimating how common a particular behaviour is, evaluating the effects of some intervention, testing a causal proposition or statistically summarising the results of a number of similar studies. No method satisfies every criterion.
- The aims and justification of the study are presented in the first section or introduction of the report or paper. All research should have clear objectives and needs clear justification for the expenditure of time and effort as well as the procedures carried out in the research.
- Hypotheses are a key component of research studies in the mainstream of psychology. Hypotheses are usually formally stated in a clear and precise form. They also need to be justified. It should be made apparent why it is important to test the hypotheses and what the basis or rationale is for them.
- Alternatively, the general aims and objectives of the research can be summarised if the research in question is not particularly conducive to presentation as a hypothesis about the relationships between small numbers of variables.
- Hypotheses are the basic building blocks of much of psychology. Some research attempts to test hypotheses, other research attempts to explore hypotheses, and yet other research seeks to generate hypotheses.
- In their simplest form, hypotheses propose that a relationship exists between a minimum of two variables.
- There is an important distinction between research hypotheses (which guide research) and statistical hypotheses (which guide statistical analyses). Research hypotheses

are evaluated by a whole range of different means, including statistics. Statistical hypothesis testing employs a very restricted concept of hypothesis.

- Of course, frequently the researcher has an idea of what the relationship is between two variables. That is, the variables are expected or predicted to be related in a particular way or direction. So wherever possible, the nature (or direction) of the relationship should be clearly stated together with the reasons for this expectation.
- The variable that is manipulated or thought to be the cause of an effect on another variable is known as the independent variable. The variable that is measured or thought to be the effect of the influence of another variable is known as the dependent variable.
- The terms 'independent' and 'dependent' are sometimes restricted to true experiments where the direction of the causal effect being investigated is clearer. However, they are frequently used in a variety of contexts and are terms which can cause confusion.
- The hypothetico-deductive method describes a dominant view of how scientists go about their work. From what has been observed, generalisations are proposed (i.e. hypotheses) which are then tested, ideally by using methods which potentially could disconfirm the hypothesis. The scientist then would either reformulate their hypothesis or test the hypothesis further depending on the outcome.

2.1 Introduction

By now, it should be clear that research is an immensely varied activity with many different objectives and purposes. In psychology, these range as widely, perhaps more so, as in any other discipline. In this chapter we will look in some detail at the keystones of most research in psychology: the aims and hypotheses underlying a study. Research is a thoughtful, rational process. It does not proceed simply by measuring variables and finding out what the relationship is between them. Instead, research is built on a sense of purpose on the part of the researcher who sees their work as fitting in with, and building on, established psychological knowledge in their chosen field. This sense of direction in research is not simply something that happens, it has to be worked at and worked towards. The idea of research is not simply to create new information or facts but to build on, expand, clarify and illuminate what is already known. To collect data without a sense of direction or purpose might be referred to, cruelly, as mindless or rampant empiricism. Simply collecting data does not constitute research.

The sense of purpose in research has to be learnt. Most of us develop it slowly as part of learning to appreciate the nature of psychology itself. That is, until one has begun to understand that psychology is more than just a few facts to learn then good research ideas are unlikely. There are a number of aspects to this:

- It is vital to understand how real psychologists (not just textbook authors) go about psychology. The only way to achieve this is to read and study in depth the writings of psychologists – especially those interested in the sorts of things that you are interested in.
- The way in which real psychologists think about their discipline, their work and the work of their intellectual colleagues has to be studied. Throughout the writings of psychologists, one will find a positive but sceptical attitude to theory and research.

There is a sense in which good psychologists regard all knowledge as tentative and even temporary – the feeling that, collectively, psychologists could always do better.

2.2 Types of study

One useful way of beginning to understand the possible aims of psychological research is to examine some broad research objectives in psychology and decide what each of them contributes. We will look at the following in turn:

- descriptive or exploratory studies;
- evaluation or outcome studies;
- meta-analytic studies.

Descriptive or exploratory studies

An obvious first approach to research in any field is simply to describe in detail the characteristics and features of the thing in question. Without such descriptive material, it is difficult for research to progress effectively. For example, it is difficult to imagine research into, say, the causes of schizophrenia without a substantial body of knowledge which describes the major features and types of schizophrenia. Descriptions require that we categorise in some way the observations we make. Curiously, perhaps perversely, psychology is not replete with famous examples of purely descriptive studies. In Part 4 of this book we discuss in detail qualitative research methods. Typical of this type of research is the use of textual material which is rich in detail and this may include descriptive analyses as well as analytic interpretations.

Case studies are reports that describe a particular case in detail. They are common in psychiatry though, once again, relatively uncommon in modern psychology. An early and often cited instance of a case study is that of 'Albert' in which an attempt was made to demonstrate that an 11-month-old boy could be taught or conditioned to become frightened of a rat when he previously was not (Watson and Rayner, 1920). Whether or not this is a purely descriptive study could probably be argued either way. Certainly the study goes beyond a mere description of the situation; for example, it could also be conceived as investigating the factors that can create fear.

In some disciplines (such as sociology and media studies), one sort of descriptive study, known as *content analysis*, is common. The main objective of this is to describe the contents of the media. So, it is common to find content analyses which report the features of television's output. For example, the types of violence contained in television programmes could be recorded, classified and counted. That is to say, the main interest of these studies lies in determining how common certain features are. Actually we have already seen a good example of content analysis. One aim of the study by Bodner (2006) mentioned in Chapter 1 was to find out the characteristics of studies published in 1999 in PsycINFO. The type of research design employed was one of the categories used by the researchers to classify the contents of the journal.

Evaluation or outcome studies

Other research has as its aim to test the effectiveness of a particular feature or intervention. Generally speaking, such studies simply concentrate on the consequences of certain activities without attempting to test theoretical propositions or ideas – that is to say, they tend to have purely empirical objectives. They often do not seek to develop theory. Good examples of an intervention into a situation are studies of the effectiveness of pscho-therapeutic treatments. Ideally in such studies participants are randomly assigned to the different treatments or conditions and, usually, one or more non-treated or control conditions. These studies are sometimes referred to as evaluation or outcome studies. When used to evaluate the effectiveness of a clinical treatment such as psychotherapy, evaluation studies are known as randomised controlled trials. Usually the purpose of the evaluation study is to assess whether the intervention taken as a whole is effective. Rarely is it possible to assess which aspects of the intervention are producing the observed changes. Nevertheless, one knows that the intervention as a whole has (or has not) achieved its desired ends. Since interventions usually take place over an extended period of time, it is much more difficult to hold other factors constant than it would in many laboratory studies that last just a few minutes.

So evaluation studies frequently seek to examine whether an intervention has had its intended effect. That is, did the intervention cause the expected change? However, explanations about why the intervention was successful are secondary or disregarded as the primary objective is not theory development.

Meta-analytic studies

A meta-analysis has the aim of statistically summarising and analysing the results of the range of studies which have investigated a particular topic. Of course, any review of studies tries to integrate the findings of the studies. Meta-analysis does this in a systematic and structured way using statistical techniques. Because it provides statistical methods for combining and differentiating between the findings of a number of data analyses, it forms a powerful integrative tool. For example, we may be interested in finding out whether cognitive behaviour therapy is more effective in treating phobias or intense fears than no treatment. If we obtain reports of studies which have investigated this question, they will contain information about the statistical trends in the findings of each of these studies. These trends may be used to calculate what is known as an *effect size*. This is merely a measure of the size of the trend in the data – depending on the measure used then this may be adjusted for the variability in the data.

There are several different measures of effect size. For example, in Chapter 35 of the companion volume *Introduction to Statistics in Psychology* (Howitt and Cramer, 2011a), we describe the procedures using the correlation coefficient as a measure of effect size. As the correlation coefficient is a common statistical measure, it is familiar to most researchers. There are other measures of effect size. For example, we can calculate the difference between the two conditions of the study and then standardise this by dividing by a measure of the variability in the individual scores. Variability can either be the standard deviation of one of the conditions (as in Glass's Δ) or the combined standard deviation of both conditions of the study (as in Cohen's d) (see Rosenthal, 1991). We can calculate the average effect size from any of these measures. Because this difference or effect size is based on a number of studies, it is more likely to give us a more clear assessment of the typical effects found in a particular area of research.

We can also see whether the effect size differs according to the ways in which the studies themselves might differ. For example, some of the studies may have been carried out on student volunteers for a study of the treatment of phobias. Because these participants have not sought professional treatment for their phobias these studies are sometimes referred to as analogue studies. Other studies may have been conducted on patients who sought professional help for their phobias. These studies are sometimes called clinical studies. We may be interested in seeing whether the effect size differs for these two types of study. It may be easier to treat phobias in students because they may be less severe. That is, the effect size will be greater for studies of the treatment of phobias using student volunteers. If there are differences in the effect size for the two kinds of study, we should be more cautious in generalising from analogue studies to clinical ones. Actually, any feature of the studies reviewed in the meta-analysis may be considered in relation to effect size even, for example, such things as the year in which it was published. The results of earlier research may be compared with later research.

When reading the results of a meta-analysis, it is important to check the reports of at least a few of the studies on which the meta-analysis was based. This will help you to familiarise yourself with specifics of the designs of these studies. Some social scientists have argued against the use of meta-analyses because they combine the results of well-designed studies with poorly designed ones. Furthermore, the results of different types of studies might also be combined. For example, they may use studies in which participants were randomly assigned to conditions together with ones in which this has not been done (Shadish and Ragsdale, 1996). Differences in the quality of design are more likely to occur in evaluation studies which are more difficult to conduct without adequate resources. However, one can compare the effect size of these two kinds of studies to see whether the effect size differs. If the effect size does not differ (as has been found in some studies), then the effect size is unlikely to be biased by the more poorly designed studies. Sometimes, meta-analytic studies have used ratings by researchers of the overall quality of each of the studies in the meta-analysis. In this way, it is possible to investigate the relationship between quality of the study and the size of the effects found. None of this amounts to a justification for researchers conducting poorly designed studies.

While few students contemplate carrying out a meta-analysis (though it is difficult to understand this reluctance), meta-analytic studies are increasingly carried out in psychology. The biggest problem with them is the need to obtain copies of the original studies from which to extract aspects of the original analysis.

2.3 Aims of research

Already it should be abundantly clear that psychological research is an intellectually highly organised and coherent activity. Research, as a whole, does not proceed willy-nilly at the whim of a privileged group of dilettante researchers. The research activities of psychologists are primarily directed at other psychologists. In this way, individual researchers and groups of researchers are contributing to a wider, collective activity. Research which fails to meet certain basic requirements is effectively excluded. Research which has no point, has a bad design, or is faulty in some other way has little chance of being published, heard about and read. The dissemination of research in psychology is subject to certain quality controls which are largely carried out by a peer review process in which experts in the field recommend whether a research report should be published or not.

Researchers have to account for the research they do by justifying key aspects of their work. Central to this is the requirement that researchers have a good, sound purpose for doing the research that they do. In other words, researchers have to specify the *aims* of their research. This is two fold:

- The researcher needs to have a coherent understanding of what purposes the research will serve and how likely it is to serve these purposes. A researcher who cannot see the point of what they are doing is likely to be a dispirited, bad researcher. Obviously, this is most likely to be the case with student researchers doing research under time pressure to meet course requirements. So clarity about the aims of research is, in the first instance, an obligation of the researcher to themselves.

- The researcher needs to be able to present the aims of their studies with enough clarity to justify the research to interested, but critical, others. This is always done in research reports, but it is also necessary, for example, in applications for research funds to outside bodies.

Clearly stated aims are essential means of indicating what the research can contribute. They also help clarify just why the research was done in the way in which it was done. By clarity, we mean a number of things. Of course it means that the aims are presented as well-written, grammatical sentences. More importantly, the aims of the research need to be clearly justified by providing their rationale. The introduction of any research report is where the case for the aims of the research is made. Justifying the aims of research can involve the following:

- Explaining the relevance of the research to what is already known about the topic. The explanation of and justification for the aims of a piece of research may include both previous theoretical and empirical advancements in the field. For many topics in psychology there may well be a great deal of previous research literature. This can be daunting to newcomers. (Chapter 7 on searching the literature describes how one can efficiently and effectively become familiar with the relevant research literature on a topic.) Examples of the sorts of reasons that can justify doing research on a particular topic are discussed in Chapter 26.

- Reference to the wider social context for research. Psychological research is often a response to the concerns of broader society as exemplified by government, social institutions such as the legal and educational system, business and so forth. Of course, there are substantial amounts of published material which emanate from these sources – government publications, statistical information, discussion documents and professional publications. These are largely not the work of psychologists but are relevant to their activities.

2.4 Research hypotheses

The use of hypotheses is far more common in psychological research than in disciplines such as sociology, economics and other related disciplines. It is a concept which derives from natural sciences such as physics, chemistry and biology which have influenced mainstream psychology more than other social and human sciences. The aims of a great deal of research in psychology (but by no means all) may be more precisely formulated in terms of one or more working suppositions about the possible research findings. These are known as hypotheses. A hypothesis does not have to be true since the point of research is to examine the empirical support or otherwise for the hypothesis. So hypotheses are working assumptions or propositions expressing expectations linked to the aims of the study.

In practice, it is not a difficult task to write hypotheses once we have clarified just what our expectations are. Since a hypothesis is merely a statement which describes the relationship expected to hold between two (or more) variables, at a minimum we need to identify what two variables we are interested in and propose that there is a relationship between the two. We could go one step further and specify the nature of that relationship. Taking the idea that we introduced in Chapter 1 that people are attracted to other people on the basis of having similar attitudes to each other, what would the hypothesis be? The two variables which derive from this might be 'attitude similarity' and 'attraction'. The hypothesised relationship between the two is that the greater the attitude similarity then the greater the attraction to the other person. Expressed as a hypothesis this could read something like: 'Higher levels of attitude similarity lead to higher levels of attraction.' However, there are many ways of writing the same thing as the following list of alternatives demonstrates:

- People with more similar attitudes will be more attracted to each other than people with less similar attitudes.
- Greater attitude similarity will be associated with greater interpersonal attraction.
- Attitude similarity is *positively* linked with interpersonal attraction.

The terms positive and negative relationship or association are fundamental concepts in research. It is important to understand their meaning as they are very commonly used phrases:

- A positive or direct association is one in which *more* of one quality (attitude similarity) goes together with *more* of another quality (interpersonal attraction).
- A negative or inverse association is one in which *more* of one quality (attitude similarity) goes together with *less* of another quality (interpersonal attraction).

An example of a negative or inverse association would be that greater attitude similarity is associated with *less* attraction. This is *not* the hypothesis we are testing, though some might consider it a reasonable hypothesis – after all there is an old saying which suggests that opposites attract. Both past research and theory have led us to the expectation that similarity leads to attraction. If that did not exist, then we would have little justification for our choice of hypothesis.

The precise phrasing of a hypothesis is guided by considerations of clarity and precision. Inevitably, different researchers will use different ways of saying more or less the same thing.

Hypotheses can be somewhat more complex than the above example. For instance, a third variable could be incorporated into our hypothesis. This third variable might be the importance of the attitudes to the individual. So it might be suggested that the more important the attitude is to the person the more they will be attracted to someone with a similar attitude. So this hypothesis actually contains three variables:

- 'attitude importance';
- 'attitude similarity';
- 'interpersonal attraction'.

In this case, the hypothesis might be expressed something like this: 'The relationship between attitude similarity and attraction will be greater when the attitudes are important.' This is quite a technically complex hypothesis to test. It requires a degree of sophistication about aspects of research design and statistical analysis. So, at this stage, we will try to confine ourselves to the simpler hypotheses that involve just two variables.

Few researchers do research which has a single aim. Usually studies involve several interrelated aims. This helps the researcher to take advantage of economies of time and other resources. A study which tests several hypotheses at the same time also potentially has more information on which to base conclusions. Another advantage is that there is a better chance that the researcher has something more interesting and more publishable. Of course, studies carried out as part of training in psychological research methods may be equally or more effective for teaching purposes if a single hypothesis is addressed.

2.5 Four types of hypothesis

The distinction between relationships and causal relationships is important. Hypotheses should be carefully phrased in order to indicate the causal nature or otherwise of the relationships being investigated.

The statement that attitude similarity is associated with attraction is an example of a non-causal hypothesis. It indicates that we believe that the two variables are interrelated but we are *not* indicating that one variable is causing the other. An association between two variables is all that we can infer with confidence when we measure two variables at the same time. Many psychologists would argue that, strictly speaking, hypotheses should be presented in a non-causal form when a non-experimental design is used. When a true experimental design is used, then the use of terms which refer directly or indirectly to a causal relationship is appropriate. True experimental designs involve the manipulation of the causal variable, participants are randomly assigned to conditions and all else is held constant. Expressing the hypothesis of a true experiment in a non-causal form fails to give credit to the main virtue of this design.

There is a range of terms which psychologists use which indicate that a causal relationship is being described or assumed. Some of these are illustrated in Figure 2.1. These phrases are so associated with questions of causality that they are best reserved for when causality is assumed to avoid confusion.

The direction of the expected relationship should be incorporated into the wording of the hypothesis if at all possible. But this is not a matter of whim and there should be good reasons for your choice. Hypotheses which indicate direction could be:

- Greater attitude similarity will lead to greater attraction.
- Greater attitude similarity will be associated with greater interpersonal attraction.

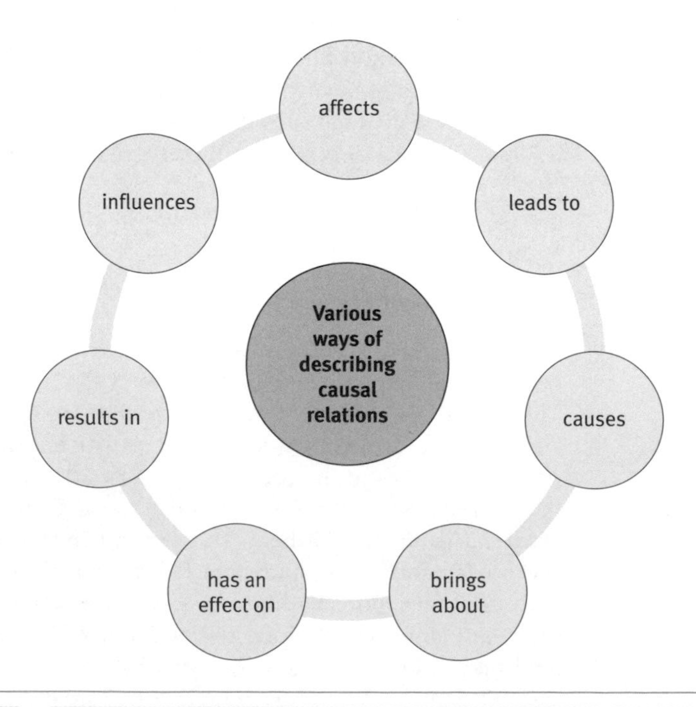

FIGURE 2.1 Alternative ways of writing about a causal relationship

Since such hypotheses indicate the direction of the relationship expected they are referred to as *directional hypotheses*. There are circumstances in which we may not be able to make predictions as to the direction of the relationship with any confidence. For example, there may be two different, but equally pertinent, theories which lead us to expect contradictory results from our study. For example, social learning theory predicts that watching aggression should lead to greater aggression whereas the idea of catharsis predicts that it should lead to less aggression. Of course, it is not always possible to pin a direction to a relationship. Sometimes hypotheses have to be stated without specifying a direction simply because there are reasonable arguments to expect either outcome and there is no strong reason to predict a particular direction of outcome. There are important issues connected with the statistical analyses of such hypotheses. These are discussed in Box 2.1.

Box 2.1 Key Ideas

Direction, hypotheses and statistical analysis

It is important to differentiate between:

- assessing the adequacy of the research hypothesis which underlies the research study in question;
- testing the null hypothesis and alternate hypothesis in significance testing (or statistical inference) as part of the statistical analysis.

These are frequently confused. The hypothesis testing model in statistical analysis deals with a very simple question: are the trends found in the data simply the consequence of chance fluctuations due to sampling? Statistical analysis in psychology is guided by the Neyman–Pearson hypothesis testing model although it is rarely referred to as such and seems to be frequently just taken for granted. This approach had its origins in the 1930s. In the Neyman–Pearson hypothesis testing model there are two statistical hypotheses offered:

- That there is no relationship between the two variables that we are investigating – this is known as the *null hypothesis*.
- That there is a relationship between the two variables – this is known as the *alternate hypothesis*.

The researcher is required to choose between the null hypothesis and the alternate hypothesis. They must accept one of them and reject the other. Since we are dealing with probabilities, we do not say that we have proven the hypothesis or null hypothesis. In effect, hypothesis testing assesses the hypothesis that any trends in the data may be reasonably explained by chance due to using samples of cases rather than all of the cases. The alternative is that the relationship found in the data represents a substantial trend which is not reasonably accountable for on the basis of chance.

To put it directly, statistical testing is only one aspect of hypothesis testing. We test research hypotheses in other ways in addition to statistically. There may be alternative explanations of our findings which perhaps fit the data even better, there may be methodological flaws in the research that statistical analysis is not intended to, and cannot, identify, or there may be evidence that the hypotheses work only with certain groups of participants, for example. So significance testing is only a minimal test of a hypothesis – there are many more considerations when properly assessing the adequacy of our research hypothesis.

Similarly, the question of direction of a hypothesis comes up in a very different way in statistical analysis. Once again, one should not confuse direction when applied to a research hypothesis with direction when applied to statistical significance testing. One-tailed testing and two-tailed testing are discussed in virtually any statistics textbook (for example, Chapter 17 of our companion statistics text *Introduction to Statistics in Psychology* (Howitt and Cramer, 2011a) is devoted to this topic). Quite simply, one-tailed testing is testing a directional hypothesis whereas two-tailed testing is for testing non-directional hypotheses. However, there are exacting requirements which need to be met before applying one-tailed testing to a statistical analysis:

- There should be very strong theoretical or empirical reasons for expecting a particular relationship between two variables.
- The decision about the nature of the relationship between the two variables should be made in ignorance of the data. That is, you do not check the data first to see which direction the data are going in – that would be tantamount to cheating.
- Neither should you try a one-tail test of significance first and then try the two-tail test of significance in its place if the trend is in the incorrect direction.

These requirements are so demanding that very little research can justify the use of one-tailed testing. Psychological theory is seldom so well developed that it can make precise enough predictions about outcomes of new research, for example. Previous research in psychology has a tendency to manifest very varied outcomes. It is notorious that there is often inconsistency between the outcomes of ostensibly similar studies in psychology. Hence, the difficulty of making precise enough predictions to warrant the use of one-tail tests.

One-tailed (directional) significance testing will produce statistically significant findings more readily than two-tailed testing – so long as the outcome is in the predicted direction. Hence the need for caution about its incorrect use since we are applying a less stringent test if these requirements are violated. Two-tailed testing should be the preferred method in all but the most exceptional circumstances as described above. The criteria for one- and two-tailed two types of significance are presented in Figure 2.2.

The distinction between a research hypothesis (which is evaluated in a multitude of ways) and a statistical hypothesis (which can be evaluated statistically only through significance testing) is very important. Any researcher who evaluates the worth of a research hypothesis merely on the basis of statistical hypothesis testing has only partially completed the task.

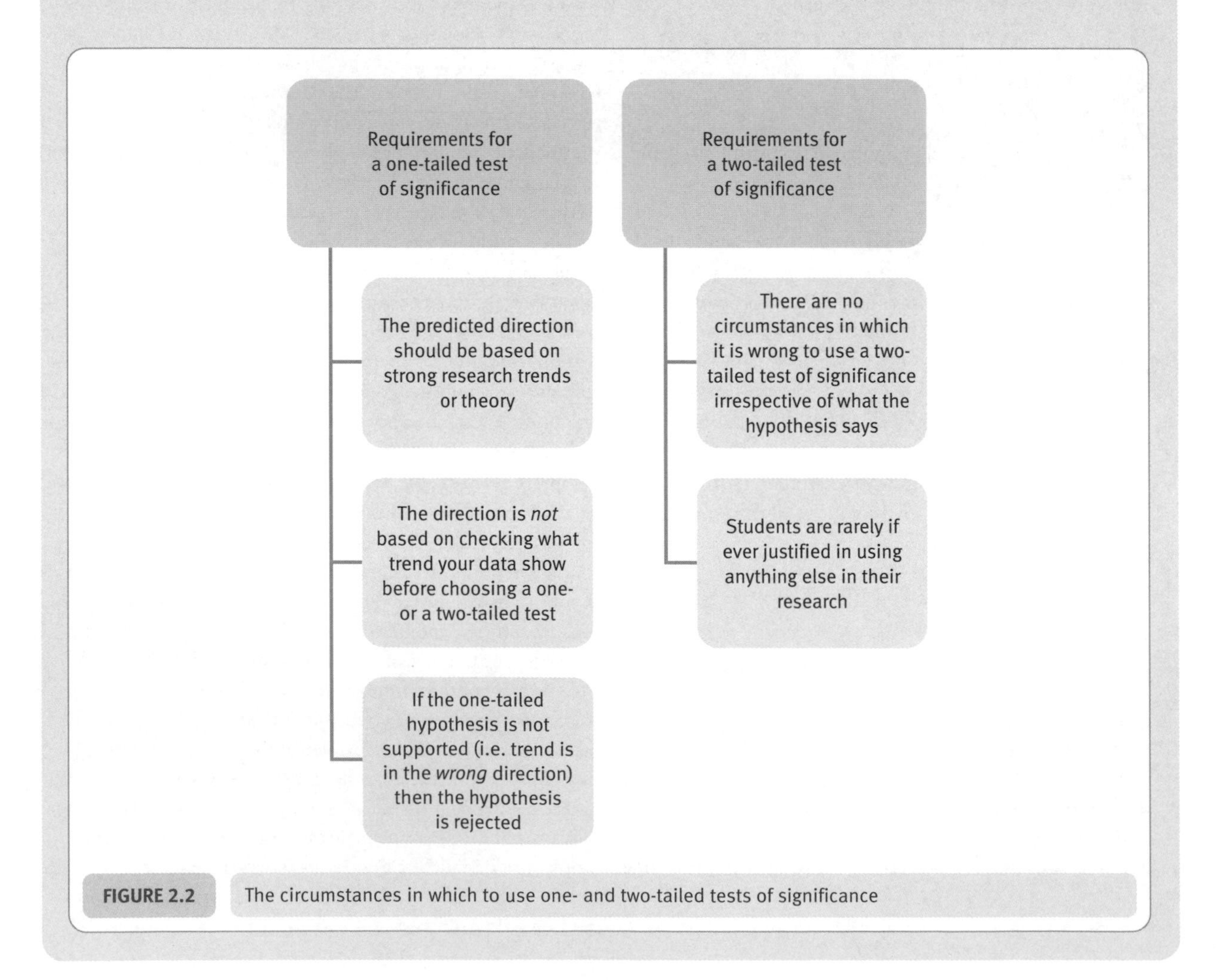

FIGURE 2.2 The circumstances in which to use one- and two-tailed tests of significance

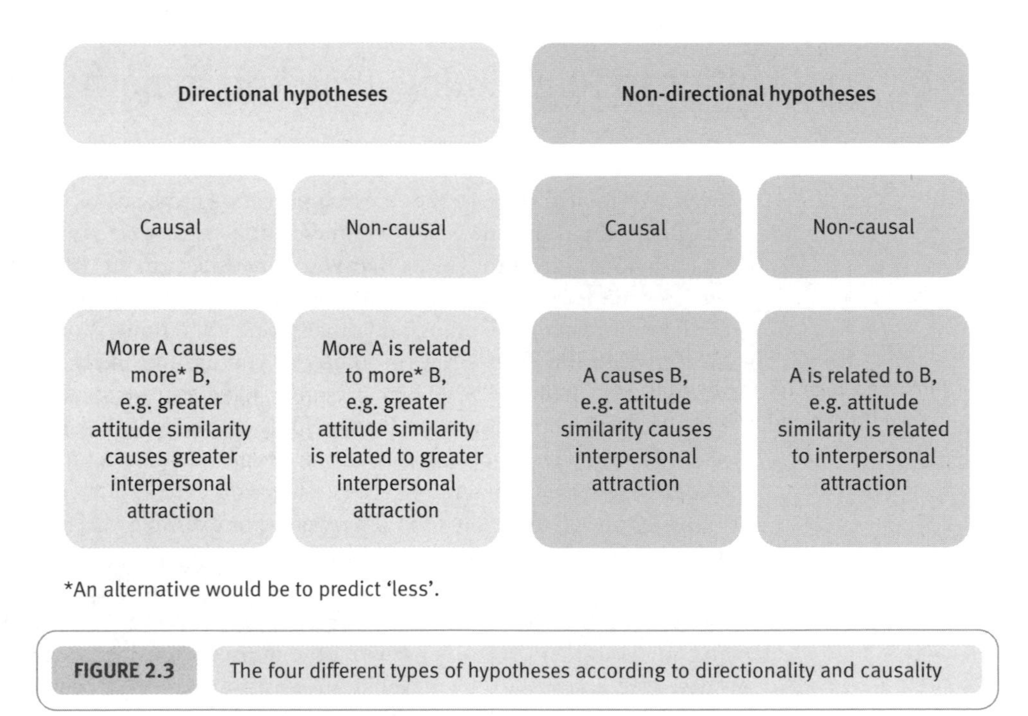

FIGURE 2.3 The four different types of hypotheses according to directionality and causality

Figure 2.3 summarises the four possible types of hypothesis which can be generated by considering the causal versus non-causal and directional versus non-directional distinctions. The letters A and B refer to the two variables. So A could be attitude similarity and B interpersonal attraction.

It should be stressed that without a rationale for a hypothesis based on theory or previous research, the case for examining the relationship between two variables is weakened. Consequently, consideration should be given to other reasons for justifying researching the relationship between two variables. Given the billions of potential variables that could be available to psychologists, why choose variable 2 743 322 and variable 99 634 187 for study? Research is *not* about data collection and analysis for its own sake. Research is part of a systematic and coordinated attempt to understand its subject matter. Until one understands the relationship between research and advancement of understanding, research methods will probably remain a mass of buzzing confusion.

The aims and hypotheses of a study are its driving force. Once the aims and hypotheses are clarified, other aspects of the research fall into place much more easily. They help focus the reading of the published literature on pertinent aspects since the aims and hypotheses help indicate what is most relevant in what we are reading. Once the past research and writings relevant to the new research study have been identified with the help of clear aims and hypotheses, the introduction can be written using more convincing and coherent justifications for them. The aims and hypotheses clarify what variables will need to be measured. Similarly, the aims and hypotheses help guide the researcher towards appropriate research design. The data will support or not support the hypotheses, either wholly or partially. Finally, the discussion of the results will primarily refer back to the aims and hypotheses. It is hardly surprising, then, to find that the aims and hypotheses of a study can be the lynchpin that holds a report together. If they are incoherent and confused then little hope can be offered about the value of the study.

2.6 Difficulties in formulating aims and hypotheses

The aims or objectives of published studies are usually well defined and clear. They are, after all, the final stage of the research process – publication. It is far more difficult to be confident about the aims and hypotheses of a study that you are planning for yourself. One obvious reason for this is that you are at the start of the research process. Refining one's crude ideas for research into aims and hypotheses is not easy – there is a lot of reading, discussing, planning and other work to be done. You will usually have a rough idea of what it is that you want to do but you are not likely to think explicitly in terms of aims and hypotheses – you probably have little experience after all. Take some comfort in personal construct theory (Kelly, 1955) which suggests that humans act like scientists and construct theories about people and the nature of the world. You may recognise yourself behaving like this when you catch yourself thinking in ways such as 'if this happens, then that should happen'. For example, 'if I send a text message then he might invite me to his party in return'.

This kind of statement is not different from saying 'if someone has the same attitude as someone else, then they will be attracted to that person' or 'the more similar someone's attitude is to that of another person, the more they will be attracted to that individual'. These are known as conditional propositions and are clearly not dissimilar from hypotheses. This kind of thinking is not always easy to recognise. Take, for example, the belief or statement that behaviour is determined by one's genes. At first sight this may not appear to be a conditional or 'if . . . , then . . .' proposition. However, it can be turned into one if we restate it as 'if someone has the same genes as another person, they will behave in the same way' or 'the more similar the genes of people are, the more similar they will behave'.

There is another fundamental thing about developing aims and hypotheses for psychological research. If people are natural scientists testing out theories and hypotheses, they also need to have a natural curiosity about people and the world. In other words, research ideas will only come to those interested in other people and society. Research can effectively be built on your interests and ideas just so long as you remember that these must be integrated with what others have done starting with similar interests.

Box 2.2 Key Ideas

Hypothetico-deductive method

The notion of the hypothesis is deeply embedded in psychological thinking and it is also one of the first ideas that psychology students learn about. However, it is a mistake to think that the testing of hypotheses is the way in which psychological research must invariably proceed. The process of hypothesis testing, however, particularly exemplifies the approach of so-called scientific psychology. Karl Popper, the twentieth-century philosopher, is generally regarded as the principal advocate and populariser of the hypothetico-deductive method, although it has its origins in the work of the nineteenth-century academic William Whewell. The foundation of the method, which is really a description of how scientists do their work, is that scientists build from the observations they make through the process of inductive reasoning. Induction refers to making generalisations from particular instances. These inductions in the scientific method are referred to as hypotheses, which comes from a Greek word meaning 'suggestion'. Thus when scientists develop hypotheses they are merely making a suggestion about what is happening in general based on their observations. Notice that induction is a creative process and that it is characteristic of much human thinking, not just that of scientists.

In the scientific method, hypotheses are tested to assess their adequacy. There are two main ways of doing this: (1) by seeking evidence which confirms the hypothesis and (2) by seeking evidence which disconfirms the hypothesis. There are problems in using confirmatory evidence since this is a very weak test of a hypothesis. For example, take Sigmund Freud's idea that hysteria is a 'disease' of women. We could seek confirmatory evidence of this by studying women and assessing them for hysteria. Each woman who has hysteria does, indeed, confirm the hypothesis. The women who do not have hysteria do not refute the hypothesis since there was no claim that all women suffer hysteria. However, by seeking confirmatory evidence, we do not put the hypothesis to its most stringent test. What evidence would disconfirm the hypothesis that hysteria is a disease of women? Well, evidence of the existence of hysteria in men would undermine the hypothesis, for example. So a scientist seeking to evaluate a hypothesis by looking for disconfirming evidence might study the incidence of hysteria in men. Any man found to have hysteria undermines the stated hypothesis. In other words, a negative instance logically should have much greater impact than any number of confirmatory instances. So, the word 'deductive' in 'hypothetico-deductive' method refers to the process of deducing logically a test of a hypothesis (which, in contrast, is based on inductive reasoning).

In this context, one of Karl Popper's most important contributions was his major proposal about what it is which differentiates scientific thinking from other forms of thinking. This is known as demarcation since it concerns what demarcates the scientific approach from non-scientific approaches. For Popper, an idea is scientific only if it is falsifiable and, by implication, a theory is scientific only if it is falsifiable. So some ideas are intrinsically non-scientific, such as the view that there is a god. It is not possible to imagine the evidence which disconfirms this so the idea is not falsifiable. Popper criticised the scientific status of the work of Sigmund Freud because Freud's theories, he argued, were often impossible to falsify.

The hypothetico-deductive method can be seen as a process as illustrated in Figure 2.4. Disconfirmation of a hypothesis should lead to an upsurge in the creative process as new hypotheses need to be developed which take account of the disconfirmation. On the other hand, finding support for the hypothesis does not imply an end to the researcher's attempts to test the hypothesis since there are many other possible ways of disconfirming the hypothesis. This is part of the reason why psychologists do not speak of a hypothesis as being proven but say that it has been supported.

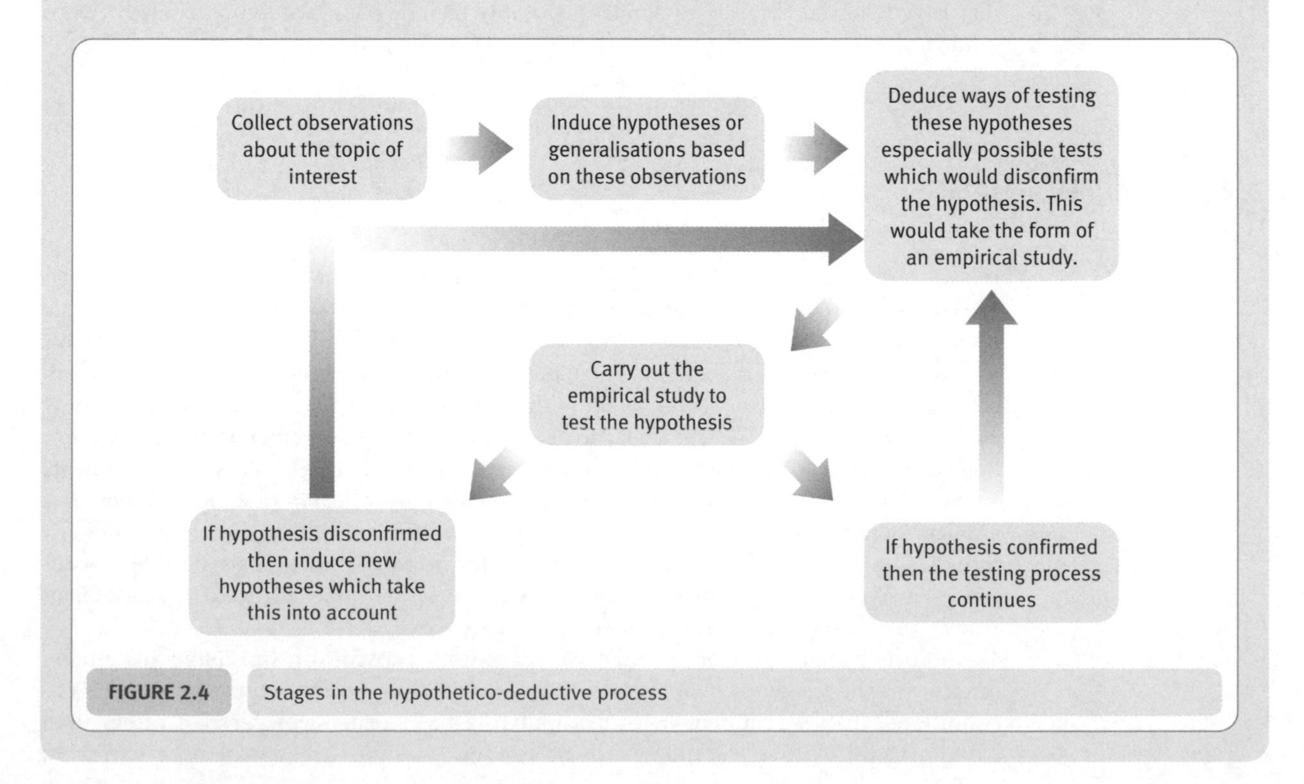

FIGURE 2.4 Stages in the hypothetico-deductive process

The comparative method

Characteristically, hypotheses in psychology imply a comparison between two or more groups. Sometimes this comparison is taken for granted in the expression of hypotheses so is not overtly stated. Unless one recognises this implicit comparison in hypotheses, it may prove difficult to formulate satisfactory ones and, furthermore, it may not be obvious that the appropriate research design should involve a comparison of two or more groups. Suppose we are interested in physical abuse in romantic relationships. One possible reason for such abuse is that one or both of the partners in the relationship are very possessive of the other person. So the violence may occur whenever a partner feels threatened by what the other person does or does not do. The research we carry out examines whether or not abusive partners are possessive. So the hypothesis is that abusive partners are possessive. We give out questionnaires measuring possessiveness to 50 people known to have physically abused their partner. Suppose we find that 70 per cent of people in abusive relationships are possessive. What can we conclude on the basis of this? Well it is certainly true that most abusive people are possessive. However, we do not know how many people not in abusive relationships are also possessive. That is, when it is suggested that abusive people are possessive, there is an implication that non-abusive people are *not* possessive.

The problem does not stem from how we have tested our hypothesis but from our understanding of the hypothesis itself. What the hypothesis implies is that abusive partners are *more* possessive than non-abusive partners. Just by looking at abusive partners we cannot tell whether they are more possessive than non-abusive partners. It could be that 70 per cent or even 90 per cent of non-abusive partners were possessive. If this were the case, abusive partners are not more possessive than non-abusive partners. They may even be less possessive than non-abusive partners.

Had the hypothesis been put as 'There is a relationship between possessiveness and abuse' then the comparison is built in but may still not be entirely obvious to those starting research for the first time. Probably the best rule of thumb is the assumption that psychological hypotheses almost invariably include or imply comparisons between groups of people.

2.7 Conclusion

It is almost a truism to suggest that the aims and hypotheses of research should be clear. This does not mean that the aims and hypotheses are obvious at the earliest stages of the research project. Since research is part of the ways in which psychological knowledge and ideas develop, it is almost inevitable that aims and hypotheses go through a developmental process. Reformulation of the aims and objectives of a study will commonly occur in the research planning stage, and sometimes after. All research is guided by aims, but hypotheses are only universal in certain types of research – especially true experiments – where it is possible to specify likely outcomes with a great deal of precision. Hypotheses are best included wherever possible since they represent the distillation of the researcher's thoughts about the subject matter. Sometimes, for non-experimental studies, the formulation of hypotheses becomes too cumbersome to be of value. Hence, many excellent studies in psychology will not include hypotheses.

The true experiment (for example, the laboratory experiment) has many advantages in terms of the testing of hypotheses – that is (a) its ability to randomise participants to conditions, (b) the requirement of manipulating the independent variable rather than using already existing variables such as gender, and (c) the control over variables.

Although we have largely discussed the testing of a single hypothesis at a time, very little research in real life is so restricted. Remember, most research studies have several aims and several hypotheses in the same study because we are usually interested in the way in which a number of different variables may be related to one another. It would also be more costly in terms of time and effort to investigate these hypotheses one at a time in separate studies.

In the penultimate section of this book on qualitative research methods, we will see that important research in psychology can proceed using a quite different approach to investigation in which the idea of specified aims and hypotheses is something of an anathema. Nevertheless, much research in mainstream psychology either overtly or tacitly subscribes to hypothesis testing as an ideal. Chapter 17 overviews the theoretical basis to these different approaches to research.

Key points

- Research studies have different general aims. Most seem to be concerned with testing causal propositions or hypotheses. Others may describe a phenomenon or intervention in detail, estimate how common a behaviour is in some population, evaluate the effects of interventions or statistically summarise the results of similar studies. The aim or aims of a study should be clearly and accurately stated.
- Studies which test causal propositions should describe clearly and accurately what these propositions are.
- The research study should make a contribution to the topic. While research usually builds on previous research in an area, the contribution of the study should be original to some extent in the sense that the particular question addressed has not been investigated in this way before.
- A hypothesis describes what the relationship is expected to be between two or more variables. The hypothesis should be stated in a causal form when the study is a true experiment. It should be stated in a non-causal form when the study is a non-experiment.
- When suggesting that variables may be related to one another, we usually expect the variables to be related in a particular way or direction. When this is the case, we should specify in the hypothesis what this direction is.
- The variable thought to be the cause may be called the independent variable and the variable presumed to be the effect the dependent variable. Some researchers feel that these two terms should be restricted to the variables in a true experiment. In non-experiments the variable assumed to be the cause may be called the predictor and the variable considered to be the effect the criterion.

ACTIVITIES

1. Choose a recent study that has been referred to either in a textbook you are reading or in a lecture that you have attended. What kind of aim or aims did the study have in terms of the aims mentioned in this chapter? What were the specific aims of this study? What kinds of variables were manipulated or measured? If the study involved testing hypotheses, were the direction and the causal nature of the relationship specified? If the hypothesis was stated in a causal form was the design a true (i.e. randomised) one?
2. You wish to test the hypothesis that we are what we eat. How could you do this? What variables could you measure?

CHAPTER 3

Variables, concepts and measures

Overview

- The variable is a key concept in psychological research. A variable is anything which varies and can be measured. There is a distinction between a concept and how it is measured.
- Despite the centrality and apparent ubiquity of the concept of variable, it was imported into psychology quite recently in the history of the discipline and largely from statistics. The dominance of 'variables' has been criticised because it tends to place emphasis on measurement rather than theoretical and other conceptual refinements of basic psychological concepts.
- In psychology, the distinction between independent and dependent variables is important. Generally, the independent variable is regarded as having an influence on the dependent variable. This is especially so in terms of experimental designs which seek to identify cause-and-effect sequences and the independent variable is the manipulated variable.
- Nominal variables are those which involve allocating cases to two or more categories. Binomial means that there are two categories, multinomial means that there are more than two categories. A quantitative variable is one in which a numerical value or score is assigned to indicate the amount of a characteristic an individual demonstrates.
- Stevens' theory of measurement suggests that variables can be measured on one of four different measurement scales – nominal, ordinal, interval and ratio. These have different implications as to the appropriate mathematical and statistical procedures which can be applied to them. However, generally in psychological research where data are collected in the form of numerical scores, the analysis tends to assume that the interval scale of measurement underlies the scores.

- Operational definitions of concepts describe concepts in terms of the procedures or processes used to measure those concepts. This is an idea introduced by psychologists from the physical sciences which attempts to avoid a lack of clarity in the definition of concepts. There is a risk that this places too great an emphasis on measurement at the expense of careful understanding of concepts.
- Mediator variables intervene between two variables and can be regarded as responsible for the relationship between those two variables. Moderator variables, on the other hand, simply show that the relationship between an independent and a dependent variable is not consistent but may be different at different levels of the moderator variable. For example, the relationship between age and income may be different for men and women. In this case, gender would be the moderator variable.
- Hypothetical constructs are not variables but theoretical or conceptual inventions which explain what we can observe.

3.1 Introduction

Variables are what we create when we try to measure concepts. So far we have used the term variable without discussing the idea in any great detail. Yet variables are at the heart of much psychological research. Hypotheses are often stated using the names of variables involved together with a statement of the relationship between the variables. In this chapter we will explore the idea of variables in some depth. A variable is any characteristic or quality that has two or more categories or values. Of course, what that characteristic or quality is has to be defined in some way by the researcher. Saying that a variable has two or more categories or values simply reminds us that a variable must vary by definition. Otherwise we call it a *constant*. Researchers refer to a number of different types of variable, as we will discuss in this chapter. Despite the apparent ubiquity of the idea of variables in psychology textbooks, the centrality of the concept of variables in much of modern psychology should be understood as applying largely to quantitative mainstream psychology. The concept is not normally used in qualitative psychology. Furthermore, historically the concept of a variable is a relatively modern introduction into the discipline, largely from statistics.

Variables are the things that we measure; they are not exactly the same thing as the concepts that we use when trying to develop theories about something. In psychological theories we talk in terms of concepts: in Freudian psychology a key concept is the ego; in social psychology a key concept is social pressure; in biological psychology a key concept might be pheromones. None of these, in itself, constitutes a variable. Concepts are about understanding things – they are not the same as the variables we measure. Of course, a major task in research is to identify variables which help us measure concepts. For example, if we wished to measure social influence we might do so in a number of different ways such as the number of people in a group who disagree with what a participant in a group has to say.

3.2 The history of the variable in psychology

The concept of *variable* has an interesting history in that psychology existed almost with no reference to variables for the first 50 or so years of the discipline's modern existence. The start of modern psychology is usually dated from the 1870s when Wilhelm Wundt (1832–1920) set up the first laboratory for psychological research at the University of Leipzig in 1879. Search through the work of early psychologists such as Sigmund Freud (1856–1939) and you find that they discuss psychological phenomena and not variables. The term 'independent variable', so familiar to all psychology students nowadays, was hardly mentioned at all in psychology publications before 1930. Most psychologists use the term without questioning the concept and it is probably one of the first pieces of psychological jargon that students come across. Psychology textbooks almost invariably discuss studies, especially laboratory experiments, in terms of *independent* and *dependent variables*; these terms are used instead of the names of the psychological phenomena or concepts that are being studied.

Variables, then, were latecomers in the history of psychology. It probably comes as no surprise to learn that the term 'variable' has its origins in nineteenth century mathematics, especially the field of statistics. It was introduced into psychology from the work of Karl Pearson (1857–1936) who originated the idea of the correlation coefficient. By the 1930s, psychologists were generally aware of statistical ideas, so familiarity with the term variable was common.

Edward Tolman (1886–1959), who is probably best remembered for his cognitive behavioural theory of learning and motivation, was the first to make extensive use of the word *variable* in the 1930s when he discussed *independent variables* and *dependent variables* together with his new idea of *intervening variables*. The significance of this can be understood better if one tries to discuss Freudian psychology, for example, in terms of these three types of variable. It is difficult to do so without losing the importance and nuances of Freudian ideas. In other words, these notions of variables tend to favour or facilitate certain ways of looking at the psychological world.

Danziger and Dzinas (1997) studied the prevalence of the term 'variables' in four major psychological journals published in 1938, 1949 and 1958. In the early journals there is some use of the term 'variable' in what Danziger and Dzinas describe as the 'softer' areas of psychology such as personality, abnormal and social psychology; surprisingly laboratory researchers were less likely to use the term. The increase in the use of the word 'variable' cannot be accounted for by an increase in the use of statistics in published articles since this was virtually universal in research published in the journals studied from 1938 onwards. The possibility that this was due to a rapidly expanding use of the term 'intervening variable' can also be dismissed on the basis that these were rarely mentioned in the context of empirical research – it was a term confined almost exclusively to theoretical discussions.

The use of the concepts of independent variable and dependent variable was being encouraged by experimental psychologists to replace the terminology of stimulus–response. Robert Woodworth (1869–1962), a prominent and highly influential author of a dominant psychology textbook of the time (Woodworth, 1938), adopted the new terminology and others followed his lead. Perhaps influenced by this, there was a substantial increase in the use of the terms independent variable and dependent variable over the time period that the journals were studied.

Danziger and Dzinas argue that the term *variable* took prominence because psychologists reconstrued psychological phenomena in terms of the *variables* familiar from statistics. In this way, psychological phenomena became mathematical entities or, at least, the distinction between the two was obscured. Thus psychologists write of personality variables when discussing aspects of personality which have not yet even been measured as they would have to be to become statistical variables. This amounts to a 'prestructuring' or construction of the psychological world in terms of variables and the consequent assumption that psychological research simply seeks to identify this structure of variables. Thus variables ceased to be merely a technical aspect of how psychological research is carried out but a statement or theory of the nature of psychological phenomena:

> When some of the texts we have examined here proceeded as if everything that exists psychologically exists as a variable they were not only taking a metaphysical position, they were also foreclosing further discussion about the appropriateness of their procedures to the reality being investigated.
>
> (Danziger and Dzinas, 1997, p. 47)

So are there any areas of psychology which do not use the concept of variable? Well it is very difficult to find any reference to variables in qualitative research, for example. Furthermore, it might be noted that facet theory (Canter, 1983; Shye and Elizur, 1994) regards the measures that we use in research simply as aspects of the world we are trying to understand. The analogy is with cutting precious stones. There are many different possible facets of a diamond depending on the way in which it is cut. Thus our measures simply reflect aspects of reality which are incomplete and less than the full picture of the psychological phenomena we are interested in. In other words, our measures are only a very limited sample of possible measures of whatever it is we are interested in theoretically. So the researcher should avoid confusing the definition of psychological concepts with how we measure them, but explore more deeply the definition of the concept at the theoretical level.

3.3 Types of variable

There are numerous different types of variable in psychology which may be indicative of the importance of the concept in psychology. Some of these are presented in Table 3.1, which indicates something of the relationship between the different types. However, the important thing about the table is that certain meanings of variable are primarily of theoretical and conceptual interest whereas others are primarily statistical in nature. Of course, given the very close relationships between psychology and statistics, many variables do not readily fall into just one of these categories. This is sometimes because psychologists have taken statistical terminology and absorbed it into their professional vocabulary to refer to slightly different things. There is an implication of the table which may not appear obvious at first sight. That is, it is very easy to fall into the trap of discussing psychological issues as if they are really statistical issues. This is best exemplified when psychologists seek to identify causal influences of one variable on another. The only way in which it is possible to establish that a relationship between two variables is causal is by employing an appropriate research design to do this. The randomised experiment is the best example of this by far. The statistical analysis employed cannot establish causality in itself.

Table 3.1 Some of the main types of variable

Type of variable	Domain – psychological or statistical	Comments
Binomial variable	Statistical	A variable which has just two possible values.
Causal variable	Psychological	It is not possible to establish cause-and-effect sequences simply on the basis of statistics. Cause and effect can be established only by the use of appropriate research designs.
Confounding variable	Psychological	A general term for variables which cause confusion as to the interpretation of the relationship between two variables of interest.
Continuous variable	Statistical	A variable for which the possible scores have every possible value with its range. So any decimal value, for example, is possible for the scores.
Dependent variable	Both	The variable assumed to be affected by the independent variable.
Discrete variable	Statistical	A variable for which the possible scores have a limited number of 'discrete' (usually whole number) values within its range – that is, not every numerical value is possible.
Dummy variable	Statistical	Used to describe the variables created to convert nominal category data to approximate score data.
Hypothetical construct	Psychological	Not really a form of variable but an unobservable psychological structure or process which explains observable findings.
Independent variable	Both	The variation in the independent variable is assumed to account for all or some of the variation in the dependent variable. As a psychological concept, it tends to be assumed that the independent variable has a causal effect on the dependent variable. This is not the case when considered as a statistical concept.
Interval variable	Statistical	Variables measured on a numerical scale where the unit of measurement is the same size irrespective of the position on the scale.
Intervening variable	Primarily psychological but also statistical	More or less the same as a mediator variable. It is a variable (concept) which is responsible for the influence of variable A on variable B. In other words it intervenes between the effect of variable A on variable B.
Mediator variable	Primarily psychological but also statistical	A variable (concept) which is responsible for the influence of variable A on variable B. In other words it mediates the effect of variable A on variable B.
Moderator variable	Statistical	A moderator variable is one which changes the character of the relationship between two variables.
Multinomial variables	Statistical	A nominal variable which has a number of values.
Nominal (category or categorical) variable	Statistical	Any variable which is measured by allocating cases to named categories without any implications of quantity.
Ordinal variable	Statistical	A variable which is measured in a way which allows the researcher to order cases in terms of the quantity of a particular characteristic. Derives from Stevens' theory of measurement.
Score variable	Statistical	Any variable which is measured using numbers which are indicative of the quantity of a particular characteristic.
Suppressor variable or masking variable	Statistical	A suppressor variable is a third variable which hides (reduces) the true relationship between two variables of interest.
Third variable	Statistical	A general term for variables which in some way influence the relationship between two variables of interest.

3.4 Independent and dependent variables

The concept of independent and dependent variables is common in psychological writings. The distinction between the two is at its clearest when we consider the true experiment and, in an ideal world, would probably best be confined to laboratory and similar true experiments. The variable which is manipulated by the researcher is known as the independent variable. Actually it is totally independent of any other variable in a true experimental design since purely random processes are used to allocate participants to the different experimental treatments. Nothing in the situation, apart from randomness, influences the level of the independent variable. The variable which is measured (rather than manipulated) is the dependent variable since the experimental manipulation is expected to influence how the participants in the experiment behave. In other words, the dependent variable is subject to the influence of the independent variable.

The concepts of independent and dependent variables would appear to be quite simple, that is, the independent variable is the manipulated variable and the dependent variable is the measured variable which is expected to be influenced by the manipulated variable. It becomes rather confusing because the distinction between independent and dependent variables is applied to non-experimental designs. For example, the term independent variable is applied to comparisons between different groups. Thus, if we were comparing men and women in terms of their computer literacy then gender would be the independent variable. Computer literacy would be the dependent variable.

Of course, gender cannot be manipulated by the researcher – it is a fixed characteristic of the participant. Variables which cannot be or were not manipulated and which are characteristic of the participant or subjects are sometimes called *subject variables*. They include such variables as how old the person is, how intelligent they are, how anxious they are and so on. All of these variables may be described as the independent variable by some researchers. However, how can a variable be the independent variable if the causal direction of the relationship between two variables is not known? In non-experiments it may be better to use more neutral terms for these two types of variable such as *predictor variable* for the independent variable and *criterion variable* for the dependent variable. In this case we are trying to predict what the value of the criterion variable is from the values of the predictor variable or variables.

Things get a little complicated since in non-experimental designs, what is the independent variable for one analysis can become the dependent variable for another and vice versa. This may be all the more reason for confining the independent–dependent variable distinction to experimental designs.

3.5 Measurement characteristics of variables

Measurement is the process of assigning individuals to the categories or values of a variable. Different variables have different measurement characteristics which need to be understood in order to plan and execute research effectively. The most important way in which variables differ is in terms of the nominal versus quantitative measurement. These are illustrated in Figure 3.1.

Nominal variables (also known as qualitative, category or categorical variables)

Nominal variables are ones in which measurement consists of categorising cases in terms of two or more named categories. The number of different categories employed is also used to describe these variables:

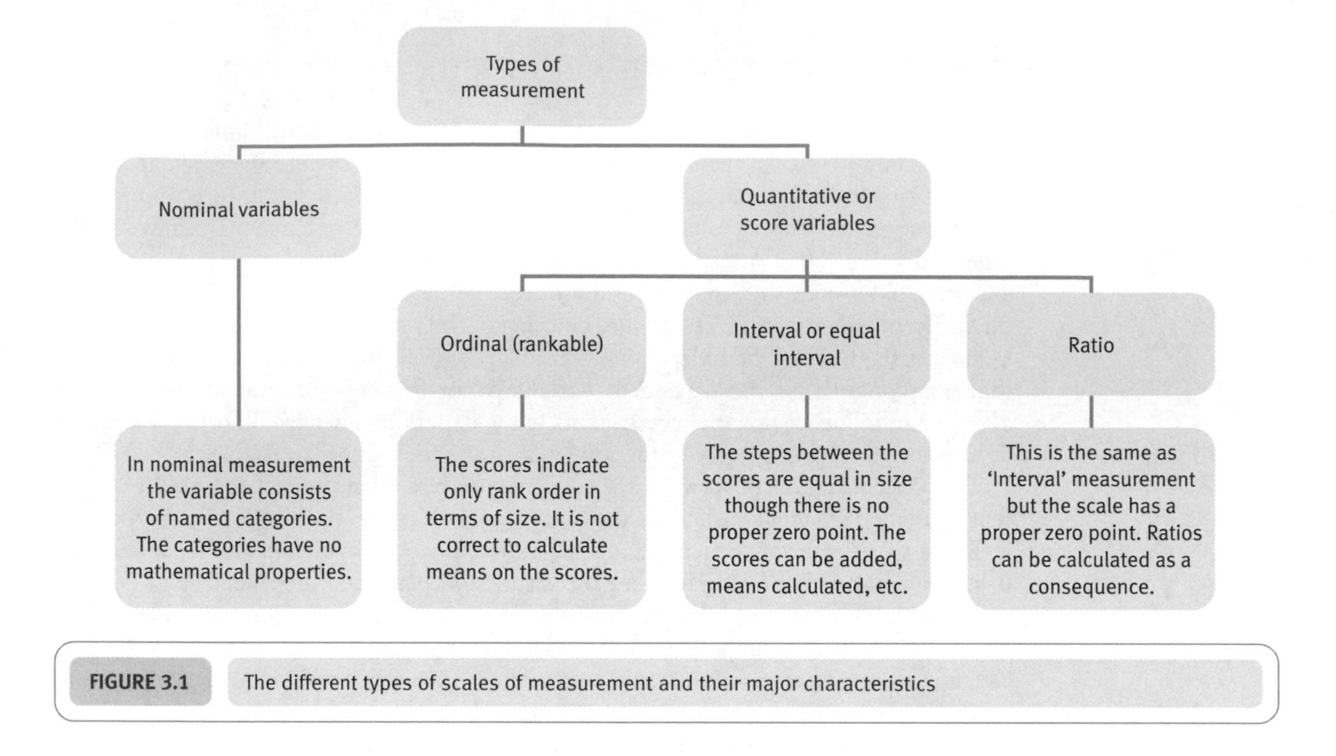

FIGURE 3.1 The different types of scales of measurement and their major characteristics

- ***Dichotomous, binomial or binary variables*** These are merely variables which are measured using just two different values. (The term dichotomous is derived from the Greek meaning equally divided or cut in two: *dicho* is Greek for apart, separate or in two parts while '-ous' is Latin for characterised by.) For example, one category could be 'friend' while the other category would be anyone else.
- ***Multinomial variables*** When a nominal variable has more than two values it is described as a multinomial, polychomous or polytomous variable (*poly* is Greek for many). We could have the four categories of 'friend', 'family member', 'acquaintance' and 'stranger'.

Each value or category of a dichotomous or multinomial variable needs to be identified or labelled. For example, we could refer to the four categories friend, family member, acquaintance and stranger as category A, category B, category C and category D. We could also refer to them as category 1, category 2, category 3 and category 4. The problem with this is that the categories named in this way have been separated from their original labels – friend, family member, acquaintance and stranger. This kind of variable may be known as a nominal, qualitative, category, categorical or frequency variable. Numbers are simply used as names or labels for the different categories. We may have to use numbers for the names of categories when analysing this sort of data on a computer. The only arithmetical operation that can be applied to dichotomous and multinomial variables is to count the frequency of how many cases fall into the different categories.

Quantitative variables

When we measure a quantitative variable, the numbers or values we assign to each person or case represent increasing levels of the variable. These numbers are known as scores since they represent amounts of something. A simple example of a quantitative variable might be social class (socio-economic status is a common variable to use in research). Suppose that social class is measured using the three different values of lower, middle and upper class. Lower class may be given the value of 1, middle class the value of 2 and upper class the value of 3. Hence, higher values represent higher social status. The important thing to remember is that the numbers here are being used to indicate different quantities of the variable social class. Many quantitative variables (such as age, income, reaction time, number of errors or the score on a questionnaire scale) are measured using many more than three categories. When a variable is measured quantitatively, then the range of arithmetic operations that can be carried out is extensive – we can, for example, calculate the average value or sum the values.

The term dichotomous can be applied to certain quantitative variables as it was to some nominal variables. For example, we could simply measure a variable such as income using two categories – poor and rich, which might be given the values 1 and 2. Quite clearly the rich have greater income than the poor so the values clearly indicate quantities. However, this is true of any dichotomous variable. Take the dichotomous variable sex or gender. For example, females may be given the value of 1 and males the value of 2. These values actually indicate quantity. A person who has the value 2 has more maleness than a person given the value 1. In other words, the distinction between quantitative and qualitative variables is reduced when considering dichotomous variables.

Box 3.1 Key Ideas

Mediator versus moderator variables

Conceptually, modern psychologists speak of the distinction between mediator and moderator variables. These present quite different views of the role of third variables in relationships among measures of two variables (e.g. the independent and dependent variables). Mediator and moderator variables are conceptual matters which are more important to research design and methodology than they are to statistical analysis as such. If we consider the relationship between the independent and dependent variable then a *mediator* variable is a variable which is responsible for this relationship. For example, the independent variable might be age and the dependent variable may be scores on a pub quiz or some other measure of general knowledge. Older people do better on the general knowledge quiz.

Independent variable
Age

Dependent variable
Scores on pub quiz

Age, itself, is not directly responsible for higher scores on the pub quiz. The reason why older people have greater general knowledge might be because they have had more time in their lives to read books and newspapers, watch television, and undertake other educational experiences. It is these learning experiences which are responsible for the relationship between age and general knowledge. In this instance, then, we can refer to these educational experiences as a mediator variable in the relationship between age and general knowledge.

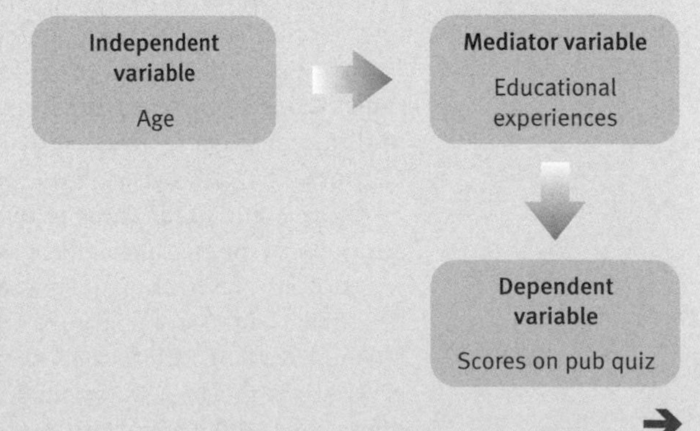

Another way of describing this is that there is an indirect effect of age on the scores on the pub quiz. Of course, there may be more than one mediator variable in the relationship between an independent and a dependent variable. There is no substantial difference between a mediator variable and an intervening variable other than that intervening variables are regarded as hypothetical variables whereas mediator variables seem not to be regarded as necessarily hypothetical.

A moderator variable is completely distinct from this. A moderator variable is one which shows that the relationship between an independent variable and a dependent variable is not consistent throughout the data. For example, imagine that a researcher investigates the relationship between age and scores on a pub quiz but adds a further dimension, that is, they consider the relationship for men and women separately. They may find that the relationship between the two is different for men and women. Perhaps there is a strong correlation of 0.7 between age and scores on the pub quiz for women but a weak one of 0.2 between age and scores on the pub quiz for men. Thus the relationship is not the same throughout the data. This implies quite different conclusions for men and for women. In other words, gender moderates the relationship between age and scores on the pub quiz. Having established that gender is a moderator variable in this case does not explain in itself why the relationship is different for men and women. One possibility is that the pub quiz in question had a gender bias such that most of the questions were about topics which are of interest to women but not to men. Consequently, we might expect that the relationship would be reduced for men.

The interpretation of moderator variables is dependent on whether or not the independent variable in the independent variable–moderator variable–dependent variable chain is a randomised variable or not. Only where it is randomised can the researcher be sure of the causal sequence. Otherwise there is uncertainty about what is causing what.

3.6 Stevens' theory of scales of measurement

The previous section describes the measurement principles underlying variables in the most useful and practical way possible. However, there is another approach which is common in textbooks. Research methods textbooks are full of ideas which appear to be the uncontroversial bedrock of the discipline. A good example of this is the theory of scales of measurement put forward by the psychologist Stanley Stevens (1906–1973) in 1946. You may have heard of other of his ideas but that of the different scales of measures already discussed in this chapter is probably the most pervasive one. Few psychologists nowadays, probably, could name Stevens as the originator of the idea of nominal, ordinal, interval and ratio scales. Remarkably, his ideas are surrounded by controversy in specialist statistical publications but one would be forgiven for thinking that they are indisputable 'facts' given the way they are uncritically presented in research methods and statistics textbooks. So in this section we will look at Stevens' theory in a somewhat critical way unlike other sources that you will find. You might be glad of the enlightenment.

By measurement, Stevens is taken to mean the allocation of a number (or symbol) to things using some consistent rule. So, for example, we could measure sweets in a number of different ways: (a) we could allocate a colour (blue, red, yellow are linguistic symbols, of course) to each sweet; (b) we could measure each sweet's weight as a number of grams; or (c) we could grade them in terms of how good they taste. Measurement, conceptually, amounts to quite a simple process as these examples illustrate. It is clear that these three ways of measurement are somewhat different in terms of how the measurement would be carried out. Stevens argued that there are four different types of measurement which have somewhat different mathematical properties. This means that the mathematical procedures that can be applied to each type of measurement differ. The mathematical operations which are appropriate for one type of measurement may be inappropriate for another. One consequence of this is that the sort of statistical analysis that is appropriate for one sort of variable may be inappropriate for a different type of variable. Choosing a statistical technique appropriate to the sort of data one has is one of the skills that has to be learnt when studying statistics in psychology.

Stevens' four types of measurement are usually put in the order nominal, ordinal, interval and ratio scales of measurement. This is actually a hierarchy from the least powerful data to the most powerful data. The later in the series the scale of measurement is in, the more information is contained within the measurement. Thus variables measured in a manner indicative of a ratio scale are at the highest level of measurement and contain more information, all other things being equal. The different measurement scales are often referred to as different levels of measurement which, of course, in itself implies a hierarchy. Let us take each of these in turn, starting with the highest level of the hierarchy.

1. ***Ratio measurement scales*** The key feature of a ratio measurement scale is that it should be possible to calculate a meaningful ratio between two things which have been measured. This is a simpler idea than it sounds because we are talking ratios when we say things like Grant is twice as tall as Michelle. So if we measure the weights of two chocolates in grams we can say that the coffee cream chocolate was 50 per cent heavier than the strawberry truffle chocolate. In order to have a ratio measurement scale it is necessary for there to be a zero point on the scale which implies zero quantity for the measurement. Weights do have a zero point – zero grams means that there is no weight – weight is a ratio measurement scale. A common measurement which does not have a proper zero (implying zero quantity) is temperature measured in terms of Celsius or Fahrenheit. To be sure, if you look at a thermometer you will find a zero point on both Celsius and Fahrenheit scales but this is not the lowest temperature possible. Zero on the Celsius scale is the freezing point of water but it can get a lot colder than that. What this means is that for temperature, it is not possible to say that something is twice as hot as something else if we measure on the temperature scales familiar to us all. Thirty degrees Celsius cannot be regarded as twice as hot as 15 degrees Celsius. Any statistical procedure can be applied to ratio data. For example, it is meaningful to calculate the mean of ratio data – as well as ratios. There is another feature of ratio measurement scales, that there should be equal intervals between points on the scale. However, this requirement of equal intervals is also a requirement of the next measurement scale – the interval measurement scale – as we shall see.
2. ***(Equal) interval measurement scale*** This involves assigning real numbers (as opposed to ranks or descriptive labels) to whatever is being measured. It is difficult to give common examples of interval measures which are not also ratio measures. Let us look on our bag of sweets: we find a sell-by date. Now sell-by date is part of a measurement scale which measures time. If a bag of sweets has the sell-by date of 22nd February then this is one day less than a sell-by date of 23rd February. Sell-by date is being measured on an equal interval measure on which each unit is a day and that unit is constant throughout the scale of measurement, of course. However, there is not a zero point on this scale. We all know that the year 0 is not the earliest year that there has ever been since it is the point where BC changes to AD. We could, if we wanted to, work out the average sell-by date on bags of sweets in a shop. The average would be meaningful because it is based on an equal-step scale of measurement. There are many statistical techniques which utilise data which are measured on interval scales of measurement. Indeed, most of the statistical techniques available to researchers can handle interval scale data. Conceptually, there is a clear difference between interval scale measurements and ratio scale measurements though in most instances it makes little difference, say, in terms of the appropriate statistical analysis – it is only important when one wishes to use ratios which, in truth, is not common in psychology.
3. ***Ordinal measurement scales*** This involves giving a value to each of the things being measured which indicates the relative order on some sort of characteristic. For example,

you might wish to order the chocolates in a box of chocolates in terms of how much you like each of them. So you could put the ones that you most like on the left-hand side of a table, the ones that you like a bit less to the right of them, and so forth until the ones that you most dislike like are on the far right-hand side of the table. In this way, the chocolates have been placed in order from the most liked to the least liked. This is like ordinal measurement in that you have ordered them from the most to the least. Nothing can really be said about how much more that you like, say, the hazelnut cluster from the cherry delight, only that one is preferred to the other because you put them at different points on the table. It is not possible to say how much more you like one of the chocolates than another. Even if you measured the distance between where you put the hazelnut cluster and where you put the cherry delight, this gives no precise indication of how much more one chocolate is liked than another. Ordinal numbers (first, second, third, fourth . . . last) could be applied to the positions from left to right at which the chocolates were placed. These ordinal numbers correspond to the rank order. The hazelnut cluster might be eighth and the cherry delight might be ninth. However, it still remains the case that although the hazelnut cluster is more liked than the cherry delight, how much more it is liked is not known. Ordinal measurements, it is argued, are not appropriate for calculating statistics such as the mean. This is certainly true for data which have been converted to ranks since the mean rank is totally determined by the number of items ranked. However, psychologists rarely collect data in the form of ranks but in the form of scores. In Stevens' measurement theory any numerical score which is not on a scale where the steps are equal intervals is defined as ordinal. Stevens argued that the mode and the median are more useful statistics for ordinal data and that the mean is inappropriate. Ordinal measurements are frequently analysed using what are known as non-parametric or distribution-free statistics. They vary, but many of these are based on putting the raw data into ranks.

4. *Nominal (or category/categorical) measurement scales* This measurement scale involves giving labels to the things being measured. It is illustrated by labelling sweets in an assorted bag of sweets in terms of their colour. Colour names are linguistic symbols but one could use letters or numbers as symbols to represent each colour if we so wished. So we could call red Colour A and blue Colour B, for instance. Furthermore, we could use numbers purely as symbols so that we could call red Colour 1 and blue Colour 2. Whether we use words, letters, numbers or some other symbol to represent each colour does not make any practical difference. It has to be recognised that if we use numbers as the code then these numbers do not have mathematical properties any more than letters would have. So things become a little more complicated when we think about what numerical procedures we can perform on these measurements. We are very restricted. For example, we cannot multiply red by blue – this is meaningless. Actually the only numerical procedure we could perform in these circumstances would be to count the number of red sweets, the number of blue sweets, and the number of yellow sweets, for example. By counting we mean the same thing as calculating the frequency of things so we are able to say that red sweets have a frequency of 10, blue sweets have a frequency of 5, and yellow sweets have a frequency of 19 in our particular bag of sweets. We can say which is the most frequent or typical sweet colour (yellow) in our bag of sweets but little else. We cannot meaningfully calculate things such as the average red sweet, the average blue sweet and the average yellow sweet based on giving each sweet a colour label. So in terms of statistical analysis, only statistics which are designed for use with frequencies could be used. Confusion can occur if the symbols used for the nominal scale of measurement are numbers. There is nothing wrong with using numbers as symbols to represent things so long as one remembers that they are merely symbols and that they are no different from words or letters in this context.

It is not too difficult to grasp the differences between these four different scales of measurement. The difficulty arises when one tries to apply what has been learnt to the vast variety of different psychological measures. It will not have escaped your attention how in order to explain interval and ratio data physical measurements such as weight and temperature were employed. This is because these measures clearly meet the requirements of these measurement scales. When it comes to psychological measures, it is much more difficult to find examples of interval and ratio data which everyone would agree about. There are a few examples but they are rare. Reaction time (the amount of time it takes to react to a particular signal) is one obvious exception, but largely because it is a physical measure (time) anyway. Examples of more psychological variables which reflect the interval and ratio levels of measurement are difficult to find. Take, for example, IQ (intelligence quotient) scores. It is certainly possible to say that a person with an IQ of 140 is more intelligent than a person with an IQ of 70. Thus we can regard IQ as an ordinal measurement. However, is it possible to say that a person with an IQ of 140 is twice as intelligent as someone with an IQ of 70, which would make it a ratio measurement? Are the units that IQ is measured in equal throughout the scale? Is the difference between an IQ of 70 and an IQ of 71 the same difference as that between an IQ of 140 and an IQ of 141? They need to be the same if IQ is being measured on an equal interval scale of measurement. One thing that should be pointed out is that one unit on our measure of IQ is in terms of the number involved being the same no matter where it occurs on the scale. The problem is that in terms of what we are really interested in – intelligence – we do not know that each mathematical unit is the same as each psychological unit. If this is not clear, then consider using electric shocks to cause pain. The scale may be from zero volts to 6000 volts. This is clearly an interval and also a ratio scale in terms of the volts, in terms of the resulting pain this is not the case. We all know that if we touch the terminals of a small battery this has no effect on us, which means that at the low voltage levels no pain is caused, but this is not true at higher voltage levels. Thus, in terms of pain this scale is not equal interval.

Not surprisingly, Stevens' theory of measurement has caused many students great consternation especially given that it is usually taught in conjunction with statistics, which itself is a difficult set of ideas for many students. The situation is that anyone using the theory – and it is a theory – will generally have great difficulty in arguing that their measures are on either an interval or ratio scale of measurement simply because there is generally no way of specifying that each interval on the scale is in some way equal to all of the others to start with. Furthermore, there are problems with the idea that a measure is ordinal. One reason for this is that psychologists rarely simply gather data in the form of ranks as opposed to some form of score. These scores therefore contain more information than that implied merely by a rank though the precise interpretation of these scores is not known. There is no way of showing that these scores are on an equal interval scale so they should be regarded as ordinal data according to Stevens' theory. Clearly this is entirely unsatisfactory. Non-parametric statistics were frequently advocated in the past for psychological data since these do not assume equality of the measurement intervals. Unfortunately, many powerful statistical techniques are excluded if one chooses the strategy of using non-parametric statistics.

One argument, which we find convincing, is that it is not the psychological implications of the scores which is important but simply the mathematical properties of the numbers involved. In other words, so long as the scores are on a numerical scale then they can be treated as if they are interval scale data (and, in exceptional circumstances where there is a zero point, as ratio scale data). The truth of the matter, and quite remarkable given the dominance of Stevens' ideas about measurement in psychology, is that researchers usually treat any measures they make involving scores as if they were interval data. This is done without questioning the status of their measures in terms of Stevens' theory. Actually there are statistical studies which partly support this in which ordinal data are subject to statistical analyses which are based on the interval scale of measurement. For many statistics, this makes little or no difference. Hence, in that sense, Stevens' theory of measurement leads to a sort of wild goose chase.

Box 3.2 Key Ideas

Hypothetical constructs

Hypothetical constructs were first defined by Kenneth MacCorquodale (1919–1986) and Paul E. Meehl (1920–2003) in 1948. They are not variables in the sense we have discussed in this chapter. Essentially a hypothetical construct is a theoretical invention which is introduced to explain more observable facts. It is something which is not directly observable but nevertheless is useful in explaining things such as relationships between variables which are found during a research study. There is a wide variety of hypothetical constructs in psychology. Self-esteem, intelligence, the ego, the id and the superego are just some examples. None of these things is directly observable yet they are discussed as explanations of any number of observable phenomena. Some of them, such as intelligence, might at first sight seem to be observable things but, usually, they are not since they are based on inferences rather than observation. In the case of the hypothetical construct of intelligence, we use observables such as the fact that a child is top of their class, is a chess champion, and has a good vocabulary to infer that they are intelligent. Perhaps more crucially, the Freudian concept of the id is not observable as such but is a way of uniting observable phenomena in a meaningful way which constitutes an explanation of those observables.

3.7 Operationalising concepts and variables

There is a crucial distinction to be made between a variable and the measure of that variable. Variables are fundamentally concepts or abstractions which are created and refined as part of the advancement of the discipline of psychology. Gender is not a tick on a questionnaire, but we can measure gender by getting participants to tick male or female on a questionnaire. The tick on the questionnaire is an *indicator* of gender, but it is not gender. Operationalisation (Bridgman, 1927) is the steps (or operations) that we take to measure the variable in question. Percy Williams Bridgman (1882–1961) was a physical scientist who, at the time that he developed his ideas, was concerned that concepts in the physical sciences were extremely poorly defined and lacked clarity. Of course, this is frequently the case in the softer discipline of psychology too. Operationalism was, however, introduced into psychology largely by Stanley Stevens, who we have earlier discussed in terms of measurement theory. For Bridgman, the solution was to argue that precision is brought to the definition of concepts by specifying precisely the operations by which a concept is measured. So the definition of weight is through describing the measurement process, for example, the steps by which one weighs something using some sort of measurement scale. Of course, operationalising concepts is not guaranteed to provide precision of definition unless the measurement process is close to the concept in question and the measurement process can be precisely defined. So, for example, is it possible to provide a good operational definition of a concept like love? By what operations can love be measured is the specific question. One possible operational definition might be to measure the amount of time that a couple spends in each other's company in a week. There are obvious problems with this operational definition which suggests that it is not wholly adequate. It has little to do with our ideas about what love is. Imagine the conversation:

'Do you love me?' 'Well I spend a lot of time with you each week, don't I?' This should quickly paint a picture of the problem with such an operational definition.

Nevertheless, some researchers in psychology argue that the best way of defining the nature of our variables is to describe how they are measured. For example, there are various ways in which we could operationalise the concept or variable of anxiety. We could manipulate it by putting participants in a situation which makes them anxious and compare that situation with one in which they do not feel anxious. We could assess anxiety by asking them how anxious they are, getting other people to rate how anxious they seem to be, or measuring some physiological index of anxiety such as their heart rate. These are different ways of operationalising anxiety. If they all reflect what we consider to be anxiety we should find that these different methods are related to one another. So we would expect participants in a situation which makes them anxious to report being more anxious, be rated as being more anxious and to have a faster heart rate than those in a situation which does not make them anxious. If these methods are not related to one another, then they may not all be measuring anxiety.

Operationalisation has benefits and drawbacks. The benefit is that by defining a concept by the steps involved in measuring it, the meaning of the concept could not be more explicit. The costs include that operationalisation places less onus on the researcher to explicate the nature of their concepts and encourages the concentration on measurement issues rather than conceptual issues. Of course, ultimately any concept used in research has to be measured using specific and specified operations. However, this should not be at the expense of careful consideration of the nature of what it is that the researcher really is trying to understand – the theoretical concepts involved. Unfortunately, we cannot escape the problem that operational definitions tend to result in a concentration on measurement in psychology rather to the detriment of the development of the ideas embedded in psychological theory.

3.8 Conclusion

Perhaps the key thing to have emerged in this chapter is that some of the accepted ideas in psychological research methods are not simply practical matters but were important philosophical contributions to the development of the methods of the discipline. A concept such as a variable has its own timeline, is not universal in the discipline, and brings to psychology its own baggage. Although some of these ideas seem to be consensually accepted by many psychologists, it does not alter the fact that they are not the only way of conceptualising psychological research, as later parts of this book will demonstrate. While the idea of operational definitions of concepts pervades much of psychology, once again it should be regarded as a notion which may have its advantages but also can be limiting in that it encourages psychologists to focus on measurement but not in the context of a thorough theoretical and conceptual understanding of what is being measured.

The concept of a variable has many different ramifications in psychology. Many of these are a consequence of the origins of the concept in statistics. Intervening variables, moderator variables, mediating variables, continuous variables, discrete variables, independent variables, dependent variables and so forth are all examples of variables but all are different conceptually. Some have more to do with statistics than others.

Measurement theory introduced the notions of nominal, ordinal, interval and ratio measurement. While these ideas are useful in helping the researcher not to make fundamental mistakes such as suggesting that one person is twice as intelligent as another when such statements require a ratio level of measurement, they are actually out of step with psychological practice in terms of the statistical analysis of research data.

Key points

- The concept of variable firmly ties psychology to statistics since it is a statistical concept at root. There are many different types of variable which relate as much to theoretical issues as empirical issues. For example, the distinction between independent and dependent variables and the distinction between moderator and mediating variables relate to explanations of psychological phenomena and not simply to empirical methods of data collection.
- Stevens' measurement theory has an important place in the teaching of statistics but is problematic in relation to the practice of research. Nevertheless, it can help prevent a researcher from making totally erroneous statements based on their data. Most research in psychology ignores measurement theory and simply assumes that data in the form of scores (as opposed to nominal data) can be analysed as if they are based on the interval scale of measurement.
- Psychologists stress operational definitions of concepts in which theoretical concepts are defined in terms of the processes by which they are measured. In the worst cases, this can encourage the concentration of easily measured variables at the expense of trying to understand the fundamental concept better at a conceptual and theoretical level.

ACTIVITIES

1. Try to list the defining characteristics of the concept love. Suggest how love can be defined using operational definitions.
2. Could Stevens' theory of measurement be applied to measures of love? For example, what type of measurement would describe classifying relationships as either platonic or romantic love?

CHAPTER 4

The problems of generalisation and decision-making in research

Chance findings and sample size

Overview

- Psychological research is based on a complex decision-making process which is irreducible to a simplistic formula or rules-of-thumb.
- A number of factors are especially influential on setting the limits to generalisation from any data. The sampling procedures used and the statistical significance of the findings are very important. At the same time, psychologists generalise because psychology has a tendency towards universalism which assumes that what is true of one group of people is true of all people.
- Psychological research (as opposed to psychological practice) is usually concerned with samples of people rather than specific individuals. This allows general trends to be considered at the expense of neglecting the idiosyncratic aspects of individuals.
- Much psychological research depends on samples being selected primarily because they are convenient for the researcher to obtain. The alternative would be random sampling from a clearly specified population which is much more expensive of time and other resources.
- Characteristically, a great deal of psychological research is concerned with studying principles of human behaviour that are assumed to apply generally. As the

generalisations being tested are assumed to be true of people in general, the necessity to ensure that the sample is representative is minimised.

- Statistical analysis is often concerned with answering the simple question of how safe it is to generalise from a particular study or sample of data. The usual model of statistical testing in psychology is based on postulating what would happen if the null hypothesis were true and then comparing this with what was actually obtained in the research.
- The probability of accepting that the results are not due to chance sampling if the null hypothesis were true is usually set at the .05 or 5 per cent level. This means that the probability of the finding being due to chance when the null hypothesis is in fact true is 5 times out of 100, or less. Results that meet this criterion are called statistically significant, otherwise they are statistically non-significant.
- The bigger the sample the more likely it is that the results will be statistically significant – all other things being equal. Consequently, it is necessary to look at the size of the result as well as its statistical significance when evaluating its importance.

4.1 Introduction

This chapter discusses in some detail the process of generalisation of research findings. Are we justified in making more general statements about the findings of our research beyond the research itself? This is a crucial step in any research. There are three important themes that need consideration:

- The lack of limitations placed on generalisation by the universalism of psychological theory.
- The limitations placed on generalisation by the sampling methods used.
- The limitations placed on generalisation by the strictures of statistical significance testing.

We will deal with each of these in turn. They are equally important but there is more to be said about qualitative analysis and generalisation in this context and so this will receive a disproportional amount of space. Each of these has a different but important influence on the question of the extent to which a researcher is wise or correct to generalise beyond the immediate setting and findings of their research study. There may be a temptation to regard statistical considerations as technical matters in research, but this is not altogether the case. Many statistical considerations are better regarded as having a bearing on important conceptual matters. For example, one might be less likely to generalise in circumstances in which your measures of concepts or variables are relatively weak or ineffective. This will tend to yield poor or low correlations between such variables and others – hence the disinclination to generalise from this finding with confidence. However, statistics can help show you such things as what the correlation would be if the measures were good and reliable. This may revise your opinion of what can be said on the basis of your data. See Figure 4.1 for a summary of some of the issues to do with generalisation.

It is important to realise that issues such as the generalisability of data are really aspects of the process of decision-making that a researcher makes throughout their research. The

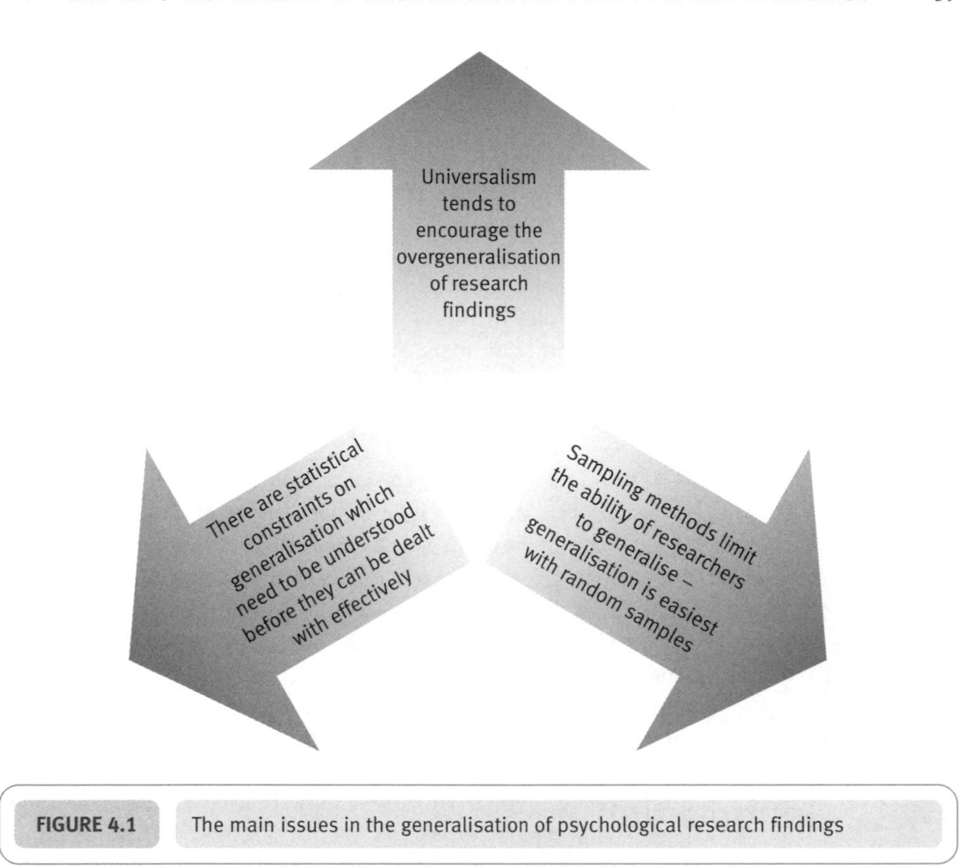

FIGURE 4.1 The main issues in the generalisation of psychological research findings

task of the researcher is to reach a balanced judgement at every stage of their research based on the information that they have in front of them and reasoned evaluations of the available choices of action available at that point in time. It is impossible to reduce this decision-making process to a few rules of thumb. It might be appealing to students to have such rules of thumb but it distorts the reality of research too much to try to reduce it to any simplistic formula. So even things such as significance testing which are often reduced to a formula in statistics textbooks turn out to be much more of a matter of judgement than that implies. Research is not simply a matter of deciding whether a hypothesis is supported by one's data or not. There are important issues such as whether it is desirable to develop further questions for further research in the area, whether an important next step is to establish whether one's findings apply in very different circumstances or with very different groups of participants or using very different methods, and the degree of confidence one should have in one's findings. There are other questions, of course, such as the desirability of abandoning this particular line of research. Again this is not simply a matter of failing to find support for a hypothesis in a particular study but a decision-making process based not simply on basic statistical outcomes but on a finer judgement as to whether the hypothesis had been given a fair chance in the research study.

4.2 Universalism

One of the characteristics of psychology is its tendency towards *universalism*. This is the fundamental assumption that the principles of psychology will not vary. Psychological findings will apply anywhere and are the same for all people irrespective of their society and their culture. So when psychologists propose a hypothesis there is an implicit assumption that it is true of all people – unless it is one of those rare cases where it is stated or implied that the principle applies only to restricted groups of people. In other words, psychologists in practice appear to be interested in making generalisations about behaviour that apply unrestrained by context and circumstances. Psychological principles are assumed to be laws of human behaviour anywhere. Increasingly psychologists question this idea of universalism and argue that a culturally specific approach to psychology is more realistic and productive (Owusu-Bempah and Howitt, 2000). Historically, many of the principles put forward by psychologists are assumed to apply not only to people but also to other animals. So it was only natural that studies of basic processes were carried out on animals and the findings applied to human beings. Examples of this include classical conditioning theory (Pavlov, 1927) and operant conditioning theory (Skinner, 1938).

While universalism is characteristic of a great deal of psychological thinking, it is rarely, if ever, stated as such in modern psychology. Nowadays psychologists are likely to be aware of the problem but, nevertheless, this awareness is not built into their practices for designing their research studies. Universalism operates covertly but reveals itself in a number of different ways – such as when university students are used unquestioningly as participants in a great deal of academic research as if what were true for university students will be true for every other grouping and sector in society. Seldom do psychologists build into their research a variety of groups of participants specifically to assess whether their findings apply throughout.

Universalism defines quantitative research in psychology much more than it does qualitative research, of course. Qualitative researchers invariably adopt a relativist perspective which rejects the idea of a single reality which can be discovered through research. Instead, qualitative researchers assume that there is a multiplicity of viewpoints on reality. This is clearly incompatible with universalism and is discussed in more detail in Part 4 of this book on qualitative research methods.

4.3 Sampling and generalisation

Many criticisms have been made of psychology for its restricted approach to sampling. As already mentioned, psychological research has sometimes been described as the psychology of psychology students or sophomores (Rosenthal and Rosnow, 1969, p. 59) (a sophomore is a second-year student in the USA). This criticism only means something if the idea of universalism in psychological research is being questioned, otherwise it would not matter since the laws of human behaviour might just as well be determined from studies using psychology students as any other group of participants. Whatever the group used, it would reveal the same universal laws. The emphasis of psychology on the processes involved in human behaviour and interaction is a strength of the discipline and not something to which sampling has anything particular to contribute from one perspective. So although sampling methods in psychology may to some extent be found to be lacking, this is not the entire story by any means.

Not all research has or needs a sample of participants. The earliest psychological research tended to use the researcher themselves as the principal or only research participant. Consequently, experimental psychologists would explore phenomena on themselves. This was extremely common in introspectionism (or structuralism) which was the dominant school of psychology at the start of modern psychology and was eventually replaced by behaviourism early in the twentieth century. Similarly, and famously, Ebbinghaus (1913) studied memory or forgetting. There are circumstances in which a single case may be an appropriate unit for study. Some psychologists still advocate using single cases or relatively few cases in order to investigate changing a particular individual's behaviour (Barlow and Hersen, 1984) and this is common in qualitative research too. A single-case experimental study in quantitative research involves applying the independent variable at different random points in time. If the independent variable is having an effect then the participant should respond differently at the points in time that the independent variable is applied than when it is not. Its obvious major advantage is that it can be helpful when the particular sort of case is very rare. For example, if a particular patient has a very unusual brain condition then such a procedure provides a way of studying the effect of that condition. Clinical researchers working with a particular patient are an obvious set of circumstances in which this style of research might be helpful.

The problems with the approach are largely to do with the high demands on the participant's time. It also has the usual problems associated with participants being aware of the nature of the design – it is somewhat apparent and obvious what is happening – which may result in the person being studied cooperating with what they see as the purpose of the study. Although this sort of 'single-case' method has never been very common in mainstream psychology and appears to be becoming less so (Forsyth *et al.*, 1999), its use questions the extent to which researchers always require substantial samples of cases in order for the research to be worthwhile or effective.

Representative samples and convenience samples

Most research studies are based on more than a few participants. The mean number of participants per study in articles published in 1988 in the *Journal of Personality and Social Psychology* was about 200 (Reis and Stiller, 1992). This is quite a substantial average number of participants. So:

- How big should a sample be in order for us to claim that our findings apply generally?
- How should samples be selected in order for us to maximise our ability to generalise from our findings?

If everyone behaved in exactly the same way in our studies, we would only need to select one person to investigate the topic in question – everyone else would behave exactly the same. The way in which we select the sample would not have any bearing on the outcome of the research because there is no variability. We only need sampling designs and statistics, for that matter, because of this variability. Psychology would also be a very boring topic to study.

Fortunately, people vary in an infinite number of different ways. Take, for example, something as basic as the number of hours people say they usually sleep a day. While most people claim to sleep between seven and eight hours, others claim that they sleep less than six hours and others that they sleep more than ten hours (Cox *et al.*, 1987, p. 129). In other words, there is considerable variation in the number of hours claimed. Furthermore, how much sleep a person has varies from day to day – one day it may be six hours and the next day eight hours. Differences between people and within a person are common just as one might expect. So our sampling methods need to be planned with the awareness of the issue of variability together with an awareness of the level of precision that we need in our estimates of the characteristics of people.

The necessary size of the samples used in research should partially reflect the consequences of the findings of the research. Research for which the outcome matters crucially may have more stringent requirements about sample size than research for which the outcome, whatever it is, is trivial. For example, what size sample would one require if the outcome of the study could result in counselling services being withdrawn by a health authority? What size sample would one require if the study is just part of the training of psychology students – a practical exercise? What size sample would one require for a pilot study prior to a major investigation? While they might disagree about the exact sample size to use, probably psychologists would all agree that larger samples are required for the study that might put the future of counselling services at risk. This is because they know that a larger sample is more likely to demonstrate a trend in the study if there is one in reality.

Also, as we have mentioned, many psychologists also tend to favour larger sample sizes because they believe that this is likely to result in greater precision in their estimates of what is being measured. For example, it is generally the case that larger samples are employed when we are trying to estimate the frequency, or typical value, of a particular behaviour or characteristic in the population. If we wanted an estimate of the mean number of reported hours of sleep in, say, the elderly, then we are likely to use a bigger sample. What this means is that it is possible to suggest that the average number of hours of sleep has a particular value and that there is only a small margin of error involved in this estimate. That is, our estimate is likely to be pretty close to what the average is in reality. On the other hand, if we want to know whether the number of hours slept is related to mental health then we may feel that a smaller sample will suffice. The reason for this is that we only need to establish that sleep and mental health are related – we are less concerned about the precise size of the relationship between the two. (If the aim is to produce an estimate of some characteristic for the population, then we will have more confidence in that estimate if the sample on which that estimate is based is selected in such a way so as to be representative of the population. The basic way of doing this is to draw samples at random. However, ways of selecting a representative sample will be discussed in Chapter 13 along with other sampling methods in some detail. Of course, the more representative we can assume our sample to be the more confidence we can have in our generalisations based on that sample.)

Probably most sampling in psychological research is what is termed *convenience samples*. These are *not* random samples of anything but groups of people that are relatively easy for the researcher to get to take part in their study. In the case of university lecturers and students, the most convenient sample typically consists of students – often psychology students. What is convenient for one psychologist may not be convenient for another, of course. For a clinical psychologist psychotherapy patients may be a more convenient sample than undergraduate students. Bodner (2006) noted that for a random sample of 200 studies selected from PsycINFO in 1999 only 25 per cent of them used college students, ranging from 5 per cent in clinical or health psychology to 50 per cent in social psychology.

Convenience samples are usually considered to be acceptable for much psychological research. Since psychological research often seeks to investigate whether there is a relationship between two or more variables, a precisely defined sample may be unnecessary (Campbell, 1969, pp. 360–2). Others would argue that this is very presumptuous about the nature of the relationship between the two variables – especially that it is consistent over different sorts of people. For example, imagine that watching television violence is related to aggressiveness in males, but inversely related to aggressiveness in females. By taking a sample of psychology students, who tend to be female, a convenience sample of university students will actually stack things in favour of finding that watching television is associated with lower levels of aggressiveness.

Whether it is possible to generalise from a sample of psychology students, or even students, to the wider population is obviously an empirical question for any one topic of research. It is also a matter of credibility since it would be scarcely credible to study post-partum depression simply on the basis of a general convenience sample of university students. There are many circumstances in which it would seem perverse to choose to study students rather than other groups. For example, if a researcher was interested in the comprehensibility of the police caution then using university students might seem less appropriate than using a sample of people with poor educational attainment. Obviously, if one is addressing an issue that is particular to a certain group such as children or psychotherapy patients, then it is important to select this group of people. The use of students as a primary group for study has its advantages in the context of their education and training as it is time-consuming to contact other groups; on the other hand it has severe difficulties for virtually any other purposes. Getting the balance right is a matter for the research community in general, not students learning to do psychology.

Often in psychological research, it is difficult to identify the population that is of concern to the researcher. Although common sense would suggest that the population is that which is represented by the actual participants in the research, this usually does not appear to be what is in the researcher's mind. Probably because psychologists tend to see research questions as general propositions about human behaviour rather than propositions about a particular type of person or specific population, they have a tendency to generalise beyond the population which would be defined by the research sample. The difficulty is, of course, just when the generalisation should stop – if ever. Similarly, there tends to be an assumption that propositions are not just true at one point in time but true across a number of points in time. That is, psychological processes first identified more than a lifetime ago are still considered relevant today. Gergen (1973) has argued for the historical relativity of psychological ideas which Schlenker (1974) has questioned.

So there appears to be a distinction between the *population of interest* and the population defined clearly by the sample of participants utilised in the research. Of course, it would be possible to limit our population in time and space. We could say that our population is all students at Harvard University in 2010. However, it is almost certain that having claimed this we would readily generalise the findings that we obtain to students at other universities, for example. We may not directly state this but we would write in a way which is suggestive of this. Furthermore, people in our research may be samples from a particular group simply because of the resource constraints affecting our options. For example, a researcher may select some, but not all, 16-year-olds from a particular school to take part in research. Within this school, participants are selected on a random basis by selecting at random from the school's list of 16-year-olds. While this would be a random sample from the school and can be correctly described as such, the population as defined by the sample would be very limited. Because of the extremely restricted nature of the initial selection of schools, the results of the study may not be seen as being more informative than a study where this random selection procedure was not used but a wider variety of research locations employed.

The question of the appropriateness of sampling methods in most psychological research is a difficult one. Psychological researchers rarely use random sampling from a clearly defined population. Almost invariably some sort of convenience sample of participants is employed – where randomisation is used it is in the form of random allocation to the conditions of an experiment or the sequence of taking part in the conditions. This is as true of the best and most influential psychological research as less auspicious and more mundane research. In other words, if precise sampling were the criterion for good research, psychology textbooks may just as well be put through the shredder. This is not to say that sampling in psychological research is good enough – there is a great deal to be desired in terms of current practices. However, given that the major justification for current practice lies in the assumed generality of psychological principles, things probably will not change materially in the near future.

Another factor needs to be considered when evaluating the adequacy of psychological sampling methods: participation rates in many sorts of research are very low. Participation rates refer to the proportion of people who take part in the research compared with the number asked to take part in the research, that is, the proportion who supply usable data. Random sampling is considerably undermined by poor participation rates; it cannot

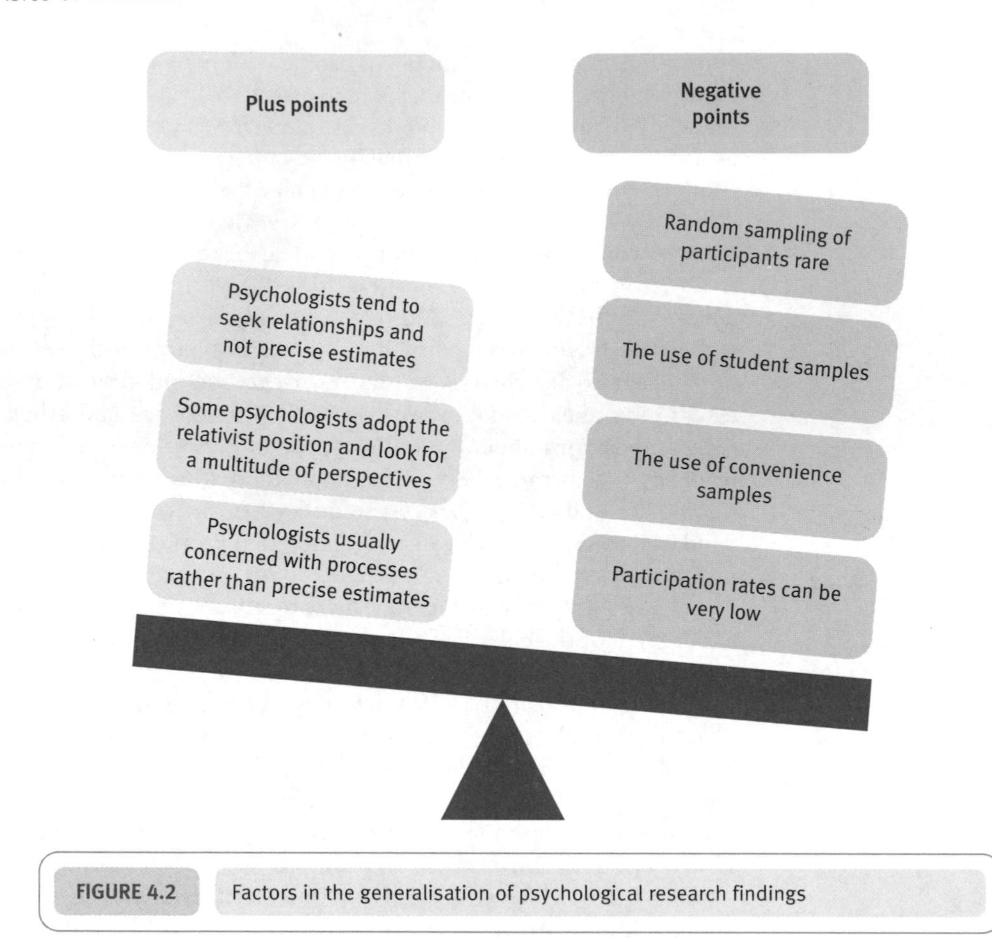
Plus points
Negative points
Random sampling of participants rare
Psychologists tend to seek relationships and not precise estimates
The use of student samples
Some psychologists adopt the relativist position and look for a multitude of perspectives
The use of convenience samples
Psychologists usually concerned with processes rather than precise estimates
Participation rates can be very low

FIGURE 4.2 Factors in the generalisation of psychological research findings

be assumed that those who do *not* participate are a random sample of the people approached. They do not participate for a variety of reasons, some of which may mean that certain sorts of participants exclude themselves. These reasons may be systematically related to the research topic – maybe potential participants are simply uninterested in the topic of the research. Alternatively, there may be more technical reasons why participation rates are low. A study which involves the completion of a questionnaire is likely to result in less literate potential participants declining to take part. In other words, issues to do with sampling require the constant attention, consideration and vigilance of researchers planning, analysing and evaluating research. The issues are complex and impossible to provide rules of thumb to deal with. The lesson is that simply using random selection methods does not ensure a random sample. In these circumstances, convenience samples may be much more attractive propositions than at first they appear to be – if poor participation rates systematically distort the sample then what is to be gained by careful sampling? Figure 4.2 displays some points about the kinds of samples typically used by psychologists.

4.4 Statistics and generalisation

Statistical analysis serves many important roles in psychology – as some students will feel they know to their cost. There are numerous statistical techniques that help researchers explore the patterns in their data, for example, which have little or nothing to do with what is taught on introductory statistics courses. Most students, however, are more familiar with what is known as 'inferential statistics' or, more likely, the concept of 'significance testing'. Significance testing is only one aspect of research but is a crucial one in terms of a researcher's willingness to generalise the trends that they find in their data. While students are encouraged to believe that statistical significance is an important criterion, it is just one of two really important things. The other is the size of the trend, difference, effect or relationship found in the research. The bigger that these are, then the more important the relationship. Furthermore, statistical significance is not the most important thing in evaluating one's research. One needs a fuller picture than just that when reaching decisions about research.

Moreover, as a consequence of the tendency of psychologists to emphasise statistical significance, they can overlook the consequences of failing to show that there is a trend in their data when, in reality, there is a trend. This can be as serious as mistakenly concluding that there is a trend when in reality there is no trend and that our sample has capitalised on chance. For example, what if the study involves an innovative treatment for autism? It would be a tragedy in this case if the researcher decided that the treatment did not work simply because the sample size used was far too small for the statistical analysis to be statistically significant. Essentially this boils down to the need to plan one's research in the light of the significance level selected, the minimum size of the effect or trend in your data that you wish to detect, and the risk that you are prepared to take of your data not showing a trend when in reality there is a trend. With these things decided, it is possible to calculate, for example, the minimum sample size that your study will need to be statistically significant for a particular size of effect. This is a rather unfamiliar area of statistics to most psychologists, which known as statistical power analysis. It is included in the new edition of the statistics textbook which accompanies this book (Howitt and Cramer, 2011a).

Chance findings and statistical significance

When investigating any research question, one decides what will be an appropriate sample size largely on the basis of the size of the effect or association expected. The bigger the effect or association, the smaller the sample can be in purely statistical terms. This is because bigger effects are more likely to be statistically significant with small sample sizes. A statistically significant finding is one that is large enough that it is unlikely to be caused by chance fluctuations due to sampling. (It should be stressed, the calculation of statistical significance is normally based on the hypothetical situation defined by the null hypothesis that there is no trend in the data.) The conventionally accepted level of significance is the 5 per cent or .05 level. This means that a finding as big as ours can be expected to occur by chance on 5 or fewer occasions if we tested that finding on 100 occasions (and assuming that the null hypothesis is in fact true). A finding or effect that is likely to occur on more than 5 out of 100 times by chance is described as being statistically non-significant or not statistically significant. Note that the correct term is non-significant, *not* that it is statistically insignificant, although authors sometimes use this term. Insignificant is a misleading term since it implies that the finding is not statistically important – but that simply is not what is meant in significance testing. The importance of a finding lies in the strength of the relationship between two variables or the size of the difference between two samples. Statistical significance testing merely refers to the question whether the trend is sufficiently large in the data so that it is unlikely that it could be the result of chance factors due to the variability inherent in sampling, that is, there is little chance that the null hypothesis of no trend or difference is correct.

Too much can be made of statistical significance if the size of the trend in the data is disregarded. For example, it has been argued that with very large samples, virtually any relationship will be statistically significant though the relationship may itself be a very small one. That is, a statistically significant relationship may, in fact, represent only a very small trend in the data. Another way of putting this is that very few null hypotheses are true *if* one deals with very large samples and one will accept even the most modest of trends in the data. What this means, though, in terms of generalisation is that small trends found in very large samples are likely not to generalise to small samples.

The difference between statistical significance and psychological significance is at the root of the following question. Which is better: a correlation of .06 which is statistically significant with a sample of 1000 participants or a correlation of .8 that is statistically significant with a sample of 6 participants?

This is a surprisingly difficult question for many psychologists to answer.

While the critical value of 5 per cent or .05 or less is an arbitrary cut-off point, nevertheless it is one widely accepted. It is not simply the point for rejecting the null hypothesis but is also the point at which a researcher is likely to wish to generalise their findings. However, there are circumstances in which this arbitrary criterion of significance may be replaced with an alternative value:

- The significance level may be set at a value other than 5 per cent or .05. If the finding had important consequences and we wanted to be more certain that our finding was not due to chance, we might set it at a more stringent level. For example, we may have developed a test that we found was significantly related at the 5 per cent level to whether or not someone had been convicted of child abuse. Because people may want to use this test to help determine whether someone had committed, or was likely to commit, child abuse, we may wish to set the critical value at a more stringent or conservative level because we would not want to wrongly suggest that someone would be likely to commit child abuse. Consequently, we may set the critical value at, say, 0.1 per cent or .001, which is 1 out of 1000 times or less. This is a matter of judgement, not merely one of applying rules.

- Where a number of effects or associations are being evaluated at the same time, this critical value may need to be set at less than the 5 per cent or .05 level. For example, if we were comparing differences between three groups, we could make a total of three comparisons altogether. We could compare group 1 with group 2, group 1 with group 3, and group 2 with group 3. If the probability of finding a difference between any two groups is set at 5 per cent or .05, then the probability of finding any of the three comparisons statistically significant at this level is three times as big, in other words 15 per cent or .15. Because we want to maintain the overall significance level at 5 per cent or .05 for the three comparisons, we could divide the 5 per cent or the .05 by 3, which would give us an adjusted or corrected critical value of 1.67 per cent (5/3 = 1.666) or .017 (.05/3 = .0166). This correction is known as a Bonferroni adjustment. (See our companion statistics text, *Introduction to Statistics in Psychology*, Howitt and Cramer, 2011a, for further information on this and other related procedures.) That is, the value of, say, the *t*-test would have to be significant at the 1.67 per cent level according to the calculation in order to be reported as statistically significant at the 5 per cent level.

- For a pilot study using a small sample and less than satisfactory measuring instruments, the 5 per cent or .05 level of significance may be an unnecessarily stringent criterion. The size of the trends in the data (relationship, difference between means, etc.) is possibly more important. For the purposes of such a pilot study, the significance level may be set at 10 per cent or .1 to the advantage of the research process in these circumstances. There may be other circumstances in which we might wish to be flexible about accepting significance levels of 5 per cent or .05. For example, in medical research, imagine that researchers have found a relationship between taking hormone replacement therapy and the development of breast cancer. Say that we find this relationship to be statistically significant at the 8 per cent or .08 level, would we willingly conclude that the null hypothesis is preferred or would we be unwilling to take the risk that the hypothesis linking hormone replacement therapy with cancer is in fact true? Probably not. The point is not that significance testing is at fault but that a whole range of factors impinge on what we do as a consequence of the test of our hypotheses. Research is an intellectual process requiring considerable careful thought in order to make what appear to be straightforward decisions on the basis of statistical significance testing.

However, students are well advised to stick with the 5 per cent or .05 level as a matter of routine. One would normally be expected to make the case for varying this and this may prove difficult to do in the typical study.

4.5 Directional and non-directional hypotheses again

The issue of directional and non-directional hypotheses was discussed in Box 2.1, but there is more that should be added at this stage. When hypotheses are being developed, researchers usually have an idea of the direction of the trend, correlation or difference that they expect. For example, who would express the opinion that there is a difference between the driving skills of men and woman without expressing an opinion as to what that difference – such as women are definitely worse drivers – is? In everyday life, a person who expresses such a belief about women's driving skills is likely to be expressing prejudices about women or joking or being deliberately provocative – they are unlikely to be a woman. Researchers, similarly, often have expectations about the likely outcome of their research – that is, the direction of the trend in their data. A researcher would not express such a view on the basis of a whim or prejudice but they would make as strong an argument as possible built on evidence suggestive of this point of view. It should also be obvious that in some cases there will be very sound reasons for expecting a particular trend in the data whereas in other circumstances no sound grounds can be put forward for such an expectation. Research works best when the researcher articulates coherent, factually based and convincing grounds for their expectations.

In other words, often research hypotheses will be expressed in a directional form. In statistical testing, a similar distinction is made between directional and non-directional tests but the justifications are required to be exacting and reasoned (see Box 2.1). In a statistical analysis, as we saw in Chapter 2, there are tough requirements before a directional hypothesis can be offered. These requirements are that there are very strong empirical or theoretical reasons for expecting the relationship to go in a particular direction and that researchers are ignorant of their data before making the prediction. It would be silly to claim to be making a prediction if one is just reporting the trend observed in the data. These criteria are so exacting that they probably mean that little or no student research should employ directional *statistical* hypotheses. Probably the main exceptions are where a student researcher is replicating the findings of a classic study, which has repeatedly been shown to demonstrate a particular trend.

The reason why directional statistical hypotheses have such exacting requirements is that conventionally the significance level is adjusted for the directional hypothesis. The directional hypothesis is referred to as one-tailed significance testing. The non-directional hypothesis is referred to as two-tailed significance testing. In two-tailed significance testing, the 5 per cent or .05 chance level is split equally between the two possibilities – that the association or difference between two variables is either positive or negative. So if the hypothesis is that cognitive behaviour therapy has an effect then this would be supported by cognitive behaviour therapy either being better in the highest 2.5 per cent or .025 of samples or worse in the lowest 2.5 per cent or .025 of samples. In one-tailed testing the 5 per cent is piled just at one extreme – the extreme which is in the direction of the one-tailed hypothesis. Put another way, a directional hypothesis is supported by weaker data than would be required by the non-directional hypothesis. The only good justification for accepting a weaker trend is that there is good reason to think that it is correct, that is, either previous research has shown much the same trend or theory powerfully predicts a particular outcome. Given the often weak predictive power of much psychological theory, the strength of the previous research is probably the most useful of the two.

If the hypothesis is directional, then the significance level is confined to just one half of the distribution – that is, the 5 per cent is just at one end of the distribution (not both) which means, in effect, that a smaller trend will be statistically significant with a directional test. There is a proviso to this and that is that the trend is in the predicted direction. Otherwise it is very bad news since even big trends are not significant if they are in the wrong direction. The problem with directional hypotheses is, then, what happens when the researcher gets it wrong, that is the trend in the data is exactly the reverse of what is suggested in the hypothesis. There are two possibilities:

- That the researcher rejects the hypothesis.
- That the researcher rejects the hypothesis but argues that the reverse of the hypothesis has been demonstrated by the data. The latter is rather like having one's cake and eating it, statistically speaking. If the original hypothesis had been supported using the less stringent requirements then the researcher would claim credit for that finding. If, on the other hand, the original hypothesis was actually substantially reversed by the data then this finding would now find favour. The reversed hypothesis, however, was deemed virtually untenable once the original directional hypothesis had been decided upon. So how can it suddenly be favoured when it was previously given no credence with good reason? The only conclusion must be that the findings were chance findings.

So the hypothesis should be rejected. The temptation, of course, is to forget about the original directional hypothesis and substitute a non-directional or reverse directional hypothesis. Both of these are totally wrong but who can say when even a researcher will succumb to temptation?

Possibly the only circumstances in which a student should employ directional *statistical* hypotheses is when conducting fairly exact replication studies. In these circumstances the direction of the hypothesis is justified by the findings of the original study. If the research supports the original direction then the conclusion is obvious. If the replication actually finds the reverse of the original findings then the researcher would be unlikely to claim that the reverse of the original findings is true since it only would apply to the replication study. The situation is one in which the original findings are in doubt as are the new findings since they are diametrically opposite.

■ One- versus two-tailed significance level

Splitting the 5 per cent or .05 chance or significance level between the two possible outcomes is usually known as the two-tailed significance level because two outcomes (directions of the trend or effect) both in a positive and a negative direction are being considered. We do this if our hypothesis is non-directional as we have not specified which of the two outcomes we expect to find. Confining the outcome to one of the two possibilities is known as the one-tailed significance level because only one outcome is predicted. This is what we do if our hypothesis is directional, where we expect the results to go in one direction.

To understand what is meant by the term 'tailed', we need to plot the probability of obtaining each of the possible outcomes that could be obtained by sampling *if the null hypothesis is assumed to be true*. This is the working assumption of hypothesis testing and reference to the null hypothesis is inescapable if hypothesis testing is to be understood. The technicalities of working out the distribution of random samples if the null hypothesis is true can be obtained from a good many statistics textbooks. The 'trick' to it all is employing the information contained in the actual data. This gives us information about the distribution of scores. One measure of the distribution of scores is the *standard deviation*. In a nutshell, this is a sort of average of the amount scores in a sample differ from the mean of the sample. It is computationally a small step from the standard deviation of scores to the standard error of the means of samples. Standard error is a sort of measure of the variation of sample means drawn from the population defined by the null hypothesis. Since we can calculate the standard error quite simply, this tells us how likely each of the different sample means are. (Standard error is the distribution of sample means.) Not surprisingly, samples very different from the outcome defined by the null hypothesis are increasingly uncommon the more different they are from what would be expected on the basis of the null hypothesis.

This is saying little more than that if the null hypothesis is true, then samples that are unlike what would be expected on the basis of this null hypothesis are likely to be uncommon.

4.6 More on the similarity between measures of effect (difference) and association

Often measures of the effect (or difference) in experimental designs are seen as unlike measures of association. This is somewhat misleading. Simple basic research designs in psychology are often analysed using the *t*-test (especially in laboratory experiments) and the Pearson correlation coefficient (especially in cross-sectional or correlational studies). The *t*-test is based on comparing the means (usually) of two samples and essentially examines the size of the difference between the two means relative to the variability in the data. The Pearson correlation coefficient is a measure of the amount of association or relationship between two variables. Generally speaking, especially in introductory statistics textbooks, they are regarded as two very different approaches to the statistical analysis of data. This can be helpful for learning purposes. However, they are actually very closely related.

A *t*-test is usually used to determine whether an effect is significant in terms of whether the mean score of two groups differ. We could use a *t*-test to find out whether the mean depression score was higher in the cognitive behaviour therapy group than in the no treatment group. A *t*-test is the mean of one group subtracted from the mean of the other group and divided by what is known as the standard error of the mean:

$$t = \frac{\text{mean of one group} - \text{mean of other group}}{\text{standard error of the mean}}$$

The standard error of the mean is a measure of the extent to which sample means are likely to differ. It is usually based on the extent to which scores in the data differ so it is also a sort of measure of the variability in the data. There are different versions of the *t*-test. Some calculate the standard error of the mean and others calculate the standard error of the difference between two means.

The value of t can be thought of as the ratio of the difference between the two means to the degree of the variability of the scores in the data. If the individual scores differ widely, then the t value will be smaller than if they do not differ much. The bigger the t value is, the more likely it is to be statistically significant. To be statistically significant at the two-tailed .05 level, the t value has to be 2.00 or bigger for samples of more than 61 cases. The t value can be slightly less than 2.00 for bigger samples. The minimum value that t has to exceed to be significant at this level is 1.96, which is for an infinite number of cases. These figures can be found in the tables in some statistics texts such as *Introduction to Statistics in Psychology* (Howitt and Cramer, 2011a).

Bigger values of t generally indicate a bigger effect (bigger difference between the sample means relative to the variability in the data). However, this is affected by the sample size so this needs to be taken into consideration as well. Bigger values of t also tend to indicate increasingly significant findings if the sample size is kept constant.

The Pearson's correlation shows the size of an association between two quantitative variables. It varies from –1 through 0 to 1:

- A negative value or correlation means that lower values on one variable go together with higher values on the other variable.
- A positive value or correlation means that higher values on one variable go together with higher values on the other variable.
- A value of zero or close to zero means that there is no relationship or no linear relationship between the two groups and the outcome measure.

Note that Pearson's correlation is typically used to indicate the association between two quantitative variables. Both variables should consist of a number of values, the frequencies of which take the shape of a bell approximately. The bell-shaped distribution is known as the normal distribution. (See the companion book *Introduction to Statistics in Psychology*, Howitt and Cramer, 2011a, or any other statistics textbook for a detailed discussion of precisely what is meant by a normal distribution.) Suppose, for example, we are interested in what the relationship is between how satisfied people are with their leisure and how satisfied they are with their work. Suppose the scores on these two measures vary from 1 to 20 and higher scores indicate greater satisfaction. A positive correlation between the two measures means that people who are more satisfied with their leisure are also more satisfied with their work. It could be that these people are generally positive or that being satisfied in one area of your life spills over into other areas. A negative correlation indicates that people who are more satisfied with their leisure are less satisfied with their work. It is possible that people who are dissatisfied in one area of their life try to compensate in another area.

A correlation of zero or close to zero shows that either there is no relationship between these two variables or there is a relationship but it does not vary in a linear way. For example, people who are the most and least satisfied with their leisure may be less satisfied with work than people who are moderately satisfied with their lives. In other words, there is a curvilinear relationship between leisure and work satisfaction. The simplest and the most appropriate way of determining whether a correlation of zero or close to zero indicates a non-linear relationship between two variables is to draw a scattergram or scatterplot as shown in Figure 4.3. Each point in the scatterplot indicates the position of one or more cases or participants in terms of their scores on the two measures of leisure and work satisfaction.

The t values which compare the means of two unrelated or different groups of cases can be converted into a Pearson's correlation or what is sometimes called a point–biserial correlation, which is the same thing. The following formula is used to convert an unrelated t value to a Pearson's correlation (which is denoted by the letter r; and n is, of course, the sample size):

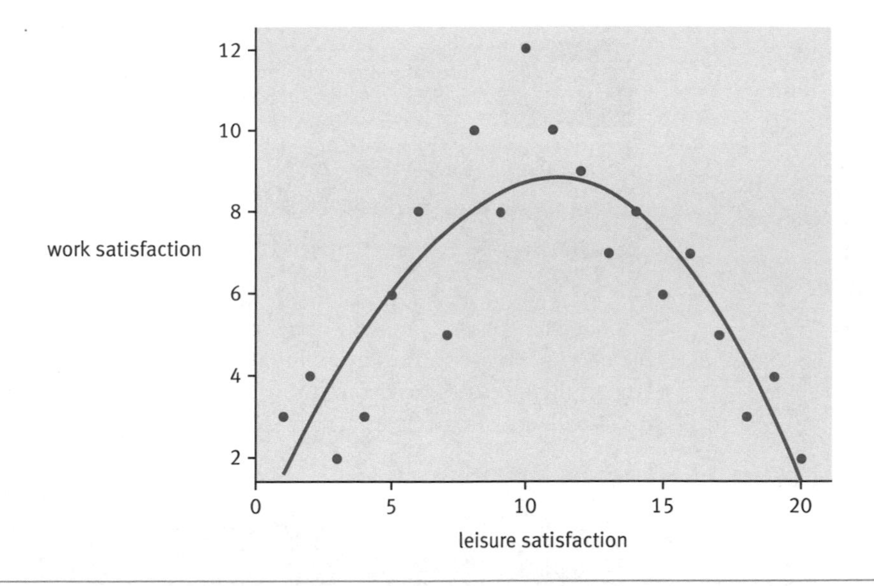

FIGURE 4.3 A scatterplot showing a non-linear relationship between leisure and work satisfaction

$$r = \sqrt{\frac{t^2}{t^2 + n - 2}}$$

Alternatively, we could calculate the value of r directly from the data. Suppose higher values on the measure of depression indicate greater depression and that the cognitive behaviour therapy group shows less depression than the no treatment group. If we code or call the no treatment group 1 and the cognitive behaviour therapy group 2, then we will have a negative correlation between the two groups and depression when we calculate the Pearson correlation value.

Because of the interchangeability of the concepts of correlation and tests of difference as shown, it should be apparent that we can speak of a difference between two groups or alternatively of a correlation or an association between group membership and another variable. This is quite a sophisticated matter and important when developing a mature understanding of research methods.

4.7 Sample size and size of association

Now that we have provided some idea as to what statistical significance is and how to convert a test of difference into a correlation, we can discuss how big a sample should be in order to see whether an effect or association is statistically significant. The bigger the correlation is, the smaller the sample can be for that correlation to be statistically significant. In Table 4.1 we have presented 11 correlations decreasing in size by .10 from ±1.00 to 0.00 and the size that the sample has to exceed to be significant at the one-tailed 5 per cent or .05 level. So, for example, if we expect the association to be about .20 or more, we would need a sample of more than 69 for that association to be statistically significant. It is unusual to have an overall sample size of less than 16. The size of an effect may be expected to be bigger when manipulating variables as in an

Table 4.1 Size of sample required for a correlation to be statistically significant at the one-tailed 5 per cent or 0.05 level

Correlation (r)	Sample size (n)	Verbal label for size of correlation	Percentage of variance shared	Verbal label for size of shared variance
±1.00		perfect	100	perfect
±.90	4		81	large
±.80	6	large, strong or high	64	
±.70	7		49	
±.60	9	modest or moderate	36	
±.50	12		25	
±.40	18		16	
±.30	32	small, weak or low	9	medium
±.20	69		4	
±.10	270		1	small
.00		none	0	none

experiment rather than simply measuring two variables as in a cross-sectional study. This is because the researcher, normally, does everything possible to reduce extraneous sources of variation in an experiment using procedural controls such as standardisation – this is another way of saying that error variance is likely to be less in a true experiment. Consequently, sample sizes are generally smaller for true experiments than non-experiments.

When dealing with two groups it is important that the two groups are roughly equal in size. If there were only a few cases in one of the groups (very disproportionate group sizes) we cannot be so confident that our estimates of the population characteristics based on these is reasonably accurate The conversion of a t value into a Pearson's correlation presupposes that the variation or variance in the two groups is also similar in size. Where there is a big disparity, the statistical outcome is likely to be somewhat imprecise. Where there are only two values, as in the case of two groups, one should ensure, if possible, that the two groups should be roughly similar in size. So it is important that the researcher should be aware of exactly what is happening in their data in respect of this.

A correlation of zero is never significant no matter how big the sample – since that is the value which best supports the null hypothesis. Correlations of 1.00 are very unusual as they represent a perfect straight-line or linear relationship between two variables. This would happen if we correlated the variable with itself. You need to have a minimum sample of three to determine the statistical significance of a correlation, though we would not suggest that you adopt this strategy since the size of the relationship would have to be rather larger than we would expect in psychological research. Thus a correlation of .99 or more would be significant at the one-tailed 5 per cent or .05 level if the sample was 3. Consequently, no sample size has been given for a correlation of 1.00. These sample sizes apply to both positive correlations and negative ones. It is the size of the correlation that matters for determining statistical significance and not its sign (unless you are carrying out a directional test).

The size of an association and its meaning

The size of a correlation is often described in words as well as in numbers. Correlations of .80 or above are usually talked of as being 'large', 'strong' or 'high'. This size of correlation may be obtained when we measure the same variable, such as depression, on two separate occasions two weeks apart. In such a case, we may say there was a strong correlation between the first and second test of depression. Correlations of .30 or less are usually spoken of as being small, weak or low. Correlations of this size are typically found when we measure different variables, such as depression and social support, on the same or different occasions. Correlations between .30 and .80 are commonly said to be moderate or modest. They are usually shown when we assess very similar measures, but which are not the same, such as (1) how supportive one sees a partner as being and (2) how satisfied one is with that relationship.

These labels may be misleading in that they may seem to be underestimating the strength of a correlation. The meaning of the size of a correlation is better understood if we square the correlation value – this gives us something called the *coefficient of determination*. So a correlation of .20 when squared gives a coefficient of determination of .04. This value represents the proportion of the variation in a variable that is shared with the variation in another variable. Technically this variation is measured in terms of a concept or formula called variance. The way to calculate variance can be found in a statistics textbook such as the companion book *Introduction to Statistics in Psychology* (Howitt and Cramer, 2011a).

A correlation of 1.00 gives a coefficient of determination of 1.00, which means that the two variables are perfectly related. A correlation of zero produces a coefficient of determination of zero, which indicates that the variables are either totally separate or they do not have a straight-line relationship such as the relationship between work and leisure satisfaction in Figure 4.3. These proportions may be expressed as a percentage, which may be easier to understand. We simply multiply the proportion by 100 so .04 becomes 4 (.04 × 100). The percentage of the variance shared by the correlations in Table 4.1 is shown in the fourth column of that table.

If we plot the percentage of variance shared against the size of the correlation as shown in Figure 4.4 we can see that there is not a straight-line or linear relationship between

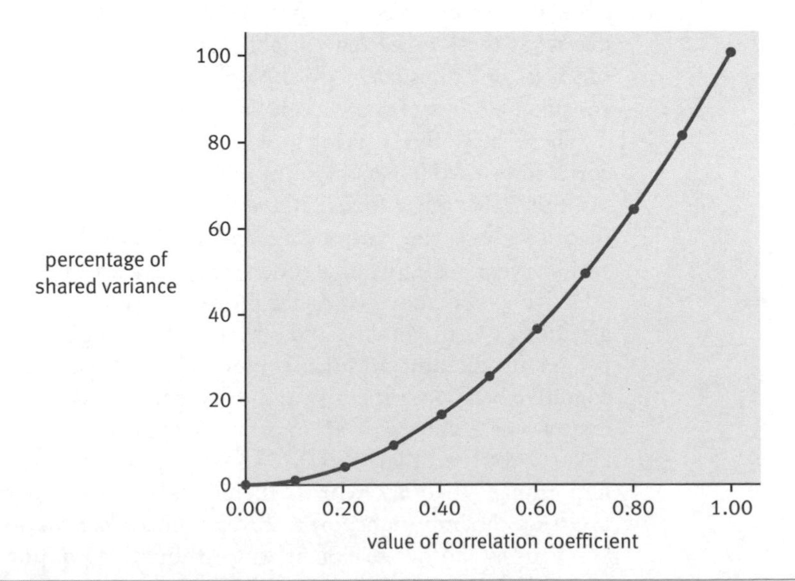

FIGURE 4.4 Relationship between correlation and percentage of shared variance

the two but what is called an exponential relationship. The percentage of variance increases at a faster rate at higher than lower correlations. As the size of a correlation doubles, the corresponding size of the percentage of shared variance quadruples. To give an example of this, a correlation of .40 is twice as big as one of .20. If we express these correlations as the percentage of shared variance we can see that a percentage of 16 is four times as big as one of 4. This should tell you that it is helpful to consider the amount of variation explained by a correlation and not simply the numerical size. A correlation of .40 is not twice as good as a correlation of .2 because in terms of the amount of variation (variance) explained, the larger correlation accounts for *four* times the amount of variation. Table 4.1 gives the figures for the amounts of variation explained.

The verbal labels generally used to describe different sizes of the shared variance have tended to differ in the research literature from those given to the correlations that correspond to them. Where the percentage of shared variance is about 1, the size of the effect or the association has been called 'small' (Cohen, 1988, pp. 24–7). Where it is about 5 it has been described as being 'medium'. Where it is more than about 10 it has been referred to as being 'large'. These judgements are obviously subjective or personal to some extent. What one psychologist considers to be a large effect, another might think of as being small. We are inclined to think that 10 may be considered medium and above 20 as large. However, such a judgement does depend on a great many factors such as what is being measured and how accurately or reliably it can be measured. One would expect lower values for the coefficient of determination if it is based on variables which cannot be measured accurately.

Justification for the use of these labels might come from considering just how many variables or factors may be expected to explain a particular kind of behaviour. Racial prejudice is a good example of such behaviour. It is reasonable to assume that racial prejudice is determined by a number of factors rather than just a single factor. The tendency towards authoritarianism has a correlation of .30 with a measure of racial prejudice (Billig and Cramer, 1990). This means that authoritarianism shares 9 per cent of its variance with racial prejudice. On the face of things, this is not a big percentage of the variance. What if we had, say, another ten variables that individually and independently explained (accounted for) a similar proportion of the variance? Then we could claim a complete account of racial prejudice. The problem in psychology is finding out what these other ten variables are – or whether they exist. Actually a correlation of .30 is not unusual in psychological research and many other variables will explain considerably less of the variance than this.

There is another way of looking at this issue. That is to ask what the value is of a correlation of .30 – a question which is meaningless in absolute terms. In the above example, the purpose of the research was basically associated with an attempt to theorise about the nature of racial prejudice. In this context, the correlation of .30 would seem to imply that one's resources would be better applied to finding more effective explanations of racial prejudice than can be offered on the basis of authoritarianism. On the other hand, what if the researcher was interested in using cognitive behaviour therapy in suicide prevention? A correlation of .30 between the use of cognitive behaviour therapy and decline in the risk of suicide is a much more important matter – it amounts to an improvement in the probability of suicide prevention from .35 to .65 (Rosenthal, 1991). This is in no sense even a moderate finding: it is of major importance. In other words, there is a case against the routine use of labels when assessing the importance of a correlation coefficient.

There is another reason why we should be cautious about the routine application of labels to correlations or any other research result. Our measures are not perfectly reliable or valid measures of what they are measuring (see Chapter 15 for a detailed discussion of reliability and validity). Because they are often relatively poor measures of what they are intended to measure, they tend usually to underestimate the true or real size of the association. There is a simple statistical procedure for taking into account the unreliability of the measures called the correction for attenuation (see the companion book *Introduction to Statistics in Psychology*, Howitt and Cramer, 2011a, Chapter 36). Basically it gives us an idealised version of the correlation between two variables as if they were perfect measures. The formula for the corrected correlation is:

$$\text{corrected correlation} = \frac{\text{correlation between measures 1 and 2}}{\sqrt{\text{measure 1 reliability} \times \text{measure 2 reliability}}}$$

If the correlation between the two measures is .30 and their reliability is .75 and .60, respectively, the corrected correlation is .45:

$$\text{corrected correlation} = \frac{.30}{\sqrt{.75 \times .60}} = \frac{.30}{\sqrt{.45}} = \frac{.30}{.67} = .45$$

This means that these two variables share about 20 per cent of their variance. If this is generally true, we would only need another four variables to explain what we are interested in. (Though this is a common view in the theory of psychological measurement, the adjustment actually redefines each of the concepts as the *stable* component of the variables. That is, it statistically makes the variables completely stable [reliable]. This obviously is to ignore the aspects of a variable which are unstable, for example, why depression varies over time, which may be as interesting and important to explain as the stable aspects of the variable.)

How do we know what size of effect or association to expect if we are just setting out on doing our research?

- Psychologists often work in areas where there has already been considerable research. While what they propose to do may never have been done before, there may be similar research. It should be possible from this research to estimate or guestimate how big the effect is likely to be.

- One may consider collecting data on a small sample to see what size of relationship may be expected and then to collect a sample of the appropriate size to ensure that statistical significance is achieved if the trend in the main study is equal to that found in the pilot study. So if the pilot study shows a correlation of .40 between the two variables we are interested in, then we would need a minimum of about 24 cases in our main study. This is because by checking tables of the significance of the correlation coefficient, we find that .40 is statistically significant at the 5 per cent level (two-tailed test) with a sample size of 24 (or more). These tables are to be found in many statistics textbooks – our companion statistics text, *Introduction to Statistics in Psychology* (Howitt and Cramer, 2011a), has all you will need.

- Another approach is to decide just what size of relationship or effect is big enough to be of interest. Remember that very small relationships and effects are significant with very large samples. If one is not interested in small trends in the data then there is little point in depleting resources by collecting data from very large samples. The difficulty is deciding what size of relationship or effect is sufficient for your purposes. Since these purposes vary widely no simple prescription may be offered. It is partly a matter of assessing the value of the relationship or effect under consideration. Then the consequences of getting things wrong need to be evaluated. (The risk of getting things wrong is higher with smaller relationships or effects, all other things being equal.) It is important not simply to operate as if statistical significance is the only basis for drawing conclusions from research.

It is probably abundantly clear by now that purely statistical approaches to generalisation of research findings are something of an impossibility. Alongside the numbers on the computer output is a variety of issues or questions that modify what we get out of the statistical analysis alone. These largely require thought about one's research findings and the need not to simply regard any aspect of research as routine or mechanical.

4.8 Conclusion

Psychologists are often interested in making generalisations about human behaviour that they believe to be true of, or apply to, people in general, though they will vary in the extent to which they believe that their generalisations apply universally. If they believe that the generalisation they are testing is specific to a particular group of people they will state what that group of people is. Because all people do not behave in exactly the same way in a situation, many psychologists believe that it is necessary to determine the extent to which the generalisation they are examining holds for a number, or sample, of people. If they believe that the generalisation applies by and large to most people and not to a particular population, they will usually test this generalisation on a sample of people that is convenient for them to use.

The data they collect to test this generalisation will be either consistent or not consistent with it. If the data are consistent with the generalisation, the extent to which they are consistent will vary. The more consistent the data are, the stronger the evidence will be for the generalisation. The process of generalisation is not based solely on simple criteria about statistical significance. Instead it involves considerations such as the nature of the sampling, the adequacy of each of the measures taken, and an assessment of the value or worth of the findings for the purpose for which they were intended.

Key points

- Psychologists are often concerned with testing generalisations about human behaviour that are thought to apply to all human beings. This is known as universalism since it assumes that psychological processes are likely to apply similarly to all people no matter their geographical location, culture or gender.
- The ability of a researcher to generalise from their research findings is limited by a range of factors and amounts to a complex decision-making process. These factors include the statistical significance of the findings, the representativeness of the sample used, participation and dropout rates, and the strength of the findings.
- Participants are usually chosen for their convenience to the researcher, for example, they are easily accessible. A case can be made for the use of convenience samples on the basis that these people are thought for theoretical purposes to be similar to people in general. Nonetheless, researchers are often expected to acknowledge this limitation of their sample.
- The data collected to test a generalisation or hypothesis will be either consistent with it or not consistent with it. The probability of accepting that the results or findings are consistent with the generalisation is set at .05 or 5 per cent. This means that these results are likely to be due to chance 5 times out of 100 or less. Findings that meet this criterion or critical value are called statistically significant. Those that do not match this criterion are called statistically non-significant.

ACTIVITIES

1. Choose a recent quantitative study that has been referred to either in a textbook you are reading or in a lecture that you have attended. What was the size of the sample used? Was a one- or a two-tailed significance level used and do you think that this tailedness was appropriate? What could the minimum size of the sample have been to meet the critical level of significance adopted in this study? What was the size of the effect or association, and do you think that this shows that the predictor or independent variable may play a reasonable role in explaining the criterion or dependent variable? Are there other variables that you think may have shown a stronger effect or association?

2. Choose a finding from just about any psychological study that you feel is important. Do you think that the principle of universalism applies to this finding? For example, does it apply to both genders, all age groups and all cultures? If not, then to which groups would you be willing to generalise the finding?

CHAPTER 5

Research reports

The total picture

Overview

- The research report is the key means of communication for researchers. Laboratory reports, projects, master's and doctoral dissertations and journal articles all use a similar and relatively standard structure.
- Research reports are more than an account of how data were collected and analysed. They describe the entire process by which psychological knowledge develops.
- Research reports have certain conventions about style, presentation, structure and content. This conventional structure aids communication once the basics have been learnt.
- The research report should be regarded as a whole entity, not a set of discrete parts. Each aspect – title, abstract, tables, text and referencing – contributes to how well the total report communicates.
- This chapter describes the detailed structure of a research report and offers practical advice on numerous difficulties.

5.1 Introduction

Research is not just about data collection and analysis. The major purpose is to advance understanding of the subject matter, that is, to develop theory, concepts and information about psychological processes. The research report describes the role that a particular study plays in this process. Research is not the application of a few techniques without rhyme or reason. Equally, the research report is not a number of unarticulated sections but a fully integrated description of the process of developing understanding and knowledge in psychology. To fully appreciate a research report requires an understanding of the many different aspects of research. Not surprisingly, writing a research report is a demanding and sometimes confusing process.

Despite there being different types of research report (laboratory report, dissertation, thesis, journal article, etc.), a broadly standard structure is often employed. Accounts of research found in psychology journals – journal articles – are the professional end of the continuum. At the other end are the research reports or laboratory reports written by undergraduate students. In between there is the final-year project, the master's dissertation and the doctoral dissertation. An undergraduate laboratory report is probably 2000 words, a journal article 5000 words, a final-year project 10 000 words, a master's dissertation 20 000–40 000 words and a doctoral dissertation in Europe 80 000 words but shorter where the programme includes substantial assessment of taught courses. Although there is a common structure which facilitates the comprehension of research reports and the absorption of the detail contained therein, this structure should be regarded as flexible enough to cope with a wide variety of contingencies. Psychology is a diverse field of study so it should come as no surprise to find conflicting ideas about what a research report should be. Some of the objections to the standard approach are discussed in the chapters on qualitative methods (Chapters 17–25).

There are two main reasons why research reports can be difficult to write:

- The research report is complex with a number of different elements, each of which requires different skills. The skills required when reviewing the previous theoretical and empirical studies in a field are not the same as those involved in drawing conclusions from statistical data. The skills of organising research and carrying it out are very different from the skills required to communicate the findings of the research effectively.

- When students first start writing research (laboratory) reports their opportunities to read other research reports – such as journal articles – are likely to have been very limited. There is a bit of a chicken-and-egg problem here. Until students have understood some of the basics of psychological research and statistics they will find journal articles very difficult to follow. At the same time, they are being asked essentially to write a report using much the same structure as a journal article. Hopefully, some of the best students will be the next generation of professional researchers writing the journal articles.

This chapter on writing the research report comes early in this book. Other books have it tucked away at the end. But to read and understand research papers it helps to understand how and why a research report is structured the way it is. Furthermore writing a report should not be regarded as an afterthought but as central to the process of doing research. Indeed, it may be regarded as the main objective of doing research. Apart from training and educational reasons, there is little point in doing research which is not communicated to others. The structure of the research report is broadly a blueprint of the entire research process though perhaps a little more organised and systematic than the actual research itself. For example, the review of previous research (literature review) would appear to be done first judging by most research reports – it follows the title and summary (abstract) after all. Nevertheless, most researchers would admit that they are still reading relevant publications even after the first draft of the report is completed. It cannot be stressed too much that the research report actually prioritises what should be done in research. In contrast, a tiny minority of researchers (see Chapter 23) reject the idea that the literature review should come first – some claim that only after the data has been collected and analysed should the previous research be examined to assess its degree of compatibility with the new findings. This is not sloppiness or laziness on their part. Instead, it is a desire to analyse data unsullied by preconceptions, but it does mean that building on previous research is not central to this alternative formulation. Put another way, a research report is largely the way it is because the methodology of psychology is the way it is. Departures from the standard practice serve to emphasise the nature and characteristics of the psychological method.

The conventional report structure, then, gives us the building blocks of conventional research. Good research integrates all of the elements into a whole – a hotchpotch of unrelated thoughts, ideas and activities is *not* required. At the same time, research reports do not give every last detail of the process but a clear synthesis of the major and critical aspects of the research process. A research report contains a rather tidy version of events, of course, and avoids the messy detail of the actual process in favour of the key stages presented logically and coherently. Writing the research report should be seen as a constructive process which can usefully begin even at the planning stage of the research. That is, the research report is *not* the final stage of the research process but integral to it. If this seems curious then perhaps you should consider what many qualitative researchers do when analysing their textual data. They begin their analysis and analytic note-taking as soon as the first data become available. Such forethought and planning are difficult to fulfil but regard them as an ideal to be aimed at.

Something as complex as a research report may be subject to a degree of inconsistency. Journals, for example, publish detailed style instructions for those submitting material to them. The American Psychological Association (APA) publishes a very substantial manual. There is no universal manual for student research reports so, not surprisingly, different lecturers and instructors have varying views of the detail of the structure and style of undergraduate student research reports. Different universities have different requirements for doctoral theses, for example. The problem goes beyond this into professional research reports. A quick search of the Internet using the search term 'psychology research report' will locate numerous websites giving advice and instruction about writing a good research report. These sites far from speak with one voice. Nevertheless, there is a great deal of overlap or unanimity of approach. It is essential for a student to understand the local rules for the research report just as it is for the professional researcher to know what is acceptable to the journal to which they submit work for possible publication. Probably students will receive advice from their lecturers and instructors giving specific requirements. In this chapter, we have opted to use the style guidelines of the American Psychological Association wherever practicable. This helps prepare students for a possible future as users and producers of psychological research. It should be remembered that research is increasingly regarded as an important skill for practitioners of all sorts and not just academic psychologists. No longer is research confined to universities; practitioner research is commonplace. Indeed the description researcher-practitioner describes the role of many psychologists outside universities.

5.2 Overall strategy of report writing

Structure

A psychological research report normally consists of the following sections:

- *Title page* This is the first page and contains the title, the author and author details such as their address, e-mail address, telephone and fax number. For a student report, this will be replaced with details such as student ID number, degree programme name and module name.
- *Abstract* This is the second page of the report and you may use the subheading 'Abstract' for clarity. The abstract is a detailed summary of the contents of the report.
- *Title* This is another new page – the title is repeated from the first page but *no* details as to authorship are provided. This is to make it easier for editors to send out the manuscript for anonymous review by other researchers.
- *Introduction* This continues on the same page but normally the subheading 'Introduction' is omitted.
- *Method* This consists of the following sections at a minimum:
  - participants,
  - materials, measures or apparatus,
  - design and procedure.
- *Results* This includes statistical analyses, tables and diagrams.
- *Discussion* This goes into a detailed explanation of the findings presented under results. It can be quite conjectural.
- *Conclusion* Usually contained within the discussion section and not a separate subheading. Nevertheless, sometimes conclusions are provided in a separate section.
- *References* One usually starts a new page for these. It is an alphabetical (then chronological if necessary) list of the sources that one has cited in the body of the text.
- *Appendices* This is an optional section and is relatively rare in professional publications. Usually it contains material which is helpful but would be confusing to incorporate in the main body of the text.

This is the basic, standard structure which underlies the majority of research reports. However, sometimes other sections are included where appropriate. Similarly, sometimes sections of the report are merged *if* this improves clarity. The different sections of the structure are presented in detail later in this chapter. Figure 5.1 gives the basic structure of a psychological report.

Although these different components may be regarded as distinct elements of the report, that they are integrated into a whole is a characteristic of a skilfully written report. In practice, this means that even the title should characterise the entire report. With only a few words at your disposal for the title this is difficult but nevertheless quite a lot can be done. Similarly the discussion needs to integrate with the earlier components such as the introduction to give a sense of completeness and coherence. The title is probably the first thing read so it is crucial to orienting the reader to the content of the report. The abstract (summary) gives an overview of all aspects of the research so clarity not

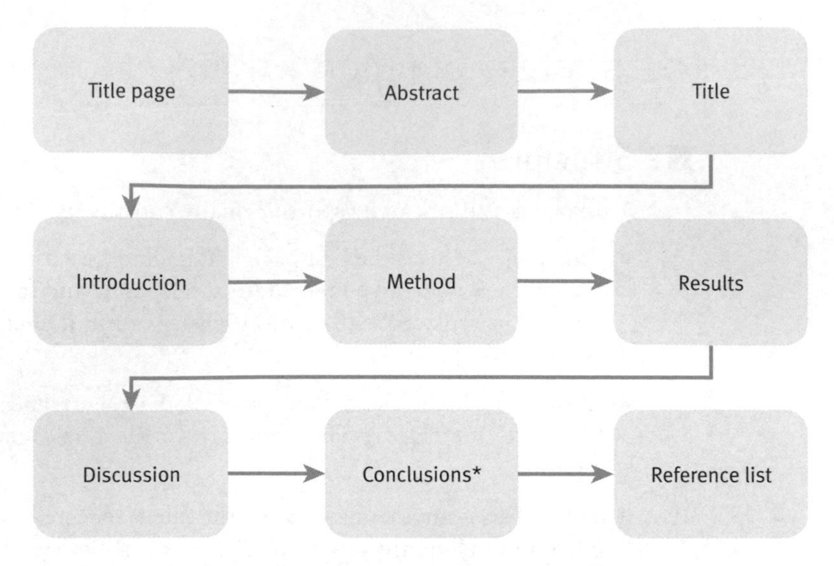

FIGURE 5.1 The basic structure of a psychological report

only creates a good impression but helps reassure the reader that the report and the research it describes are of a high quality.

Overall writing style

Clarity is essential since there is a great deal of information contained within a research report. The material contained in the report should be geared to the major theme of the report. This is particularly the case with the introduction in which the research literature is reviewed. It is a bad mistake to simply review research in the chosen field and fail to integrate your choice with the particular aspects addressed by your research.

A number of stylistic points (as summarised in Figure 5.2) should be remembered:

- Keep sentences short and as simple as possible. Sentences of eight to ten words are probably the optimum. Check carefully for sentences over 20 words in length. The reader will have forgotten the beginning of the sentence by the time the end is reached! With modern word processing it is possible to check for sentence length. In Microsoft Word, for example, the spelling and grammar checker does this and it is possible to get readability statistics by selecting the appropriate option. If you have the slightest problem about sentence length then you should consider using this facility. (Readability statistics are based on such features as average sentence length. As such they provide a numerical indication of stylistic inadequacies.)

- Paragraphing needs care and thought. A lack of paragraphing makes a report difficult to read. Probably a paragraph should be no more than about half a printed page. Equally, numerous one-sentence paragraphs make the report incoherent and unreadable. Take a good look at your work as bad paragraphing looks odd. So always check your paragraphs. Break up any very long paragraphs. Combine very short paragraphs, especially those of just one or two sentences in length.

- It is useful to use subheadings (as well as the conventional headings). The reason for this is that subheadings indicate precisely what should be under that subheading – and what should not be. *Even if you delete the subheadings before submitting the final report*, you will benefit by having a report in which the material is in meaningful order. If you think that your draft report is unclear, try to put in subheadings. Often this will help you spot just where the material has got out of order. Then it is a much easier job to put it right.

- Make sure that your sentences are in a correct and logical order. It is easy to get sentences slightly out of order. The same is true for your paragraphing. You will find that subheadings help you spot this.
- It is normally inappropriate to use personal pronouns such as 'I' and 'we' in a research report. However, care needs to be taken as this can lead to lengthy passive sentences. In an effort to avoid 'We gave the participants a questionnaire to complete.' the result can be the following passive sentence: 'Participants were given a questionnaire to complete.' It would be better to use a more active sentence structure such as 'Participants completed a questionnaire.' This is shorter by far. In the active sentence it is the subject that performs the action; for example, 'We [SUBJECT] wrote [VERB] the report [OBJECT]'. In a passive sentence the subject suffers the action, as in 'The report [SUBJECT] was written [VERB]'.
- The dominant tense in the research report is the past tense. This is because the bulk of the report describes completed activities in the past (for example, 'The questionnaire measured two different components of loneliness.'). That is, the activities completed by the researcher in the process of collecting, analysing and interpreting the data took place in the past and are no longer ongoing. Other tenses are, however, sometimes used. The present tense is often used to describe the current beliefs of researchers (for example, 'It is generally considered that loneliness consists of two major components . . .'). Put this idea into the past tense and the implications are clearly different (for example, 'It was generally considered that loneliness consists of two major components . . .'). The future tense is also used sometimes (for example, 'Clarification of the reasons for the relationship between loneliness and lack of social support will help clinicians plan treatment strategies.').
- Remember that the tables and diagrams included in the report need to communicate as clearly and effectively as the text. Some readers will focus on tables and diagrams before reading the text since these give a quick overview of what the research and the research findings are about. Too many tables and diagrams are not helpful and every table and diagram should be made as clear as possible by using headings and clear labels.
- Avoid racist and sexist language, and other demeaning and otherwise offensive language about minority groups. The inclusion of this in a professional research report may result in the rejection of the article or substantial revision to eliminate such material (see Box 5.1).
- Numbers are expressed as 27, 3, 7, etc. in most of the text except where they occur as the first words of the sentence. In this case, we would write; 'Twenty-seven airline pilots and 35 cabin crew completed the alcoholism scale.'
- It is a virtue to keep the report reasonably compact. Do not waffle or put in material simply because you have it available. It is not desirable to exceed word limits so sometimes material has to be omitted. It is not uncommon to find that excess length can be trimmed simply by judicious editing of the text. A quarter or even a third of words can be edited out if necessary.
- Do *not* include quotations from other authors except in those cases where it is undesirable to omit them. This is particularly the case when one wishes to dispute what a previous writer has written. In this instance, only by quoting the origin can its nuances be communicated.

Box 5.1 Talking Point

Avoiding bias in language

Racism, sexism, homophobia and hostility to minorities such as people with disabilities are against the ethics of psychologists. The use of racist and sexist language and other unacceptable modes of expression are to be avoided in research reports. Indeed, such language may result in the material being rejected for publication. We would stress that the avoidance of racist and sexist language cannot fully be reduced to a list of dos and don'ts. The reason is that racism and sexism can manifest themselves in a multiplicity of different forms and those forms may well change with time. For example, Howitt and Owusu-Bempah (1994) trace the history of racism in psychology and how the ways it is manifest have changed over time. While it is easy to see the appalling racism of psychology from a century ago, it is far harder to understand its operation in present day psychology. For detailed examples of how the writings of psychologists may reinforce racism see Owusu-Bempah and Howitt (1995) and Howitt and Owusu-Bempah (1990).

Probably the first step towards the elimination of racism and sexism in psychological research is for researchers to undergo racism and sexism awareness training. This is increasingly available in universities and many work locations. In this way, not only will the avoidance of offensive language be helped but, more important, the inadvertent propagation of racist and sexist ideas through research will be made much more difficult.

A few examples of avoidable language use follow:

- Writing things like 'the black sample . . .' can readily be modified to 'the sample of black people . . .' or, if you prefer, 'the sample of people of colour . . .'. In this way, the most important characteristic is drawn attention to: the fact that you are referring to people first and foremost who also happen to be black. You might also wish to ask why one needs to refer to the race of people at all.
- Avoid references to the racial (or gender) characteristics of participants which are irrelevant to the substance of the report. For example, 'Female participant Y was a black lone-parent . . .'. Not only does this contain the elements of a stereotypical portrayal of black people as being associated with father absence and 'broken families', but the race of the participant may be totally irrelevant to what the report is about.
- Do not refer to man, mankind or social man, for example. These terms do not make people think of man and woman but of men only. Words like 'people' can be substituted. Similarly referring to 'he' contributes to the invisibility of women and so such terms should not be used.

Of course, the use of demeaning and similar language is not confined to race and gender. Homophobic language and writings are similarly to be avoided. Equally, careful thought and consideration should be given when writing about any disadvantaged or discriminated against group. So people with disabilities should be treated with dignity in the choice of language and terms used. So, for example, the phrase 'disabled people' is not acceptable and should be replaced with 'people with disabilities'.

The website of the American Psychological Association contains in-depth material on these topics – race and ethnicity, gender and disabilities. Should your report touch on any of these, you are well advised to consult the Association's guidance. The following location deals with various aspects of APA style: http://www.apastyle.org/index.aspx

- Generally introductions are the longest section of a research report. Some authorities suggest about a third of the available space should be devoted to the introduction. Of course, adjustments have to be made according to circumstances. Research which collects data on numerous variables may need to devote more space to the results section.
- A rule of thumb is to present the results of calculations to no more than two decimal places. There is a danger of spuriously implying a greater degree of accuracy than psychological data usually possess. Whatever you do, be consistent. You need to understand how to round to two decimals. Basically, if the original number ends with a figure of 5 or above then we round up, otherwise we round down. So 21.4551 gives 21.46 rounded whereas 21.4549 gives 21.45 rounded.

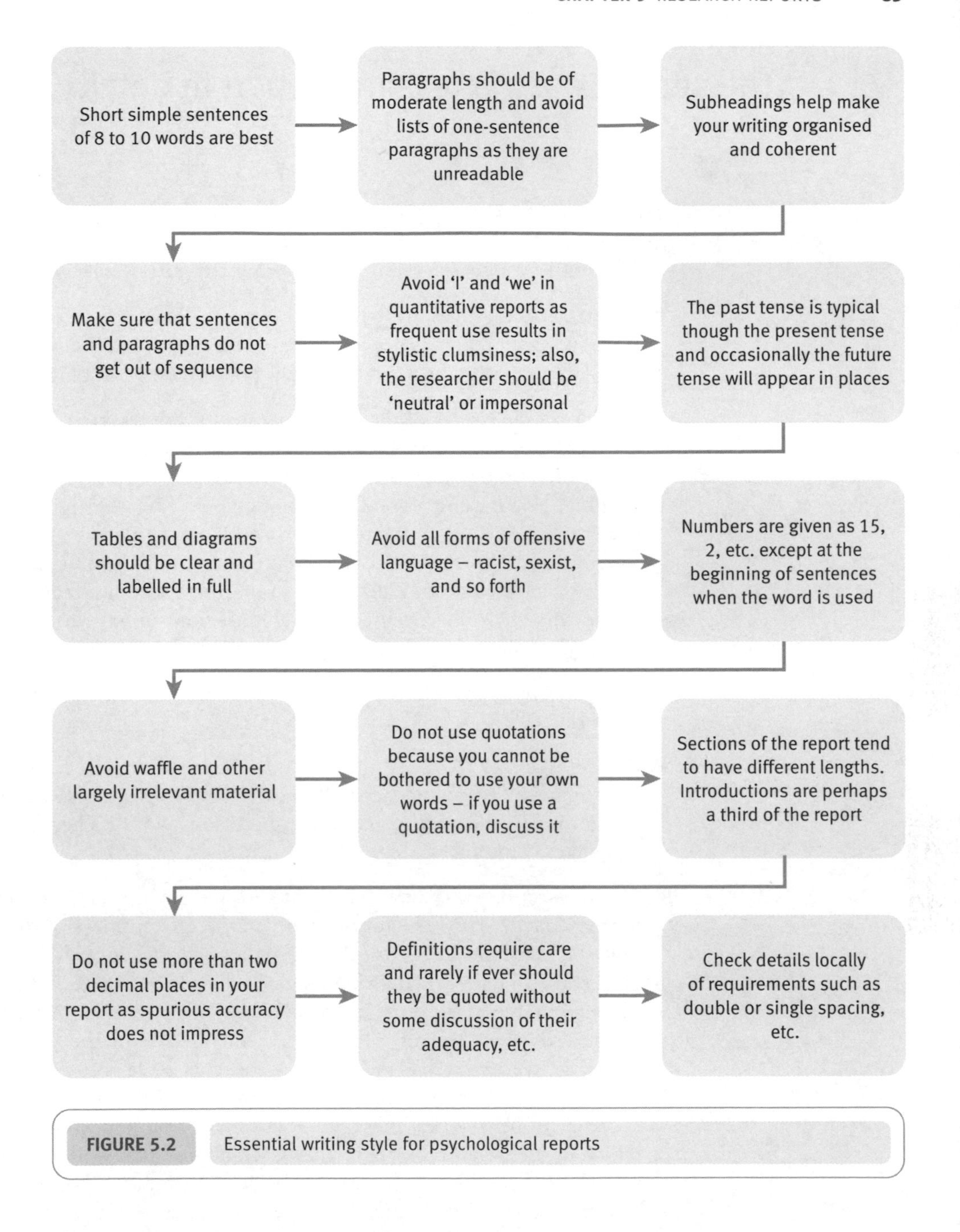

FIGURE 5.2 Essential writing style for psychological reports

- Psychological terms may not have a standard definition which is accepted by all researchers. Consequently, you may find it necessary to define how you are using terms in your report. Always remember that definitions in psychology are rarely definitive and they are often problematic in themselves.
- Layout: normally the recommendation is to double space your work and word-process it. However, check local requirements on this. Leave wide margins for comments. Use underlining or bold for headings and subheadings. The underlying assumption behind this is that the report is being reviewed by another person. A report that will not be commented upon might not require double spacing. Check the local rules where you are studying.

5.3 The sections of the research report in detail

Title

The title is *not* used as a heading or subheading. Often it is given twice – once on the title page and again just before the introduction.

The title of a research report serves two main purposes:

- To attract the attention of potential readers. This is especially the case for professional research reports since the potential reader probably comes across the title either in a database or by browsing through the contents of a journal.
- To inform the reader of the major features of the research paper. In other words, it amounts to a summary of the contents of the research report in no more than about 12 words (although 20 words might be used if necessary). This includes any subheading. You require a good understanding of your research before you can write a good title. It is a good discipline to try to write a title even before you have finished the research. This may need honing into shape since initial attempts are often a little clumsy and too wordy. Enlist the help of others who are familiar with your research as they may be able to help you to rephrase your initial efforts. The key thing is that the reader gains a broad idea of the contents of the report from the title.

The following suggestions may help you write a clear and communicative title for your work:

- Phrases such as 'A study of . . .', 'An investigation into . . .' and 'An experiment investigating . . .' would normally not be included in well-written titles since they are not really informative and take up precious words. They should be struck out in normal circumstances.
- Avoid clever or tricky titles which, at best, merely attract attention. So titles like 'The journey out of darkness' for a report on the effectiveness of therapy for depression fails the informativeness test. It may be a good title for a novel but not a psychological research report. Better titles might include 'Effectiveness of cognitive behavioural therapy in recovery from depression during long-term imprisonment'. This title includes a great deal of information compared with the previous one. From our new title, we know that the key dependent variable is depression, that the population being researched is long-term prisoners, and that the key independent variable is cognitive behavioural therapy. Occasionally you may come across a title which is tricky or over-clever but nevertheless communicates well. For example, one of us published a paper with the title 'Attitudes do predict behaviour – in mails at least'. The first four words clearly state the overriding theme of the paper. The last four words do not contain a misprint but an indication that the paper refers to letter post. Another example from a recent study is 'Deception among pairs: "Let's say we had lunch and hope they will swallow it!"' All in all, the best advice is to avoid being this smart.
- If all else fails, one can concentrate on the major hypothesis being tested in the study (if there is one). Titles based on this approach would have a rudimentary structure something like 'The effects of (variable A) on (variable B)'. Alternatively, 'The relationship between (variable A) and (variable B)'. A published example of this is 'The effects of children's age and delay on recall in a cognitive or structured interview'. The basic structure can be elaborated as in the following published example: 'Effects of pretrial juror bias, strength of evidence and deliberation process on juror decisions:

New validity evidence of the juror bias scale scores'. The phrase 'Effects of . . .' may create some confusion. It may mean 'the causal effect of variable A on variable B' but not necessarily so. Often it means 'the relationship between variable A and variable B'. 'The effects of' and 'the relationship between' are sometimes used interchangeably. It is preferable to restrict 'Effects of' to true experiments and 'Relationships between' to non-experiments.

Abstract

The abstract is best given this heading in a student's report although the title 'Abstract' is not normally included in psychology journals. Since the abstract is a summary of many aspects of the research report, normally it is written after the main body of the report has been drafted. This helps prevent the situation in which the abstract refers to things *not* actually in the main body of the report. Since the abstract is crucially important, expect to write several drafts before it takes its final shape. The brevity of the abstract is one major reason for the difficulty.

The key thing is that the abstract is a (fairly detailed) summary of *all* aspects of the research report. It is usually limited to a maximum number of words. This maximum may vary, but limits of 100 to 200 words are typical. With space available for only 10 to 20 short sentences, inevitably the summary has to be selective. Do *not* cope with the word limit by concentrating on just one or two aspects of the whole report, for example, the hypotheses and the data collection method used would be insufficient on their own. When writing an abstract you should take each of the major sections of the report in turn and summarise the key features of each. There is an element of judgement in this but a well-written abstract will give a good overview of the contents of the report.

It is increasingly common to find 'structured abstracts'. The structure may vary but a good structure is four subheadings:

- Purpose
- Methods
- Results
- Conclusions.

This structure ensures that the abstract covers the major components of the research. You could use it to draft an abstract and delete these headings after they have served their purpose of concentrating your mind on each component of the research.

Although this does not apply to student research reports, the abstract (apart from the title) is likely to be all that potential readers have available in the first instance. Databases of publications in psychology and other academic disciplines usually include just the title and the abstract together, perhaps, with a few search terms. Hence, the abstract is very important in a literature search – it is readily available to the researcher whereas obtaining the actual research report may require some additional effort. Most students and researchers will be able to obtain abstracts almost instantly by using Internet connections to databases. A badly written abstract may deter some researchers from reading the original research report and may cause others to waste effort obtaining a report which is not quite what they expected it to be. The clearer and more comprehensive the information in the abstract, the more effective will be the decision of whether or not to obtain the original paper for detailed reading.

The other function of the abstract is that it provides a structure for when one is reading the entire paper. In other words, the reader will know what to expect in the report having read the abstract, and this speeds up and simplifies the task of reading. Since first impressions are important, writing the abstract should not be regarded as a drudgery but an opportunity to establish the value of your research. Get it wrong, and the reader may get the impression that you are confused and muddled – bad news if that person is giving you a grade or possibly considering your work for possible publication.

You will find examples of abstracts in any psychology journal. Figure 5.3 shows the components of a report to be summarised in the abstract.

Box 5.2 Practical Advice

Important points to summarise in the abstract

Ideally, the following should be outlined in the abstract. Normally subheadings are *not* used except in structured abstracts though this rule may be broken if necessary. They are given here simply for purposes of clarity. They relate to the major subheadings of the report itself:

- *Introduction* This is a brief statement justifying the research and explaining the purpose, followed by a short statement of the research question or the main hypotheses. The justification may be in terms of the social or practical utility of the research, its relevance to theory, or even the absence of previous research. The research question or hypotheses will also be given. Probably no more than 30 per cent of the abstract will be such introductory material.
- *Method* This a broad orientation to the type of research that was carried out. Often a simple phrase will be sufficient to orient the reader to the style of research in question. So phrases like 'Brain activity was studied using PET (positron emission tomography) and FMRI (functional magnetic resonance imaging) . . .', 'A controlled experiment was conducted . . .', 'The interview transcripts were analysed using discourse analysis . . .' and 'A survey was conducted . . .' suggest a great deal about the way in which the research was carried out without being wordy.
- *Participants* This will consist of essential detail about the sample(s) employed. For example, 'Interview data from an opportunity sample consisting of young carers of older relatives was compared with a sample of young people entering the labour market for the first time, matched for age'.
- *Procedure* This should identify the main measures employed. For example, 'Loneliness was assessed using the shortened UCLA loneliness scale. A new scale was developed to measure social support'. By stipulating the important measures employed one also identifies the key variables. For an experiment, in addition it would be appropriate to describe how the different conditions were created (i.e. manipulated). For example, 'Levels of hunger were manipulated by asking participants to refrain from eating or drinking for 1 hour, 3 hours and 6 hours prior to the experiment'.
- *Results* There is no space in an abstract for elaborate presentations of the statistical analyses that the researcher may have carried out. Typically, however, broad indications are given of the style of analysis. For example, 'Factor analysis of the 20-item anxiety scale revealed two main factors', 'The groups were compared using a mixed-design ANOVA' or 'Binomial logistic regression revealed five main factors which differentiated men and women'. Now these statistical techniques may be meaningless to you at the moment but they will not be to most researchers. They refer to very distinct types of analysis so the terms are very informative to researchers. In addition, the major findings of the statistical analysis need to be reported. Normally this will be the important, statistically significant features of the data analysis. Of course, sometimes the lack of significance is the most important thing to draw attention to in the abstract. There is no need and normally no space to use the succinct methods of the reporting of statistics in the abstract. So things like ($t = 2.43$, df $= 17$, $p < 0.05$) are rare in abstracts and best omitted.
- *Discussion* In an abstract, the discussion (and conclusions) need to be confined to the main things that the reader should take away from the research. As ever, there are a number of ways of doing this. If you have already stated the hypothesis then you need do little other than confirm whether or not this was supported, given any limitations you think are important concerning your research, and possibly mention any crucial recommendations for further research activity in the field.

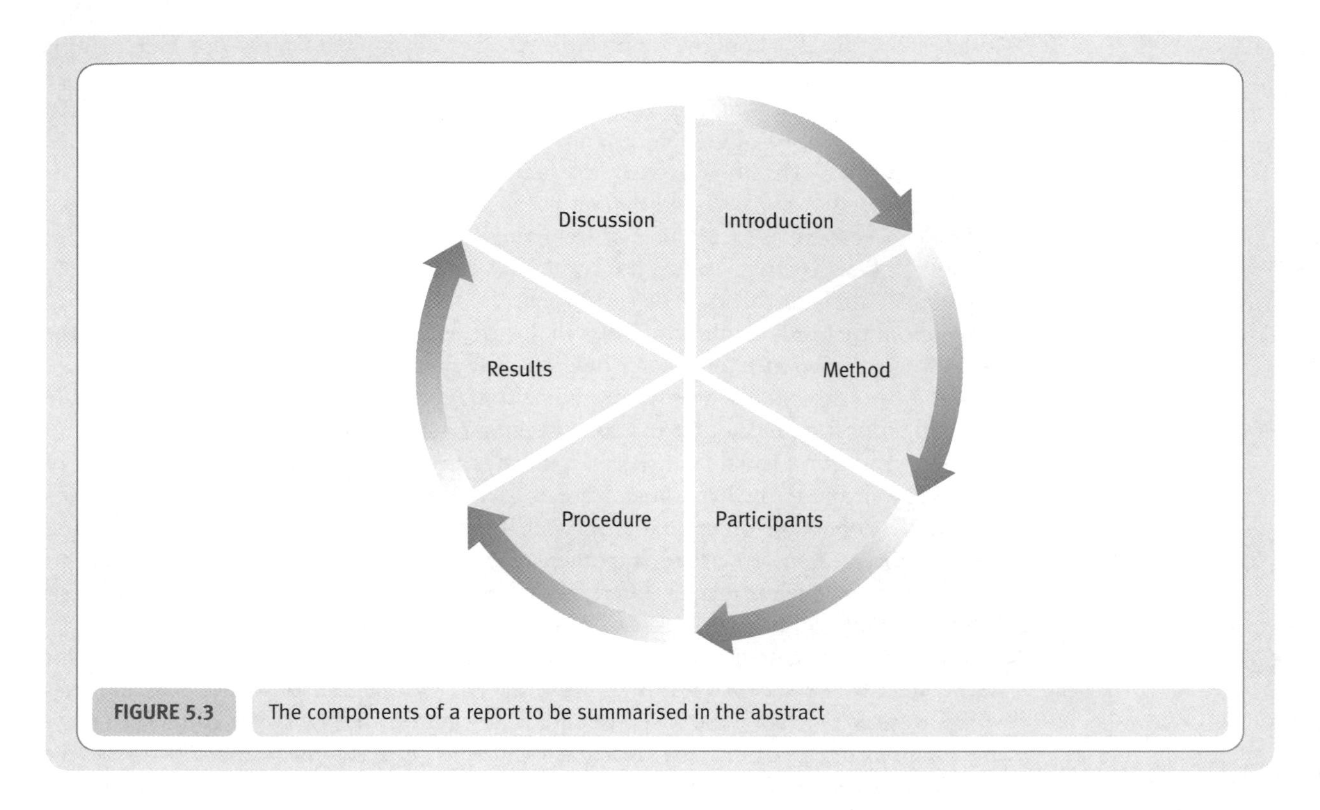

FIGURE 5.3 The components of a report to be summarised in the abstract

Introduction

Usually, the introduction to a report is *not* given a heading or subheading – to do so merely states the obvious. The introduction sets the scene for the research, the analysis and discussion which follow it. In effect, it is an explanation of why your chosen research topic deserved researching and the importance of the particular aspect of the topic you have chosen to focus on.

Explanations or justifications for a particular research topic include the following:

- There is a need to empirically test ideas that have been developed in theoretical discussions of the topic. In other words, the advancement of theory may be offered as full or partial reasons for engaging in research.
- There is a pressing concern over a particular issue which can be informed by empirical data. Often social research is carried out into issues of public concern but the stimulus may, instead, come from the concerns of organisations such as health services, industry and commerce, the criminal justice system, and so forth.
- There is an unresolved issue arising out of previous research on a topic which may be illuminated by further research – especially research which constructively replicates the earlier research.

Just being interested in the topic is not a good or sufficient reason for doing research in academic terms. You should make the intellectual case for doing the research, not the personal case for doing so.

The introduction contains a *pertinent* review of previous research and publications on the topic in question, partly to justify the new research. One explains the theoretical issues, the problems with previous research, or even the pressing public interest by reference to what other researchers have done and written before. Research is regarded as part of a collective enterprise in which each individual contribution is part and builds on the totality. The keyword is *pertinent* – or relevant – previous research and publications.

Just writing vaguely about the topic of the research using any material that is to hand is not appropriate. The literature that you need to incorporate is that which most directly leads to the research that you have carried out and are about to describe later in the report. In other words, the literature review needs to be in tune with the general thrust of your research. That it is vaguely relevant is not a good reason for the inclusion of anything and you may well find that an unfocused review is counterproductive.

Students sometimes face problems stemming from their use of secondary sources, for example, a description given in a textbook. This may contain very little detail about, say, a particular study or theory. As a consequence, the student lacks information when they write about the study or theory. Often they introduce errors because they read into the secondary source things that they would not if they had read the original source. There is no easy way around this other than reading sources that cover the topic in depth. Ideally this will be the original source but some secondary sources are better than others.

The introduction should consistently and steadily lead to a statement of the aims of your research and to the hypotheses (though there is some research for which hypotheses are either not possible or are inappropriate). There is a difference between the aims and the hypotheses. Aims are broad, hypotheses more specific. For example, 'the aim of this study was to investigate gender differences in conversation' might include the hypothesis 'Males will interrupt more than females in mixed-gender dyads'.

It is usually suggested that the introduction should be written in the past tense. This may generally be the case but is not always so. The past perfect tense describes activities which were completed in the past. I ran home, the car broke down, the CD finished playing are all examples of the past tense. Unfortunately, it is not always possible to use the past tense – sometimes to do so would produce silly or confusing writing. For example, the sentence 'The general consensus among researchers is that loneliness is a multifaceted concept' is in the present tense. The past tense cannot convey the same meaning. 'The general consensus among researchers was that loneliness is a multifaceted concept' actually implies that this is no longer the general consensus. Hence one needs to be aware of the pitfalls of the present tense for communicating certain ideas. However, since most of the material in the introduction refers to completed actions in the past then most of it will be in the past tense. Sentences like, 'Smith and Hardcastle (1976) showed that intelligence cannot be adequately assessed using motor skills alone' refer to past events. Similarly, 'Haig (2004) argued for the inclusion of "waiting list" control groups in studies of the effectiveness of counselling' cannot be expressed well using a different tense.

Putting all of this together, a typical structure of an introduction is as follows:

- A brief description of the topic of the research.
- Key concepts and ideas should be explained in some detail or defined if this is possible.
- Criticisms of aspects of the relevant research literature together with synthesis where the literature clearly leads to certain conclusions.
- A review of the most important and relevant aspects of the research literature:
  - Theoretical matters pertinent to the research
  - Describe and discuss as necessary the key variables to be explored in your research.

List your aims and hypotheses as summary statements at the end of the introduction.

Method

This is a major heading. The method section can be tricky to write since the overall strategy is to provide sufficient information for another researcher to replicate your study precisely. At the same time, the minutiae of the procedures that you carried out are not included. Clearly getting the balance between what to include and what to omit is difficult. Too much detail and the report becomes unclear and difficult to read. Too little detail and significant aspects of the research may be difficult for other researchers to reproduce. Really the task is to describe your methods in sufficient detail that the reader has a clear understanding of what you did – they could probably replicate the research, more or less. In a sense it is rather like a musical score: broadly speaking the musicians will know what and how to play, but they will have to fill in some of the detail themselves. Nevertheless, it is inevitable that the method section contains a greater density of detail than most other sections, which tend to summarise and take an overview.

Box 5.3 Talking Point

Abbreviations

Abbreviations should be used with caution in a research report. Their main advantage is brevity. Before the days of computerisation, typesetting was very expensive, and the use of abbreviations saved some considerable expense. This is no longer really the case. Student writing rarely benefits from the use of abbreviations. If they are used badly or inappropriately, then they risk confusing the reader who may then feel that the student is confused.

The major disadvantage of abbreviations is that they hamper communications and readability. Ss, DV, *n*, SD and SS are examples of abbreviations that sometimes were included in reports. The trouble is that we assume that the reader knows what the abbreviations refer to. If the reader is not familiar with the abbreviation then their use hampers rather than aids communication and clarity. The problem is not solved simply by stating the meaning of the abbreviation early in the report (for example, 'The dependent variable (DV) was mental wonderment (MW)'). The reader may not read your definition or they may forget the meaning of the abbreviation the next time it appears. Acronyms for organisations can similarly tax readers unnecessarily. This is because acronyms are commonly used by those involved with an organisation but an outside reader may be unfamiliar with their use. We recommend using abbreviations only in exceptional circumstances and where their use is conventional – as when you succinctly report your statistical findings (for example, $t(27) = 2.30$, $p = .05$).

Box 5.4 Key Ideas

Independent and dependent variables

In student research reports, it is common to identify what variable(s) is the independent variable (IV) and what variable(s) is the dependent variable (DV). This is much less common in professional research reports. In general, it is probably a useful thing for students new to research to do. However, it is something that few professional researchers do and, consequently, might best be left out as soon as possible. The problem in professional research is that the independent variables and dependent variables are often interchangeable. This is especially the case in regression analyses. The independent variable is the variable which is expected to affect the value of the dependent variable. There is no necessary assumption that there is a causal link at all. In controlled experiments (see Chapter 9) the independent variable is always the variable(s) defined by the experimental manipulation(s). So if the level of anger in participants is manipulated by the experimenter then anger level is the independent variable. In an experiment, any variables which might be expected to be affected by varying the level of anger or any other independent variable is called a dependent variable. The dependent variable is the one for which we calculate the means and standard deviations, etc. In a controlled or randomised experiment, the effect of the independent variable on the dependent variable is regarded as causal. In this case, the independent variable is expected to have a direct effect on the scores of the dependent variable. In non-randomised research, all that is established is that there is an association or relationship between the independent and dependent variable.

The method section may include as many as six or more sections if the procedures are especially complex. These include:

- participants (sometimes archaically referred to as subjects);
- materials or apparatus (or even both);
- procedure (always);
- design (possibly);
- stimuli (if these require detailed description);
- ethical considerations (recommended).

Of these, design and stimuli would only be used if they are not simple or straightforward in these respects; it is a matter of judgement whether to include them or not. One approach would be to include them in the early draft of your report. If they seem unnecessary or if you write very little under them (just a few sentences) then they can be combined quickly with a few mouse clicks on the word processor. An ethical considerations section is becoming increasingly common and is required in some professional writings. While students should not carry out any research which does not use well-established, ethically sound procedures, including an ethics section demonstrates that the ethical standing of the research has been considered.

Normally the methods heading is given as a major title, perhaps centred and underlined or in bold. The section headings are subheadings and are aligned to the left margin. It is usual to go straight from the methods heading to the participants subheading. No preamble is required.

Participants

This section should contain sufficient information so that the reader knows how many participants you had in total, how many participants were in each condition of the study, and sufficient detail about the characteristics of the participants to make it clear what the limitations on generalising the findings are likely to be. Given that much student research is conducted on other students, often the description will be brief in these cases. It is old-fashioned (and misleading; see Box 5.5) to refer to those taking part in your research as subjects. Avoid doing so. Once in a while the word subjects will occur in relation to particular statistical analyses which traditionally used the term (e.g. related-subjects analysis of variance). It is difficult to change this usage.

We would normally expect to include the following information to describe the participants:

- The total number of participants.
- The numbers of participants in each of the groups or conditions of the study.
- The gender distribution of the participants.
- The average age of the participants (group by group if they are not randomly allocated to conditions) together with an appropriate measure of the spread of the characteristic (for example, standard deviation, variance or standard error – these are equally understandable to most researchers and more-or-less equivalent in this context).
- Major characteristics of the participants or groups of participants. Often this will be university students but other research may have different participant characteristics, for example, preschool children, visitors to a health farm, etc. These may also be presented in numerical form as frequencies.

Box 5.5 Talking Point

No subjects – just participants

One of the most misleading terms ever in psychology was the concept of 'subject'. Monarchs have subjects, psychologists do not. In the early days of psychological research, terms such as reactors, observers, experimentees and individuals under experiment were used. Danziger (1985) points out that these early research studies used other professors, students and friends which may explain the lack of use of the slightly hostile term 'subjects'. Although the term subjects had a long history in psychological writing, it is inadequate because it gives a false picture of the people who take part in research. The term implies a powerful researcher and a manipulated subject. Research has long since dispelled this myth – people who take part in research are not passive but actively involve themselves in the research process. They form hypotheses about the researcher's hypotheses, for example. As a consequence they must be regarded as active contributors in the research process. In the 1990s, psychological associations such as the American Psychological Association and the British Psychological Society recommended/insisted on the modern terminology for their journals.

- How the participants were contacted or recruited initially.
- How the participants were selected for the research. Rarely are participants formally sampled using random sampling in psychological research. Convenience sampling is much more common (see Chapter 13).
- It is good practice, but not universal, to give some indication of refusal rates and dropout rates for the participants. Refusal rates are the numbers who are asked to take part in the research but say no or otherwise fail to do so. Dropout rates are the numbers who initially take part in the research but for some reason fail to complete all of the stages. Sometimes this is known alarmingly as 'the mortality rate' or the 'experimental mortality'.
- Any inducements or rewards given to participants to take part in the study. So, for example, giving the participants monetary rewards or course credits would be mentioned.

Materials/apparatus

The materials or apparatus section describes psychological tests, other questionnaire measures, laboratory equipment and other such resources which are essential components of the research. Once again the idea is to supply enough detail so that another researcher could essentially replicate the study and the reader gains a clear impression of the questionnaire, tests and equipment used. These requirements allow for a degree of individual interpretation, but experience at writing and considering other people's writings in research journals, especially, will help improve your style and hone the level of detail that you include. Remember that the degree of detail you go into should be sufficient to help the reader contextualise your research but not so detailed that the wood cannot be seen for the trees. So trivial detail should be omitted. Details such as the make of the stopwatch used, the colour of the pen given to participants, and the physical dimensions of the research setting (laboratory, for example) would not normally be given unless they were especially pertinent. If the study were, for example, about the effects of stress on feelings of claustrophobia then the physical dimensions of the laboratory would be very important and ought to be given. Provide the name and address of suppliers of specialised laboratory equipment, for example, but not for commonplace items such as stopwatches. When using computers in psychological research, details of the software used to present experimental stimuli, for example, would normally be given.

It is also usual to give details of any psychological tests and measures employed. This would probably include the official name of the test, the number of items used in the measure, broadly what sorts of items are employed, the response format, basic information that is available about its reliability and validity, and any adjustments or alterations you may have made to the psychological test. Of course, you may be developing your own psychological measures (see Chapter 14), in which case more detail should be provided about the items included in the questionnaire and so forth. It is unlikely that you would include the full test or measure in the description though this may be included in an appendix if space is available. The method section is normally the place where one outlines how questionnaires etc. were quantified (scored) or coded.

Remember that the materials and apparatus are related to the variables being measured and the procedures being employed. Hence, it is important to structure this section clearly so that the reader will gain a good understanding of the key aspects of the study. In other words, an organised description of the materials and apparatus communicates key features of the research procedures and design. Jumble this section and your reader will have trouble in understanding the research clearly.

Procedure

The *procedure* subsection describes the essential sequence of events through which you put participants in your research. The key word is sequence and this implies a chronological order. It is a good idea to list the major steps in your research before writing the procedure. This will help you get the key stages in order – and space allocated proportional to their importance. The instructions given to participants in the research should be given and variations in the contents of these instructions (e.g. between experimental and control group) should be described. Do not forget to include debriefing and similar aspects of the research in this section.

Also you should mention any randomisation that was involved – random allocation to conditions, for example, or randomisation of the order of presentation of materials or stimuli. An experimental design will always include randomisation.

It is difficult to recommend a length for the procedure section since studies vary enormously in their complexity in this respect. A complex laboratory experiment may take rather more space to describe than a simple study of the differences in mathematical ability in male and female psychology students. Of course, a great deal of research uses procedures which are very similar to those of previous research studies in that field. By checking through earlier studies you should gain a better idea of what to include and exclude – this is not an invitation to plagiarise the work of others.

Finally, it may be appropriate to describe broadly the strategy of the statistical analysis especially, for example, if specialised computer programs or statistics are employed. This is in order to give an overview should one seem necessary.

Design

The additional subheading *Design* might be included if the research design is not simple. A diagram may be appropriate if it is difficult to explain in words alone. The design subheading may be desirable when one has a between-subjects or repeated measures design, for example (see Chapters 9 and 10).

■ Results

The results section also has a main heading. Like many other aspects of the report, it is largely written in the past tense. The results section is mainly about the outcome of the statistical analysis of your data. Statistical outcomes of your research are *not* the same as the psychological interpretation of your statistical findings. Statistical analysis is rather more limited than psychological analysis, which involves the development of psychological knowledge and theory. Thus the outcomes of your analyses should normally be reported without conjecture about what they mean – you simply say what the findings *are*. That is, say that there is a significant difference in the means of two groups or a correlation between two variables. Draw no further inferences than that there is a relationship or a difference. Since the statistical analysis is often related to the hypotheses, it is perfectly appropriate to present the results hypothesis by hypothesis. It is also appropriate to indicate whether the statistical evidence supports or fails to support each of the hypotheses.

The results section will vary in length according, in part, to the numbers of variables and type of statistical analysis employed. It is not usual to go into a detailed description of the statistical tests used – just the outcome of applying them. There is no need to mention how the computations were done. If you used a statistical package such as SPSS Statistics then there is no need to mention this fact – or even that you did your calculations using a calculator. Of course, there may be circumstances in which you are using unusual software or highly specialised statistical techniques. In this case then essential details should be provided. Sometimes this would be put in the methods section but not necessarily so.

One common difficulty occurs when the standard structure is applied to research which does not involve hypothesis or theory testing. That is, largely, when we are not carrying out laboratory experiments. Sometimes it is very difficult to separate the description of the results of the study from a discussion of those results (i.e. the next section). Ultimately, clarity of communication has to take precedence over the standard conventions and some blurring of the boundaries of the standard structure for reports may be necessary.

Statistical analyses, almost without exception in psychology, consist of two components:

- Descriptive statistics which describe the characteristics of your data. For example, the means and standard deviations of all of your variables, where appropriate.
- Inferential statistics which indicate whether your findings are statistically significant in the sense that they are unlikely to be due to chance. The correlation coefficient and *t*-test are examples of the sorts of inferential statistics that beginning researchers use.

Both descriptive and inferential statistics should be included in the results section though not to the extent that they simply cloud the issues. The statistical analysis is not a description of everything that you did with the data but the crucial steps in terms of reaching the conclusions that you draw.

Conventionally, the raw, unprocessed data are *not* included in the results section. The means, standard deviations and other characteristics of the variables are given instead. This convention is difficult to justify other than on the basis of the impracticality of including large data sets in a report. Students should always consider including their raw data in an appendix. All researchers should remember the ethical principle of the APA which requires that they should make their data available to any other researcher who has a legitimate reason to verify their findings.

Tables and diagrams are common in the results section. They are not decorations. They are a means of communicating one's findings to others. We suggest a few rules-of-thumb for consideration:

- Keep the number of tables and diagrams to a minimum. If you include too many they become confusing. Worse still, they become irritating. For example, giving separate tables for many very similar variables can exhaust the reader. Far better to have fewer tables and diagrams but ones which can allow comparisons, say, between the variables in your study. In the analysis of your data, you may have produced many different tables, graphs and the like. But those were for the purpose of exploring and analysing the data and many of them will be of little interest to anyone other than the researcher. So do not try to include tables and diagrams which serve no purpose in the report.

- Always take care about titling and labelling your tables and diagrams. If the title and labels are unclear then the whole table or diagram becomes unclear. It is easy to use misleading or clumsy titles and labels, so check them and revise them if necessary.
- Some readers will look at tables and diagrams before reading your text. So for them the quality of the tables and diagrams is even more important. Those used to statistical analyses will be drawn to such tables as they often are quicker to absorb than what you say about the data in the text.
- Tables and diagrams should be numbered in the order they appear.
- Tables and diagrams *must* be referred to in your text. At the very least you need to say what the table or diagram indicates – the important features of the table or diagram.
- Tables and diagrams are readily created in SPSS Statistics, Excel and other computer programs. The difficulty is that the basic versions (the default options) of tables and diagrams are often unclear or in some other way inadequate. They need editing to make them sufficiently clear. Tables may need simplifying to make them more accessible to the reader. Much student work is spoilt by the use of computer-generated tables and diagrams without modification. Think very carefully before using any unmodified output from a computer program in your report.

A glance at any psychology research journal will indicate that relatively little space is devoted to presenting the key features of each statistical analysis. Succinct methods are used which provide the key elements of the analysis simply and clearly. These are discussed in detail in the companion book *Introduction to Statistics in Psychology*, Chapter 16 (Howitt and Cramer, 2011a). Basically the strategy is to report the statistical test used, the sample size or the degrees of freedom, and the level of statistical significance. So you will see things like:

$t(14) = 2.37, p = .05$
$(t = 2.37, N = 16, p < .05)$
$(t = 2.37, df = 14, p = 5\%)$

All of these are much the same. They give the statistic used, an indication of the sample size(s) or degrees of freedom (*df*), and the level of statistical significance.

Discussion

This is a main heading. The past tense will dominate the discussion section but you will also use the present and future tenses from time to time. It all depends on what you are writing about. The material in the discussion should not simply rehash that in the introduction. You may need to move material between the introduction and the discussion sections to ensure this. A previous study may be put in the discussion rather than the introduction. A research report is no place to be repetitive.

If your study tested a hypothesis, then the discussion is likely to begin with a statement indicating whether or not your hypothesis was supported by the statistical analysis (i.e. the results). Remember that most statistical analyses are based on samples and so the findings are probabilistic, not absolute. Consequently, researchers can only find their hypotheses to be supported or not supported. Research based on samples (i.e. most research) cannot definitely establish the hypothesis's truth or falsity since a different sample might produce different outcomes. Consequently it grates to read that the hypothesis was 'proved' or 'not proved'. It suggests that the writer does not fully understand the nature of statistical inference.

Your research findings should be related to those of previous research. They may completely, partially or not support those of previous studies. The differences between the current research and past research should be described as well as the similarities. Sometimes, previous research findings may cast light on the findings of your study. Where there is a disparity between the new findings and previous findings, attempts should be made to explain the disparity. Different types of sample, for example, may be the explanation though it may not be as simple as that. In the discussion section one has new findings as well as older ones from previous research. The task is to explain how our knowledge is extended, enhanced or complicated by the new findings. It may be that the new findings tend to support one interpretation of implications of the previous research rather than another. This should be drawn out.

Of course there may be methodological features which may explain the disparity between the new and older findings. Notice that we use the term methodological features rather than methodological flaws. The influence of methodological differences cuts both ways – your research may have problems and strengths but the previous research may have had other problems and strengths. Try to identify what these may be. Accuracy in identifying the role of methodological features is important since vague, unspecified suggestions leave the reader unclear as to what you regard as the key differences between studies. Try to identify the role of methodological features as accurately and precisely as you can – merely pointing out that the samples are different is not very helpful. Better to explain the ways in which the sample differences could produce the differences in the findings. That is, make sure that it is clear why methodological factors might affect the findings differentially. And, of course, it is ideal if you can recommend improvements to your methodology.

Try not to include routine commonplaces. The inclusion of phrases such as 'A larger sample size may result in different findings' is not much of a contribution, especially when you have demonstrated large trends in your data.

The discussion should not simply refer back to the previous research, it should include the theoretical implications which may be consequent on your findings. Perhaps the theory is not quite so rigorous as you initially thought. There may be implications of your findings. Perhaps you could suggest further research or practical applications.

Finally, the discussion should lead to your conclusion. This should be the main thing that the reader should take away from your research. It is not typically the case that a separate heading is included for the conclusions. It can seem clumsy or awkward to do so in non-experimental research especially. But this is a matter of choice – and judgement.

References

This a major heading and should start on a fresh page. Academic disciplines consider it important to provide evidence in support of all aspects of the argument that you are making. This can clearly be seen in research reports. The research evidence itself and your interpretation of it is perhaps the most obvious evidence for your argument. However, your report will make a lot of claims over and above this. You will claim that the relevant theory suggests something, that previous studies indicate something else, and so forth. In order for the reader of your report to be in a position to check the accuracy or truth of your claims, it is essential to refer them to the sources of information and ideas that you use. Simply suggesting that 'research has found that' or 'it is obvious that' is not enough. So it is necessary to identify the source of your assertions. This includes two main components:

- You cite your sources in the text as, say, (Donovan & Jenkins, 1963). 'Donovan & Jenkins' gives the name of the authors and 1963 is the date of publication (dissemination) of the work. There are many systems in use for giving citations, but in psychology it is virtually universal to use this author–date system. It is known as the Harvard system but there are variants of this and we will use the American Psychological Association's version which is the basis of those employed throughout the world by other psychological associations.

- You provide an alphabetical list of references by the surname of the first author. There is a standard format for the references though this varies in detail according to whether it is a book, a journal article, an Internet source and so forth. The reference contains sufficient information for a reader to track down and, in most cases, obtain a copy of the original document.

The citation

While the citation in the text seems to be very straightforward, there are a number of things that you need to remember:

- The citation should be placed adjacent to the idea which it supports. Sometimes confusion can be caused because the citation is placed at the end of a sentence which contains more than one idea. In these circumstances, the reader may be misled about which of the ideas the citation concerns. For that reason, think very carefully about where you insert the citation. You may choose to put it part-way through the sentence or at the end of the sentence. The decision can be made only in the light of the structure of the sentence and whether things are clear enough with the citation in a particular position.

- Citations should be to *your* source of information. So if you read the information in Smith (2004) then you should cite this as the source really. The trouble is that Smith (2004) may be a secondary source which is explaining, say, Freud's theory of neurosis, Piaget's developmental stages, or the work of some other theorist or researcher. Students rarely have time or resources to read all of the original publication from which the idea came although it is a good idea to try to read some of it. So although uncommon in professional report writing, student research reports will often contain citations such as (Piaget, 1953, cited in Atkinson, 2005). In this way the ultimate source of the idea is acknowledged but the actual source is also given. To attribute to Atkinson ideas which the reader might recognise as those of Piaget might cause some confusion and would be misleading anyway. In your reference list you would list Atkinson (2005) and also Piaget (1953).

- Citations with several authors are given as Smith et al. (1976) with the et al followed by a full stop to indicate an abbreviation. In the past, et al. was in italics to indicate a foreign word or phrase but, increasingly, this is not done. APA style does not italicise et al.

- Student writing (especially where students have carried out a literature review) can become very stilted because they write things like 'Brownlow (1989) argued that children's memories are very different from those of adults. Singh (1996) found evidence of this. Perkins and Ottaway (2002) confirmed Singh's findings in a group of 7-year-olds.' The problem with this is that the sentence structure is repetitive and the person who you are citing tends to appear more important than their ideas. It is a good idea to keep this structure to a minimum and bring their contributions to the fore instead. For example: 'Memory in children is different from that of adults (Brownlow, 1989). This was initially confirmed in preschool children (Singh, 1996) and extended to 7-year-olds by Perkins and Ottaway (2002).' Both versions contain similar information but the second illustrates a greater range of stylistic possibilities.

- When you cite several sources at the same time (Brownlow, 1989; Perkins & Ottaway, 2002; Singh, 2002, 2003) do so in alphabetical order (and then date order if necessary).
- Sometimes an author (or authors) has published more than one thing in a particular year and you want to cite all of them. There may be two papers by Kerry Brownlow published in 2009. To distinguish them, the sources would be labelled Brownlow (2009a) and Brownlow (2009b). The order of the title of the books or articles in the report determines which is 'a' and which is 'b'. In the references, we include the 'a' and 'b' after the date so that the sources are clearly distinguished. Remember to do this as soon as possible to save you from having to re-read the two sources later to know which contains what. If you were citing both sources then you would condense things by putting (Brownlow, 2009a, b).
- It is important to demonstrate that you have read up-to-date material. However, do not try to bamboozle your lecturers by inserting citations which you have not read in a misguided attempt to impress. Your lecturer will not be taken in. There are many tell-tale signs such as citing obscure sources which are not in your university library or otherwise easily obtainable.

Box 5.6 Talking Point

Citing what you have not actually read!

The basic rules of citations are clear. You indicate the source of the idea in the text just like this (Conqueror, 1066) and then put where the information was found in the reference list. This is all very well in theory but in practice causes problems in relation to student work. The difficulty is that the student may only have read textbooks or other secondary sources and they may not be able to get their hands on Conqueror (1066). Now one could simply cite the textbook from which the information came but this has problems. If one cites the secondary source it reads like the secondary source was actually responsible for the idea which they were not. So what does the student do?

There are three, probably equally acceptable, ways of doing this in student work:

- In the main body of the text give the original source first followed by 'cited in' then the secondary source (Conqueror, 1066, cited in Bradley, 2004). Then in the reference list simply list Bradley (2004) in full in the usual way. This has the advantage of keeping the reference list short.
- In the main body of the text give the original source (Conqueror, 1066). Then in the reference list insert

  Conqueror, W. (1066). Visual acuity in fatally wounded monarchs. *Journal of Monocular Vision Studies*, 5 (3), 361–72. Cited in Bradley, M. (2004). *Introduction to Historical Psychology* (Hastings: Forlorn Hope Press).

  This method allows one to note the full source of both the primary information and the secondary information.
- In the main body of the text give the original source (Conqueror, 1066). Then in the reference list insert

  Conqueror, W. (1066). Cited in Bradley, M. (2004). *Introduction to Historical Psychology* (Hastings: Forlorn Hope Press).

  This is much the same as the previous version but the full details of the original source are not given.

It might be wise to check which version your lecturer/supervisor prefers. Stick to one method and do not mix them up in your work.

Some of the problems associated with citations and reference lists can be avoided by using a reference and citation bibliographic software such as RefWorks and Endnote (see Chapter 7). These are basically databases in which you enter essential details such as the authors, the date of publication, the title of the publication, and the source of the publication. One can also insert notes as to the work's contents. It is also possible to download details of publications directly from some Internet databases of publications which can save a lot of typing. If you do this properly and systematically, it is possible to use the program in conjunction with a word processing program to insert citations at appropriate places and to generate a reference list. Even more useful is that the software will do the citations and reference list in a range of styles to suit different journal requirements, etc. The researcher can easily change the references to another style should this be necessary.

The main problem with these programs may be their cost. Bona fide students may get heavily discounted rates. Increasingly universities have site licences for this sort of bibliographic software so check before making any unnecessary expenditure.

Reference list

References will be a main heading at the end of the report. There is a difference between a list of references and a bibliography. The reference list only contains the sources which you cite in the main body of your report. Bibliographies are *not* usually included in research reports. A bibliography lists everything that you have read which is pertinent to the research report – even if it is not cited in the text. Normally just include the references you cite unless it is indicated to you to do otherwise.

Items in reference lists are *not* numbered in the Harvard system (APA style), they are merely given in alphabetical order by surname of the author.

One problem with reference lists is that the structure varies depending on the original source. The structure for books is different from the structure for journal articles. Both are different from the structure for Internet sources. Unpublished sources have yet another structure. In the world of professional researchers, this results in lengthy style guides for citing and referencing. Fortunately, the basic style for references boils down to just a few standard patterns. However, house styles of publishers differ. The house style of Pearson Education, for example, as seen in the references at the end of this book differs slightly from that recommended here. We would recommend that you obtain examples of reference lists from journal articles and books which correspond to the approved style. These constitute the most compact style guide possible.

Traditionally journal names were underlined as were book titles. This was a printer's convention to indicate that emphasis should be added. In the final printed version, it is likely that what was underlined appeared in italics. If preparing a manuscript for publication, this convention is generally no longer followed. If it is a report or thesis then it is appropriate for you to use italics for emphasis instead of the underline. Do not use underlining in addition to italics. The use of italics has the advantage of being less cluttered and has positive advantage for readers with dyslexia as underlining often makes the text harder to read.

The following is indicative of the style that you should adopt for different kinds of source.

Books

Author family name, author initials or first name, year of publication in brackets, stop/period, title of book in lower case except for the first word (or words – where the first letter is usually capitalised) as in example, stop/period, place of publication, colon, publisher details.

Howitt, D. (2002). *Forensic and criminal psychology*. Harlow: Pearson.

Journal articles

Author family name, author initials or first name, year of publication in brackets, stop/period, journal article title in lower case except for first word, stop/period, title of journal in italics or underlined, with capitals on first letter of first word and significant parts of the title, comma, volume of journal in italics, comma, pages of journal in italics. The latest version of the APA's Publication Manual recommends that the number that is used to identify the electronic source of an article (a DOI or Digital Object Identifier) should be presented at the end of the reference where it is available. However, this may not be necessary for student work.

> Schopp, R. F. (1996). Communicating risk assessments: Accuracy, efficacy, and responsibility. *American Psychologist*, *9*, 939–944.

Web sources

For a journal on the World Wide Web then simply add the source:

> Schopp, R. F. (1996). Communicating risk assessments: Accuracy, efficacy, and responsibility. *American Psychologist*, *9*, 939–944. Retrieved from PsyKnow database.

This is a developing area and the style depends a little on the source. Search the American Psychological Association site for further information and details.

Box 5.7 Talking Point

The use of quotations

The most straightforward approach to quotations is *never* to use them. It is generally best to put things in your own words. The use of quotations tends to cause problems because they are often used as a substitute for explaining and describing things yourself. The only legitimate use of quotations, we would suggest, is when the wording of the quotation does not lend itself to putting in your own words for some reason. Sometimes the nuances of the wording are essential. The use of a quotation really should always be accompanied by some commentary of your own. This might be a critical discussion of what the quoted author wrote.

Quotations should always be clearly identified as such by making sure that they appear in quotation marks and indicating just where they appear in the original source. There are two ways of doing this. One is simply to put the page number into the citation: (Smith, 2004: 56). This means that the quotation is on page 56 of Smith (2004). Alternatively the pages can be indicated at the end of the quotation as (pp. 45–6) or (p. 47). This latter style is favoured by the APA.

Appendix/appendices

In some types of publication appendices are a rarity. This is because they are space consuming. The main reason for using appendices is to avoid cluttering up the main body of the report with overlong detail which may confuse the reader and hamper good presentation. So, for example, it may be perfectly sensible to include your 50 item questionnaire in your report but common sense may dictate that it is put at the very end of the report in the section for appendices. In this case, it would be usual to give indicative examples of questions under materials. Similar considerations would apply to the multitude of tables that the statistical analysis may generate but which are too numerous to include in the results section. These too may be confined to the appendices. Remember:

- to refer to the relevant appendix in the main text where appropriate;
- to number and title the appendices appropriately in order to facilitate their location;
- that you may be partly evaluated on the basis of the contents of your appendices. It is inappropriate simply to place a load of junk material there.

5.4 Conclusion

It should be evident that research reports require a range of skills to be effective. That is why they are among the most exacting tasks that any researcher can undertake. There seems to be a great deal to remember. In truth, few professional researchers would have all of the detail committed to memory. Not surprisingly, details frequently need to be checked. The complexity, however, can be very daunting for students who may feel overwhelmed by having so much to remember. It will clearly take time to become skilled at writing research reports. The key points are as follows:

- Make sure that the text is clearly written with attention paid to spelling and grammar.
- Keep to the conventional structure (title, abstract, introduction, method, etc.) as far as is possible at the initial stages.
- Ensure that you cite your sources carefully and include all of them in the list of references.
- Carefully label and title all tables and diagrams. Make sure that they are helpful and that they communicate effectively and efficiently.
- Remember that reading some of the relevant research literature will not only improve the quality of your report but also quickly familiarise you with the way in which professionals write about their research.

Box 5.8 Practical Advice

Some essentials of the research report at a glance

Title

- This is normally centred and is often emphasised in bold.
- Should be informative about study.
- Usually no more than 12 words but sometimes longer.
- Avoid uninformative phrases such as 'A study of'.
- A good title will orient the reader to the contents of the research report.

Abstract or summary

- Usually 100 to 200 words long but this may vary.
- The abstract is a summary of *all* aspects of the report. It should include key elements from the introduction, method, findings and conclusions.
- The abstract is crucial in providing access to your study and needs very careful writing and editing.

Introduction

- This is not normally given a heading in research reports, unless it is a very long thesis.
- It should be a focused account of why the research was needed. All material should be pertinent and lead to the question addressed by the research.
- It should contain key concepts and ideas together with any relevant theory.
- Avoid using quotations unless their content is to be discussed in detail. Do not use them as a substitute for writing in your own words.
- Consider using subheadings to ensure that the flow of the argument is structured. They can easily be removed once the report is complete.
- Make sure that the argument leads directly to the aims of your research.

Method

- This is a centred, main heading.
- Sections should include participants, materials or apparatus, procedure, design where not simple, stimuli and (recommended) ethical considerations.
- It is difficult to judge the level of detail to include. Basically the aim is to provide enough detail that another researcher could replicate the study in its essence.
- Do not regard the structure as too rigid. It is more important to communicate effectively than to include all sections no matter whether they apply or not.

Results

- This is a centred, main heading.
- The results are intended to be the outcomes of the statistical analysis of the data. Quite clearly, this is not appropriate for many qualitative studies.
- Do not evaluate the results or draw general conclusions in the results section.
- Remember that tables and diagrams are extremely important and need to be very well done. They help provide a structure for the reader. So good titles, labelling and general clarity are necessary.
- Do not leave the reader to find the results in your tables and diagrams. You need to write what the results are – you should not leave it to the reader to find them for themselves.

Discussion

- The discussion is the discussion of the results. It is not the discussion of new material except in so far as the new material helps in understanding the results.
- Do not regurgitate material from the introduction here.
- Ensure that your findings are related back to previous research findings.
- Methodological differences between your study and previous studies which might explain any disparities in the findings should be highlighted. Explain why the disparities might explain the different outcomes.
- The discussion should lead to the conclusions you may wish to draw.

References

- This is a centred, main heading.
- It is an alphabetical list of the sources that you cite in the text of the report.
- A bibliography is *not* normally given. This is an alphabetical list of *all* the things that you have read when preparing the report.
- The sources are given in a standard fashion for each of journal articles, books, reports, unpublished sources and Internet sites. Examples of how to do this are given on pages 98 and 99.
- Multiple references by the same author(s) published in a single year are given letters, starting with 'a', after the date to distinguish each (for example, 2004a, 2004b).

If you wish to cite sources which you have only obtained from secondary sources (i.e. you have not read the original but, say, read about it in textbooks) then you must indicate this. Box 5.6 gives several ways of doing this.

Appendix

- Appendix (or appendices) is a centred, main heading.
- These are uncommon in published research reports largely because of the expense. However, they can be a place for questionnaires and the like. For student reports, they may be an appropriate place for providing the raw data.
- The appendices should be numbered (Appendix 1, Appendix 2, etc.) and referred to in the main text. For example, you may refer to the appropriate appendix by putting in brackets (see Appendix 5).

- As much work should go into the appendices as other components of the report. They should be clear, carefully structured and organised.

Tables and diagrams

- These should be placed in the text at appropriate points and their presence indicated in the text (with phrases such as 'see Table 3'). In work submitted for publication tables and diagrams are put on separate pages. Their approximate location in the text is indicated by the phrase 'Insert Table 5 about here' put in the text and centred.
- Tables and diagrams are key features of an effectively communicating report. There should be a balance between keeping their numbers low and providing sufficient detail.
- They should be numbered and given an accurate and descriptive title.
- All components should be carefully labelled, for example, axes given titles, frequencies indicated to be frequencies and so forth.
- Avoid using a multitude of virtually identical tables by combining them into a clear summary table or diagram.
- Remember that well-constructed tables and diagrams may be helpful to the reader as a means of giving an overview of your research.

Key points

- The research report draws together the important features of the research process and does not simply describe the details of the empirical research. As such, it brings the various aspects of the research process into an entirety. It is difficult to write because of the variety of different skills involved.
- There is a basic, standard structure that underlies all research reports, which allows a degree of flexibility. The detail of the structure is too great to remember in detail and style manuals are available for professional researchers to help them with this.
- Although quality of writing is an important aspect of all research reports, there are conventions which should be followed in all but exceptional circumstances. For example, most of the report is written in the past tense, avoids the personal pronoun, and uses the active, not passive, voice.
- A research report needs to document carefully the sources of the evidence supporting the arguments made. Citations and references are the key to this and should correspond to the recommended format.
- All parts of the report should work to communicate the major messages emerging from the empirical research. Thus the title and abstract are as important in the communication as the discussion and conclusions.

ACTIVITIES

1. Photocopy or print an article in a current psychology journal held in the library. Draw up a list of any disparities between this article and the conventional structure described in this chapter. Why did the author(s) make these changes?

2. Find a recent practical report that you have written. Using the material in this chapter, list some of the ways in which your report could be better.

CHAPTER 6

Examples of how to write research reports

Overview

- Writing up a research study is a complex business which takes time to master. It needs thought and practice since it involves the full range of knowledge and skills employed by research psychologists. So there is a limit to how much one can short-cut the process by reducing it to a set of 'rules' to follow. Nevertheless, this chapter is intended to provide easy-access practice in thinking about report writing.
- A fictitious laboratory report is presented of a study which essentially replicates Loftus and Palmer's (1974) study of the effect of questioning on the memory of an eye-witnessed incident. This is a classic in psychology and illustrates the influence of questioning on memory for witnessed events.
- This should be read in conjunction with Chapter 5 which explains the important features of a good write-up of research. Box 5.8 may be especially useful to refer to. Chapter 5 takes the components of the research report in turn and describes good practice and pitfalls. So you may wish to check back as you read through the research study written up in this chapter.
- This chapter presents a short laboratory report which is evaluated in terms of the presence or absence of important features, its logical structure and the numerous aspects of a good laboratory report. Of course, there are many other styles of research but, almost without exception, the basic structure of the laboratory report can be modified to provide a satisfactory structure for any style of research.
- Looking at psychological research journals will add to your understanding of how psychological research is written up. Indeed, having a published article for comparison is a very useful guide as to the important elements of any report you are writing. This is really what your lecturers would like you to be able to emulate so using a journal

article as a template for your own work is not a bad idea. Just make sure that the article is from a core psychology journal so that the 'psychology' style of doing things is used.

- There is a 'model' write-up given of the same study. This is not intended as 'perfection' but as a way of indicating some of the features of better than average work. This write-up gives a clear impression of a student who is on top of the research that they conducted, understands the basics of report writing, and can accurately communicate ideas.

6.1 Introduction

The bottom line is that it is not easy to write an entirely satisfactory research report, as we saw in the previous chapter. Each new study carried out brings up new difficulties often quite different from ones previously encountered. We, like many other researchers, still find it difficult to write the reports of our own research simply because of the complexity of the task of putting all of the elements of the research into one relatively brief document. Not surprisingly, then, newcomers who perhaps have never even read a psychological research report will find report writing a problem. Although everyone will get better with practice there will always be errors and criticisms no matter how sophisticated one becomes. Furthermore, a novice researcher looking at the reports of research in psychological journals will almost certainly be daunted by what they find. These reports are usually the work of seasoned professionals and have been through quality-checking procedures of the peer-review system in which other researchers comment upon manuscripts submitted to journals. This means that the work has been reviewed by two or three experts in that field of research who will identify problems in the report – they may also insist that these difficulties are corrected. The work of students is largely their unaided effort and usually has not been reviewed by their lecturers before it is submitted for marking.

In this chapter, there is a sample research report which contains numerous errors and inadequacies but also some good features for you to learn by. Your task is to spot the good and bad elements. You may identify more than we mention – there is always a subjective aspect to the assessment of any academic work. Of course, there is a problem in supplying a sample research report since this is a learning exercise, not an exercise in marking. Although we could provide an example of the real work of students, such a research report is unlikely to demonstrate a sufficiently wide range of problems. So, instead, we have written a report which features problems in various areas to illustrate the kinds of error that can occur as well as some of the good points. We then ask you to identify what these problems are and to make suggestions about how to correct them. We have indicated many problem areas in the research report by the use of highlighted numbers which may serve as a clue as to where we think that there are problems. You may well find problems which we have failed to notice. Our ideas as to how the report could be improved follow the report. It is unlikely that your own research reports will have such detailed feedback as we have provided for this example, so do not assume that if the assessor of your report has not commented on aspects of it that these parts cannot be improved. Assessors cannot be expected to remark on everything that you have written.

One of the most famous studies in modern psychology is Elizabeth Loftus's study of memory (Loftus and Palmer, 1974) in which participants were shown a video of a vehicle accident and then asked one of a variety of questions such as 'About how fast were the cars going when they smashed each other?' Other participants were given words such as hit, collided, bumped or contacted instead of smashed. Participants gave different estimates according to the particular version of the question asked. Those who were asked about the speed when the cars 'contacted' gave an average estimate of 31 miles per hour but those asked about the cars which 'smashed' each other estimated a speed 10 miles per hour faster than this on average. The argument is, of course, that this study demonstrates that memory can be modified by the kind of question asked after the event. We have decided to write up a fictional study which replicates Loftus and Palmer's study but with some variations. The report is brief compared with, say, the length a journal article would be, and in parts it is truncated as a consequence. Nevertheless, it is about 2000 words in length, which is probably in the middle of the range of word-lengths that lecturers demand. Of course, it would be too short for a final-year project/dissertation. Nevertheless, many of the points we make here would apply to much more substantial pieces of writing.

It would be wise to familiarise yourself with the contents of Chapter 5 on writing research reports before going further. Then read through the following practical report, carefully noting what you believe to be the problems and the good qualities of the report. You should then make suggestions about how the report could be improved. It is easier to spot problems than identify good elements so you will find that the former dominates in our comments. Remember that the highlighted numerals shown at various points of the report roughly indicate the points at which we have something to comment on. Do not forget that there is likely to be some variation in how your lecturers expect you to write up your research. This is common in psychology. For example, different journals may insist on slightly different formats for manuscripts. So you may be given specific instructions on writing up your research which differ slightly from our suggestions. If so, bear this advice in mind alongside our comments.

Notice that our research report introduces a novel element into the study which was not part of the Loftus and Palmer original. It is a variation on the original idea which might have psychological implications. Depending on the level of study, the expectation of originality for a student's work may vary. Introductory level students are more likely to be given precise instructions about the research that they are to carry out whereas more advanced students may be expected to introduce their own ideas into the research that they do. We have taken the middle ground in which the student has been encouraged to replicate an earlier study, introducing some relevant variation in the design. This is often referred to as a constructive replication.

You will find it helpful to have a copy of a relevant journal article to hand when you write your own reports. You may find such an article helpful when studying our fictitious report. It would be a good idea to get hold of Palmer and Loftus's original report, though this may have variations from what is now the accepted style of writing and presenting research reports. Of course, whatever journal article you use should reflect the mainstream psychology style of writing reports. So make sure that you use an article from a core psychology journal as other disciplines often have different styles. Remember too, that research reports in the field that you carry out your research will be good guides for your future research reports. Of course, the important thing is to use them as 'guides' or 'models' – do not copy the material directly as this is bad practice which is likely to get you into trouble for plagiarism (see Chapter 8).

6.2 A poorly written practical report

Particular issues are flagged with numbers in the report and then explained in detail in the analysis that follows.

Practical Report

A Smashing Study: Memory for Accidents 1

Ellie Simms

Abstract

This was an investigation of the way in which people remember accidents after a period of time has elapsed. **2** Seventy-six subjects took part in the study in which they were shown a video of a young man running down the street and colliding with a pushchair being pushed by a young woman. **3** Following this, the participants were given one of two different questionnaires. In one version the participants were asked a number of questions, one of which they were asked was 'How fast was the young man running when he injured the baby in the pushchair?' and in the other condition subjects were asked 'How fast was the young man running when he bumped into the pushchair? **4** Participants were asked to estimate the speed of the runner in miles per hour. The data was analysed **5** using the SPSS Statistics computer program which is a standard way of carrying out statistical analyses of data. The data estimated speeds were put in one column of the SPSS Statistics spreadsheet. **6** The difference between the conditions was significant at the 5 per cent level of significance with a sample size of 76 participants. **7** So the null hypothesis was disproven **6** and the alternate hypothesis proved. **8**

Introduction

I wanted to carry out this study because eyewitness evidence is notoriously unreliable. There are numerous cases where eyewitness evidence has produced wrong verdicts. It has been shown that most of the cases of false convictions for crimes have been established by later DNA evidence involved eyewitness testimony. **9** Loftus and Palmer (1974) carried out a study in which they asked participants questions about an accident they had witnessed on a video. The researchers found that the specific content of questioning subsequently had an influence on how fast the vehicle in the accident was going at the time of the collision. **10** Much higher speeds were reported when the term 'smashed' was used than when the term 'contacted' was used. Numerous other researchers have replicated these findings (Adamson et al., 1983; Wilcox and Henry (1982); Brown, 1987; Fabian, 1989). **11** However, there have been a number of criticisms of the research such as the artificial nature of the eyewitness situation which may be very different from witnessing a real-life accident which is likely to be a much more emotional experience. Furthermore, it is notoriously difficult to judge the speed of vehicles. **12** In addition, participants may have been given strong clues as to the expectations of the researchers by the questions used to assess the speed of the impact. While Loftus and Palmer conclude that the content of the questions affected memory for the collision, it may be that memory is actually unaffected and that the influence of the questioning is only on the estimates given rather than the memory trace of the events.

13 Rodgers (1987) argued that the Loftus and Palmer study had no validity in terms of eyewitness research. Valkery and Dunn (1983) stated that the unreliability of eyewitness testimony reflects personality characteristics of the eyewitness more than the influence of questioning on memory. Eastwood, Marr and Anderson, 1985, stated that memory is fallible under conditions of high stress. Myrtleberry and Duckworth 1979 recommend that student samples are notoriously unrepresentative of the population in general and should not be used in research into memory intrusions in order to improve ecological validity. Pickering (1984) states that 'Loftus and Palmer have made an enormous contribution to our understanding of memory phenomenon in eyewitness research.' **14**

Loftus and Palmer's study seems to demonstrate that the wording of a question can influence the way in which memories are reported. **15** In order to make the research more realistic, it was decided to replicate their study but with a different way of influencing recall of the events. It was reasoned that memory for events such as accidents may be influenced by the consequence of an accident such as whether or not someone was injured in the accident. Does the consequence of an accident influence the way in which it was perceived? **16** This was believed to be a more realistic aspect of eyewitness behaviour than the rather unsubtle questioning manipulation employed in the Loftus and Palmer's research. **17**

It was hypothesised that an accident which results in injury to an individual will be regarded as involving more dangerous behaviour. The null hypothesis states that accidents which result in injury to individuals will be regarded as involving less dangerous behaviour. **18**

Participants 19

76 students at the University were recruited to participate in the study using a convenience sampling method. Those who agreed to take part were allocated to either the experimental ($n = 29$) or control condition ($n = 47$). **20**

Materials and apparatus

A specially prepared video of an accidental collision between a running man and a woman pushing a pushchair with what appeared to be a baby in it. The video lasted 2 minutes and shows the man initially walking down a street but then he begins to run down what is a fairly crowded street. Turning a corner, he collides with the woman pushing the pushchair. The video was filmed on a digital video camera by myself with the help of other students. A Pananony S516 camera was used which features a 15× zoom lens and four mega-pixels image resolution. It was mounted on a Jenkins Video Tripod to maximise the quality of the recording.

The participants were given a short self-completion questionnaire including two versions of the critical question which comprised the experimental manipulation. The first version read 'How fast do you think the man was running in miles per hour when he collided with the woman with the pushchair and the baby was injured?'. The second version read 'How fast do you think the man was running in miles per hour when he collided with the woman with the pushchair and baby?' The questionnaire began with questions about the gender of the participant, their age, and what degree course they were taking. The critical questions were embedded in a sequence of five questions which were filler questions designed to divert the participant's attention from the purpose of the study. These questionnaires were 'What colour was the man's shirt?', 'How many people saw the collision occur?', 'What was the name of the shop outside of which the accident occurred?', 'Was the running man wearing trainers?', and 'Was the woman with the pushchair wearing jeans?'.

Procedure

Participants were recruited from psychological students on the university campus. **21** They were recruited randomly. **22** It was explained to them that the research was for an undergraduate project and that participation was voluntary and that they could withdraw from the study at any stage they wished. The participants in the research were offered a bonus on their coursework of 5 per cent for taking part in three different pieces of research but that does not appear to have affected their willingness to take part in the research. Students failing to participate in research are referred to the Head of Department as it is part of their training. Participants were taken to a small psychological laboratory in the Psychology Department. A data projector was used to show them the short video of the running man and his collision with the woman with a pushchair. The video was two minutes long and in colour. After the video had been shown, the participants were given the questionnaire to complete. Finally they were thanked for their participation in the research and left.

Ethics

The research met the current British Psychological Society ethical standards and complied with the University Ethical Advisory Committee's requirements. **23** The participants were free to withdraw from the research at any time and they were told that their data would be destroyed if they so wished. All participants signed to confirm that they agreed to these requirements.

➔

Results

Group Statistics **24**

	group	N	Mean	Std Deviation	Std Error Mean
speed	1.00	29	4.7138	1.66749	.30964
	2.00	47	3.1500	1.37161	.20007

The scores on ERS **25** were compared between the two groups using the Independent Samples t-Test on SPSS Statistics. **26** SPSS Statistics is the internationally accepted computer program for the analysis of statistical data. **27** The t-test is used where there are two levels of the independent variable and where the dependent variable is a score. **28**

The mean scores for the two groups are different with the scores being higher where the baby was injured in the collision. **29** The difference between the two means was statistically significant at the .000 **30** level using the t-test. **31**

t = 4.443 **32**, df = 74, p = .000 **33**

Thus the hypothesis was proven and the null hypothesis shown to be wrong. **34**

Independent Samples Test **35**

		Levene's Test for Equality of Variances		t-test for Equality of Means				
		F	Sig.	t	df	Sig. (2-tailed)	Mean Difference	Std Error Difference
Speed	Equal variances assumed	.784	.379	4.443	74	.000	1.56379	.35195
	Equal variances not assumed			4.242	50.863	.000	1.56379	.36866

Discussion and conclusions

This study supports the findings of Loftus and Palmer (1974) in that memory is affected by being asked questions following the witnessed incident. **36** Memory can be changed by events following the incident witnessed. Everyone will be affected by questions which contain information relevant to the witnessed event and their memories will change permanently. **37** In addition, it is clear that memory is not simply affected by asking leading questions of the sort used by Loftus and Palmer, but perceptions of the events leading to the incident are affected by the seriousness of the consequences of those actions.

It is not clear why serious consequences should affect memory in this way but there are parallels with the Loftus and Palmer research. When asked questions about vehicles smashing into each other then this implies a more serious consequence than if the vehicles had only bumped. This is much the same as the present research in which memories of events are affected by the injury to the baby which is an indication of the seriousness of the accident. The faster the man ran then the more likely it is that someone would get hurt.

The study provides support for the view that eyewitness evidence is unreliable and cannot be trusted. Many innocent people have spent years in prison for crimes that they did not commit because judges have not paid attention to the findings of Loftus and Palmer and many other researchers. **38**

There are a number of limitations on this study. In particular, the use of a more general sample of participants than university students would be appropriate and would provide more valid data. **39** A larger sample of participants would increase the validity of the research findings. **40** It is suggested that a further improvement to the research design would be to add a neutral condition in which participants simply rated the speed of the runner with no reference to the accident. This could be achieved by having a separate group estimate the speed of the runner.

It was concluded that eyewitness testimony is affected by a variety of factors which make its value difficult to assess. **41**

References

Adamson, P. T. & Huthwaite, N. (1983). Eyewitness recall of events under different questioning styles. Cognitive and Applied Psychology, 18, 312–321. **42**

Brown, I. (1987). *The gullible eyewitness*. Advances in Criminological Research. **43**

Myrtleberry, P. I. E. & Duckworth, J. (1979). The artificiality of eyewitness research: Recommendations for improving the fit between research and practice. *Critical Conclusions in Psychology Quarterly*, 9, 312–319. **44**

Eastwood, A., Marr, W. & Anderson, P. (1985). The fallibility of memory under conditions of high and low stress. Memory and Cognition Quarterly, 46, 208–224. **45**

Fabian (1989). *The Fallible Mind*. London: University of Battersea Press. **46**

Howitt, D. & Cramer, D. (2011). *Introduction to SPSS Statistics in Psychology* (4th edn). Harlow: Pearson. **47**

Loftus, E. F. & Palmer, J. C. (1974). Reconstruction of automobile destruction: An example of the interaction between language and memory. *Journal of Verbal Learning and Verbal Behaviour*, 13, 585–589.

Pickering, M. (1984). Elizabeth Loftus: An Appreciation. *Genuflecting Psychology Review*, 29, 29–43. **48**

Rodgers, T. J. (1987). The Ecological Validity of Laboratory-based Eyewitness Research. *Critical Psychology and Theory Development*, 8 (1), 588–601. **49**

Valkery, Robert O., Dunn, B. W. (1983). The unreliable witness as a personality disposition. *Personality and Forensic Psychology*, 19 (4), 21–39. **50**

Wilcox, A. R. and Henry, Z. W. (1982). *Two hypotheses about questioning style and recall*. Unpublished paper, Department of Psychology, University of Northallerton. **51**

6.3 Analysis of the report

Title

1 The title is clever but not very informative as to the research you are describing. Make sure that your title contains as much information as possible about the contents of your report. Loftus and Palmer, themselves, entitled their study 'Reconstructions of automobile destruction: An example of the interaction between language and memory'. This is more informative but probably could be improved on since all of the useful information is in the second part of the title. A better title might be 'The relation between memory for witnessed events and later suggestive interviewing'. This would also be a good title for our study though it might be better as 'The influence of later suggestive interviewing on recall of witnessed events: A constructive replication of Loftus and Palmer's classic study'. An alternative title might be 'The effect of question wording on eyewitness recall'. Using the term 'effect' suggests that the study employs a true or randomised design. Note that it is not necessary to preface this title with a phrase such as 'An experimental study of' or 'An investigation into' because we can assume that the report is of a study and so this phrase is redundant.

Abstract

2 The description of the purpose of the study is not accurate and precise enough. The study is one on the influences of leading questioning on memory for events not on recall of eye-witnessed events over a period of time, as such. This may confuse the reader as it is inconsistent with what is described in the rest of the report.

3 *Subjects* is an old-fashioned and misleading term for participants, which is the modern and accurate way of characterising those who take part in research, though you sometimes still see it. Also note that the abstract says little about who the participants were. Finally, often there are tight word limits for abstracts so shortening sentences is desirable where possible. Changing the sentence to read 'Seventy-six undergraduates watched a video of a young man running down the street and colliding with a pushchair' corrects these three main errors and so is more acceptable.

4 The wording of these two sentences could be improved. At present the second sentence could be read to suggest that participants given the second version of the questionnaire were only asked one question, which was not the case. One way of rewriting these two sentences is as follows: 'They were then given a questionnaire consisting of six questions in which the wording of a question about how fast the man was running was varied. In one version the baby was described as injured while in the other version there was no mention of this'.

5 The word 'data' is plural so this should read 'data were'.

6 There is a lot of unnecessary detail about SPSS Statistics and data entry which adds nothing to our understanding of the research. This could be deleted without loss. By doing so, space would be freed for providing more important information about the study which is currently missing from the abstract. This information would include a clear description of what the hypothesis was.

7 The findings of the study are not very clear from this sentence and the reader would have to guess what was actually found. A better version would be 'The speed of the runner was estimated to be significantly faster when the baby was injured. $t(74) = 4.43$, $p < .001$'. This contains more new information and presents the statistical findings in a succinct but professional manner. However, statistical values such as t and probability levels are not usually presented in abstracts mainly because of the shortness of abstracts. It is important to state whether the results being presented are statistically significant and this can be done by using the adjective 'significant' or the adverb 'significantly'. Note that the sample size is mentioned twice in the original which is both repetitive and wastes words which are tight in an abstract.

8 Hypotheses can be supported, confirmed or accepted but they cannot be proven (or disproved for that matter). It would be better to say that the research provided support for the hypothesis. But notice that the writer has not said what the hypothesis was so how meaningful to the reader is this part of the write-up? Ideally the main aim or hypothesis of a study should be described earlier on in the abstract. If this had been done, then we would know what the hypothesis was, in which case it would not be necessary to repeat it here. Also, the reader is left without a clear idea of what the researcher has concluded

from this study. Leaving it as simply a test of a hypothesis fails to place the study in its wider context. While significance testing is usually taught in terms of the null and the alternate hypothesis, the results of research are generally more simply described in terms of whether the (alternate) hypothesis was confirmed or not. In other words, it is not necessary to mention the null hypothesis. If the alternate hypothesis is confirmed, then we can take as read that the null hypothesis has been disconfirmed.

Introduction

9 There are a number of problems with these first few lines of introduction: (a) it is not the convention to write in the first person; (b) what has been written is not particularly relevant to the research that was actually carried out and so is something of a waste of space; (c) the sources of the statement are not cited so the reader does not know where the information came from; and (d) these statements mislead the reader into thinking that the research to be described is about the fallibility of eyewitness evidence. It is not. Of course, the extent to which relevance is an issue depends on the space available and what comes later in the report anyway. If the researcher is arguing that most of the fallibility of eyewitnesses is because of the intrusive effects of later questioning then these introductory lines become more relevant.

10 This sentence is inaccurately expressed. The question asked did not affect the actual speed of the car. It affected the estimated speed of the car.

11 There are a number of problems with these citations: (a) they are not in alphabetical order; (b) they are not separated consistently by a semi-colon; (c) 'et al' should have a full stop after 'al', but also this is the first occurrence of the citation and it would be more usual to list the authors in full; (d) the citations are quite dated and the impression is created that the student is relying on a fairly elderly textbook for the information; and (e) since it appears that the writer has not read the sources that they are citing it would be better to be honest and cite one's actual source, for example, by writing something like (Fabian, 1989, cited in Green, 1997).

12 Not only is this comment not documented with a citation, but it is not clear what the argument is. If we assume that the comment is true, in just what way does it imply a criticism of the original study? As it stands, the comment seems irrelevant to the point being made. It would, therefore, be better deleted.

13 It is good to indent the first line of every paragraph. This gives a clear indication as to the paragraph structure of the report. Without these indentations, it is not always clear where one paragraph ends and another begins, which makes the report harder to read. This is especially a problem where one paragraph ends at the right-hand margin at the very bottom of one page and the new paragraph begins at the start of the new page. Without the indentation the division between the two paragraphs is not clear.

14 This entire paragraph is stylistically clumsy since it consists of the name of a researcher followed by a statement of what they did, said, wrote or thought. It is then succeeded by several sentences using exactly the same structure. The entire paragraph needs rewriting so that the issues being discussed are the focus of the writing. Generally, avoid citing an authority at the beginning of sentences as this poor style is the result. Discuss their idea and then cite them at the end. But there are other problems. The paragraph seems to be a set of unconnected notes which have been gleaned from some source or other without much attempt to process the material into something coherent. Just how each point contributes to the general argument being made in the report is unclear. Furthermore, be very careful when using direct quotations. The writer, in this case, has failed to give the page or pages from where the quotation was obtained.

Also, it needs to be questioned just what purpose using the quotation serves. The writer could probably have said it just as clearly in their own words and there is obviously no discussion of the quotation – it is just there and serves no particular purpose. Quotations may be used but there needs to be a good reason for them such as where a report goes on to discuss, question or criticise what is in the quote in some way. A more minor point is that the reference to Eastwood, Marr and Anderson and to Myrtleberry and Duckworth should have the date or year in brackets.

15 Why is the phrase 'seems to' used? If the study does not demonstrate what the researchers claim of it then its validity should be questioned as part of the discussion of the material.

16 This is an interesting idea but is it really the case that there is no relevant research to bring in at this point? Certainly no citations are given to research relevant to this point. One would look for previous research on whether the consequences of events are taken into account when assessing the seriousness of the behaviours which led to these events.

17 It is not clear how the questions used by Loftus and Palmer were unsubtle. The ways in which their question manipulation is unsubtle needs to be explained. It is good to see a critical argument being used but the presentation of the argument could be clearer.

18 The hypothesis is not sufficiently clearly or accurately stated. It might be better to begin by describing the hypothesis more generally as 'It was hypothesised that memory for a witnessed event will be affected by later information concerning the consequences of that event'. It might be preferable to try to formulate it in a way which relates more closely to what was actually tested. For example, we might wish to hypothesise that 'The estimated speed of an object will be recalled as faster the more serious the impact that that object is later said to have had'. The null hypothesis reveals a misunderstanding of the nature of what a null hypothesis is. A null hypothesis is simply a statement that there is no relationship between two variables. It does not imply a relationship between the two variables in question. The null hypothesis is not usually presented in reports as it is generally assumed that the reader knows what it is. It is also better to describe the hypothesis in terms of the independent and dependent variables and to state what the direction of the results are expected to be if this can be specified. For example, we could say that 'Participants who were informed that the baby was injured were predicted to give faster estimates of running time than those who were not told this'.

Method

19 The main section following the Introduction is the Method section and should be titled as such. This overall title is missing from the report. The Method section is broken down into subsections which in this case starts with Participants and ends with Ethics. You may wish to make the distinction between sections and subsections clearer by centring the titles of the sections.

20 (a) Numbers such as 76 would be written in words if they begin a sentence. (b) The way in which participants were allocated to the conditions should be clearly stated. It is not clear whether this was done randomly, systematically or in some other way. (c) What condition is the experimental one and what condition is the control should be described clearly as this has not been previously done. To work out what is the experimental and control group requires that we search elsewhere in the report, which makes the report harder to read. (d) More information should be given about the participants. It should be indicated whether the sample consisted of both men and women and, if so, what numbers of each gender there were. In addition, the mean age of the sample should be given as well as some indication of its variability. Variability is usually described in terms of standard deviation but other indices can be used, such as the minimum and maximum age.

21 The students were not psychological. They were psychology students.

22 In the Participants subsection it was stated that participants were a convenience sample which means that they were recruited at the convenience of the researcher, not randomly or systematically. In psychology the term 'random' or 'randomly' has a specific technical meaning and should be used only when a randomised procedure is employed. If such a procedure had been used, it is necessary to describe what the population consisted of (for example, all psychology undergraduates at that university), what the target sample was (for example, 10 per cent of that population) and what the selection procedure was (for example, numbering the last name of each psychology undergraduate alphabetically, generating 100 numbers randomly and then approaching the students given those numbers). When a random or systematic procedure has not been used, as seems to be the case in this study, it is not necessary to describe the selection procedure in any detail. It may be sufficient simply to state that 'An e-mail was sent to all psychology undergraduates inviting them to participate in the study' or 'Psychology undergraduates in practical classes were invited to take part in the study'.

23 This sentence is inaccurate as the research did not meet all the ethical requirements of the BPS. For example, it is not stated that they were debriefed at the end of their participation which they should have been. The previous paragraph describes several ethically dubious procedures which the writer does not appear to acknowledge.

Results

24 (a) All tables need to be given the title 'Table' followed by a number indicating the order of the tables in the report. As this is the first table in the report it would be called Table 1. (b) The table should have a brief label describing its contents. For example, this table could be called 'Descriptive statistics for speed in the two conditions'. (c) In general, it is not a good idea to begin the Results section with a table. It is better to present the table after the text which refers to the content of the table. In this case this would be after the second paragraph where the mean scores are described. (d) SPSS Statistics tables should not be pasted into the report because generally they contain too much statistical information and are often somewhat unclear. For example, it is sufficient to describe all statistical values apart from probability levels to two decimal places. Values for the standard error of the mean need not be presented. Notice that the table does not identify what group 1.00 is and what 2.00 is. (e) Tables should be used only where doing so makes it easier to understand the results. With only two conditions it seems preferable to describe this information as text. For example, we could report the results very concisely as follows: 'The mean estimated running speed for those told the baby was injured ($M = 4.71$, $SD = 1.67$) was significantly faster/slower, unrelated $t(74) = 4.43$, 2-tailed $p < .001$, than those not told this ($M = 3.15$, $SD = 1.37$)'. (In order to be able to write this, we need to know which group is Group 1.00 and which group is Group 2.00 in the table.) The Results section will be very short if we report the results in this way but this concise description of the results is sufficient for this study. There is no need to make it longer than is necessary. If our sample included a sufficient number of men and women we could have included the gender of the participants as another variable to be analysed. If this were the case we could have carried out a 2 (condition) × 2 (gender) unrelated analysis of variance (ANOVA).

25 It is generally better not to use abbreviations to refer to measures when writing up a study. If you do use abbreviations, you need to give the full unabbreviated name first followed by the abbreviation in brackets when the measure is first mentioned. In this

case, we have to guess that ERS refers to 'estimated running speed' as it has not previously been mentioned.

26 As we have discussed under number **24**, we can describe the results of this study in a single sentence. The term 'unrelated *t*' refers to the unrelated *t*-test which SPSS Statistics calls the Independent Samples t-Test. In our summary sentence, it is clear that we are using the unrelated *t*-test to compare the mean of the two groups so it is not necessary to state this again. We do not have to state how the *t*-test was calculated. We generally do not need to mention the type of statistical software used for this kind of analysis. We may only need to do this if we were carrying out some more specialist statistics such as structural equation modelling or hierarchical linear modelling which employs less familiar software than SPSS Statistics which is unlikely for student work.

27 As we discussed under number **26**, we do not usually need to mention the statistical software we used so this sentence should be omitted. The sentence basically gives superfluous detail anyway.

28 As the *t*-test should be very familiar to psychologists, there is no need to describe when it should be used.

29 If it was thought advisable to present a table of the mean and standard deviation of running speed for the two groups, we need to refer to this table in the text. At present, the table is not linked to the text. We could do this here by starting this sentence with a phrase such as 'As shown in Table 1' or 'As can be seen from Table 1'. It would also be more informative if the direction of the difference was described in this sentence rather than simply saying that the means differ. Notice that this is the first sentence in the report which enables us to identify which condition is associated with the highest scores.

30 The level of statistical significance or probability can never be zero. Some readers would see a probability of .000 as being zero probability. This figure is taken from the output produced by SPSS Statistics which gives the significance level to three decimal places. What this means is that the significance level is less than .0005. For most purposes it is sufficient to give the significance level to three decimal places in which case some psychologists would round up the third zero to a 1. In other words the significance level here is .001. Strictly speaking the significance level is equal to or less than a particular level such as .001, although the symbol for this (≤) is rarely used.

31 We have previously stated that the *t*-test was used so it is not necessary to state it again.

32 It is sufficient to give statistical values other than the significance level to two decimal places. In this case we can write that $t = 4.44$.

33 As presently displayed, these statistical values appear to hang on their own and seem not to be part of a sentence. They should be clearly incorporated into a sentence. We have already shown under number **24** how this can be concisely done by placing these values in brackets.

34 As discussed under number **8** hypotheses cannot be proven or shown to be wrong. Saying that a hypothesis has been proved or disproved implies that other results are not possible. Another study might find that there is no difference between the two conditions or that the results are in the opposite direction to that found here. Consequently, when the results are consistent with the hypothesis it is more accurate to describe them as being confirmed, accepted or supported rather than proved. When the results are not consistent with the hypothesis it is better to describe them as not confirmed, not accepted or not supported rather than disproved. It is not necessary to mention the null hypothesis. If we know the results for the alternate hypothesis, we will know the results for the null hypothesis.

35 As previously mentioned under number **24**, we need to label any tables to indicate what they refer to. Also, we should not paste in tables from SPSS Statistics output. We have been able to describe succinctly the results of our analysis in a single sentence which includes the essential information from this table, so there is no need to include the table. If you want to show the results of your SPSS Statistics output, then it is better to append this to your report rather than include it in the Results section.

Discussion

36 It is usual to begin the Discussion section by reporting what the main findings of the study are. In this case it would be saying something along the lines that 'The hypothesis that the memory of a witnessed event is affected by later information about the consequences of that event was supported in that participants who were informed that the baby was injured estimated the speed of the man running into the pushchair as significantly faster than those not told this'.

37 This assertion is not supported by any evidence. No information is presented to show that everyone was affected let alone will be affected or that the change was permanent.

38 No evidence is cited to support this statement. It would be difficult to test this assertion. How can we determine whether someone is innocent when the evidence is often circumstantial? How can we show that these wrong convictions were due to the way in which the witnesses were questioned? These are not easy questions to test or to carry out research on. It would be much easier to find out how familiar judges were with research on eyewitness testimony and whether this knowledge affected the way in which they made their judgements. Unless evidence can be found to support this assertion, it would be better to describe it as a possibility rather than a fact. In other words, we could rewrite this sentence as follows: 'Many innocent people may have spent years in prison for crimes that they did not commit because judges may have not paid attention to the findings of Loftus and Palmer and many other researchers'.

39 It would be better to say in what way the data would be more valid if a more general sample of participants was used. For example, we might say that the use of a more general sample would determine the extent to which the findings could be replicated in a more diverse group of people.

40 It is not stated how a larger sample would improve the validity of the findings. This would not appear to be the case. As the size of sample used in this study produced significant findings, we do not have to use a larger sample to determine the replicability of the results. However, it would be more informative if we could suggest a further study which would help our understanding of why this effect occurs or the conditions under which it occurs rather than simply repeat the same study.

41 No evidence is presented in the discussion of a variety of factors affecting eyewitness testimony so this cannot be a conclusion. As it stands, it is a new idea introduced right at the end of the report. It is also unclear what the term 'value' means so the writer needs to be more specific. If eyewitness testimony has been shown to be affected by various factors it is unclear why this makes its value difficult to assess. One or more reasons need to be given to support this assertion.

References

42 It is important to use a consistent style when giving full details of the references. It is usual to italicise the title and volume number of journals. As this is done for the other journals, this should be done here.

43 Although this may not be immediately obvious, there seem to be three errors here. It is usual to italicise the names of books, journals and the titles of unpublished papers. The name of a book is followed by the place of publication and the name of the publisher. As this was not done here, it implies that the subsequent title refers to the name of a journal. If this is the case, then (a) the title of the paper should not be italicised, (b) the title of the journal should be italicised, and (c) the volume number and the page numbers of the journal should be given for that paper.

44 This reference has not been placed in the correct alphabetical order of the last name of the first author. This reference should come after Loftus and Palmer.

45 Underlining is usually used to indicate to publishers that the underlined text is to be printed in italics. This convention was developed when manuscripts were written with manual typewriters. As it is easy to italicise text in electronic documents there is less need for this convention. As the journal titles of the other references have not been underlined, this title should not be underlined and should be italic.

46 The initial or initials of this author are missing. The titles of books are often presented in what is called in Microsoft Word 'sentence case'. This means that the first letter of the first word is capitalised but the first letters of the subsequent words are not unless they refer to a name.

47 Although we hope that the student has used this book to help them analyse their results, they have not cited it and so it should not be listed as a reference.

48 The titles of journal papers are usually presented in sentence case. The first letter of 'Appreciation' should be small case.

49 The title of this journal paper should be in sentence case as most of the titles of other papers are. It is considered important that you are consistent in the way you present references. The number of the issue in which the paper was published is given. This is indicated by the number in brackets after the volume number. This is not usually done and in this paper is not generally done, so decide what you want to do and do it consistently.

50 The ampersand sign (&) indicating 'and' is missing between the two authors. This is usually placed between the initials of the penultimate author and the last name of the last author. First names of authors are not usually given and have not been given for the other authors listed here, so should not be shown here.

51 'and' is used instead of the & to link the names of the two authors. Once again, you need to be consistent in how you do this. The American Psychological Association uses '&' while some other publishers use 'and'. This book uses 'and' since this is part of the standard style of Pearson Education, its publisher.

6.4 An improved version of the report

While a lot may be learnt by studying examples of below-par work, it is helpful to have a good example to hand, so a better write-up of the study follows. This is not to say that the report is perfect – you may well find problems remain – but that it is an improvement over the previous version. Look through this version and see what you can learn from it. We would suggest that you take note of the following:

- Notice how the reader is given clear information in both the title and the abstract. These summarise the research very well and give the reader a good idea of what to expect in the main body of the text. Put another way, they give a good impression of the competence of the writer.

- This version of the report is a big improvement since a fairly coherent argument runs all the way through it. The writing is not episodic but fairly integrated throughout.
- Just the right amount of information is provided in a logical and coherent order.
- While the results section is very short, it contains all the detail that the reader needs. At no point is it unclear quite what the writer is referring to. This is achieved by carefully stating the results in words, using succinct statistical reporting methods, and by making sure that any table included is a model of clarity.
- All of the arguments are justified throughout the report and the discussion and conclusions section makes pertinent points throughout which have a bearing on the research that had been carried out.
- Finally, the reference section is well ordered and consistent. The writer has found up-to-date work relevant to the new study which creates the impression of a student who is actively involved in their studies rather than someone who simply does what is easiest. A good touch is that the writer shows honesty by indicating where they have not read the original publication. This has been done by indicating the actual source of the information.

Practical Report

The Effect of Later Suggestive Interviewing on Memory for Witnessed Events

Ellie Simms

Abstract

The influence of leading questioning on memory for events was investigated in a constructive replication of the Loftus and Palmer (1974) study. It was hypothesised that memory for a witnessed event will be affected by later information concerning the consequences of that event. Thirty four male and 42 female undergraduates watched a video of a young man running down the street and colliding with a pushchair. They were then given a questionnaire consisting of six questions in which the wording of a question about how fast the man was running was varied. In one version the baby was described as injured while in the other version there was no mention of this. It was found that the estimated running speed was significantly greater where the consequences of the action was more serious. Thus the hypothesis was supported.

Introduction

This study explored the effect of the seriousness of the consequences of events on memory for those events. In a classic study, Loftus and Palmer (1974) investigated the influences of later questioning on memory for events that had been witnessed. Participants in their research were asked questions about an accident they had witnessed on a video. The researchers found that the specific content of questioning influences estimates of how fast the vehicle in the accident was going at the time of the collision. Much higher average speeds were reported when the term 'smashed' was used than when the term 'contacted' was used. Numerous other researchers have replicated these findings (Adamson & Huthwaite, 1983; Brown, 1987; Edmonson, 2007; Fabian, 1989; Jacobs, 2004; Wilcox & Henry, 1982). However, there have been a number of criticisms of the research such as the artificial nature of the eyewitness situation which may be very different from witnessing

➔

a real-life accident, which is likely to be a much more emotional experience (Slatterly, 2006). Furthermore, it is notoriously difficult to judge the speed of vehicles (Blair & Brown, 2007). In addition, participants may have been given strong clues as to the expectations of the researchers by the questions used to assess the speed of the impact. While Loftus and Palmer conclude that the content of the questions affected memory for the collision, it may be that memory is actually unaffected and that the influence of the questioning is only on the estimates given rather than the memory trace of the events (Pink, 2001).

The validity of Loftus and Palmer's research in terms of its relevance to eyewitness evidence in real-life situations has been questioned by Rodgers (1987) who argues that the study has poor validity. Furthermore, student populations may be unrepresentative of the more general population and should be avoided to improve the ecological validity of research in this field (Myrtleberry & Duckworth, 1979). These views are not shared by all researchers, thus Pickering (1984) writes of the important contribution that the Loftus and Palmer study has made to our understanding of eyewitness memory.

Loftus and Palmer demonstrated that the form of questioning following a witnessed event can influence the way in which those events are later recalled. However, there is evidence that evaluations of crime are influenced by the consequences of the crime rather than the criminal actions involved (Parker, 2001). So, for example, a burglary which results in the victim having subsequent psychological problems is judged to be more serious than an identical crime which led to no serious consequence. It was reasoned that memory for events such as accidents may be influenced by the consequence of an accident such as whether or not someone was injured in the accident. Does the consequence of an accident influence the way in which the events leading up to the accident are recalled?

Based on this, it was hypothesised that memory for a witnessed event will be affected by later information concerning the consequences of that event. In particular, it was predicted that where the consequences of the event were more severe the events leading up to the accident would be perceived as more extreme than when the consequences were less severe.

Method

Participants

Thirty four male and 42 female psychology students at the University were recruited to participate in the study using a convenience sampling method which involved inviting them in lectures and elsewhere to participate in the research. Those who agreed to take part were randomly allocated to either the experimental ($n = 29$) or control condition ($n = 47$). There were 15 male and 14 female participants in the experimental group and 19 male and 28 female participants in the control group. The mean age of participants was 20.38 years ($SD = 1.73$).

Materials and apparatus

A specially prepared two-minute video of an accidental collision between a running man and a woman pushing a pushchair with what appeared to be a baby in it was shown. Initially, the young man is seen walking down a street but then he begins to run down what is a fairly crowded street. Turning a corner, he collides with the woman pushing the pushchair. The video was filmed using a good-quality digital video camera mounted on a tripod with a high degree of image resolution by myself with the help of other students. A data projector was used to show the video.

The participants were given a short self-completion questionnaire including two versions of the critical question which comprised the experimental manipulation. The first version which was given to the experimental group read 'How fast do you think the man was running in miles per hour when he collided with the woman with the pushchair and the baby was injured?' The second version which was given to the control group read 'How fast do you think the man was running in miles per hour when he collided with the woman with the pushchair and baby?' These questions were embedded in the questionnaire among other questions which started with ones concerning the gender of the participant, their age and what degree course they were taking. The critical questions were placed at the end of five questions which were filler questions designed to divert the participants' attention from the purpose of the study. These questions were 'What colour was the man's shirt?', 'How many people saw the collision occur?', 'What was the name of the shop outside of which the accident occurred?', 'Was the running man wearing trainers?', and 'Was the woman with the pushchair wearing jeans?'

Design and procedure

The study employed an experimental design in which participants were randomly assigned to these conditions on the basis of the toss of a coin. The experimental group witnessed events which led to the serious consequence of an injury to a baby and the control group witnessed the same events but with no serious consequence.

Participants took part in the study individually in a small psychology laboratory on the University campus. Prior to showing the video, it was explained to them that the research was for an undergraduate project and that participation was voluntary and that they could withdraw from the study at any stage if they wished. Psychology students are encouraged to volunteer as participants in other students' studies for educational reasons though there is no requirement that they should do so. A data projector was used to show them the short video of the running man and his collision with the woman with a pushchair. The video was two minutes long and in colour. After the video had been shown, the participants were given one of the two different versions of the questionnaire to complete. Finally, they were thanked for their participation in the research and debriefed about the study and given an opportunity to ask questions. Participants were asked if they wished to receive a brief summary of the research findings when these were available.

Ethics

The research met the current British Psychological Society ethical standards and complied with the University Ethical Advisory Committee's requirements. In particular, the voluntary nature of participation was stressed to participants and care was taken to debrief all participants at the end of their involvement in the study. All data were recorded anonymously. All participants signed to confirm that they had been informed of the ethical principles underlying the research.

Results

Table 1 gives the mean estimates of the running speed in the video for the serious consequence and the no-consequence conditions. The mean estimated running speed for those told the baby was injured ($M = 4.71$, $SD = 1.67$) was significantly faster, $t(74) = 4.43$, 2-tailed $p < .001$, than those not told this ($M = 3.15$, $SD = 1.37$).

Table 1 Descriptive statistics on estimated running speed in the two conditions

Condition	Sample size	*M*	*SD*
Serious consequence	29	4.71	1.67
No consequence	47	3.15	1.37

This finding supports the hypothesis that memory for a witnessed event will be affected by later information concerning the consequences of that event.

Discussion and conclusions

This study supports the findings of Loftus and Palmer (1974) in that memory was affected by the nature of questions asked following the witnessed incident. Memory can be changed by events following the incident witnessed. There was a tendency for those who believed that the incident had led to a serious injury to estimate that the runner who was responsible for the accident was running faster than did members of the

➔

control group. This is important because it illustrates that the consequences of an action may influence the perceptions of the characteristics of that action.

However, the study does not explain why the serious consequences of an incident should affect memory in this way but there are parallels with the Loftus and Palmer research which may be helpful. Asking questions about vehicles 'smashing' into each other implies a more serious consequence than if the vehicles had only 'bumped'. This is much the same as the present research in which memories of events were affected by the injury to the baby, which is an indication of the seriousness of the accident. The faster the man ran then the more likely it was that someone would get hurt.

There are implications of the study for the interviewing of witnesses. In particular, the research raises the question of the extent to which the police should give additional information unknown to the witness during the course of an interview. In real life, an eyewitness may not know that the victim of an accident had, say, died later in hospital. Is it appropriate that the police should provide this information in the light of the findings of the present study?

There are a number of limitations on this study. In particular, the use of a more representative sample of the general population would provide an indication of the generalisability of the findings of the present sample. A further improvement would be to add a neutral condition in which participants simply rated the speed of the runner with no reference to the accident. This could be achieved by having a separate group estimate the speed of the runner without any reference to a collision in the question. Finally, the speed of the runner is not the only measure that could be taken. For example, questions could be asked about the reason why the man was running, whether he was looking where he was running, and whether the woman pushing the pushchair was partly responsible for the collision.

It is concluded that memory for eyewitnessed events is affected by information about the consequences of those events. This may have implications for police interviews with eyewitnesses and the amount of information that the police supply in this context.

References

Adamson, P. T., & Huthwaite, N. (1983). Eyewitness recall of events under different questioning styles. *Cognitive and Applied Psychology*, *18*, 312–321.

Blair, A., & Brown, G. (2007). Speed estimates of real and virtual objects. *Traffic Psychology*, *3*, 21–27.

Brown, I. (1987). The gullible eyewitness. *Advances in Criminological Research*, *3*, 229–241.

Edmonson, C. (2007). Question content and eye-witness recall. *Journal of Criminal Investigations, 5*, 31–39. Cited in D. Smith (2008), *Introduction to cognition*. Lakeside, UK: Independent Psychology Press.

Fabian, G. (1989). *The fallible mind*. London: University of Battersea Press.

Jacobs, D. (2004). Eyewitness evidence and interview techniques. *Forensic Investigation Quarterly*, *11*, 48–62.

Loftus, E. F., & Palmer, J. C. (1974). Reconstruction of auto-mobile destruction: An example of the interaction between language and memory. *Journal of Verbal Learning and Verbal Behaviour*, *13*, 585–589.

Myrtleberry, P. I. E., & Duckworth, J. (1979). The artificiality of eyewitness research: Recommendations for improving the fit between research and practice. *Critical Conclusions in Psychology Quarterly*, *9*, 312–319.

Parker, V. (2001). Consequences and judgement. *Applied Cognitive Behavior*, *6*, 249–263.

Pickering, M. (1984). Elizabeth Loftus: An appreciation. *Genuflecting Psychology Review*, *29*, 29–43.

Pink, J. W. (2001). What changes follow leading interviews: The memory or the report of the memory? *Essential Psychology Review*, *22*, 142–151.

Rodgers, T. J. (1987). The ecological validity of laboratory-based eyewitness research. *Critical Psychology and Theory Development*, *8*, 588–601.

Slatterly, O. (2006). Validity issues in forensic psychology. *Criminal and Forensic Research, 2*, 121–129.

Wilcox, A. R., & Henry, Z. W. (1982). *Two hypotheses about questioning style and recall*. Unpublished paper, Department of Psychology, University of Northallerton.

6.5 Conclusion

It is not easy to write a good research report. You need to provide a strong and convincing argument for what you have done. To be convincing the argument has to be clear otherwise the reader will not be able to follow it. It also has to be accurate. You should try to ensure that what you write is an accurate description of what you are writing about. When you refer to the work of others, it is important that you are familiar with their work so you know in what way their work is relevant to your own. In writing your report it is important to check it carefully sentence by sentence to make sure that it makes sense and is clearly and accurately articulated. It is sometimes difficult for us to evaluate our own work because we often interpret what we have written in terms of what we know but which we have not mentioned in the report itself. Although we may be able to find other people to check what we have written, we cannot always be sure how thoroughly or critically they will do this. They may not want to offend us by being critical or they may not be sufficiently interested in having a thorough grasp of what we have done and to question what we have written. Consequently, we need to check what we have written ourselves. It is often useful to leave the work for a few days and to return to it when we are less familiar with it. It may then be easier to spot anything that is not as clear as it should be.

To help you become more skilled in evaluating your own report writing we have presented you with a report which contains a number of examples of poor practice. We hope you have been able to spot many of these errors and will not make them when writing your own reports.

Key points

- Writing research reports is a complex task. We have to have a clear idea of what we want to say and to say it clearly. We need to present a strong and convincing argument as to why we believe our research is important and how exactly it makes a contribution to what is already known about the topic we are studying.
- A research report consists of the following major parts: a title, an abstract, an introduction, a method section, a results section, a discussion section and a list of references. All of these components of a report are important and deserve careful consideration. It is a pity to spoil an otherwise good report with a clumsy title or an insufficiently detailed abstract.
- It is useful to bear in mind what the main aims or hypotheses of your study are and to use these to structure your report. These should be clearly stated. They are generally stated and restated in various parts of the report. When doing this it is important to make sure that you do not change what they are as this will cause confusion. They should be most fully described in the introduction and should form the basis of the discussion section. They should be briefly mentioned in the abstract and it should be clear in the results section how they were analysed and what the results of these analyses were.
- We need to describe the most relevant previous research that has been carried out on the topic and to show how this work is related to what we have done and how it has not addressed the question or questions that we are interested in.
- The abstract is written last. It may be useful to have a working title which you may change later on so that it captures in as few words as possible what your study is about. ➔

- Your views on what you write may change as you think more thoroughly about what you have done and as you become more familiar with your research and your understanding of it improves. You should expect to have to revise what you have already written in terms of what you choose to say in a later draft subsequently. For example, writing about your findings in the discussion section may lead you to carry out further analyses or to change or add material to the introduction.
- If you are having difficulty in writing any part of your report, look at how authors of published research have handled this part of their report. This is usually best done by looking at journal articles which are the most relevant or most strongly related to your own research.
- Writing the report is ultimately your own responsibility. You need to read and re-read it carefully a number of times. It is often a good idea to let some time elapse before re-reading your report so that you can look at it again with a fresh mind.

ACTIVITY

You might like to offer to read and to provide constructive criticism of a report written by one of your fellow students. Where appropriate, you could ask them to clarify or to better substantiate what they are saying or to suggest an alternative way of saying it.

CHAPTER 7

The literature search

Overview

- The literature search is an integral part of the research process. Pertinent research studies and theoretical papers obtained in this way provide the researcher with an overview of thinking and research in a particular area. Although time-consuming, it is essential to developing ideas about the issues to be addressed and what more needs to be explored.
- The literature search is best seen as a process of starting broadly but moving as rapidly as possible to a more narrow and focused search. One common strategy is to focus, first, on the most recent research and writings on a topic. These contain the fruits of other researchers' literature searches as well as up-to-date information of where current research has taken us. The current literature is likely to alert us to what still needs to be done in the field. Of course, the major limitation of starting with current publications is that important ideas from earlier times can become ignored and neglected without justification.
- Computers and computerised databases are the modern, highly efficient way of searching the research literature through electronic databases such as Web of Science and PsycINFO. Among a great deal of information of various sorts, they provide a brief abstract or summary of the publication. The abstract in the Web of Science is that of the article itself whereas this may not be the case in PsycINFO.
- Abstracts in research reports, if well-written, contain a great deal of information which will provide a degree of detail about the research in question and the theoretical context. Usually abstracts contain enough information to help the reader decide whether or not to obtain the original article or report. Almost certainly local college and university libraries are unlikely to have anything other than a small fraction of these publications in stock although the use of electronic versions of journals by libraries is changing that situation. Consequently, it is necessary to obtain the article by some means from elsewhere. There are various ways of doing this, including visiting other libraries, e-mailing the author for a copy of the article, or getting the library to obtain a copy or photocopy of the article in question.

→

- There are a number of reasons why it may be essential to obtain the original article. For example, it is the only way of obtaining an overview of the methods and procedures employed. Sometimes one may be suspicious of how sound the conclusions of the study are and may wish to evaluate, say, the statistical analysis carried out or consider possible flaws in the method.
- You should keep a careful record of the publications that you consider important to your research. Although initially this is time-consuming it is far less frustrating in the long run. There are a number of ways of doing this, including computerised databases (such as RefWorks or EndNote), simple hand-written index cards or computer files to which material may be copied and pasted.

7.1 Introduction

How the literature search is conducted depends a little on one's starting point. A professional researcher with a well-established reputation will have much of the previous research and theory readily at their command. A beginning student will have little or no knowledge. If one knows very little about a topic, then a sensible first stage is to read some introductory material such as that found in textbooks. A relatively recent textbook is likely to cover fairly recent thinking on the topic although in brief overview form. At the same time, the textbook is likely to provide a fairly rapid access to a field in general. This is especially useful for students doing research for the first time in practical classes. Because it is readily to hand, material in the college or university library or accessible from the Internet will be your first port of call. Getting material from elsewhere may take a little time and cause problems in managing your time – and getting your assignments in before the deadline. Of course, professional researchers regularly keep up to date, perhaps searching the new literature on a monthly basis.

It cannot be stressed too much that professional researchers are part of complex networks of individuals and groups of individuals sharing ideas and interests. As such, information flows through a variety of channels and few researchers would rely exclusively on the sources described in this chapter. For one thing, no matter how efficient the system – and it is impressive – there is always a delay between research being done and the final report being published. This can be a year or two in many cases. So if one needs to be more up to date than that then one needs to rely on conferences and other sources of contact and information. This would be characteristic of the activities of most researchers.

Searching one's college or university library usually involves using its electronic catalogue system via computer terminals. As there are a number of such systems, these may differ across universities. Many British university libraries use the OPAC system (Online Public Access Catalogue). Leaflets about using the local system are likely to be available from the library, there may be induction or training sessions for new library users, or you may simply seek help from members of the library staff. If you have, say, a specific book in mind then its title and author will quickly help you discover where it is located in the library. However, if you simply are searching with a general keyword such as 'memory' or 'intelligence' then you are likely to find more entries or hits. Perhaps too many. Sometimes it may be quicker to go to the section of the library where items with particular keywords are likely to be held, though this is less systematic and others on the course may have beaten you there. Library classification systems need to be understood in general if one is to use this sort of method.

7.2 Library classification systems

There are two main systems for classifying and arranging non-fiction (mostly) books in a library. It is sufficient to know how to find the books in the library without having a detailed knowledge of the system used by your library.

- One scheme is the Dewey Decimal Classification (DDC) system developed by Melvil Dewey in 1876 which is reputedly the world's most widely used library classification system, although not necessarily in university libraries (Chan and Mitchell, 2003; Dewey Services, n.d.). Each publication is given three whole numbers followed by several decimal places as shown in Figure 7.1. These numbers

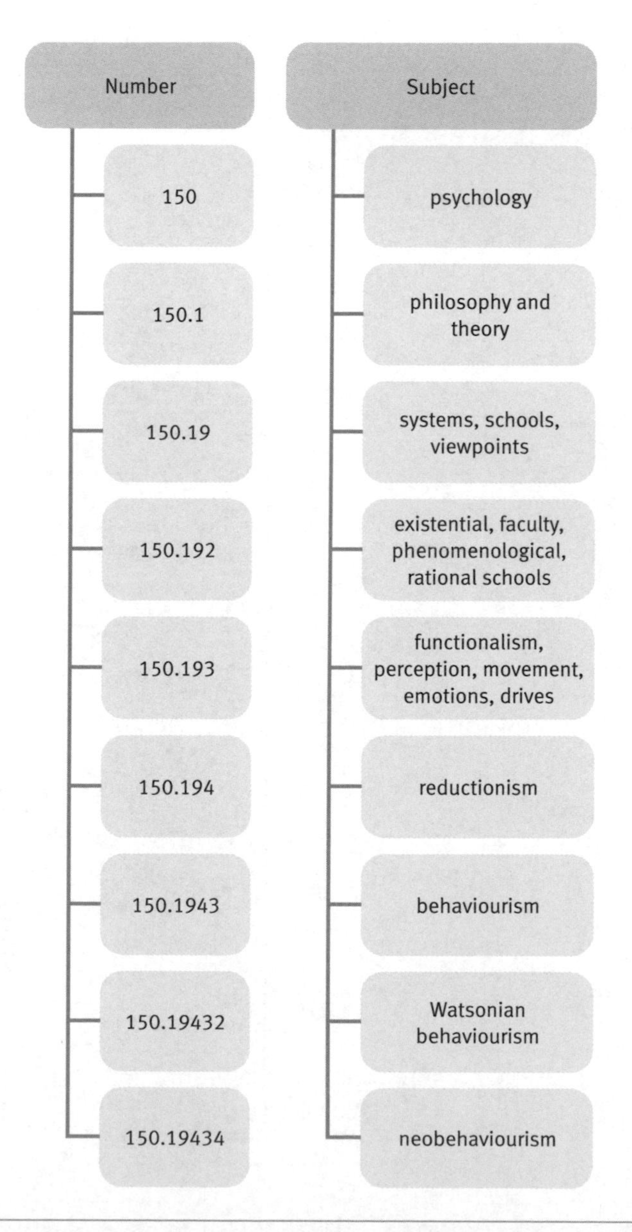

Number
Subject
150
psychology
150.1
philosophy and theory
150.19
systems, schools, viewpoints
150.192
existential, faculty, phenomenological, rational schools
150.193
functionalism, perception, movement, emotions, drives
150.194
reductionism
150.1943
behaviourism
150.19432
Watsonian behaviourism
150.19434
neobehaviourism

FIGURE 7.1 Some psychology subcategories in the Dewey Decimal Classification

are known as *call numbers* in both systems. The first of the three whole numbers indicates the classes of which there are 10 as shown in Figure 7.2. So psychology mainly comes under 1 _ _ although certain areas fall into other classes. For example, abnormal or clinical psychology is classified under 6 _ _. The second whole number shows the divisions. Much of psychology comes under 1 5 _. The third whole number refers to the section. The decimal numbers indicate further subdivisions of the sections.

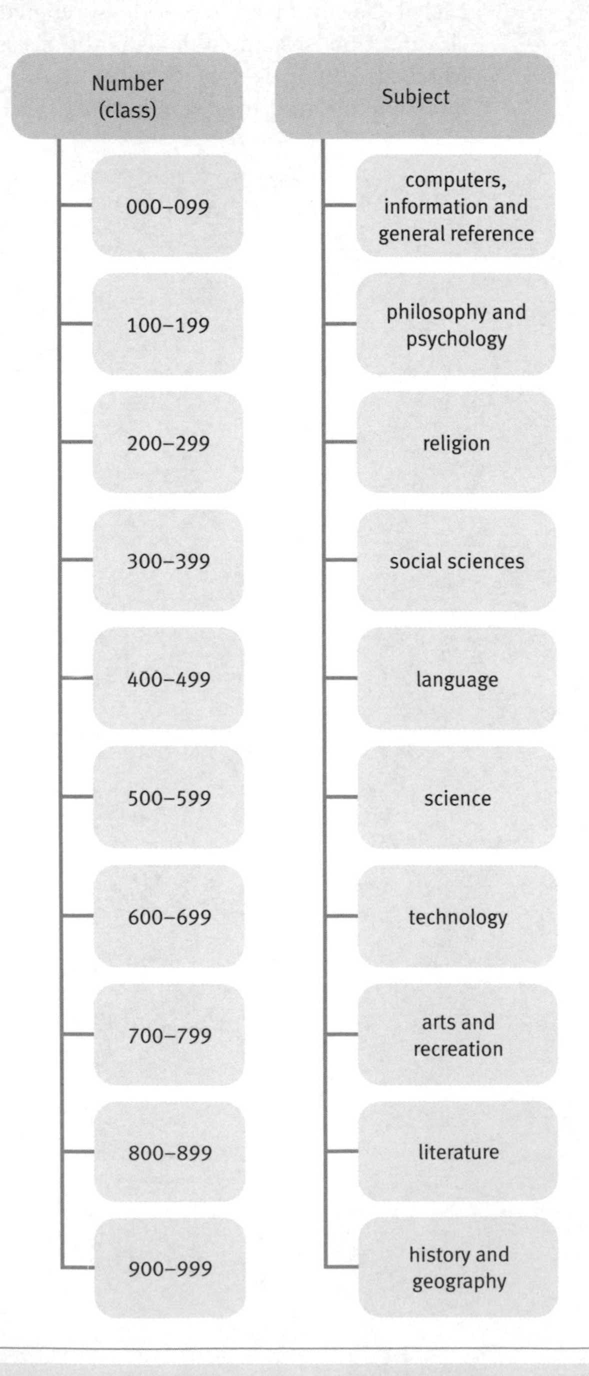

Number
(class)
Subject
000–099
computers, information and general reference
100–199
philosophy and psychology
200–299
religion
300–399
social sciences
400–499
language
500–599
science
600–699
technology
700–799
arts and recreation
800–899
literature
900–999
history and geography

FIGURE 7.2 Ten major classes of the Dewey Decimal Classification for cataloguing library material

- The other main system for organising non-fiction material in a library is the Library of Congress classification system (Chan, 1999; Library of Congress Classification Outline, n.d.) which was developed by that library in the United States. Each publication is assigned one or two letters, signifying a category, followed by a whole number between 1 and 9999. There are 21 main categories labelled A to Z but excluding I, O, W, X and Y. These categories are shown in Table 7.1. Psychology largely comes under BF. Some of the numbers and categories under BF are presented in Table 7.2.

Table 7.1 Twenty-one major categories of the Library of Congress classification system for cataloguing non-fiction

Letter (category)	Subject	Letter (category)	Subject
A	General works	M	Music
B	Philosophy, psychology, religion	N	Fine arts
C	Auxiliary sciences of history	P	Language and literature
D	History: (general) and history of Europe	Q	Science
E	History: America	R	Medicine
F	History: America	S	Agriculture
G	Geography, anthropology, recreation	T	Technology
H	Social sciences	U	Military science
J	Political science	V	Naval science
K	Law	Z	Bibliography, library science, information resources
L	Education		

Table 7.2 Some psychology subcategories in the Library of Congress classification system

Letter (category)	Number (subcategory)	Subject
BF	1–1999	Psychology, parapsychology, occult sciences
	1–990	Psychology
	38–64	Philosophy, relation to other topics
	173–175.5	Psychoanalysis
	176–176.5	Psychological tests and testing
	180–198.7	Experimental psychology
	203	Gestalt psychology
	207–209	Psychotropic drugs and other substances
	231–299	Sensation, aesthesiology
	309–499	Consciousness, cognition

Box 7.1 Practical Advice

The psychology of the literature search

Carrying out literature searches can be daunting especially as a newcomer to a research area. Professional researchers and academics are more likely to update a literature search than carry out a completely new one. Students sometimes seem overwhelmed when their attempts at a literature search do not initially go well. Furthermore, carrying out a literature search can involve a lot of time and there is no absolute certainty that it will be fruitful. The following may be of help to keep the task of searching the literature manageable:

- As a student you will have only a limited amount of time to devote to the literature search. It is better to concentrate on material published in recent years since this demonstrates that you are aware of up-to-date research, is more easily obtainable, and is likely to contain references to the older material some of which remains important in your field of interest or has been forgotten but perhaps warrants reviving.
- Literature searches on topics rising out of a student's general knowledge of psychology are less likely to present problems than where the idea being pursued is not already based on reading. Where the student has an idea based on their novel experiences then difficulties tend to arise. For one reason, the appropriate terminology may elude the student because their common-sense terminology is not what is used in the research literature. For example, a common-sense term may be remembering whereas the appropriate research literature might refer to reminiscence. Finding appropriate search terms can be difficult and often many different ones will need to be tried.
- Many students avoid carrying out a literature search because the cost of obtaining the material seems prohibitive. However, nowadays authors can be contacted by e-mail and many of them are delighted to e-mail you copies of their articles. This is without cost of course and often very quick. This does not apply to obtaining copies of books for obvious reasons.
- Most databases count the number of 'hits' that your search terms have produced. If this number is large (say more than 200) and they appear to be largely irrelevant then you need to try more restricted search terms which produce a manageable number of pertinent publications. If the database yields only a small number of articles (say fewer than ten) then this can be equally problematic especially where they are not particularly relevant to your interests. You need to formulate searches which identify more papers.
- There may be research areas that you are interested in which have only a rudimentary or non-existent research base. In these cases, your search may well be fruitless. The trouble is that it can take some time to determine whether or not this is the case since it could be your search terms which are at fault.
- Most databases contain advanced search options which can be used to maximise the number of 'hits' you make on relevant material and may reduce the number of times you find irrelevant material. Usually it is advantageous to confine your search to the titles or abstracts sections rather than to, say, anywhere on the database.
- When you find articles which are relevant to your interests, examine their database entries in order to get clues as to the sorts of keywords or terms you should be searching for to find articles like the one you are looking for. Furthermore, as many databases now include the full reference lists from the original article, these should be perused as they are likely to contain other references pertinent to your interests.
- The articles which the database identifies on the basis of your search need to be examined in order to decide whether they are actually pertinent. It is probably best to restrict this on-screen to the article's title. Reading a lot of abstracts in one sitting on the computer can become very tiring. Once potential articles have been selected on the basis of their title, you should then save the details of the article including the abstracts for later perusal. This can be done often by ticking a box on screen and e-mailing the text to yourself or by employing a cut-and-paste procedure to put the text into a file. It is then much easier to carefully select the articles which you wish to follow up.

7.3 Electronic databases

Using the library catalogue in this way is clearly a rather haphazard process. It is totally dependent on what books (and other publications) are actually in the library. Consequently you may prefer to go directly to electronic databases (as opposed to catalogues). There are a number of different electronic databases that contain information relevant to psychology. Generally libraries have to pay for access to these but research and scholarship would be severely hampered without them. The web pages for your library will usually provide details as to what is available to you at your university or college. For those databases that are only available on the web, you will generally need to have a username and password. This is obtained from your library or computer centre. Two databases that you may find especially useful are called *ISI Web of Science* and *PsycINFO*. ISI stands for Institute for Scientific Research though we will refer to this database as just the Web of Science. PsycINFO is short for Psychological Information and is produced by the American Psychological Association. Each of these has its advantages and disadvantages and it is worth becoming familiar with both of them (and others) if they are available. Both PsycINFO and Web of Science are essentially archives of summaries of research articles and other publications. These summaries are known as abstracts. Apart from reading the full article, they are probably the most complete summary of the contents of journal articles.

PsycINFO is more comprehensive in its coverage of the content of psychology journals than Web of Science. It includes what was formerly published as Psychological Abstracts and contains summaries of the content of psychology books and sometimes individual chapters but primarily PsycINFO is dominated by journal articles. For books, it goes back as far as 1840. Abstracts of books and chapters make up 11 per cent of its database (PsycINFO Database Information, n.d.). However, it also includes abstracts of postgraduate dissertations called *Dissertation Abstracts International* which constitute a further 12 per cent of the data. These abstracts are based on postgraduate work which has not been specially reviewed for publication unlike the vast majority of published research reports. The dissertations themselves are often the length of a short book and rather difficult to get access to – certainly they are difficult to obtain in a hurry. Consequently, their use is problematic when normal student submission deadlines are considered.

Web of Science contains the abstracts of articles published since 1945 for science articles and since 1956 for social science articles. It covers only those in journals that are thought to be the most important in a discipline (Thomson Reuters, n.d.). It ranges through a number of disciplines and is not restricted to psychology. Moreover, like PsycINFO it may be linked to the electronic catalogue of journals held by your library. If so, it will inform you about the availability of the journal in your library. This facility is very useful as there are a large number of psychology journals and your library will subscribe to only some of them.

Using Web of Science

Web of Science is currently accessed through Web of Knowledge. This may appear as in Figure 7.3. It is not possible in the limited space available here to show you all its different facilities. Once you have tried out one or two basic literature searches you may wish to explore its other capabilities. Find out from your college library whether you can access it and, if so, how to do so. Once you are logged into Web of Knowledge and have selected ISI Web of Science, the home page shown in Figure 7.4 will appear.

Quick Search is sufficient in most cases. To restrict your search you may wish to de-select the *Arts & Humanities Citation Index* and, possibly, the *Science Citation Index*

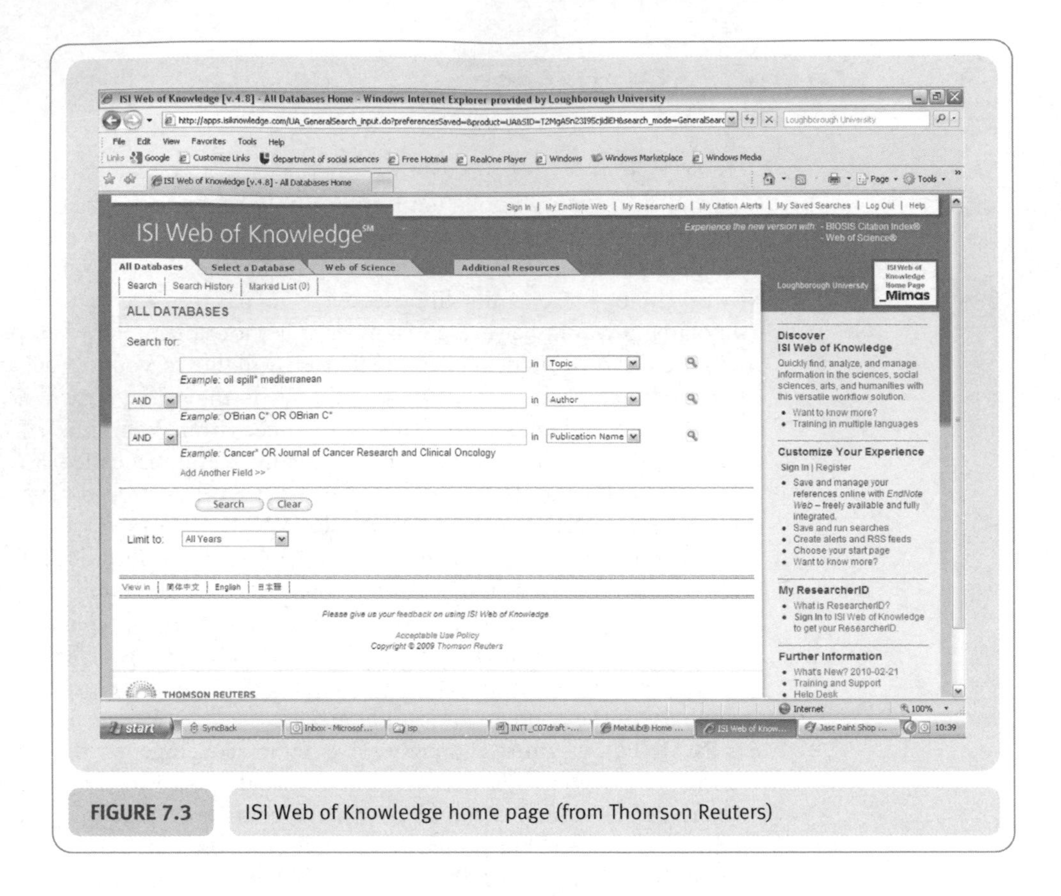

FIGURE 7.3 ISI Web of Knowledge home page (from Thomson Reuters)

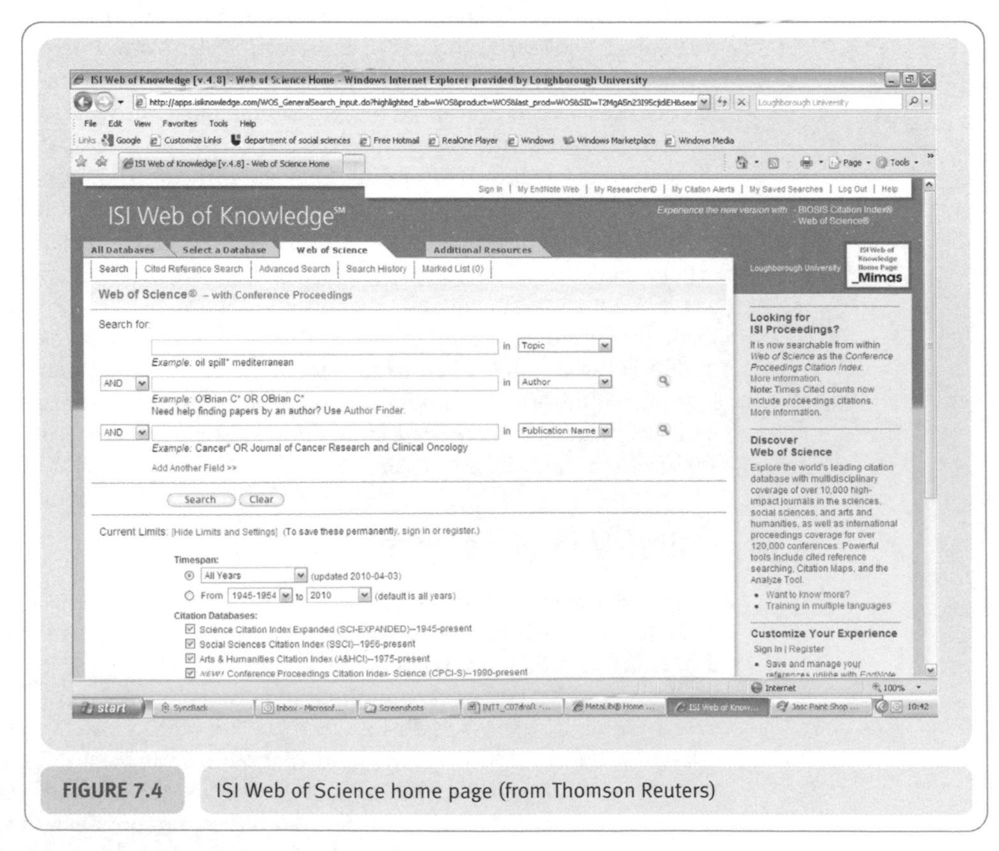

FIGURE 7.4 ISI Web of Science home page (from Thomson Reuters)

Expanded by clicking on the box containing the tick mark. If too many inappropriate references come up when the number of databases is not restricted, then go back and limit your search.

Enter the key words or terms that describe the topic that you want to conduct the search on. If too many references are found, limit your search by adding further keywords. There is a help facility if you want more information on what to do. Suppose you want to find out what articles there are on the topic of interpersonal attraction and attitude similarity. You type in these terms in the box provided combining them with the word or search operator 'and'. Then press the *Return* key or select the *Search* option.

The first part of the first page of the *Summary* of the results of this search is shown in Figure 7.5. Of course, if you search using these terms now you will get newer publications than these as this example was done in April 2010. Articles are listed in order of the most recent ones unless you have selected them in order of the highest relevance of the keywords. This option is shown in the *Sort by* box in Figure 7.5. With this option articles containing more of these terms and presenting them closer together are listed first.

Four kinds of information are provided for each article listed in the summary:

- the family name of the authors and their initials;
- the title of the article;
- the name of the journal together with the volume number, the issue number in parentheses, the first and last page numbers of the article, and the month and the year the issue was published; and
- the number of times the article has been cited by other papers.

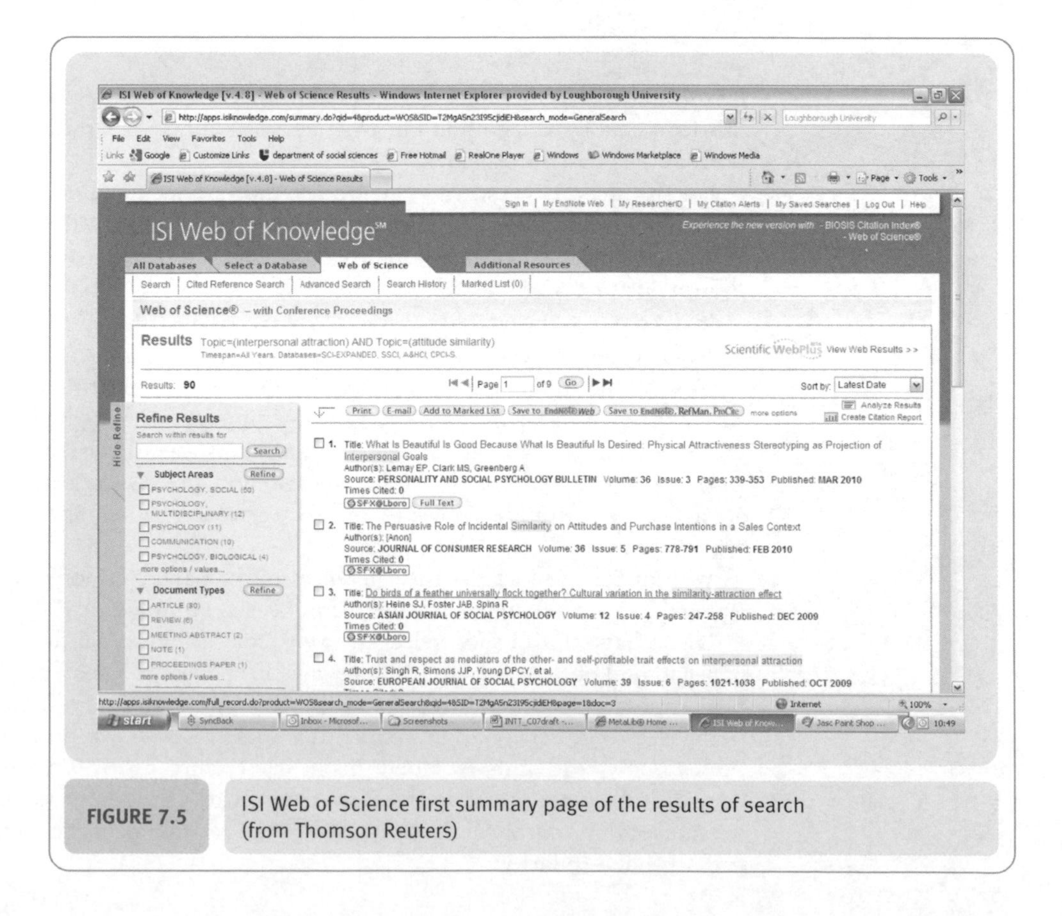

FIGURE 7.5 ISI Web of Science first summary page of the results of search (from Thomson Reuters)

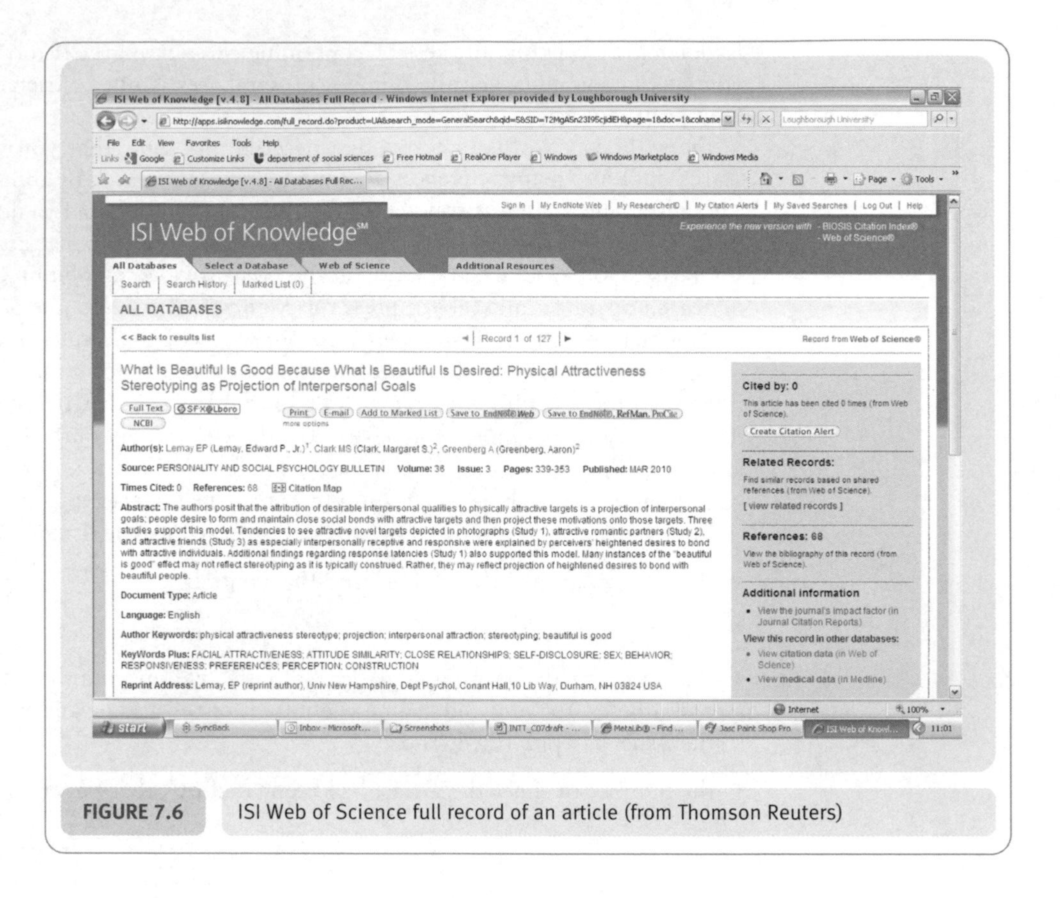

FIGURE 7.6 ISI Web of Science full record of an article (from Thomson Reuters)

If your library has this software, just below this last entry may be the *SFX* icon. Selecting this icon enables you to find out whether your library has this journal. The use of this procedure is described below.

For the first article shown in Figure 7.5, the authors are Lemay, Clark and Greenberg. The title of the article is 'What Is Beautiful Is Good Because What Is Beautiful Is Desired: Physical Attractiveness Stereotyping as Projection of Interpersonal Goals'. The journal is *Personality and Social Psychology Bulletin.*

It does not seem possible to tell from the title of this article whether it is directly relevant to our topic as the title does not refer to attraction. To see whether the article is relevant and to find out further details of it we select the title which produces the *Full Record* shown in Figure 7.6. The keyword 'interpersonal attraction' is listed as one of the *Author Keywords* and 'attitude similarity' as one of the *KeyWords Plus.* From the Abstract it would seem that this paper is not directly concerned with interpersonal attraction and attitude similarity and so we would be inclined to look at some of the other references.

Web of Science includes the references in the paper. To look at the references select *References* near the top of the full record as shown in Figure 7.7.

If you have this facility, select the *SFX* icon in Figure 7.6 just below the title of the paper to find out whether your library has this paper. *SFX* may produce the kind of web page shown in Figure 7.8. We can see that Loughborough University Library has access to the electronic version of this paper. If we select *Go* the window in Figure 7.9 appears. We can now read and download this article (Figure 7.10). We can search for our two keywords by typing them in the *Find* box towards the top of the screen. If we do this we can see that this paper does not look at the relation between interpersonal attraction and attitude similarity.

FIGURE 7.7 ISI Web of Science cited references of an article (from Thomson Reuters)

There are several ways of saving the information on Web of Science. Perhaps the easiest method is to move the cursor to the start of the information you want to save, hold down the left button of the mouse and drag the cursor down the page until you reach the end of the information you want to save. The address is useful if you want to contact the authors. This area will be highlighted. Select the *Edit* option on the bar at the top of the screen which will produce a dropdown menu. Select *Copy* from this menu. Then paste this copied material into a Word file.

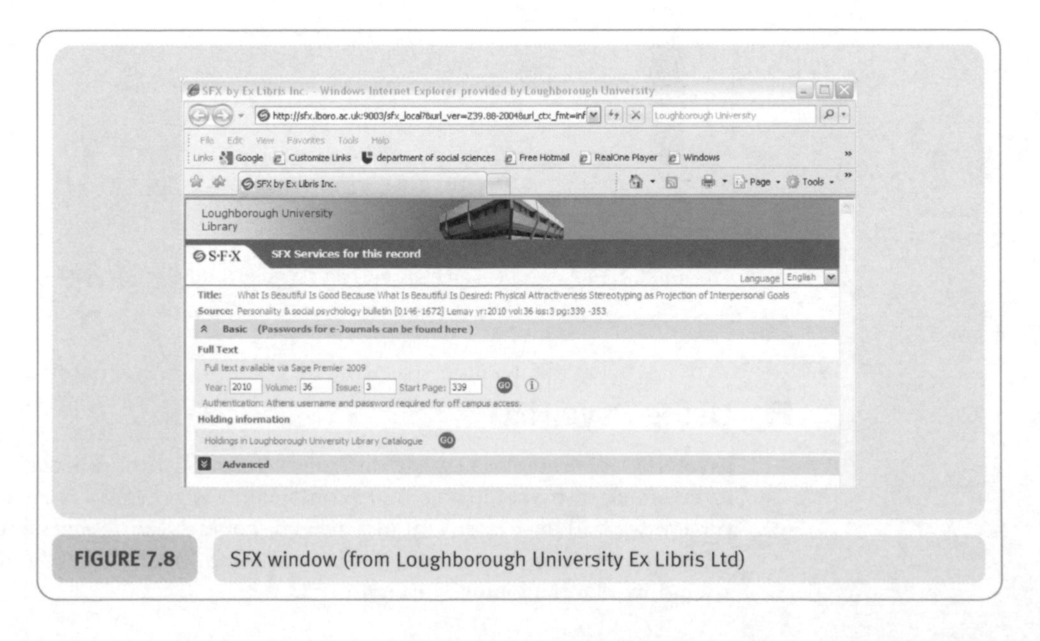

FIGURE 7.8 SFX window (from Loughborough University Ex Libris Ltd)

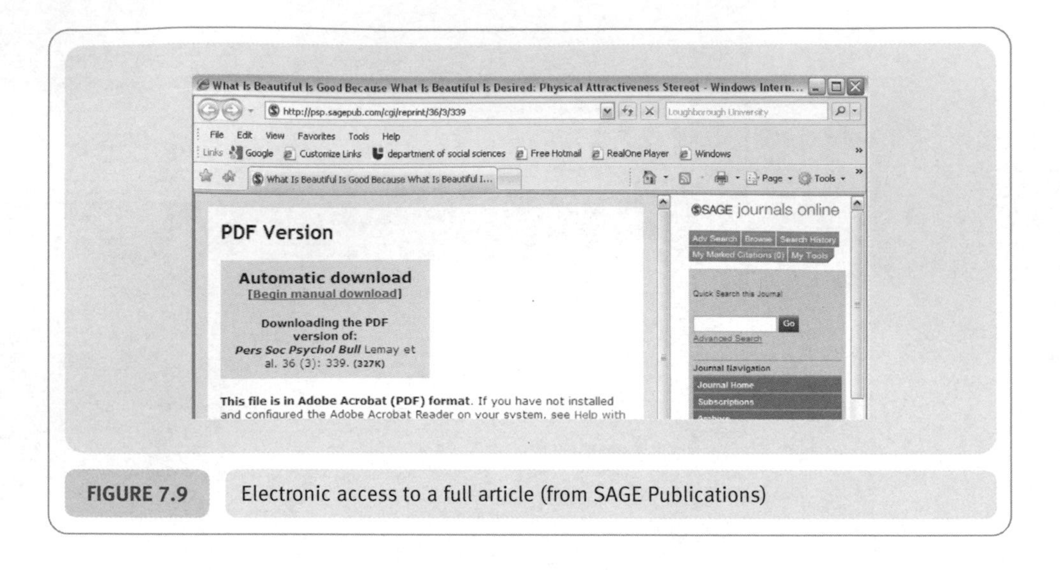

FIGURE 7.9 Electronic access to a full article (from SAGE Publications)

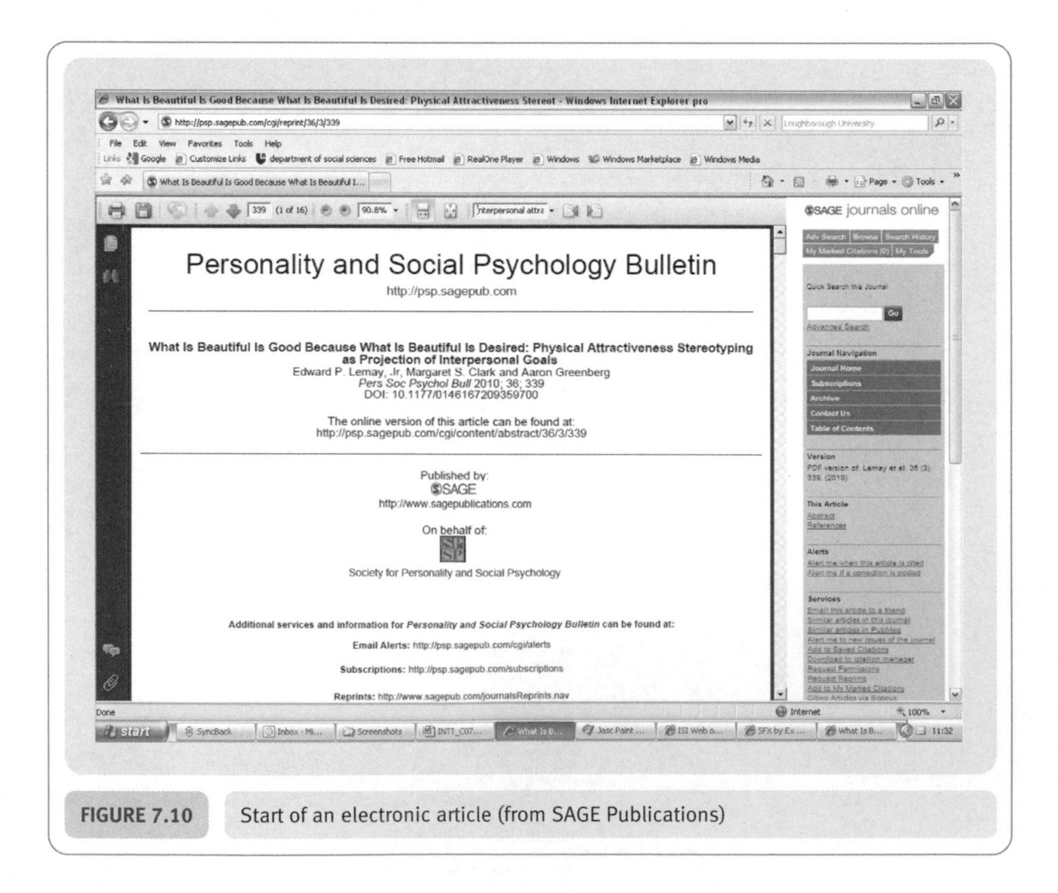

FIGURE 7.10 Start of an electronic article (from SAGE Publications)

Using PsycINFO

PsycINFO operates somewhat differently from Web of Science. Its use is essential as its coverage of psychology is more complete. It is generally accessed online which is the version we will illustrate. You may need to contact your college library to see if you can access it and, if so, how. After you have selected PsycINFO, you may be presented with a window like that shown in Figure 7.11.

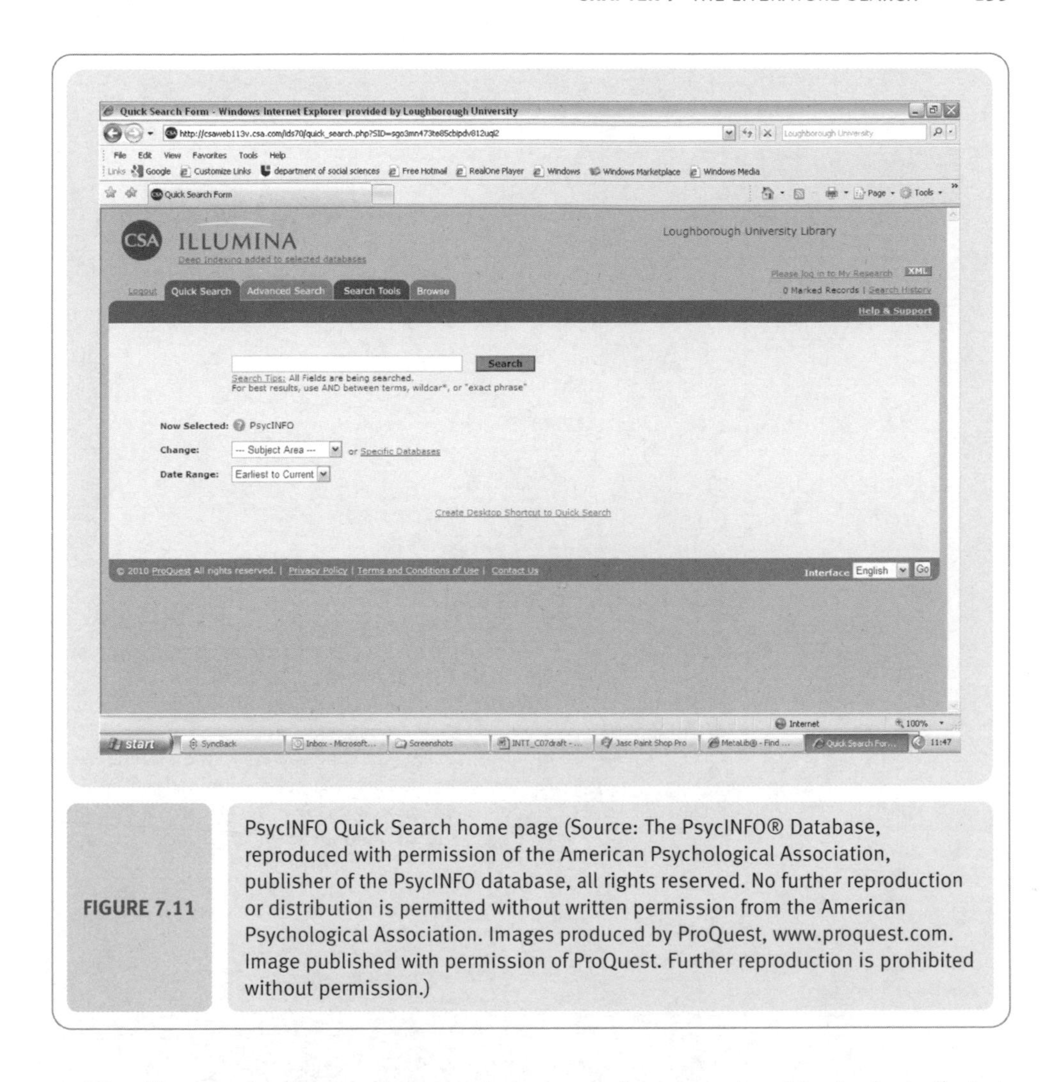

FIGURE 7.11 PsycINFO Quick Search home page (Source: The PsycINFO® Database, reproduced with permission of the American Psychological Association, publisher of the PsycINFO database, all rights reserved. No further reproduction or distribution is permitted without written permission from the American Psychological Association. Images produced by ProQuest, www.proquest.com. Image published with permission of ProQuest. Further reproduction is prohibited without permission.)

We will again search using the terms 'interpersonal attraction' and 'attitude similarity'. It is better to use the *Advanced Search* option so that you can specify that you only want material which has these keywords in their Abstract. Otherwise you will be presented with material which has these keywords elsewhere such as in the references listed at the end of a paper. The *Advanced Search* option is shown in Figure 7.12 with the keywords in the two boxes connected by the operator 'and' and restricted to be found in the 'Abstract'. To select 'Abstract', select the button on the relevant row of the rightmost box when a menu will appear as shown in Figure 7.13. 'Abstract' is the sixth keyword on this list. Then press the *Return* key or select *Search*. This will produce the kind of list shown in Figure 7.14. Your list will, of course, be more up to date if you follow these steps now. Note that this list is somewhat different from that for Web of Science shown in Figure 7.5. However, the same three kinds of information are provided for each record or reference – the title of the reference, the authors and where it was published. Also shown are the first few lines of the Abstract. You can restrict what publications are listed by selecting *Peer-Reviewed Journals*.

If you want to see the full abstract for an item, select *View Record*. The first part of the complete record for the peer-reviewed paper by Singh and colleagues is presented in Figure 7.15. Note that the record also contains the references included in the original article. To keep a copy of the details of a search it is probably easiest to select and copy the information you want and then paste it into a Word file as described for Web of Science.

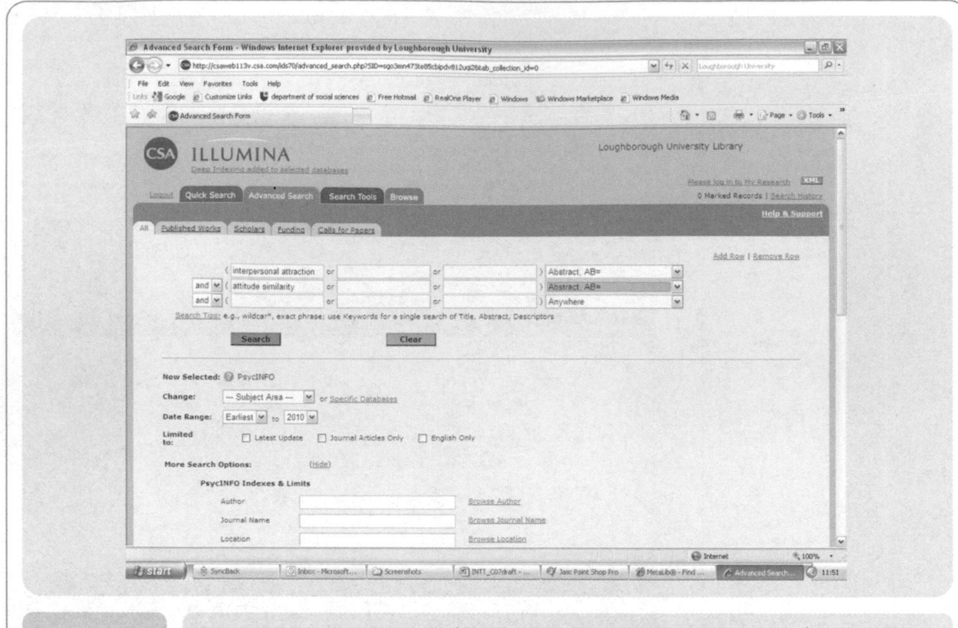

FIGURE 7.12

PsycINFO Advanced Search (Source: The PsycINFO® Database, reproduced with permission of the American Psychological Association, publisher of the PsycINFO database, all rights reserved. No further reproduction or distribution is permitted without written permission from the American Psychological Association. Images produced by ProQuest, www.proquest.com. PsycINFO is a registered trademark of the American Psychological Association (APA). The PsycINFO Database content is reproduced with permission of the APA. The CSA Illumina internet platform is the property of ProQuest LLC. Image published with permission of ProQuest. Further reproduction is prohibited without permission.)

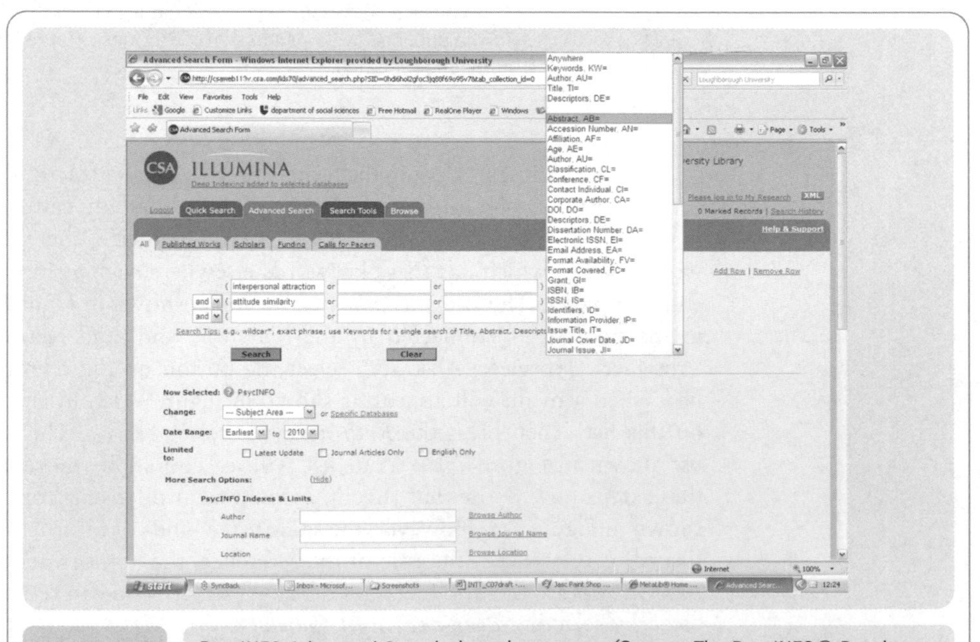

FIGURE 7.13

PsycINFO Advanced Search drop-down menu (Source: The PsycINFO® Database, reproduced with permission of the American Psychological Association, publisher of the PsycINFO database, all rights reserved. No further reproduction or distribution is permitted without written permission from the American Psychological Association. Images produced by ProQuest, www.proquest.com. Image published with permission of ProQuest. Further reproduction is prohibited without permission.)

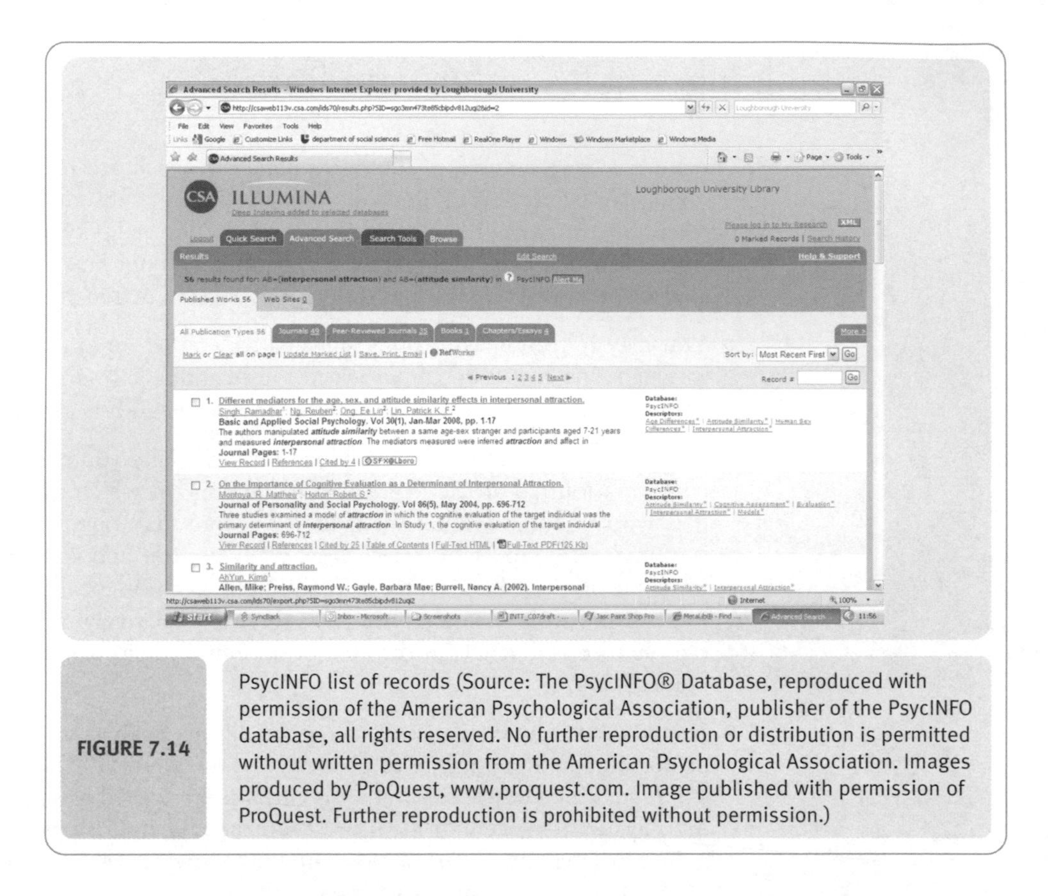

FIGURE 7.14 PsycINFO list of records (Source: The PsycINFO® Database, reproduced with permission of the American Psychological Association, publisher of the PsycINFO database, all rights reserved. No further reproduction or distribution is permitted without written permission from the American Psychological Association. Images produced by ProQuest, www.proquest.com. Image published with permission of ProQuest. Further reproduction is prohibited without permission.)

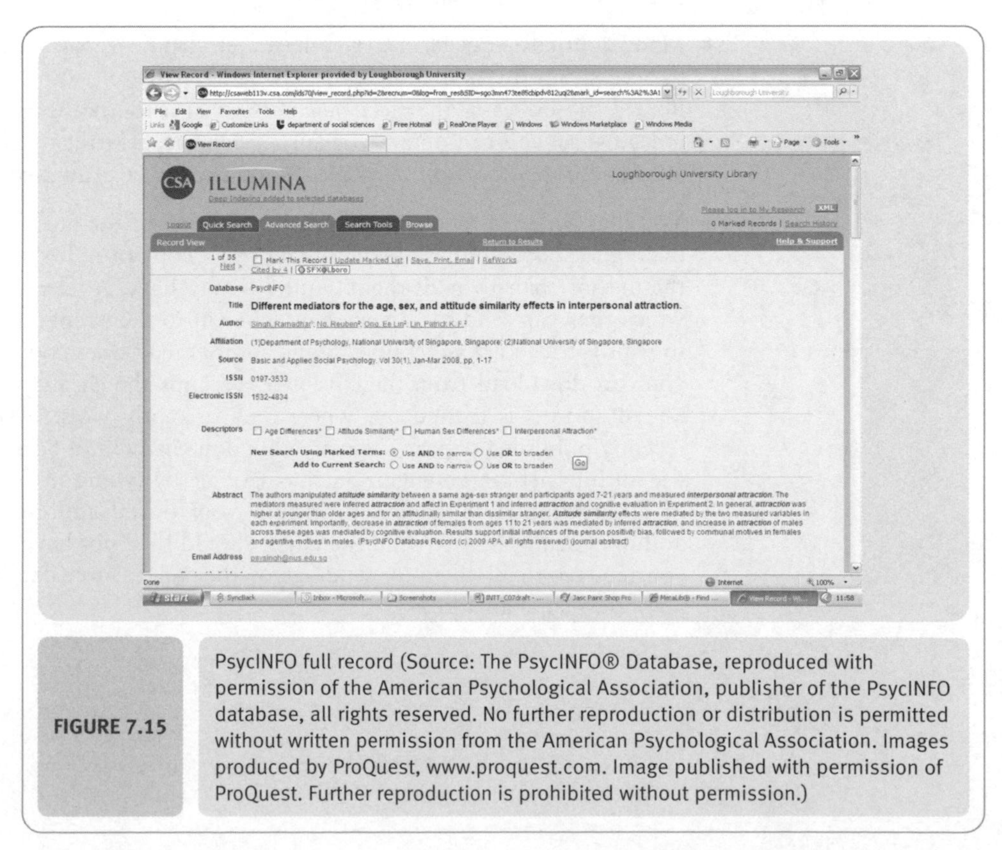

FIGURE 7.15 PsycINFO full record (Source: The PsycINFO® Database, reproduced with permission of the American Psychological Association, publisher of the PsycINFO database, all rights reserved. No further reproduction or distribution is permitted without written permission from the American Psychological Association. Images produced by ProQuest, www.proquest.com. Image published with permission of ProQuest. Further reproduction is prohibited without permission.)

7.4 Obtaining articles not in your library

There are a large number of journals published which are relevant to psychology. As the budgets of libraries are limited, your library will subscribe to only some of them. Furthermore, it may not have a complete set of a journal. Consequently, it is likely that some of the articles you are interested in reading will not be available in your library. If this occurs, there are at least five different courses of action you can take:

- You may have friends at other universities or there may be other universities near to where you live. You can check in their catalogue to see if they have the journal volume you need using the web pages of your local library in many cases.
- Libraries provide an inter-library loan service where you can obtain either a photocopy of a journal article or a copy of the issue in which the article was published. It is worth obtaining the issue if there is more than one article in that issue which is of interest to you. This service is generally not free and you may be expected to pay for some or all of these loans. You will need to check locally what arrangements are in place for using such a service. This service is relatively quick and you may receive the photocopied article in the post or an e-mailed electronic copy within a week of requesting it.
- Sometimes it may be worth travelling to a library such as the British Lending Library at Boston Spa in Yorkshire to photocopy these yourself. The number of articles you can request in a day at this library is currently restricted to 16 if you order five working days in advance and a further 8 on the day itself. It is worth checking before going. You will find contact details on the following website: http://www.york.ac.uk/library/libraries/britishlibrary/#bspa
- Many journals now have a web-based electronic version which you may be able to access. Information about this may be available on your library's website. If your library subscribes to the electronic version of a journal then this is very good news indeed since you can obtain virtually instant access to it from your computer which allows you to view it on the screen, save a copy, or print a copy.
- You may write to or e-mail the author (or one of the authors if there is more than one) of the paper and ask them to send you a copy of it. It should be quicker to e-mail the author than to mail them. Authors may have an electronic copy of the paper which they can send to you as an attachment to their reply. Otherwise you will have to wait for it to arrive by post. Some authors may have copies of their papers which you can download from their website. Perhaps the easiest way to find an author's e-mail address is to find out where they currently work by looking up their most recently published paper. This is readily done in Web of Science or PsycINFO. Then use an Internet search engine such as Google by typing in their name and the name of the institution. You need to include your postal address in your e-mail so that authors know where to send the paper should they not have an electronic copy. It is courteous to thank them for sending you the paper. Some databases routinely provide an author's e-mail address.

Box 7.2 Talking Point

Judging the reputation of a publication

There is a form of pecking-order for research journals in all disciplines and that includes psychology. To have an article published in *Nature* or *Science* signals something of the importance of one's work. Researchers are attracted to publishing in the most prestigious journals for professional advancement. Virtually every journal has a surfeit of material submitted to it so very good material may sometimes be rejected. The rejection rate of articles submitted for publication in journals is relatively high. In 2008, rejection rates varied from 35 per cent for *Experimental and Clinical Pharmacology* to 89 per cent for *Teaching of Psychology* with an average of 69 per cent across the non-divisional journals published by the leading American Psychological Association (American Psychological Association, 2009). Not unexpectedly, agreement between referees or reviewers about the quality of an article may be considerably less than perfect (Cicchetti, 1991). Quality, after all, is a matter of judgement. Authors may well find that an article rejected by one journal will be accepted by the next journal they approach.

The impact factor is a measure of the frequency with which the average article in a journal has been cited within a particular period. This may be regarded as a useful indicator of the quality of the journal. More prestigious journals should be more frequently cited than less prestigious ones. The Institute for Scientific Information which produces the Web of Science also publishes *Journal Citation Reports* annually in the summer following the year they cover. The impact factor of a particular journal may be found using these reports either online, on CD-ROM or on microfiche. The period looked at by the *Journal Citation Reports* is the two years prior to the year being considered (Institute for Scientific Information, 1994). For example, if the year being considered is 2011, the two years prior to that are 2009 to 2010. The impact factor of a journal in 2011 is the ratio of the number of times in 2011 that articles published in that journal in 2009 and 2010 were cited in that and other journals to the number of articles published in that journal in 2009 and 2010:

$$\textit{journal's impact factor } 2011 = \frac{\textit{citations in } 2011 \textit{ of articles published in journal in } 2009\text{–}2010}{\textit{number of articles published in journal in } 2009\text{–}2010}$$

So, for example, if the total number of articles published in 2009 and 2010 was 200 and the number of citations of those articles in 2011 was 200, the impact factor is 1.00. The impact factor excludes what are called self-citations where authors refer to their previous articles.

Taking into account the number of articles published in a particular period controls for the size of the journal. If a journal publishes more articles than another journal, then that journal is more likely to be cited simply for that reason if all else is equal. This correction may not be necessary as it was found by Tomer (1986) that the corrected and the uncorrected impact factor correlates almost perfectly (0.97).

The impact factors for a selection of psychology journals for the years 2004 to 2008 are presented in Table 7.3. The impact factor varies across years for a journal. For example, for the *British Journal of Social Psychology* it decreased from 1.99 in 2007 to 1.71 in 2008. It also differs between journals. For these journals in 2008, the highest impact factor is 5.04 for the *Journal of Personality and Social Psychology* and the lowest is 0.59 for *The Journal of Psychology*. An impact factor of about 1.00 means that the average article published in that journal was cited about once in the previous two years taking into account the number of articles published in that journal in those two years. The Web of Science includes only those journals that are considered to be the most important (Thomson Reuters, n.d.).

However, even the Institute for Scientific Information which introduced the impact factor measure says that the usefulness of a journal should not be judged only on its impact factor but also on the views of informed colleagues or peers (Institute for Scientific Information, 1994). The impact factor is likely to be affected by a number of variables such as the average number of references cited in a journal or the number of review articles that are published by a journal. The relationship between the citation count of a journal and the subjective judgement of its standing by psychologists has not been found to be strong.

For example, Buss and McDermot (1976) reported a rank-order correlation of 0.45 between the frequency of citations for 64 psychology journals in the period 1973–1975 and a five-point rating made of those journals

Table 7.3 Impact factors for some psychology journals for 2008 to 2004

Journal	2008	2007	2006	2005	2004
British Journal of Social Psychology	1.71	1.99	1.42	2.11	1.59
Journal of Personality and Social Psychology	5.04	4.51	4.22	4.21	3.63
The Journal of Psychology	0.59	0.54	0.59	0.53	0.42
Journal of Social and Personal Relationships	1.10	0.87	0.99	0.72	0.82
The Journal of Social Psychology	0.73	0.86	0.66	0.60	0.60
Personality and Social Psychology Bulletin	2.46	2.58	2.42	2.09	1.90
Social Psychology Quarterly	1.14	2.07	1.30	1.06	1.40

by the chairs or heads of 48 psychology departments in the United States in an earlier study by Mace and Warner (1973). This relationship was stronger at 0.56 when it was restricted to the ten most highly cited journals. In other words, agreement was higher when the less highly cited journals were excluded. Rushton and Roediger (1978) found a Kendall's tau correlation of 0.45 between the ratings of these journals by these departmental heads and their impact factor. Chairs of departments are an influential group of people in that they are often responsible for selecting, giving tenure and promoting academic staff. However, it is possible that nowadays chairs are more aware of the impact factor and so the relationship between the impact factor and the rating of the journal may be higher.

There appears not to be a strong relationship between the number of times a published paper is cited by other authors and either the quality or the impact of the paper as rated by about 380 current or former editors, associate editors and consulting editors of nine major psychology journals who had not published papers in those journals (Gottfredson, 1978). Because the distribution of the number of citations was highly skewed with most articles not being cited, the logarithm of the citation number was taken. The correlation between this transformed number was 0.22 for the quality scale and 0.36 for the impact scale. The number of times a paper is cited is given by Web of Science just below the journal title as shown in Figure 7.5. If you select the number after *Times Cited* (provided that it is not zero), you will see details of the papers that have cited this reference.

The lack of agreement about the quality of published papers was dramatically illustrated in a study by Peters and Ceci (1982) in which 12 papers which had been published in highly regarded American psychology journals were resubmitted to them 18 to 32 months later using fictitious names and institutions. Of the 38 editors and reviewers who dealt with these papers, only three realised that they were resubmissions. Of the nine remaining papers, eight of these previously published papers were rejected largely on the grounds of having serious methodological flaws. This finding emphasises the importance of the reader being able to evaluate the worth of a paper by themselves and not relying entirely on the judgements of others.

7.5 Personal bibliographic database software

There is much bibliographic database software which enables you to quickly store the details of references of interest to you from electronic databases such as Web of Science and PsycINFO. These include EndNote, RefMan, ProCite and RefWorks. If you look at the Web of Science screenshots in Figure 7.5 or 7.6 you will see that there is an option to *Save to* EndNote, RefMan and ProCite. In the PsycINFO screenshot in Figure 7.14 there is a RefWorks icon which if you select will permit you to save references to RefWorks. For example, we could save the details of the reference in Figure 7.14 by Singh and colleagues in RefWorks as shown in Figure 7.16. You can also use this kind of software to write out the references that you cite in your work in a particular style,

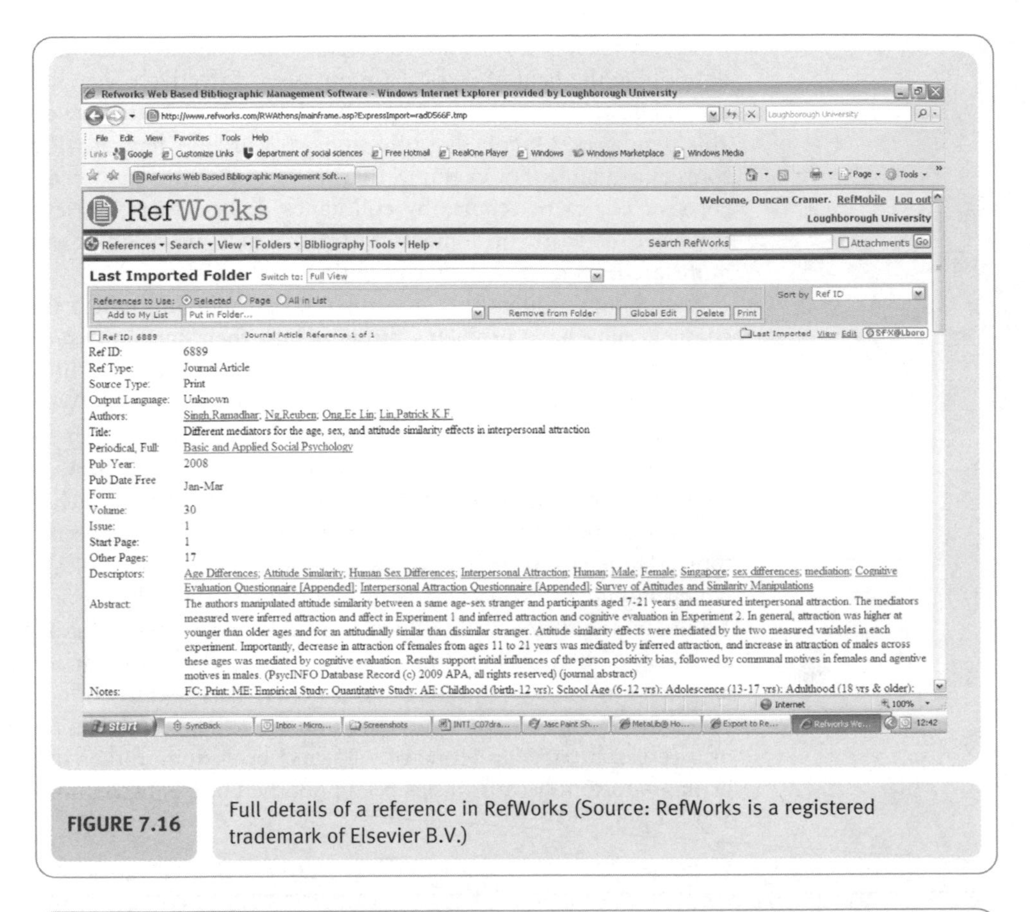

FIGURE 7.16 Full details of a reference in RefWorks (Source: RefWorks is a registered trademark of Elsevier B.V.)

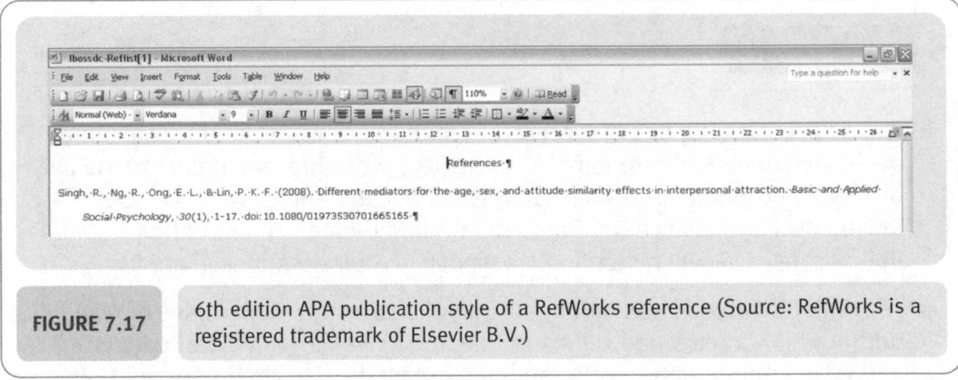
References

Singh, R., Ng, R., Ong, E. L., & Lin, P. K. F. (2008). Different mediators for the age, sex, and attitude similarity effects in interpersonal attraction. *Basic and Applied Social Psychology*, *30*(1), 1-17. doi:10.1080/01973530701665165

FIGURE 7.17 6th edition APA publication style of a RefWorks reference (Source: RefWorks is a registered trademark of Elsevier B.V.)

such as that recommended by the American Psychological Association. For example, we could format the Singh reference in terms of 6th edition of the APA Publication manual as presented in Figure 7.17. This does not mean that we do not have to familiarise ourselves with the details of this particular style, as we still have to check whether the software and our use of it has presented the references in the appropriate style.

7.6 Conclusion

The development of any discipline is the collective effort of numerous researchers acting to a degree independently. It is necessary for researchers to communicate their findings and ideas in publications such as journal articles. Similarly, researchers need to be able to access the work of other researchers in order to make an effective contribution to developing the field of research in question. Effectively searching the literature involves a number of skills. In this chapter we have concentrated on efficient searches of available databases. Of course, professional researchers have a wider variety of information sources available. For example, they go to conferences and hear of new work there, they get sent copies of reports by colleagues doing research elsewhere, and they have an extensive network of contacts through which news of research elsewhere gets communicated. Students have fewer options at first.

Searching the literature on a topic takes time and unless this is taken into account, students may have problems fitting it into their schedule. No one can expect that all materials will be available in their university or college library. There are ways of obtaining material which are increasingly dependent on the World Wide Web. If you are unfamiliar with the research on a particular topic, it may be helpful to find a recently published book which includes an introduction to that topic to give you some idea of what has been done and found. Electronic databases such as Web of Science and PsycINFO are a very convenient way to find out what has been published on a particular topic. These electronic databases provide short abstracts or summaries of publications which should give you a clearer idea of whether the publication is relevant to your needs. You may not always realise the importance or relevance of a paper when you first come across it. Consequently, it may be better to make a note of a paper even if it does not appear immediately relevant to your needs. This is easily done with the copy and paste functions of the computer software you are using. You need to learn to judge the value of a research paper in terms of what has been done rather than simply accepting it as being important because it has been published or has been published in what is reputedly a good journal.

Key points

- The key to carrying out a successful literature research in any discipline lies in using the various available sources of information. Of these, modern research is most heavily dependent on the use of electronic databases such as Web of Science and PsycINFO. These are generally available in universities and elsewhere. Students will find them useful but often the materials available via their library catalogue will take priority because of their ease of availability.
- The major databases essentially consist of abstracts or summaries of research publications including both journal articles and books. An abstract gives a fairly detailed summary of the research article and is an intermediate step to help the reader decide whether or not the complete article or book is required. In addition, these databases frequently contain enough information to enable the author to be contacted – often this goes as far as including an e-mail address.
- Databases are not identical and one may supplement another. Furthermore, there may well be other sources of information that some researchers can profitably refer to. For example, the fields of biology, medicine, sociology and economics might provide essential information for researchers in some fields of psychology. Knowledge of these builds up with experience.
- Abstracts and other information may be copied and pasted on your computer. In this way it is possible to build up a record of the materials you feel will be useful to you.
- There are numerous ways of obtaining published research. The Internet and e-mail are increasingly rich sources. It can be surprisingly easy to get in touch with academics all over the world. Many are quite happy to send copies of their work either in the mail or electronically.

ACTIVITIES

1. If you do not already know this, find out what electronic databases are available in your library and how to access them. The best way of checking how good a system is and how to use it is to try it out on a topic that you are familiar with. It should produce information you are already aware of. If it does not do this, then you can try to find out how to locate this information in this system. Try out a few systems to see which suits your purposes best.

2. Many university libraries provide training in the use of their resources and systems. These are an excellent way of quickly learning about the local situation. Enquire at your library and sign up for the most promising. Afterwards try to turn your effort into better grades by conducting a more thorough search as preparation for your essays, practical reports and projects. Using information sources effectively is a valuable skill, and should be recognised and rewarded.

CHAPTER 8

Ethics and data management in research

Overview

- Psychological ethics are the moral principles that govern psychological activity. Research ethics are the result of applying these broader principles to research. Occasions arise when there is a conflict between ethical principles – ethical dilemmas – which are not simply resolved.
- Psychology's professional bodies (for example, the American Psychological Association and the British Psychological Society) publish detailed ethical guidelines. They overlap significantly. This chapter is based on recent revisions of the ethical principles of these bodies.
- Deception, potential harm, informed consent and confidentiality are commonly the focus of the debate about ethics. However, ethical issues stretch much more widely. They include responsibilities to other organisations, the law and ethical committees, circumstances in which photos and video-recording are appropriate, and the publication of findings, plagiarism and fabricating data.
- Significantly, ethical considerations are the responsibility of all psychologists including students in training.
- It is increasingly the norm that a researcher obtains formal consent from their participants that they agree to take part in the research on an informed basis.
- Data management refers to the ways that you may need to store and handle the personal data which you collect in research in order to maintain confidentiality. Data items which are anonymous are not included in the requirements of the Data Protection Act in the UK.

8.1 Introduction

Quite simply, ethics are the moral principles by which we conduct ourselves. Psychological ethics, then, are the moral principles by which psychologists conduct themselves. It is wrong to regard ethics as being merely the rules or regulations which govern conduct. The activities of psychologists are far too varied and complex for that. Psychological work inevitably throws up situations which are genuinely dilemmas which no amount of rules or regulations could effectively police. Ethical dilemmas involve conflicts between different principles of moral conduct. Consequently psychologists may differ in terms of their position on a particular matter. Ethical behaviour is not the responsibility of each individual psychologist alone but a responsibility of the entire psychological community. Monitoring the activities of fellow psychologists, seeking the advice of other psychologists when ethical difficulties come to light and collectively advancing ethical behaviour in their workplace are all instances of the mutual concern that psychologists have about the conduct of the profession. Equally, psychological ethics cannot be entirely separated from personal morality.

The American Psychological Association's most recent ethical code was first published in 2002. It came into effect on 1 June 2003. It amounts to a substantial ethical programme for psychological practitioners, not just researchers. This is important since unethical behaviour reflects on the entire psychological community. The collective strength of psychology lies largely in the profession's ability to control and monitor all aspects of the work of psychologists. The code, nevertheless, only applies to the professional activities of the psychologist – their scientific, professional and educational roles. For example, it requires an ethical stance in psychology teaching – so that there is a requirement of fidelity in the content of psychology courses such that they should accurately reflect the current state of knowledge. These newest ethical standards do not simply apply to members of the American Psychological Association but also to student affiliates/members. Ignorance of the relevant ethical standards is not a defence for unethical conduct and neither is failure to understand the standards properly. Quite simply, this means that all psychology students need a full and mature understanding of the ethical principles which govern the profession. It is not something to be left until the student is professionally qualified. Whenever scientific, professional and educational work in psychology is involved so too are ethics, irrespective of the status. We have chosen to focus on the American Psychological Association's ethical guidelines as they are the most comprehensive available, considering rather wider issues than any others. As such, they bring to attention matters which otherwise might be overlooked. We believe that it is no excuse to disregard them simply because they are not mentioned by one's own professional ethics, for example.

What is the purpose of ethics? The answer to this may seem self-evident, that is, psychologists ought to know how to conduct themselves properly. But there is more to it than that. One of the characteristics of the professions (medicine being the prime example) is the desire to retain autonomy. The history of the emergence of professions such as medicine during the nineteenth century illustrates this well (Howitt, 1992a). Autonomy implies self-regulation of the affairs of members by the professional body. It is not possible to be autonomous if the activities of members are under the detailed control of legislation. So professions need to stipulate and police standards of conduct. There is another important reason why psychological work should maintain high ethical standards. The good reputation of psychology and psychologists among the general public, for example, is essential for the development of psychology. If psychologists collectively enjoyed a reputation for being dishonest, exploitative, prurient liars then few would employ their services or willingly participate in their research. Trust in the profession is essential.

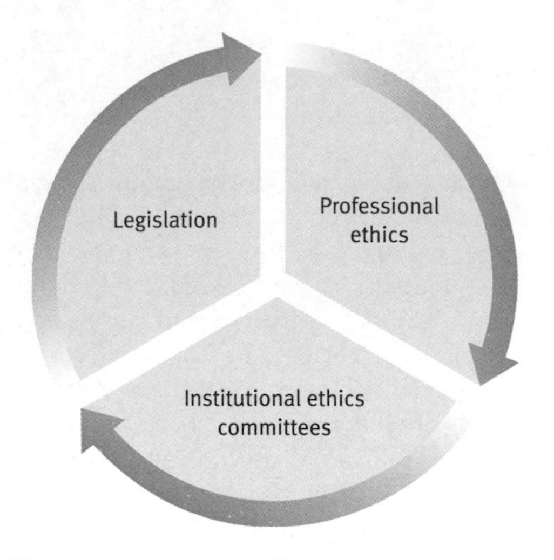

FIGURE 8.1 The ethical environment of psychology

Failure to adhere to sound ethical principles may result in complaints to professional bodies such as the American Psychological Association, British Psychological Society and so forth. Sanctions may be imposed on those violating ethical principles. Ultimately the final sanction is ending the individual's membership of the professional body, which may result in the individual being unable to practise professionally. Many organisations including universities have ethics committees that both supervise the research carried out by employees but also that of other researchers wishing to do research within the organisation. While this does provide a measure of protection for all parties (including the researcher), it should not be regarded as the final guarantee of good ethical practices in research.

However, no matter the role of professional ethics in psychology, this is not the only form of control on research activities (see Figure 8.1). Probably the more direct day-to-day influence on research are the ethical committees of major public institutions such as universities and health services. These have a more immediate impact since they review the research proposals of researchers planning most forms of research. Legislation is also relevant, of course, and the particular impact of data protection legislation constitutes the best example of this.

8.2 APA ethics: The general principles

The APA ethics are based on five general principle:

- *Principle A: Beneficence and non-maleficence* Psychologists seek to benefit and avoid harm to those whom they engage with professionally. This includes the animals used in research. Psychologists should both be aware of and guard against those factors which may result in harm to others. The list of factors is long and includes financial, social and institutional considerations.
- *Principle B: Fidelity and responsibility* Psychologists are in relationships of trust in their professional activities. They are thus required to take responsibility for their actions, adhere to professional standards of conduct, and make clear exactly their role

and obligations in all aspects of their professional activities. In relation to research and practice, psychologists are not merely concerned with their own personal activities but with the ethical conduct of their colleagues (widely defined). It is worthwhile quoting word for word one aspect of the professional fidelity ethic: 'Psychologists strive to contribute a portion of their professional time for little or no compensation or personal advantage.'

- *Principle C: Integrity – accuracy, honesty, truthfulness* Psychologists are expected to manifest integrity in all aspects of their professional work. One possible exception to this is circumstances in which the ratio of benefits to harm of using deception is large. Nevertheless, it remains the duty of psychologists even in these circumstances to seriously assess the possible harmful consequences of the deception including the ensuing distrust. The psychologist has a duty to correct these harmful consequences. The problem of deception is discussed in more detail later.
- *Principle D: Justice – equality of access to the benefits of psychology* This means that psychologists exercise careful judgement and take care to enable all people to experience just and fair psychological practices. Psychologists should be aware of the nature of their biases (potential and actual). They should not engage in, or condone, unjust practices and need to be aware of the ways in which injustice may manifest itself.
- *Principle E: Respect for people's rights and dignity* According to the American Psychological Association, individuals have the rights of privacy, confidentiality and self-determination. Consequently, psychologists need to be aware of the vulnerabilities of some individuals that make it difficult for them to make autonomous decisions. Children are an obvious example. The principle also requires psychologists to be aware of and respect differences among cultures, individuals and roles. Age, culture, disability, ethnicity, gender, gender identity, language, national origin, race, religion, sexual orientation and socio-economic status are among these differences. Psychologists should avoid and remove biases related to these differences while being vigilant for, and critical of, those who fail to meet this standard.

Detailed recommendations about ethical conduct are provided on the basis of these principles. In this chapter we will concentrate on those issues which are especially pertinent to research.

8.3 Research ethics

Ethical issues are presented by the American Psychological Association's documentation in the order in which they are likely to be of concern to the researcher. Hence the list starts with the preparatory stages of planning research and culminates with publication.

Institutional approval

Much research takes place in organisations such as the police, prisons, schools and health services. Many, if not all, of these require formal approval before the research may be carried out in that organisation or by members of that organisation. Sometimes this authority to permit research is the responsibility of an individual (for example, a headteacher) but, more likely, it will be the responsibility of a committee which considers ethics. In addition, in universities the researcher is usually required to obtain permission to carry out their research from their school, department or an ethics committee such as an Institutional Review Board (IRB). It is incumbent on the researcher to obtain approval for their planned research. Furthermore, the proposal they put forward should be transparent in the sense that the information contained in the documentation and any other communication should accurately reflect the nature of the research. The organisation should be in a position to understand precisely what the researcher intends on the basis of the documentation provided by the researcher and any other communications. So any form of deceit or sharp practice such as lies, lying by omission and partial truths is unacceptable. Finally, the research should be carried out strictly in accordance with the protocol for the research as laid down by the researcher in the documentation. Material changes are not permissible and, if unavoidable, may require additional approval to remain ethical.

The next set of ethical requirements superficially seem rather different from each other. Nevertheless, they all indicate that participation in research should be a freely made decision of the participant. Undue pressure, fear, coercion and the like should not be present or implied. In addition, participants need to understand just what they are subjecting themselves to by agreeing to be part of the research. Without this, they may inadvertently agree to participate in something which they would otherwise decline to do.

Informed consent to research

The general principle of informed consent applies widely and would include assessment, counselling and therapy as well as research. People have the right to have prior knowledge of just what they are agreeing to before agreeing to it. Only in this way is it possible for them to decide not to participate. Potential participants in research need to have the nature of the research explained to them in terms which they could reasonably be expected to understand. So the explanation given to a child may be different from that given to a university student. According to the ethical principles, sometimes research may be conducted without informed consent if it is allowed by the ethical code or where the law and other regulations specifically permit. (Although one might question whether research is ethical merely because the law permits it.)

The main provisions which need to be in place to justify the claim of informed consent are as follows:

- The purpose, procedures and approximate duration of the research should be provided to potential participants.
- Participants should be made aware that they are free to refuse to take part in the research and also free to withdraw from the research at any stage. Usually researchers accept that this freedom to withdraw involves the freedom to withdraw any data provided up to the point of withdrawal. For example, the shredding of questionnaires and the destruction of recordings are appropriate ways of doing this if the withdrawing participant wishes. Or they may simply be given to the participant to dispose of as they wish.
- The participant should be made aware of the possible outcomes or consequences of refusing to take part in the research or withdrawing. Frequently, there are no consequences but this is not always the case. For example, some organisations require that clients take part in research as part of the 'contract' between the organisation and client. Failure to take part in research might be taken as an indicator of non-cooperation. The sex offender undergoing treatment who declines to take part in the research might be regarded as lacking in contrition. Such judgements may have implications for the future disposal of the client. The researcher cannot be responsible for the original contract but they should be aware of the (subtle) pressure to participate and stress the voluntary nature of participation.

- The participants should be informed of those aspects of the research which might influence their decision to participate. These include discomforts, risks and adverse outcomes. For example, one might include features of the study which might offend the sensibilities of the participant. Research on pornography in which pornographic images will be shown may offend the moral and/or social sensibilities of some participants.
- Similarly, the participant should be informed of the benefits that may emerge from the research. A wide view of this would include benefits for academic research, benefits for the community, and even benefits for the individual participant. In this way, the potential participant is provided with a fuller picture of what the research might achieve which otherwise might not be obvious to them.
- Participants should be told of any limits to the confidentiality of information provided during the research. Normally, researchers ensure the anonymity of the data that they collect and also the identity of the source of the data. But this is not always possible: for example, if one were researching sex offenders and they disclosed other offences of which authorities were unaware. It may be a requirement placed on the researcher that such undisclosed offences are reported to the authorities. In these circumstances, the appropriate course of action might be to indicate to the participant that the researcher would have to report such previously undisclosed offences to the authorities.
- Participants should be informed of the nature of any incentives being made to participate. Some participants may agree to take part as an act of kindness or because they believe that the research is important. If they are unaware of a cash payment, they may feel that their good intentions for taking part in the research are compromised when the payment is eventually offered.
- Participants should be given contact details of someone whom they may approach for further details about the research and the rights of participants in the research. This information allows potential participants to ask more detailed questions and to obtain clarification. Furthermore, it has the benefit of helping to establish the bona fides of the research. For example, if the contact is a professor at a university, then this would help establish the reputability of the research.

Special provisions apply to experimental research involving potentially beneficial treatments which may not be offered to all participants (see Box 8.1).

Informed consent for recordings and photography

Taking voice recordings, videos or photographs of participants is subject to the usual principle of informed consent. However, exceptions are stipulated in the ethical code:

- Informed consent is not necessary if the recording or photography takes place in a public place and is naturalistic (that is, there is no experimental intervention). This is ethical only to the extent that there is no risk of the inadvertent participants being identified personally or harmed by the recording or photography.
- If the research requires deception (and that deception is ethical) then consent for using the recording may be obtained retrospectively during the debriefing session in which the participant is given information about the research and an opportunity to ask questions. Deception is discussed below.

Circumstances in which informed consent may not be necessary

The ethical guidelines do not impose an invariant requirement of informed consent. They suggest circumstances in which it may be permissible to carry out research without prior

Box 8.1 Talking Point

Informed consent in intervention experiments

When a psychologist conducts *intervention research* there may be issues of informed consent. This does not refer to every experiment but those in which there may be significant advantages to receiving the treatment and significant disadvantages in *not* receiving the treatment. The treatment, for example, might be a therapeutic drug, counselling or therapy. Clearly, in these circumstances many participants would prefer to receive the treatment rather than not receive the treatment. If this were medical research it would be equivalent to some cancer patients being in the control group and dying because they are not given the newly developed drug that the experimental group benefits from. In psychological research, someone may be left suffering depression simply because they are allocated to the control group not receiving treatment. Owing to these possibilities, the researcher in these circumstances should do the following:

- The experimental nature of the treatments should be explained at the outset of the research.
- It should be made clear the services or treatments which will *not* be available should the participant be allocated to the control condition.
- The method of assignment to the experimental or the control conditions should be explained clearly. If the method of selection for the experimental and control conditions is random then this needs to be explained.
- The nature of the services or treatments available to those who choose *not* to take part in the research should be explained.
- Financial aspects of participation should be clarified. For example, the participant may be paid for participation, but it is conceivable that they may be expected to contribute to the cost of their treatment.

The classic study violating the above principles is known as the Tuskegee Experiment (Jones, 1981). Significantly, the study involved only black people as participants. They were suffering from syphilis at a time when this was a killer disease. The researchers, unbeknown to the participants, allocated them to experimental and control conditions. Hence those in the control had effective treatment withheld, so were at a serious health risk as a consequence. This may have been bad enough but there was worse. Even when it had become clear that the treatment was effective, the members of the control group were left to suffer from the disease because the researchers also wished to study its natural progression!

consent of this sort. The overriding requirement is that the research could not be expected to (i.e. can be regarded as *not* likely to) cause distress or harm to participants. Additionally, at least *one* of the following should apply to the research in question:

- The study uses anonymous questionnaires or observations in a natural setting or archival materials – even then such participants should not be placed at risk of harm of any sort (even to their reputation) and confidentiality should be maintained.
- The study concerns jobs or related organisational matters in circumstances where the participant is under no risk concerning employment issues and the requirements of confidentiality are met.
- The study concerns 'normal educational practices, curricula or classroom management methods' in a context of an educational establishment.

The ethics also permit research not using informed consent *if* the law or institutional regulations permit research without informed consent. This provision of the ethical principles might cause some consternation. Most of us probably have no difficulty with the principle that psychologists should keep to the law in terms of their professional activities. Stealing from clients, for example, is illegal as well as unethical. However, there is a distinction to be made between what is permissible in law and what is ethical. A good example of this is the medical practitioner who has consenting sex with a patient. This may not be illegal and no crime committed in some countries. However, it is unethical for a doctor to do so and the punishment imposed by the medical profession is severe: possibly removal from the register of medical practitioners. There may be a potential conflict between ethics and the law. It seems to be somewhat lame to prefer the permission of the law rather than the constraints of the ethical standards.

Research with individuals in a less powerful/subordinate position to the researcher

Psychologists are often in a position of power relative to others. A university professor of psychology has power over his or her students. Clients of psychologists are dependent on the psychologists for help or treatment. Junior members of research staff are dependent on senior research staff and subordinate to them. It follows that some potential research participant may suffer adverse consequences as a result of refusing to take part in research or may be under undue pressure to participate simply because of this power differential. Any psychologist in such a position of power has an ethical duty to protect these vulnerable individuals from such adverse consequences. Sometimes, participation in research is a requirement of particular university courses or inducements may be given to participate in the form of additional credit. In these circumstances, the ethical recommendation is that fair alternative choices should be made available for individuals who do not wish to participate in research.

Inducements to participate

Financial and other encouragement to participate in research are subject to the following requirements:

- Psychologists should not offer unreasonably large monetary or other inducements (for example, gifts) to potential participants in research. In some circumstances such rewards can become coercive. One simply has to take the medical analogy of offering people large amounts of money to donate organs in order to understand the undesirability of this. While acceptable levels of inducements are not stipulated in the ethics, one reasonable approach might be to limit payments where offered to out-of-pocket expenses (such as travel) and a modest hourly rate for time. Of course, even this provision is probably out of the question for student researchers.

- Sometimes professional services are offered as a way of encouraging participation in research. These might be, for example, counselling or psychological advice of some sort. In these circumstances, it is essential to clarify the precise nature of the services, including possible risks, further obligations and the limitations to the provision of such services. A further requirement, not mentioned in the APA ethics, might be that the researcher should be competent to deliver these services. Once again, it is difficult to imagine the circumstances in which students could be offering such inducements.

The use of deception in research

The fundamental ethical position is that deception should *not* be used in psychological research procedures. There are *no* circumstances in which deception is acceptable if there is a reasonable expectation that physical pain or emotional distress will be caused. However, it is recognised that there are circumstances in which the use of deception may be justified. If the proposed research has 'scientific, educational or applied value' (or the prospect of it) then deception may be considered. The next step is to establish that no effective alternative approach is possible which does not use deception. These are not matters on which individual psychologists should regard themselves as their own personal arbiters. Consultation with disinterested colleagues is an appropriate course of action.

If the use of deception is the only feasible option, it is incumbent on the psychologist to explain the deception as early as possible. This is preferably immediately after the data have been collected from each individual, but it may be delayed until all of the data from all of the participants have been collected. The deceived participant should be given the unfettered opportunity to withdraw their data. Box 8.2 discusses how deception has been a central feature of social psychological research. The ethics of the British Psychological Society indicate that a distinction may be drawn between deliberate lies and omission of particular details about the nature of the research that the individual is participating in. This is essentially the distinction between lying by omission and lying by commission. You might wonder if this distinction is sufficient justification of anything. The British Psychological Society indicates that a key test of the acceptability is the response of participants at debriefing when the nature of the deception is revealed. If they express anger, discomfort or otherwise object to the deception then the deception was inappropriate and the future of the project should be reviewed. The BPS guidelines do not specify the next step, however.

Box 8.2 Talking Point

Deception in the history of social psychology

The use of deception has been much more characteristic of the work of social psychologists than any other branch of psychological research. Korn (1997) argues that this was the almost inevitable consequence of using laboratory methods to study social phenomena. Deception first occurred in psychological research in 1897 when Leon Solomons studied how we discriminate between a single point touching our skin and two points touching our skin at the same time. Some participants were led to believe that there were two points and others that there was just one point. Whichever they believed made a difference to what they perceived. Interestingly, Solomons told his participants that they might be being deceived before participation.

In the early history of psychology deception was indeed a rare occurrence, but so was any sort of empirical research. There was a gradual growth in the use of deception between 1921 and 1947. The *Journal of Abnormal and Social Psychology* was surveyed during this period. Fifty-two per cent of articles involved studies using misinformation as part of the procedure while 42 per cent used 'false cover stories' (Korn, 1997). Little fuss was made about deception at this time. According to a variety of surveys of journals, the use of deception increased between 1948 and 1979 despite more and more questions about psychological ethics being asked. Furthermore, there appears to be no moderation in the scale of the sort of deceptions employed during this period. Of course, the purpose of deception in many cases is simply to hide the true purpose of the experiment. Participants threatened with a painful injection might be expected to behave differently if they believe this is necessary for a physiological experiment than if they know that the threat of the injection is simply a way of manipulating their stress levels.

Many of the classic studies in social psychology – ones still discussed in textbooks – used deceit of some sort or another. These did not necessarily involve trivial matters. In Milgram's studies of obedience in which participants were told that they were punishing a third party with electric shock, it appeared at one stage that the victim of the shock had been severely hurt. All of this was a lie and deception (Milgram, 1974). Milgram tended to refer to his deceptions as 'technical illusions' but this would appear to be nothing other than a euphemism. In studies by other researchers, participants believed that they were in an emergency situation when smoke was seeping into a laboratory through a vent – again a deliberate deception (Latane and Darley, 1970). Deception was endemic and routine in social psychological research. It had to be, given the great stress on laboratory experimentation in the social psychology of the time. Without the staging of such extreme situations by means of deception, experimental social psychological research would be difficult if not impossible.

Sometimes the deceptions seem relatively trivial and innocuous. For example, imagine that one wished to study the effects of the gender of a student on the grades that they get for an essay. Few would have grave concerns about taking an essay and giving it to a sample of lecturers for marking, telling half of them that the essay was by a male and the other half the essay was by a woman. It would seem to be important to know through research whether or not there is a gender bias in marking which favours one gender over the other. Clearly there has been a deception – a lie if one prefers – but it is one which probably does not jeopardise in any way the participants' psychological well-being though there are circumstances in which it could. Believing that you have endangered someone's life by giving them dangerous levels of electric shock is not benign but may fundamentally affect a person's ideas about themselves. In some studies, participants have been deceitfully abused about their abilities or competence in order to make them angry (Berkowitz, 1962).

How would studies like these stand up to ethical scrutiny? Well, deception as such is not banned by ethical codes. There are circumstances in which it may be justifiable. Deception may be appropriate when the study has, or potentially has, significant 'scientific, educational or applied value' according to the APA ethical principles. Some might question what this means. For example, if we wanted to study the grieving process, would it be right to tell someone that the university had been informed that their mother had just died? Grief is an important experience and clearly it is of great importance to study the phenomenon. Does that give the researcher carte blanche to do anything?

Deception is common in our society. The white lie is a deception, for example. Does the fact that deception is endemic in society justify its use in research? Psychologists are professionals who as a group do not benefit from developing a reputation as tricksters. The culture of deception in research may lead to suspicion and hostility towards participation in the research.

Debriefing

As soon as the research is over (or essential stages are complete), debriefing should be carried out. There is a mutual discussion between researcher and participant to fully inform the participant about matters such as the nature of the result, the results of the research and the conclusions of the research. The researcher should try to correct the misconceptions of the participant that may have developed about any aspect of research. Of course, there may be good scientific or humane reasons for withholding some information – or delaying the main debriefing until a suitable time. For example, it may be that the research involves two or more stages separated by a considerable interval of time. Debriefing participants after the first stage may considerably contaminate the results at the second stage.

Debriefing cannot be guaranteed to deal effectively with the harm done to participants by deception. Whenever a researcher recognises that a particular participant appears to have been (inadvertently) harmed in some way by the procedures then reasonable efforts should be made to deal with this harm. It should be remembered that researchers are not normally qualified to offer counselling, and other forms of help and referral to relevant professionals may be the only appropriate course of action. There is a body of research on the effects of debriefing (for example, Epley and Huff, 1998; Smith and Richardson, 1983).

Box 8.3 Talking Point

Ethics and research with animals

Nowadays, few students have contact with laboratory animals during their education and training. Many universities simply do not have any facilities at all for animal research. However, many students have active concerns about the welfare of animals and so may be particularly interested in the ethical provision for such research in psychology. It needs to be stressed that this is an area where the law in many countries has exacting requirements that may be even more stringent than those required ethically. The first principle is that psychologists involved in research with animals must adhere to the pertinent laws and regulations. This includes the means by which laboratory animals are acquired, the ways in which the animals are cared for, the ways in which the animals are used, and the way in which laboratory animals are disposed of or retired from research.

Some further ethical requirements are as follows:

- Psychologists both experienced and trained in research methods with laboratory animals should adopt a supervisory role for all work involving animals. Their responsibilities include consideration of the 'comfort, health and humane treatment' of animals under their supervision.
- It should be ensured that all individuals using animals have training in animal research methods and the care of animals. This should include appropriate ways of looking after the particular species of animal in question and the ways in which they should be handled. The supervising psychologist is responsible for this.
- Psychologists should take appropriate action in order that the adverse aspects of animal research should be minimised. This includes matters such as the animals' pain, comfort, freedom from infection and illnesses.
- While in some circumstances it may be ethically acceptable to expose animals to stress, pain or some form of privation of its bodily needs, this is subject to requirements. There must be no alternative way of doing the research. Furthermore, it should be done only when it is possible to justify the procedures on the basis of its 'scientific, educational or applied value'.
- Anaesthesia before and after surgery is required to minimise pain. Techniques which minimise the risk of infection are also required.
- Should it be necessary and appropriate to terminate the animal's life, this should be done painlessly and as quickly as possible. The accepted procedures for doing so should be employed.

One suspects that many will regard this list as inadequate. The list makes a number of assumptions – not the least being that it is ethically justifiable to carry out research on animals in certain conditions. But is this morally acceptable? Some might question whether cruelty to animals (and the unnecessary infliction of pain is cruel) is defensible in any circumstances. Others may be concerned about the lack of clarity in terms of when animal research is appropriate. Isn't any research defensible on the grounds of scientific progress? What does scientific progress mean? Is it achieved by publication in an academic psychology journal?

8.4 Ethics and publication

The following few ethical standards for research might have particular significance for student researchers.

Ethical standards in reporting research

It is ethically wrong to fabricate data. Remember that this applies to students. Of course, errors may inadvertently be made in published data. These are most likely to be computational or statistical error. The researcher, on spotting the error, should take reasonable efforts to correct it. Among the possibilities are corrections or retractions in the journal in question.

Plagiarism

Plagiarism is when the work of another person is used without acknowledgement and as if it was one's own work. Psychologists do *not* plagiarise. Ethical principles hold that merely occasionally citing the original source is insufficient to militate against the charge of plagiarism. So, copying chunks of other people's work directly is inappropriate even if they are occasionally cited during this procedure. Of course, quotations clearly identified as such by the use of quotation marks, attribution of authorship, and citation of the source are normally acceptable. Even then, quotations should be kept short and within the limits set by publishers, for example.

Proper credit for publications

It is ethically inappropriate to stake a claim on work which one has not actually done or in some way contributed to substantially. This includes claiming authorship on publications. The principal author of a publication (the first-named) should be the individual who has contributed the most to the research. Of course, sometimes such a decision will be arbitrary where contributions cannot be ranked. Being senior in terms of formal employment role should not be a reason for principal authorship. Being in charge of a research unit is no reason for being included in the list of authors. There are often circumstances in which an individual makes a contribution but less than a significant one. This should be dealt with by a footnote acknowledging their contribution or some similar means. Authorship is not the reward for a minor contribution of this sort.

It is of particular importance to note that publications based on the dissertations of students should credit the student as principal (first) author. The issue of publication credit should be raised with students as soon as practicable by responsible academics.

Publishing the same data repeatedly

When data are published for the second or more time then the publication should clearly indicate the fact of republication. This is acceptable. It is not acceptable to repeatedly publish the same data as if for the first time.

The availability of data for verification

Following the publication of the results of research, they should be available for checking or verification by others competent to do so. This is not carte blanche for anyone to take another person's data for publication in some other form – that would require agreement. It is merely a safeguard for the verification of substantive claims made by the original researcher. The verifying psychologist may have to meet the costs of supplying the data for verification. Exceptions to this principle of verification are:

- circumstances in which the participants' confidentiality (e.g. anonymity) cannot be ensured;
- if the data may not be released because another party has proprietary rights over the data which prevent their release.

8.5 Obtaining the participant's consent

It is commonplace nowadays that the researcher both provides the potential participant with written information about the nature of the study and obtains their agreement or consent to participation in the study. Usually these include a statement of the participant's rights and the obligations of the researcher. The things which normally would go into this sort of documentation are described separately for the information sheet/ study description and the consent form. It is important that these are geared to your particular study so what follows is a list of things to consider for inclusion rather than a ready-made form to adopt.

The information sheet/study description

The information sheet or study description should be written in such a way that it communicates effectively to those taking part in the study. It should therefore avoid complex language and, especially, the use of jargon which will be meaningless to anyone not trained in psychology. The following are the broad areas which should be covered in what you write. Some of these things might be irrelevant to your particular study:

- The purpose of the study and what it aims to achieve.
- What the participant will be expected to do in the study.
- Indications of the likely amount of time which the participant will devote to the study.
- The arrangements to deal with the confidentiality of the data.
- The arrangements to deal with the privacy of any personal data stored.
- The arrangements for the security of the data.
- A list of who would have access to the data.
- The purposes for which the data will be used.
- Whether participants will be personally identifiable in publications based on the research.
- Participation is entirely voluntary.
- It is the participants right to withdraw themselves and the data from the study without giving a reason or explanation (possibly also a statement that there will be no consequences of doing so such as the withdrawal of psychological services if the context of the research requires this).
- What benefits might participation in the research bring the individual and others.
- Any risks or potential harm that the research might pose to those participating.
- If you wish to contact the participant in future for further participation in the research, it is necessary to get their permission to do so at this stage. If you do not, you cannot contact them in the future under the terms of the British Data Protection Act.
- Give details for the research team or your supervisor if you are a student from which the participant can obtain further information if necessary and the contact details of the relevant Ethics Committee in case of issues which cannot be dealt by the research team or the supervisor.

The consent form

The consent form provides an opportunity for the participants to indicate that they understand the arrangements for the research and give their agreement to take part in the research in the light of these. The typical consent form probably should cover the following points, though perhaps modified in parts:

- The title of the research project.
- I have been informed about and understand the nature of the study. Yes/No
- Any questions that I had were answered to my satisfaction. Yes/No
- I understand that I am free to withdraw myself and my data from the research at any time with no adverse consequences. Yes/No
- No information about me will be published in a form which might potentially identify me. Yes/No
- My data, in an anonymous form, may be used by other researchers. Yes/No
- I consent to participate in the study as outlined in the information sheet. Yes/No
- Space for the signature of the participant, their name in full, and the date of the agreement.

8.6 Data management

Data management includes some issues very closely related to ethical matters; however, it is different. Ethical matters, as we have seen, are not driven primarily by legislation whereas data management issues have a substantial basis in legislation. Data protection, in European countries, is required by legislation to cover all forms of recorded information whether it is digitally stored on a computer, for example, or in hard copy form in filing cabinets. The university or college that you study at should have a data protection policy. The department that you study in is also likely to have its own policy on data protection. Now data protection is not mainly or substantially about data in research; it is far wider than that. Data protection covers any personal data which are held by an organisation for whatever purpose. There are exemptions but the legislation is likely to apply to anything that you do professionally and even as a student of psychology. It covers things such as application forms, work and health records, and much more – anything which involves personal data period. So it is vital to understand data management in relation to your professional work in psychology in the future since you will almost certainly collect information from clients and others which comes under the legislation. Research is treated positively in data protection legislation in the UK.

The good news is that data protection legislation does not apply if the personal data are in anonymous form. Essentially this means that the data should be anonymous at the point of collection. This could be achieved, for example, by not asking those completing a questionnaire to give their name or address or anything like that. It might be wise to avoid other potentially identifiable information in order to be on the safe side – for example, just ask for their year of birth rather than the precise date if the latter risks identifying participants. All of this needs some thought. It obviously imposes some limits on what you can do – for example, you could not contact the participant to take part in a follow-up to the study and you cannot supplement the data that you have with additional information from other sources. But most of the time you would not want to do these things anyway.

Of course, some data inevitably will allow for the identification of a research participant. Just because they are not named does not mean that they are not identifiable. For example, videoed research participants may well be identifiable and individuals with a particular job within an organisation may also be identifiable by virtue of that fact. So it is possible that data protection legislation applies. It is immaterial in what form the data are stored – hard copy, digital recording media, or what-have-you: if the data are personal and the person is identifiable then the act applies. What follows will be familiar from parts of the previous section. Data protection requires that the researcher must give consideration to the safe keeping of identifiable personal data. So it includes the question of which people have access to the data. Probably this is all that you need to know about data protection but organisations will have their own data protection officers from whom you may seek advice if necessary.

8.7 Conclusion

Research ethics cover virtually every stage of the research process. The literature review, for example, is covered by the requirements of fidelity and other stages of the process have specific recommendations attached to them. It is in the nature of ethics that they do not simply list proscribed behaviours. Frequently they offer advice on what aspects of research require ethical attention and the circumstances in which exceptions to the generally accepted standards may be considered. They impose a duty on all psychologists to engage in consideration and consultation about the ethical standing of their research as well as that of other members of the psychological community. Furthermore, the process does not end prior to the commencement of data collection but requires attention and vigilance throughout the research process since new information may indicate ethical problems where they had not been anticipated.

One important thing about ethics is that they require a degree of judgement in their application. It is easy for students to seek rules for their research. For example, is it unethical to cause a degree of upset in the participants in your research? What if your research was into experiences of bereavement? Is it wrong to interview people about bereavement knowing that it will distress some of them? Assume that you have carefully explained to participants that the interviews are about bereavement. Is it wrong then to cause them any distress in this way? What if the research was just a Friday afternoon practical class on interviewing? Is it right to cause distress in these circumstances? What if it were a Friday workshop for trainee clinical psychologists on bereavement counselling? Is it any more acceptable? All of this reinforces the idea that ethics are fine judgements, not blanket prohibitions for the most part. Of course, ethics committees may take away some of this need for fine judgement from researchers.

The consideration of ethics is a fundamental requirement of the research process that cannot be avoided by any psychologist – including students at any level. It starts with not fiddling the data and not plagiarising. And what if your best friend fiddles the data and plagiarises?

Key points

- Psychological associations such as the American Psychological Association and the British Psychological Society publish ethical guidelines to help their members behave morally in relation to their professional work. Self-regulation of ethics is a characteristic of professions.
- Ethics may be based on broad principles, but frequently advice is provided in guidelines about their specific application, for example, in the context of research. So one general ethical principle is that of integrity, meaning accuracy, honesty and truthfulness. This principle clearly has different implications to the use of deception in research from those when reporting data.
- Informed consent is the principle that participants in research should willingly consent to taking part in research in the light of a clear explanation by the researcher about what the research entails. At the same time, participants in research should feel in a position to withdraw from the research at any stage with the option of withdrawing any data that have already been provided. There are exceptions where informed consent is not deemed necessary – especially naturalistic observations of people who might expect to be observed by someone since they are in a public place.
- Deception of participants in research is regarded as problematic in modern psychology despite being endemic in some fields, particularly social psychology. Nevertheless, there is no complete ban on deception, only the requirements that the deception is absolutely necessary since the research is important and there is no effective alternative deception-free way of conducting it. The response of participants during debriefing to the deception may be taken as an indicator of the risks inherent in that deception.
- The publication of research is subject to ethical constraints. The fabrication of data, plagiarism of the work of others, claiming the role of author on a publication to which one has only minimally contributed, and the full acknowledgement by first authorship of students' research work are all covered in recent ethical guidelines.
- Increasingly there are more formal constraints on researchers such as those coming from ethics committees and the increased need to obtain research participant's formal consent. Although data protection legislation can apply to research data, data in an anonymous/unidentifiable form are exempt from the legislation.

ACTIVITIES

Are any principles of ethical conduct violated in the following examples? What valid arguments could be made to justify what occurs? These are matters that could be debated. Alternatively, you could list the ethical pros and cons of each before reaching a conclusion.

(a) Ken is researching memory and Dawn volunteers to be a participant in the research. Ken is very attracted to Dawn and asks for her address and mobile phone number, explaining that she may need to be contacted for a follow-up interview. This is a lie as no such interviews are planned. He later phones her up for a date.

(b) A research team is planning to study Internet sex offenders. They set up a bogus Internet pornography site – 'All tastes sex'. The site contains a range of links to specialised pages devoted to a specific sexual interest – bondage, mature sex, Asian women and the like. Visitors to the site who press these links see mild pornographic pictures in line with the theme of the link. The main focus of the researchers is on child pornography users on the Internet. To this end

they have a series of links labelled '12-year-olds and under', 'young boys need men friends', 'schoolgirls for real', 'sexy toddlers' and so forth. These links lead nowhere but the researchers have the site programmed such that visitors to the different pages can be counted. Furthermore, they have a 'data miner' which implants itself onto the visitor's computer and can extract information from that computer and report back to the researchers. They use this information in order to send out an e-mail questionnaire concerning the lifestyle of the visitor to the porn site – details such as their age, interests, address and so forth as well as psychological tests. To encourage completion, the researchers claim that in return for completing the questionnaire, they have a chance of being selected for a prize of a Caribbean holiday. The research team is approached by the police who believe that the data being gathered may be useful in tracking down paedophiles.

(c) A student researcher is studying illicit drug use on a university campus. She is given permission to distribute questionnaires during an introductory psychology lecture. Participants are assured anonymity and confidentiality, although the researcher has deliberately included questions about demographic information such as the participants' exact date of birth, their home town, the modules they are taking and so forth. However, the student researcher is really interested in personality factors and drug taking. She gets another student to distribute personality questionnaires to the same class a few weeks later. The same information about exact date of birth, home town, place of birth and so forth is collected. This is used to match each drug questionnaire with that same person's personality questionnaire. However, the questionnaires are anonymous since no name is requested.

(d) Professor Green is interested in fascist and other far-right political organisations. Since he believes that these organisations would not permit a researcher to observe them, he poses as a market trader and applies for and is given membership of several of these organisations. He attends the meetings and other events with other members. He is carrying out participant observation and is compiling extensive notes of what he witnesses for eventual publication.

(e) A researcher studying sleep feels that a young man taking part in the research is physically attracted to him. She tries to kiss him.

(f) Some researchers believe that watching filmed violence leads to violence in real life. Professor Jenkins carries out a study in which scenes of extreme violence taken from the film *Reservoir Dogs* are shown to a focus group. A week later, one of the participants in the focus group is arrested for the murder of his partner on the day after seeing the film.

(g) A discourse analyst examines President Bill Clinton's television claim that he did not have sexual intercourse with Monica Lewinsky in order to assess discursive strategies that he employed and to seek any evidence of lying. The results of this analysis are published in a psychology journal.

(h) 'Kitty Friend complained to an ethics committee about a psychologist she read about in the newspaper who was doing research on evoked potentials in cat brains. She asserted that the use of domesticated cats in research was unethical, inhumane, and immoral' (Keith-Spiegel and Koocher, 1985, p. 35). The ethics committee chooses not to consider the complaint.

(i) A psychology student chooses to investigate suicidal thoughts in a student population. She distributes a range of personality questionnaires among her friends. Scoring the test she notices that one of her friends, Tom, has scored heavily on a measure of suicide ideation and has written at the end of the questionnaire that he feels desperately depressed. She knows that it is Tom from the handwriting, which is very distinctive.

(j) Steffens (1931) describes how along with others he studied the laboratory records of a student of Wilhelm Wundt, generally regarded as the founder of the first psychological laboratory. This student went on to be a distinguished professor in America. Basically the student's data failed to support aspects of Wundt's psychological writings. Steffens writes that the student

> must have thought . . . that Wundt might have been reluctant to crown a discovery which would require the old philosopher [Wundt] to rewrite volumes of his lifework. The budding psychologist solved the ethical problem before him by deciding to alter his results, and his papers showed how he did this, by changing the figures item by item, experiment by experiment, so as to make the curve of his averages come out for instead of against our school. After a few minutes of silent admiration of the mathematical feat performed on the papers before us, we buried sadly these remains of a great sacrifice to loyalty, to the school spirit, and to practical ethics.
>
> (p. 151)

PART 2

Quantitative research methods

CHAPTER 9

The basic laboratory experiment

Overview

- The laboratory experiment has a key role in psychological research in that it allows the investigation of causal relationships between variables. In other words, it identifies whether one variable affects another in a cause and effect sequence.
- Essentially an experiment involves systematically varying the level of the variable that is thought to be causal then measuring the effect of this variation on the measured variable while holding all other variables constant.
- The simplest experimental design used by psychologists consists of two conditions. One condition has a higher level of the manipulated variable than the other condition. The former condition is sometimes known as the experimental condition while the condition having the lower level is known as the control condition. The two experimental conditions may also be referred to as the independent variable. The researcher assesses whether the scores on the measured variable differ between the two conditions. The measured variable is often referred to as the dependent variable.
- If the size of the effect differs significantly between the two conditions and all variables other than the manipulated variable have been held constant, then this difference is most likely due to the manipulated variable.
- There are a number of ways by which the researcher tries to hold all of the variables constant other than the independent and dependent variables. These include: randomly assigning participants to the different conditions (which ensures equality in the long run), carrying out the study in the controlled setting of a laboratory where hopefully other factors are constant, and making the conditions as similar as possible for all participants except as far as they are in the experimental or control condition.

- In *between-subjects designs* participants are randomly assigned to just *one* of the conditions of the study. In *within-subjects designs* the same participants carry out all conditions. Rather than being randomly assigned to just one condition, they are randomly assigned to the different orders in which the conditions are to be run.
- Random assignment of participants only guarantees that in the long run the participants in all conditions start off similar in all regards. For any individual study, random assignment cannot ensure equality. Consequently, some experimental designs use a prior measure of the dependent variable (the measured variable) which can be used to assess how effective the random assignment has been. The experimental and control groups should have similar (ideally identical) mean scores on this pre-measure. This prior measure is known as a pre-test while the measurement after the experimental manipulation is known as a post-test. The difference between the pre-test and post-test provides an indication of the change in the measured variable.
- A number of disadvantages of the basic laboratory experiment should be recognised. The artificiality of the laboratory experiment is obvious so it is always possible that the experimental manipulation and the setting fail to reflect what happens in more natural or realistic research settings. Furthermore, the number of variables that can be manipulated in a single experiment is limited, which can be frustrating when one is studying a complex psychological process.

9.1 Introduction

When used appropriately, the randomised laboratory experiment is one of the most powerful tools available to researchers. This does not mean that it is always or even often the ideal research method. It simply means that the laboratory is an appropriate environment for studying many psychological processes – particularly physiological, sensory or cognitive processes. The use of the laboratory to study social processes, for example, is not greeted with universal enthusiasm. Nevertheless, many studies in psychology take place in a research laboratory using true or randomised experimental designs. Any psychologist, even if they never carry out a laboratory experiment in their professional career, needs to understand the basics of laboratory research. Otherwise a great deal of psychological research will pass over their heads.

It is essential to be able to differentiate between two major sorts of research designs (see Figure 9.1):

- Experiments in which different participants take part in different conditions are known variously as between-subjects, between-participants, independent-groups, unrelated-groups or uncorrelated-groups designs. A diagram of a simple between-subjects design with only two conditions is shown in Figure 9.2a.
- Designs in which the same participants take part in all (or sometimes some) of the various conditions are called within-subjects, within-participants, repeated-measures, dependent-groups, related-groups or correlated-groups designs. A diagram of a simple within-subjects design with only two conditions is presented in Figure 9.2b.

An example of a between-subjects design would be a study that compared the number of errors made entering data into a computer spreadsheet for a sample of people listening

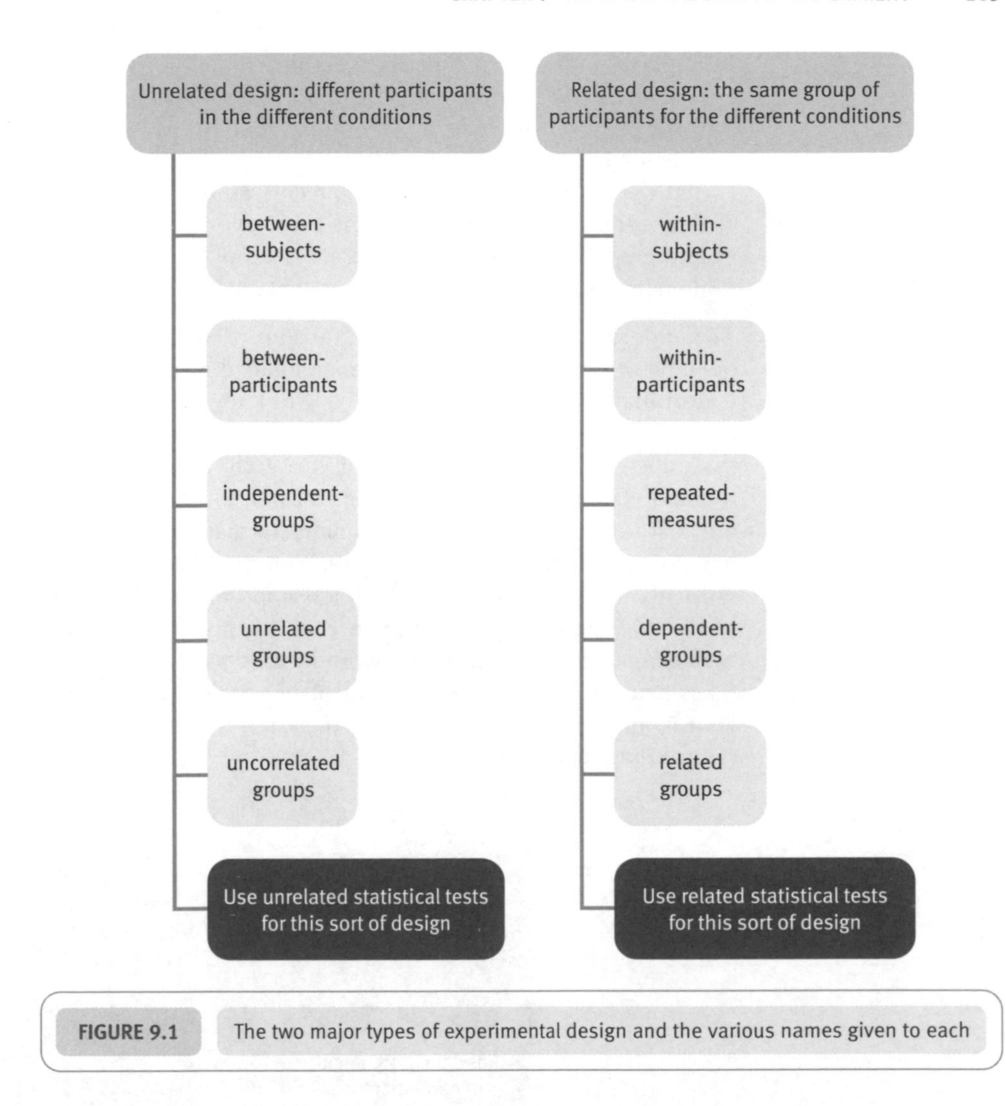
Unrelated design: different participants in the different conditions
between-subjects
between-participants
independent-groups
unrelated groups
uncorrelated groups
Use unrelated statistical tests for this sort of design
Related design: the same group of participants for the different conditions
within-subjects
within-participants
repeated-measures
dependent-groups
related groups
Use related statistical tests for this sort of design

FIGURE 9.1 The two major types of experimental design and the various names given to each

to loud popular music with the number of errors made by a different control sample listening to white noise at the same volume. That is to say, two different groups of people are compared. An example of a within-subjects design would be a study of the number of keyboard errors made by a group of 20 secretaries, comparing the number of errors when music is being played with when music is not being played. That is to say, the performance of one group of people is compared in two different circumstances.

One of the reasons why it is important to distinguish between these two broad types of design is that they use rather different methods of analysis. For example, they use different statistical tests. The first design which uses different groups of participants for each condition would require an unrelated statistical test (such as the unrelated or uncorrelated *t*-test). The latter design in which the same group of participants take part in every condition of the experiment would require a related statistical test (such as the related or correlated *t*-test).

Although it is perfectly feasible to do experiments in a wide variety of settings, a custom-designed research laboratory is usually preferred. There are two main reasons for using a research laboratory:

- *Practicalities* A study may require the use of equipment or apparatus that may be too bulky or too heavy to move elsewhere, or needs to be kept secure because it may be expensive or inconvenient to replace.

- *Experimental control* In an experiment it is important to try to keep all factors constant other than the variable or variables that are manipulated. This is obviously easier to do in a room or laboratory that is custom designed for the task and in which all participants take part in the study. The lighting, the temperature, the noise and the arrangement of any equipment can all be kept constant. In addition, other distractions such as people walking through or talking can be excluded. You will often find great detail about the physical set up of the laboratory where the research was done in some reports of laboratory experiments.

The importance of holding these extraneous or environmental factors constant depends on how strong the effect of the manipulated variable is on the measured variable. Unfortunately, the researcher is unlikely to know in advance what their effect is in some cases. To the extent that the extraneous variable seems not to, or has been shown not to, influence the key variables in the research, one may consider moving the research to a more appropriate or convenient research location. For example, if you are carrying out a study in a school, then it is unlikely that that school will have a purpose-built psychology laboratory for you to use. You may find yourself in a small room which is normally used for other purposes. If it is important that you control the kind or the level of the noise in the room, then you could be able to do this by playing what is called 'white noise' through earphones worn by each participant. If it is essential that the study takes place in a more carefully controlled setting, then the pupils will need to come to your laboratory. The essential point to realise is that the setting of the study may be less critical than its design.

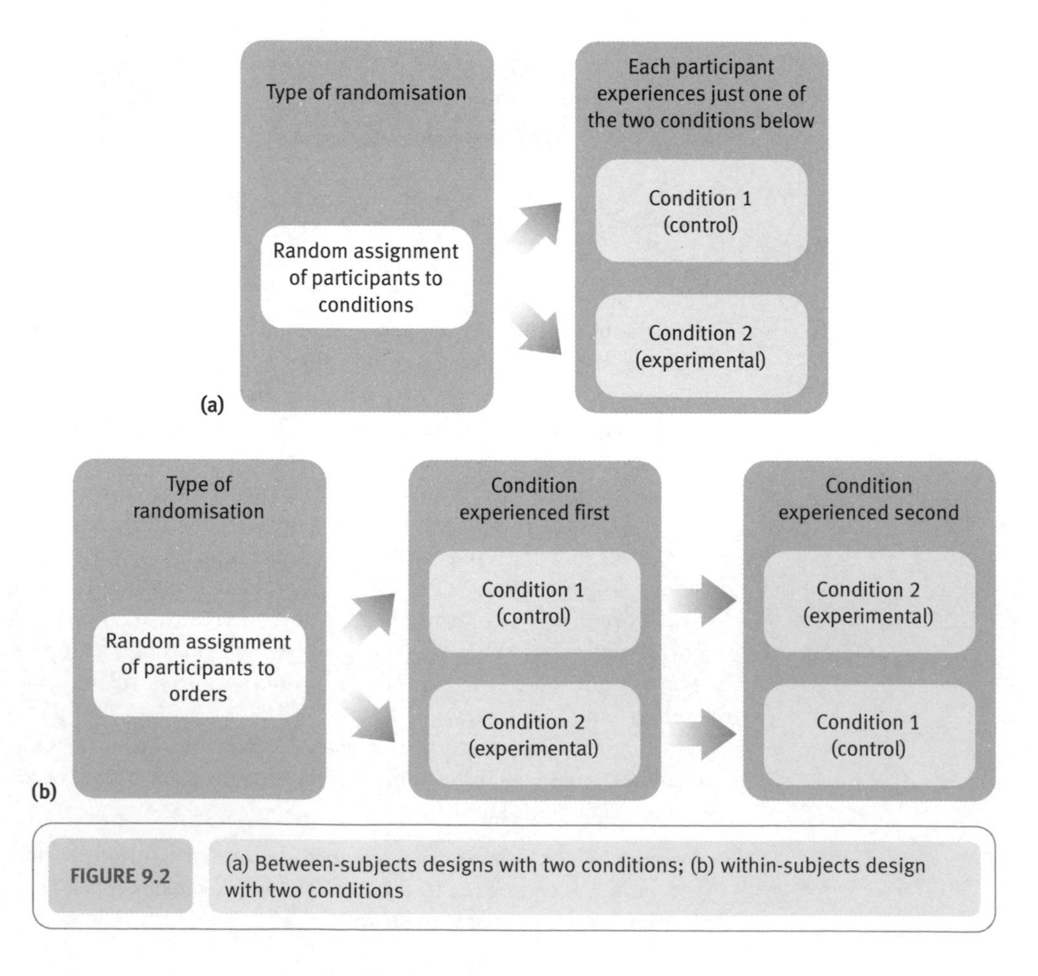

FIGURE 9.2 (a) Between-subjects designs with two conditions; (b) within-subjects design with two conditions

It should be stressed that while researchers may aim for perfection, most research is to a degree a compromise between a number of considerations. The perfect research study has probably never been designed and is probably an oxymoron – that should not stop you trying for the best possible, but it should be remembered when your research seems not to be able to reach the highest standard.

Because of the important role it plays in psychological research, we introduced the concept of a true experiment in Chapter 1 on the role and nature of research in psychology. As we saw, studies employing a true or randomised experimental design were the most common kind of study published in a random selection of psychology journals in 1999 (Bodner, 2006). They constituted 41 per cent of all the studies sampled in that year. The proportion of studies using a true or randomised experimental design was higher in areas such as learning, perception, cognition and memory, which have traditionally been called experimental psychology because of the use of this design.

9.2 Characteristics of the true or randomised experiment

The most basic laboratory experiment is easily demonstrated. Decide on an experimental and control group and allocate participants to one or other on the basis of a coin toss. Put the experimental group through a slightly different procedure from that of the control group. This is known as the experimental manipulation and corresponds to a variable which the researcher believes might affect responses on another variable called the dependent variable. After the data have been collected, the researcher examines the average score for the two conditions to see whether or not there is a substantial difference between them. For many purposes, such a simple design will work well.

In order to be able to run experiments satisfactorily one needs to understand the three essential characteristics of the true or randomised experiment – and more (see Figure 9.3):

- Experimental manipulation.
- Standardisation of procedures – that is the control of all variables other than the independent variable.
- Random assignment to conditions or order.

We will consider each of these in turn.

Experimental manipulation

Only the variable that is assumed to cause or affect another variable is manipulated (varied). This manipulated variable is often referred to as the *independent variable* because it is assumed to be varied independently of any other variable. If it is not manipulated independently, then any effect that we observe may be due to those other variables. An example of an independent variable is alcohol. If we think that alcohol increases the number of mistakes we make and we are interested in seeing whether this is the case, alcohol (or more strictly speaking, the level or amount of alcohol) is the independent variable which we manipulate. The variable that is presumed to be affected by the independent or manipulated variable is called the *dependent variable* because it is thought to be dependent on the independent variable. It is the variable that we measure. In this example, the dependent variable is the number of mistakes made in a task such as walking along a straight line.

Experimental manipulation	Standardisation of procedures	Randomisation
In an experiment, the experimental manipulation essentially 'creates' the independent variable	Ideally the only differences between the experimental and control conditions should be the experimental manipulation	Randomisation is primarily used for allocating participants to the experimental and control groups in unrelated designs
In an experiment, the experimental manipulation is believed likely to have an effect on the dependent variable	The way in which the experimenter interacts with the participants should be standardised – some have gone to the extent of playing recordings of the instructions	Randomisation is in terms of the order that the experimental conditions are run through in related designs
The manipulation must have two different levels (versions) or more	Random allocation to the sequence helps avoid inadvertently running one group sooner than the others. Time of day effects may be the consequence of not randomising	Participants can sometimes be matched in 'pairs' and then randomly allocated to the two conditions as a pair
It is easy to inadvertently manipulate more than one thing and care should be taken to avoid this	Failure to standardise procedures increases the amount of uncontrolled variation – your analysis is less likely to be statistically significant	If matching is used, then the design becomes a related one. As a consequence statistical tests for related designs should be used
Checks on the effectiveness of the experimental manipulation should be carried out – and questions asked at debriefing	Systematic but unintended differences in the way that the experimental and control groups are treated can lead to erroneous conclusions	However, matching should be done on variables which are related to the dependent variable. Otherwise it has no effect

FIGURE 9.3 Essential features in a simple experimental design summarised

In the most basic true or randomised experiment, we would only have two *conditions* (also known as *levels of treatment* or *groups*). In one condition, a lower amount of alcohol would be given to the participant while in the other condition the participant receives a higher level of alcohol. The amount of alcohol given would be standard for all participants in a condition. So, in condition 1 the amount might be standardised as 8 millilitres (ml). This is about the amount of alcohol in a glass of wine. In the second condition the amount may be doubled to 16 ml. If the size of the effect varies directly with the amount of alcohol given, then the more the two groups differ the bigger the effect.

Why did we choose to give both groups alcohol? After all, we could have given the experimental group alcohol but the control group no alcohol at all. Both groups are given alcohol in the hope that by doing so participants in both groups are aware that they have consumed alcohol. Participants receiving alcohol probably realise that they have been given alcohol. Unless the quantity of alcohol was very small, participants are likely to detect it. So two things have happened – they have been given alcohol and they are also aware that they have been given alcohol. However, if one group received no alcohol then, unless we deliberately misled them, members of this group will not believe that they have taken alcohol. So we would not know whether the alcohol or the belief that they had been given alcohol was the key causal variable. Since the effects of alcohol are well known, participants believing that they have taken alcohol may behave accordingly. By giving both groups alcohol, both groups will believe that they have taken alcohol. The only thing that varies is the key variable of the amount of alcohol taken. In good experimental research the effectiveness of the experimental manipulation is often evaluated. This is discussed in Box 9.1. In this case, participants in the experiment might be asked about whether or not they believed that they had taken alcohol in a debriefing interview at the end of the study.

The condition having the lower quantity of alcohol is referred to as the *control condition*. The condition having the higher quantity of alcohol may be called the *experimental condition*. The purpose of the control condition is to see how participants behave when they receive less of the variable that is being manipulated.

Box 9.1 Key Ideas

Checks on the experimental manipulation

It can be a grave mistake to assume that simply because an experimental manipulation has been introduced by the researcher that the independent *variable* has actually been effectively manipulated. It might be argued that if the researcher finds a difference between the experimental and control conditions on the dependent variable that the manipulation must have been effective. Things are not that simple.

Assume that we are investigating the effects of anger on memory. In order to manipulate anger, the researcher deliberately says certain pre-*scripted* offensive comments to the participants in the experimental group whereas nice things are said to the participants in the control group. It is very presumptuous to assume that this procedure will work effectively without subjecting it to some test.

For example, the participants might well in some circumstances regard the offensive comments as a joke rather than an insult so the manipulation may make them happier rather than angrier. Alternatively, the control may find the nice comments of the experimenter to be patronising and become somewhat annoyed or angry as a consequence. So there is a degree of uncertainty whether or not the experimental manipulation has actually worked. One relatively simple thing to do in this case would be to get participants to complete a questionnaire about their mood containing a variety of emotions, such as angry, happy and sad, which the participant rates in terms of their own feelings. In this way it would be possible to assess whether the experimental group was indeed angrier than the control group following the anger manipulation. Alternatively, at the debriefing session following participation in the experiment, the participants could be asked about how they felt after the experimenter said the offensive or nice things. This check would also demonstrate that the manipulation had had a measurable effect on the participants' anger levels.

Sometimes it is appropriate, as part of pilot work trying out one's procedures prior to the study proper, to establish the effectiveness of the experimental manipulation as a distinct step in its own right. Researchers need to be careful not to assume that simply because they obtain statistically significant differences between the experimental and control conditions that this is evidence of the effectiveness of their experimental manipulation. If the experimental manipulation has had an effect on the participants but not the one intended, it is vital that the researcher knows this. Otherwise, the conceptual basis for their analysis may be inappropriate. For example, they may be discussing the effects of anger when they should be discussing the effects of happiness.

In our experience, checks on the experimental manipulation are relatively rare in published research and are, probably, even rarer in student research. Yet such checks would seem to be essential. As we have seen, the debriefing session can be an ideal opportunity to interview participants about this aspect of the study along with its other features. The most thorough researchers may also consider a more objective demonstration of the effectiveness of the manipulation as above when the participants' mood was assessed.

In theory, if not always in practice, the experimental and control conditions should be identical in every way but for the variable being manipulated. It is easy to overlook differences. So, in our alcohol experiment participants in both groups should be given the same quantity of liquid to drink. But it is easy to overlook this if the low alcohol group are, say, given one glass of wine and the high alcohol group two glasses of wine. If the participants in the control condition are not given the same amount to drink, then alcohol is not being manipulated independently of all other factors. Participants in the experimental condition would have been given more to drink while participants in the control condition would have been given less to drink. If we find that reaction time was slower in the experimental condition than in the control condition, then we could not be certain whether this difference was due to the alcohol or the amount of liquid drunk. This may seem very pernickety in this case but dehydration through having less liquid may have an effect on behaviour. The point of the laboratory experiment is that we try to control as many factors as possible but, as we have seen, this is not as easy as it sounds. Of course, variations in the volume of liquid drunk could be introduced into the research design in order to discover what the effect of varying volume is on errors.

Standardisation of procedures

A second essential characteristic of the true experiment is implicit in the previous characteristic. That is, all factors should be held constant apart from the variable(s) being investigated. This is largely achieved by standardising all aspects of the procedures employed. Only the experimental manipulation should vary. We have already seen the importance of this form of control when we stressed in our alcohol study that the two conditions should be identical apart from the amount of alcohol taken. So participants in both conditions were made aware that they will be given alcohol and that they are given the same amount of liquid to drink.

There are other factors which we should try to hold constant which are not so obvious. The time of day the study is carried out, the body weight of the participants, the amount of time that has lapsed since they have last eaten and so forth are all good examples of this. Such standardisation is not always easy to achieve. For example, what about variations in the behaviour of the experimenter during an experiment? If the experimenter's behaviour differs systematically between the two groups then experimenter behaviour and the effects of the independent variable will be confounded. We may confuse the variability in the behaviour of the experimenter with the effects of different quantities of alcohol. There have been experiments in which the procedure is automated so that there is no experimenter present in the laboratory. For example, tape-recorded instructions to the participants are played through loudspeakers in the laboratory. In this way, the instructions can be presented identically in every case and variations in the experimenter's behaviour eliminated.

Standardisation of procedures is easier said than done but remains an ideal in the laboratory experiment. It is usual for details such as the instructions to participants to be written out as a guide for the experimenter when running the experiment. One of the difficulties is that standardisation has to be considered in relation to the tasks being carried out by participants, so it is impossible to give advice that would apply in every case. Because of the difficulty in standardising all aspects of the experiment, it is desirable to randomly order the running of the experimental and control conditions. For example, we know that cognitive functions vary according to the time of day. It is very difficult to standardise the time of day that an experiment is run. Hence it is desirable to decide randomly which condition the participant who arrives at 2.30 p.m. will be in, which condition the participant who arrives next will be in, and so forth. In this way there will be no systematic bias for one of the conditions of the experiment to be run at different times of the day from the other conditions.

Random assignment

Random assignment is the third essential feature of an experiment. There are two main procedures according to the type of experimental design:

- Participants are put in the experimental or control condition at random using a proper randomisation procedure (which may be simply the toss of a coin – heads the experimental group, tails the control group). There are other methods of randomisation as we will see. Random assignment to conditions is used when the participants only take part in one condition.
- Alternatively, if participants undertake more than one condition (in the simplest case both the experimental and control conditions), then they are randomly assigned to the different orders of those two conditions. With just two conditions there are just two possible orders – experimental condition first followed by control condition second, or control condition first followed by experimental condition second. Of course, with three or more conditions there is a rapidly increasing numbers of possible orders.

It is easy to get confused about the term 'random'. Random assignment requires the use of a proper random procedure. This is not the same thing at all as a haphazard or casual choice. By random we mean that each possible outcome has an equal chance of being selected. We do not want a selection process which systematically favours one outcome rather than another. (For example, the toss of a coin is normally a random process but it would not be if the coin had been doctored in some way so that it lands heads up most of the time.) There are a number of random procedures which may be employed. We have already mentioned the toss of a coin but will include it in the list of possibilities again as it is a good and simple procedure:

- With two conditions or orders you can toss a coin where the participant will be assigned to one of them if the coin lands 'heads' up and to the other if it lands 'tails' up.
- Similarly, especially if there are more than two conditions, you could throw a die.
- You could write the two conditions or orders on two separate index cards or slips of paper, shuffle them without seeing them and select one of them.
- You could use random number tables where, say, even numbers represent one condition or order and odd numbers represent the other one.
- Sometimes a computer can be used to generate a sequence of random numbers. Again you could use an odd number for the experimental group and an even number for allocating the participant to the control group (or vice versa).

One can either go through one of these randomisation procedures for each successive participant or you can draw up a list in advance for the entire experiment. However, there are two things that you need to consider:

- You may find that you have 'runs' of the same condition such as six participants in sequence all of which are in, say, the control group. If you get runs like this you may find, for example, that you are testing one condition more often at a particular time of day. That is, despite the randomisation, the two conditions are not similar in all respects. For example, these six participants may all be tested in the morning rather than spread throughout the day.
- Alternatively, you may also find that the number of participants in the two conditions or orders is very different. There is even a remote possibility that all your participants are assigned to one condition or order.

Randomisation only equates things in the long run. In the short run, it merely guarantees that there is no systematic bias in the selection. In the short term, chance factors may nevertheless lead to differences between the conditions.

There is no need to go into technical details, but if it is possible to have equal numbers in each condition of an experimental design then you should try to do so. Most statistical tests work optimally in these circumstances. If equal numbers are impossible then so be it. However, there are ways in which one can ensure that there are equal numbers of participants in each condition. For example, one can employ *matched* or *block randomisation*. That is, the first participant of every pair of participants is assigned at random using a specified procedure while the second participant is assigned to the remaining condition or order. So, if the first participant has been randomly assigned to the control condition, the second participant will be allocated to the experimental condition. If you do this, you will end up with equal numbers of participants in the two conditions or orders if you have an equal number of participants. Box 9.2 discusses how you can pair off participants to ensure that the different groups have similar characteristics.

In the between-subjects design (in which participants serve in just one condition of the experiment), any differences between the participants are usually controlled by random assignment. The prime purpose of this is to avoid systematic biases in the allocation of participants to one or other condition. If the experimenter merely decided on the spot which group a participant should be in then all sorts of 'subconscious' factors may influence this choice and perhaps influence the outcome of the experiment as a consequence. For example, without randomisation it is possible that the researcher allocates males to the experimental group and females to the control group – and does not even notice what they have done. If there is a gender difference on the dependent variable, the results of the experiment may confuse the experimental effect with the bias in participant selection.

Box 9.2 Key Ideas

Matching

One way of ensuring that the participants in the experimental and control group are similar on variables which might be expected to affect the outcome of the study is to use matching. Participants in an experiment will vary in many ways so there may be occasions when you want to ensure that there is a degree of consistency. For instance, some participants may be older than others unless we ensure that they are all the same age. However, it is difficult, if not impossible, to control for all possible individual differences. For example, some participants may have had less sleep than others the night before or gone without breakfast. Some might have more familiarity with the type of task to be carried out than others and so on.

We could try to hold all these factors constant by making sure, for example, that all participants were female, aged 18, weighed 12 stone (76 kilograms), had slept 7 hours the night before and so on. But this is far from easy. It is generally much more practicable to use random assignment. To simplify this illustration, we will think of all these variables as being dichotomous or only having two categories such as female/male, older/younger, heavier/lighter and so on. If you look at Table 9.1, you will see that we have arranged our participants in order and that they fall into sets of individuals who have the same pattern of characteristics on these three variables. For example, the first three individuals are all female, older and heavier. This is a matched set of individuals. We could choose one of these three at random to be in the experimental condition and another at random to be in the control condition. The third individual would not be matched with anyone else so they cannot be used in our matched study in this case. We could then move on to the next set of matched participants and select one of them at random for the experimental condition and a second for the control condition.

You might like to try this with the rest of the cases in Table 9.1 which consists of information about the gender, the age and the weight of 24 people who we are going to randomly assign to two groups.

Matching is a useful tool in some circumstances. There are a few things that have to be remembered if you use matching as part of your research design:

- The appropriate statistical tests are those for related data. So a test like the related *t*-test or the Wilcoxon matched pairs test would be appropriate.
- Variables which correlate with both the independent and dependent variables are needed for the matching variables. If a variable is unrelated to either one or both of the independent or dependent variables then there is no point in using it as a matching variable. It could make no difference to the outcome of the study.
- The most appropriate variable to match on is most probably the dependent variable measured at the start of the study. This is not unrelated to the idea of pre-testing though in pre-testing participants have already been allocated to the experimental and control conditions. But pre-testing, you've guessed it, also has its problems.

Table 9.1 Gender, age and weight details for 24 participants

Number	Gender	Age	Weight
1	female	older	heavier
2	female	older	heavier
3	female	older	heavier
4	female	older	lighter
5	female	older	lighter
6	female	older	lighter
7	female	younger	heavier
8	female	younger	heavier
9	female	younger	heavier
10	female	younger	lighter
11	female	younger	lighter
12	female	younger	lighter
13	male	older	heavier
14	male	older	heavier
15	male	older	heavier
16	male	older	lighter
17	male	older	lighter
18	male	older	lighter
19	male	younger	heavier
20	male	younger	heavier
21	male	younger	heavier
22	male	younger	lighter
23	male	younger	lighter
24	male	younger	lighter

9.3 More advanced research designs

We have stressed that there is no such thing as a perfect research design that can be used irrespective of the research question and circumstances. If there were such a thing then not only would this book be rather short but research would probably rank in the top three most boring jobs in the world. Research is intellectually challenging because it is problematic. The best research that any of us can do is probably a balance between a wide range of different considerations. In this chapter we are essentially looking at the simplest laboratory experiment in which we have a single independent variable. But even this basic experimental design gathers levels of complexity as we try to plug the holes in the simple design. The simplest design, as we are beginning to see, has problems. One of these problems is that if a single study is to be relied on, then the more that we can be certain that the experimental and control conditions are similar prior to the experimental manipulation the better. The answer is obvious: assess the two groups prior to the experimental manipulation to see whether they are similar on the dependent variable. This is a good move but, as we will see, it brings with it further problems to solve. It should be stressed that none of what you are about to read reduces the importance of using random allocation procedures for participants in experimental studies.

Pre-test and post-test sensitisation effects

The pre-test is a way of checking whether random assignment has, in fact, equated the experimental and control groups prior to the experimental manipulation. It is crucial that the two groups are similar on the dependent variable prior to the experimental manipulation. Otherwise it is not possible to know whether the differences following the experimental manipulation are due to the experimental manipulation or to pre-existing differences between the groups on the dependent variable.

The number of mistakes is the dependent variable in our alcohol-effects example. If members of one group make more mistakes than do members of the other group before drinking alcohol, then they are likely to make more mistakes after drinking alcohol. For example, if the participants in the 8 ml alcohol condition have a tendency to make more errors regardless of whether or not they have had any alcohol, then they may make more mistakes after drinking 8 ml of alcohol than the participants who have drunk 16 ml.

This situation is illustrated in Figure 9.4. In this graph the vertical axis represents the number of mistakes made. On the horizontal axis are two marks which indicate participants' performance before drinking alcohol and after drinking alcohol. The measurement of the participants' performance before receiving the manipulation is usually called the pre-test

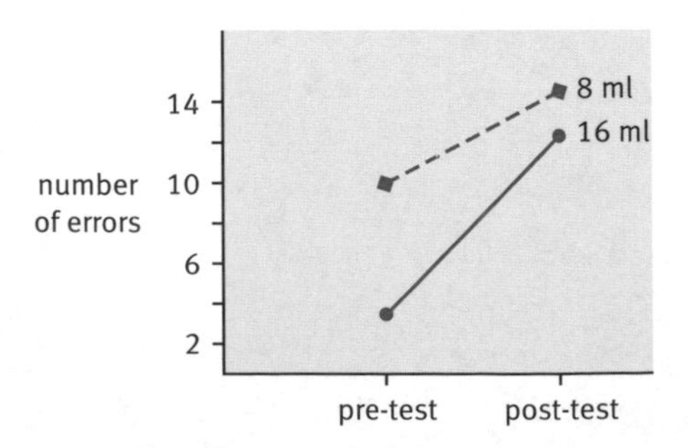

FIGURE 9.4 Performance differences before the manipulation

and the measurement after receiving the manipulation the post-test. The results of the post-test are usually placed after those of the pre-test in graphs and tables as time is usually depicted as travelling from left to right.

Without the pre-test measure, there is only the measure of performance after drinking alcohol. Just looking at these post-test measures, people who drank 8 ml of alcohol made more mistakes than those who drank 16 ml. In other words drinking more alcohol seems to have resulted in making fewer mistakes (and not more mistakes as we might have anticipated). This interpretation is incorrect since, by chance, random assignment to conditions resulted in the participants in the 8 ml condition being those who tend to make more mistakes. Without the pre-test we cannot know this, however.

It is clearer to see what is going on if we calculate the difference between the number of mistakes made at pre-test and at post-test (simply by subtracting one from the other). Now it can be seen that the increase in the number of mistakes was greater for the 16 ml condition (12 – 4 = 8) than for the 8 ml condition (14 – 10 = 4). In other words, the increase in the number of mistakes made was greater for those drinking more alcohol.

We can illustrate the situation summarised in Figure 9.4 with the fictitious raw data in Table 9.2 where there are three participants in each of the two conditions. Each participant is represented by the letter P with a subscript from 1 to 6 to indicate the six different participants. There are two scores for each participant – the first for the pre-test and the second for the post-test. These data could be analysed in a number of different ways. Among the better of these would be the mixed-design analysis of variance. This statistical test is described in some introductory statistics texts such as the companion book *Introduction to Statistics in Psychology* (Howitt and Cramer, 2011a). However, this requires more than a basic level of statistical sophistication. Essentially, though, you would be looking for an interaction effect. A simpler way of analysing the same data would be to compare the differences between the pre-test and post-test measures for the two conditions. An unrelated *t*-test would be suitable for this.

Experimental designs which include a pre-test are referred to as a pre-test–post-test design while those without a pre-test are called a post-test-only design. There are two main advantages of having a pre-test:

- As we have already seen, it enables us to determine whether randomisation has worked.

Table 9.2 Fictitious data for a pre-test–post-test two-group design

		Pre-test	Post-test
Condition 1	P_1	9	13
	P_2	10	15
	P_3	11	14
Sum		**30**	**42**
Mean		**30/3 = 10**	**42/3 = 14**
Condition 2	P_4	3	12
	P_5	4	11
	P_6	5	13
Sum		**12**	**36**
Mean		**12/3 = 4**	**36/3 = 12**

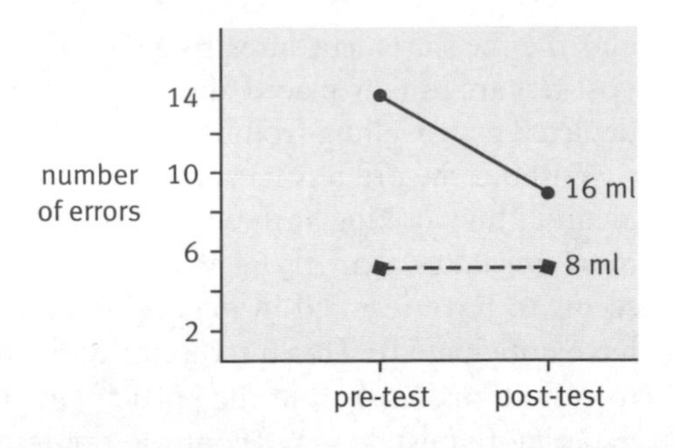

FIGURE 9.5 Change in performance between pre-test and post-test

- It allows us to determine whether or not there has been a change in performance between pre-test and post-test. If we just have the post-test scores, we cannot tell whether there has been a change in those scores and what that change is. For example, the post-test scores may show a decline from the pre-test. Without the pre-test, we may suggest incorrectly that the independent variable is increasing the scores on the dependent variable.

Look at the data shown in the graph in Figure 9.5. Concentrate on the post-test scores and ignore the pre-test. That is, pretend that we have a post-test-only design for the moment. Participants who had drunk 16 ml of alcohol made more errors than those who had drunk 8 ml. From these results we may conclude that drinking more alcohol increases the number of mistakes made. If the pre-test number of errors made were as shown in Figure 9.5, this interpretation would be incorrect. If we know the pre-test scores we can see that drinking 16 ml of alcohol decreased the number of errors made ($10 - 14 = -4$) while drinking 8 ml of alcohol had no effect on the number of errors ($6 - 6 = 0$). Having a pre-test enables us to determine whether or not randomisation has been successful and what, if any, was the change in the scores. (Indeed, we are not being precise if we talk of the conditions in a post-test-only study as increasing or decreasing scores on the dependent variable. All that we can legitimately say is that there is a difference between the conditions.)

Whatever their advantages, pre-tests have disadvantages. One common criticism of pre-test designs is that they may alert participants as to the purpose of the experiment and consequently influence their behaviour. That is, the pre-test affects or sensitises participants in terms of their behaviour on the post-test (Lana, 1969; Solomon, 1949; Wilson and Putnam, 1982). Again we might extend our basic research design to take this into account. We need to add to our basic design groups which undergo the pre-test and other groups which do not. Solomon (1949) called this a four-group design since at a minimum there will be two groups (an experimental and control group) that include a pre-test and two further groups that do not have a pre-test as shown in Figure 9.6.

One way of analysing the results of this more sophisticated design is to tabulate the data as illustrated in Table 9.3. This contains fictitious post-test scores for three participants in each of the four conditions. The pre-test scores are *not* given in Table 9.3. Each participant is represented by the letter P with a subscript consisting of a number ranging from 1 to 12 to denote there are 12 participants.

The analysis of these data involves combining the data over the two conditions. That is, we have a group of six cases which had a pre-test and another group of six cases which did not have the pre-test. The mean score of the group which had a pre-test is 8 whereas the mean score of the group which had no pre-test is 2. In other words we are ignoring the effect of the two conditions at this stage. We have a pre-test sensitisation effect

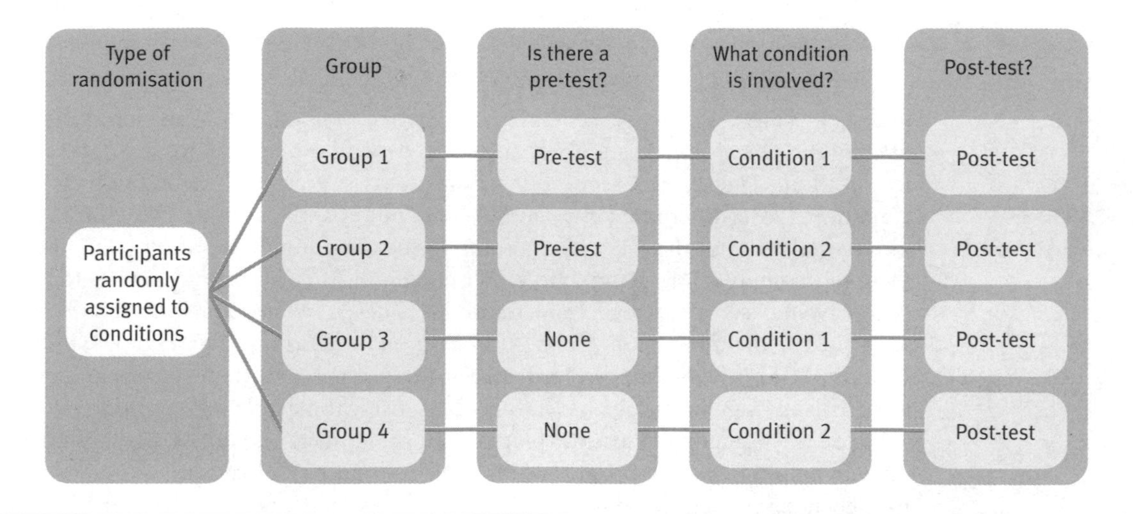

FIGURE 9.6 Solomon's (1949) four-group design

Table 9.3 Fictitious post-test scores for a Solomon four-group design

	Had pre-test	Had no pre-test	Row means
Condition 1	$P_1 = 4$	$P_7 = 0$	
	$P_2 = 5$	$P_8 = 1$	
	$P_3 = 6$	$P_9 = 2$	
Sum	**15**	**3**	**18**
Mean	**15/3 = 5**	**3/3 = 1**	**18/6 = 3**
Condition 2	$P_4 = 10$	$P_{10} = 2$	
	$P_5 = 11$	$P_{11} = 3$	
	$P_6 = 12$	$P_{12} = 4$	
Sum	**33**	**9**	**42**
Mean	**33/3 = 11**	**9/3 = 3**	**42/6 = 7**
Column sums	**48**	**12**	
Column means	**48/6 = 8**	**12/6 = 2**	

if the means for these two (combined) conditions differ significantly. In our example, there may be a pre-test sensitisation effect since the mean score of the combined two conditions with the pre-test is 8 which is higher than the mean score of 2 for the two conditions without the pre-test combined. If this difference is statistically significant we have a pre-test sensitisation effect. (The difference in the two means could be tested using an unrelated *t*-test. Alternatively, one could use a two-way analysis of variance. In this case we would look for a pre-test–no pre-test main effect.)

Of course, it is possible that the pre-test sensitisation effect is different for the experimental and control conditions (conditions 1 and 2):

- For condition 1 we can see in Table 9.3 that the difference in the mean score for the group with the pre-test and the group without the pre-test is 5 − 1 = 4.

- For condition 2 we can see that the difference in the mean score for the group with the pre-test and the group without the pre-test is 11 – 3 = 8.

In other words, the mean scores of the two conditions with the pre-test and the two conditions without the pre-test appear to depend on, or interact with, the condition in question. The effect of pre-test sensitisation is greater for condition 1 than for condition 2. The difference between the two in our example is quite small, however. This differential effect of the pre-test according to the condition in question would be termed a pre-test/condition interaction effect. (We could test for such an interaction effect using a two-way analysis of variance. How to do this is described in some introductory statistics texts such as the companion volume *Introduction to Statistics in Psychology*, Howitt and Cramer, 2011a. If the interaction between the pre-test and the experimental condition is statistically significant, we have a pre-test sensitisation interaction effect.)

If you are beginning to lose the plot in a sea of numbers and tables, perhaps the following will help. Pre-test sensitisation simply means that participants who are pre-tested on the dependent variable tend to have different scores on the post-test from the participants who were not pre-tested. There are many reasons for this. For example, the pre-test may simply coach the participants in the task in question. However, the pre-test/condition interaction means that the effect of pre-testing is different for the experimental and the control conditions. Again there may be many reasons why the effects of pre-testing will differ for the experimental group. For example, participants in the experimental group may have many more clues as to what the experimenter is expecting to happen. As a consequence, they may change their behaviour more in the experimental condition than in the control condition.

Pre-test sensitisation in itself may not be a problem whereas if it interacts with the condition to produce different outcomes it is problematic:

- A pre-test sensitisation interaction effect causes problems in interpreting the results of a study. We simply do not know with certainty if the effect is different in the different conditions. Further investigation would be needed to shed additional light on the matter. If we are interested in understanding this differential effect we need to investigate it further to find out why it has occurred.

- A pre-test sensitisation effect without a pre-test sensitisation interaction effect would not be a problem if we are simply interested in the relative effect of an independent variable and not its absolute effect. For example, it would not be a problem if we just wanted to know whether drinking a greater amount of alcohol leads to making more errors than drinking a smaller amount. The size of the difference would be similar with a pre-test to without one. On the other hand, we might be interested in the absolute number of errors made by drinking alcohol. For example, we may want to recommend the maximum amount of alcohol that can be taken without affecting performance as a driver. In these circumstances it is important to know about pre-test sensitisation effects if these result in greater errors. In this case we would base our recommendation on the testing condition which resulted in the greater number of errors.

If one wishes to use a pre-test but nevertheless reduce pre-test sensitisation effects to a minimum then there are techniques that could be used:

- Try to disguise the pre-test by embedding it in some other task or carrying it out in a different context.

- Increase the length of the interval between the pre-test and the manipulation so that the pre-test is less likely to have an effect on the post-test. So if the pre-test serves as a practice for the post-test measure, then a big interval of time may result in a reduced practice effect.

- If the effects of the manipulation were relatively short-lived we could give the 'pre-test' after the post-test. For example, if we were studying the effects of alcohol on errors then we could test the participants a couple of hours later when the effects of alcohol would have worn away. The two groups could be tested to see if they made similar numbers of errors once the effects of alcohol had dissipated.

While there are many studies which use a pre-test and a post-test measure fruitfully, the same is not true of the Solomon four-groups design. Such studies are scarce. That is, while it is important to be aware of pre-test sensitisation effects, we know of very few published studies which have tested for pre-test sensitisation effects.

Within-subjects design

Where the same participants take part in all conditions, this effectively controls for many differences between participants. For example, we may have a participant who makes numerous errors irrespective of condition. Because this person is in every condition of the experiment, the pre-existing tendency for them to make a lot of errors will apply equally to every condition of the experiment. In other words, they would make more errors in every condition. The effects, say, of alcohol will simply change the number of errors they make differentially. The advantage of the within-subjects design is that it provides a more sensitive test of the difference between conditions because it controls for differences between individuals. Having a more sensitive test and having the same participants take part in all conditions means that, ideally, fewer participants can be used in a within-subjects than in a between-subjects design.

The extent to which this is the case depends on the extent to which there is a correlation between the scores in the experimental and control conditions. Many of the statistical tests appropriate for within-subjects designs will give an indication of this correlation as well as a test for the significance of the difference between the two conditions. This is discussed in more detail in Chapter 12 of our companion statistics text, Introduction to Statistics in Psychology *(Howitt and Cramer, 2011a). It is also discussed in Box 9.3 below. So long as the correlation is significant then there is no problem. If it is not significant then the test of the difference between the two means may well be not very powerful.*

In a within-subjects design the effects that may occur as a result of doing the conditions in a particular order must be controlled. In a design consisting of only two conditions, these order effects are dealt with by counterbalancing the two orders so that both orders occur equally frequently. This counterbalancing is important since the data may be affected by any of a number of effects of order. The main ones are as follows:

- *Fatigue or boredom* Participants may become progressively more tired or bored with the task they are performing. So the number of mistakes they make may be greater in the second than in the first condition, regardless of which condition, because they are tired. An example of a fatigue effect is illustrated in the bar chart in Figure 9.7 for the effect of two different amounts of alcohol on the number of mistakes made. The vertical axis shows the number of errors made. The horizontal axis shows the two conditions of 8 ml and 16 ml of alcohol. Within each condition, the order in which the conditions were run is indicated. So '1st' means that that condition was run first and '2nd' that that condition was run second. We can see that there is a similar fatigue effect for both conditions. More errors are made when the condition is run second than when it is run first. In the 8 ml condition 6 errors are made when it is run second compared with 4 when it is run first. The same difference holds for the 16 ml condition where 12 errors are made when it is run second compared with 10 when it is run first.

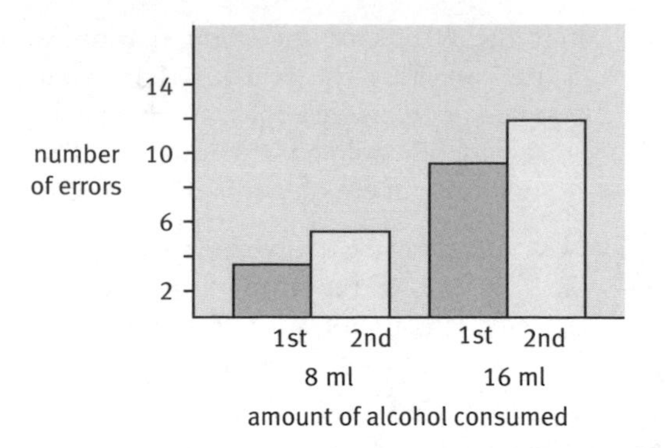

FIGURE 9.7 Fatigue effect in within-subjects design

- *Practice* Participants may become better at the task they are carrying out. So the number of mistakes they make may be less in the second than in the first condition, regardless of which condition, because they have learnt to respond more accurately. Sometimes the term 'practice effect' is used to cover both learning as described here and fatigue or boredom.

- *Carryover, asymmetrical transfer or differential transfer* Here the effect of an earlier condition affects a subsequent one but not equally for all orders. (One can refer to this as an interaction between the conditions and the order of the conditions.) For example, if the interval between the two alcohol conditions is close together, the carryover effect of drinking 16 ml of alcohol first may be greater on the effect of drinking 8 ml of alcohol second than the carryover effect of drinking 8 ml of alcohol first on the effect of drinking 16 ml of alcohol second. This pattern of results is illustrated in Figure 9.8. When the 8 ml condition is run second the number of mistakes made is much greater (12) than when it is run first (4) and is almost the same as the number of mistakes made when the 16 ml condition is run first (10). When the 16 ml condition is run second the number of mistakes made (13) is not much different from when it is run first. This asymmetrical transfer effect reduces the overall difference between the 8 and 16 ml conditions. If one finds such an asymmetrical transfer effect then it may be possible to make adjustments to the research design to get rid of them. In the alcohol example, one could increase the amount of time

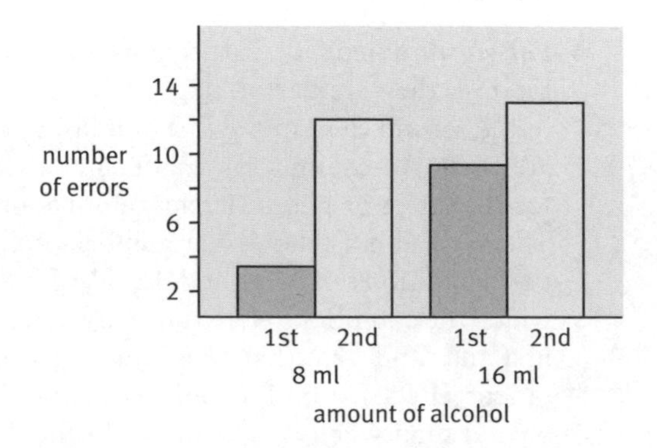

FIGURE 9.8 Asymmetrical transfer effect in a within-subjects design

between the two conditions. In this way, the alcohol consumed in the first condition may have worked its way out of the blood system. Of course, this has disadvantages. It might involve participants returning to the laboratory at a later time rather than the entire study being run at the same time. This increases the risk that participants may not return. Worse still, you may find that participants in one of the conditions may fail to return at different rates from participants in the other condition.

Of course, this implies that counterbalanced designs are not always effective at balancing any effects of order. They clearly do balance out order effects in circumstances in which there is no significant interaction between the conditions and the order in which the conditions are run. If there is a significant interaction between the conditions and the order in which they are run, we need to describe what this interaction is. We can illustrate the interaction summarised in Figure 9.8 with the fictitious raw data in Table 9.4 where there are three participants who carry out the two different orders in which the two conditions are run. Each participant is signified by the letter P with a subscript consisting of two whole numbers. The first number refers to a particular participant and varies from 1 to 6 as there are six participants. The second number represents the two conditions. We could analyse these data with a mixed analysis of variance. This statistical test is described in some introductory statistics texts such as the companion text *Introduction to Statistics in Psychology* (Howitt and Cramer, 2011a).

With counterbalancing, it is obviously important that equal numbers of participants are included in each condition. The random assignment of participants to different orders in a within-subjects design is necessary to ensure that the orders are exactly the same apart from the order in which the conditions are run. For example, in our study of the effects of alcohol on the number of errors made it is important that the proportion of people who are inclined to make more errors is the same in the two different orders. So if the proportion is not the same, then there may be a difference between the two orders which may result in a significant interaction effect.

If there is a significant interaction effect in a counterbalanced design, the analysis becomes a little cumbersome. Essentially, one regards the study as having two different parts – one part for each different order. The data from each part (order) are then analysed to see what is the apparent effect of the experimental treatment. If the same conclusions are reached for the different orders, then all is well as far as one's findings are concerned. Things become difficult when the conclusions from the different orders

Table 9.4 Fictitious scores for a within-subjects design with two conditions

	Condition 1	Condition 2
Condition 1 first	$P_{1,1} = 3$	$P_{1,2} = 11$
	$P_{2,1} = 4$	$P_{2,2} = 10$
	$P_{3,1} = 5$	$P_{3,2} = 9$
Sum	12	30
Mean	12/3 = 4	30/3 = 10
Condition 1 second	$P_{4,1} = 11$	$P_{4,2} = 13$
	$P_{5,1} = 12$	$P_{5,2} = 12$
	$P_{6,1} = 13$	$P_{6,2} = 14$
Sum	36	39
Mean	36/3 = 12	39/3 = 13

are not compatible with each other. Also note that this also effectively reduces the maximum sample size and so takes away the advantage of a within-subjects design.

Of course, researchers using their intelligence would have anticipated the problem for a study such as this in which the effects of alcohol are being studied. To the extent that one can anticipate problems due to the order of running through conditions then one would be less inclined to use a within-subjects design. This is a case where noting problems with counterbalanced designs identified by researchers investigating similar topics to one's own may help decide whether a within-subjects design should be avoided.

Stable individual differences between people are controlled in a within-subjects design by requiring the same participants to carry out all conditions. Nevertheless, it remains a requirement that assignment to different orders is done randomly. Of course, this does not mean that the process of randomisation has not left differences between the orders. This may be checked by pre-testing participants prior to the different first conditions to be run. One would check whether the pre-test means were the same for those who took condition 1 first as for those who took condition 2 first. In other words, it is possible to have a pre-test–post-test within-subjects design. It would also be possible to extend the Solomon four-group design to investigate not only the effects of pre-testing but also the effects of having participants undertake more than one condition.

Box 9.3 Key Ideas

Statistical significance

Statistical significance is one of those ideas that many students have difficulty with. So it can be usefully returned to from time to time so that the ideas are reinforced. We can explain the concept of statistical significance with the experiment on the effects of alcohol on errors. Suppose that we find in our study of the effects of alcohol on making errors that the participants who drink less alcohol make fewer errors than those who drink more alcohol. We may find that we obtain the results shown in Table 9.5. All the participants who drink less alcohol make fewer errors than those who drink more alcohol.

The mean number of errors made by the participants who drink less alcohol is 2 compared with a mean of 5 for those who drink more alcohol. The absolute difference between these two means, which ignores the sign of the difference, is 3 ($2 - 5 = 3$). (To be precise, this should be written as ($|2 - 5| = 3$), which indicates that the absolute value of the difference should be taken.) Can we conclude from these results that drinking more alcohol causes us to make more mistakes?

We cannot draw this conclusion without determining the extent to which we may find this difference simply by chance as it is possible to obtain a difference of 3 by chance. If this difference has a probability of occurring by chance of 5 per cent or .05 or less we can conclude that this difference is quite unusual and unlikely to be due to chance. It represents a real difference between the two conditions. If this difference has a probability of occurring by chance of more than 5 per cent or .05, we would conclude that this difference could be due to chance and so does not represent a real difference between the two conditions. It needs to be stressed that the 5 per cent or .05 significance level is just a conventional and generally accepted figure. It indicates a fairly uncommon outcome if differences between groups were simply due to chance factors resulting from sampling.

We can demonstrate the probability of this difference occurring by chance in the following way. Suppose that it is only possible to make between 1 and 6 mistakes on this task. (We are using this for convenience; it would be more accurate in terms of real statistical analysis to use the same figures but arranged in a bell-shaped or normal distribution in which scores of 3 and 4 are the most common and scores of 1 and 6 were the most uncommon.) If we only have three participants in each condition and the results were simply determined by chance, then the mean for any group would be the mean of the three numbers selected by chance. We could randomly select these three numbers

Table 9.5 Fictitious data for a two-group between-subjects post-test only design

	Post-test
Condition 1	$P_{1,1} = 1$
	$P_{2,1} = 2$
	$P_{3,1} = 3$
Sum	**6**
Mean	**6/3 = 2**
Condition 2	$P_{4,2} = 4$
	$P_{5,2} = 5$
	$P_{6,2} = 6$
Sum	**15**
Mean	**15/3 = 5**

in several ways. We could toss a die three times. We could write the six numbers on six separate index cards or slips of paper, shuffle them, select a card or slip, note down the number, put the card or slip back, shuffle them again and repeat this procedure three times. We could use a statistical package such as SPSS Statistics. We would enter the numbers 1 to 6 in one of the columns. We would then select *Data*, *Select Cases . . .* , *Random sample of cases*, *Sample . . .* , *Exactly*, and then enter one case from the first six cases. We would note down the number of the case selected and repeat this procedure twice.

As we have two groups we would have to do this once for each group, calculate the mean of the three numbers for each group and then subtract the mean of one group from the mean of the other group. We would then repeat this procedure 19 or more times. The results of our doing this 20 times are shown in Table 9.6.

The mean for both the first two groups is 2.00 so the difference between them is zero. As the six numbers are equiprobable, the mean of three of these numbers selected at random is likely to be 3.5. This value is close to the mean for the 40 groups which is 3.52. However, the means can vary from a minimum of 1 [(1 + 1 + 1)/3 = 1] to a maximum of 6 [(6 + 6 + 6)/3 = 6].

The distribution of the frequency of an infinite number of means will take the shape of an inverted U or bell as shown by the normal curve in Figure 9.9 which has been superimposed onto the histogram of the means in Table 9.6. Of these means, the smallest is 1.67 and the largest is 5.67. The distribution of these means approximates the shape of an inverted U or bell as shown in the histogram in Figure 9.9. The more samples of three scores we select at random, the more likely it is that the distribution of the means of those samples will resemble a normal curve. The horizontal width of each rectangle in the histogram is 0.50. The first rectangle, which is on the left, ranges from 1.50 to 2.00 and contains two means of 1.67. The last rectangle which is on the right, varies from 5.50 to 6.00 and includes two means of 5.67.

If the means of the two groups tend to be 3.5, then the difference between them is likely to be zero. They will vary from a difference of −5 (1 − 6 = −5) to a difference of 5 (6 − 1 = 5) with most of them close to zero as shown by the normal curve superimposed on the histogram in Figure 9.10. Of the 20 differences in means in Table 9.6 the lowest is −2.33 and the highest is 3.00. If we plot the frequency of these differences in means in terms of the histogram in Figure 9.10 we can see that its shape approximates that of a normal curve. If we plotted an infinite number or a very large number of such differences then the distribution would resemble a normal curve. The horizontal width of each rectangle in this histogram is 1.00. The first rectangle, which is on the left, ranges from −3.00 to −2.00 and contains two differences in means of −2.33. The last rectangle which is on the right, varies from 2.00 to 3.00 and includes two differences in means of 2.67 and 3.00.

We can see that the probability of obtaining by chance a difference as large as −3.00 is quite small. One test for

Table 9.6 Differences between the means of three randomly selected numbers varying between 1 and 6

	Condition 1				Condition 2				
	P11	P21	P31	Mean	P42	P52	P62	Mean	Difference
1	1	3	2	**2.00**	2	3	1	**2.00**	0.00
2	5	1	2	**2.67**	5	5	4	**4.67**	−2.00
3	6	1	4	**3.67**	6	1	6	**4.33**	−0.66
4	6	3	6	**5.00**	5	6	6	**5.67**	−0.67
5	1	5	3	**3.00**	6	3	1	**3.33**	−0.33
6	1	2	4	**2.33**	1	3	6	**3.33**	−1.00
7	3	6	2	**3.67**	6	3	5	**4.67**	−1.00
8	1	2	5	**2.67**	5	5	4	**4.67**	−2.00
9	6	2	3	**3.67**	2	4	4	**3.33**	0.34
10	6	1	1	**2.67**	6	3	3	**4.00**	−1.33
11	4	4	6	**4.67**	2	3	1	**2.00**	2.67
12	1	6	2	**3.00**	2	5	2	**3.00**	0.00
13	2	5	5	**4.00**	5	3	6	**4.67**	−0.67
14	2	1	2	**1.67**	6	5	1	**4.00**	−2.33
15	4	6	3	**4.33**	6	5	6	**5.67**	−1.34
16	2	4	5	**3.67**	6	1	3	**3.33**	0.34
17	2	5	4	**3.67**	5	1	5	**3.67**	0.00
18	2	2	2	**2.00**	1	6	6	**4.33**	−2.33
19	2	1	2	**1.67**	3	1	5	**3.00**	−1.33
20	6	3	6	**5.00**	2	2	2	**2.00**	3.00

determining this probability is the unrelated *t*-test. This test is described in most introductory statistics texts including our companion volume *Introduction to Statistics in Psychology* (Howitt and Cramer, 2011a). If the variances in the scores for the two groups are equal or similar, the probability of the unrelated *t*-test will be the same as a one-way analysis of variance with two groups. If we had strong grounds for thinking that the mean number of errors would be smaller for those who drank less rather than more alcohol, then we could confine our 5 per cent or .05 probability to the left tail or side of the distribution which covers this possibility. This is usually called the one-tailed level of probability. If we did not have good reasons for predicting the direction of the results, then we are saying that the number of errors made by the participants drinking less alcohol could be either less or more than those made by the participants drinking more alcohol. In other words, the difference between the means could be either negative or positive in sign. If this was the case, the 5 per cent or .05 probability level would cover the two tails or sides of the distribution. This is normally referred to as the two-tailed level of probability. To be significant at the two rather than the one-tailed level, the difference in the means would have to be bigger as the 5 per cent or .05 level is split between the two tails so that it covers a more extreme difference. If the difference between the two means is statistically significant, which it is for the scores in Table 9.5, we could conclude that drinking less alcohol results in making fewer errors.

Our companion statistics text, *Introduction to Statistics in Psychology* (Howitt and Cramer, 2011a), presents a more extended version of this explanation for the correlation coefficient and the *t*-test.

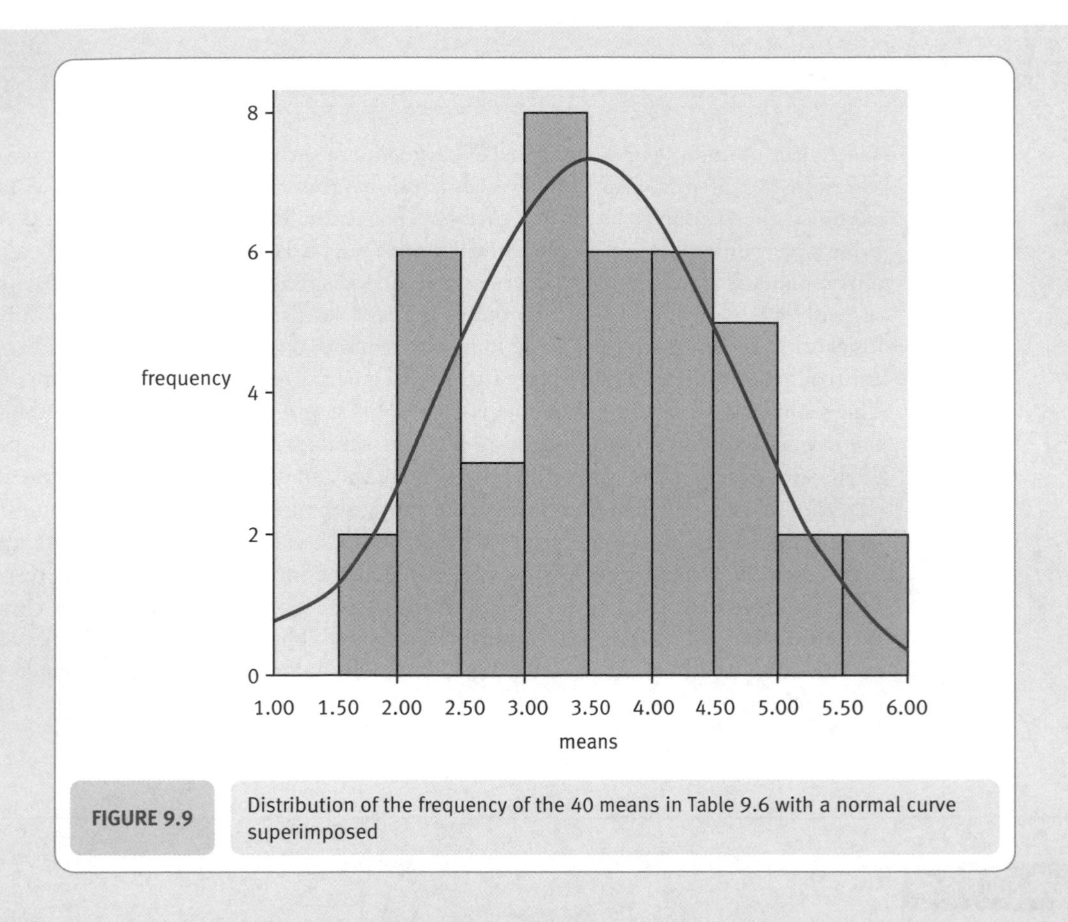

FIGURE 9.9 Distribution of the frequency of the 40 means in Table 9.6 with a normal curve superimposed

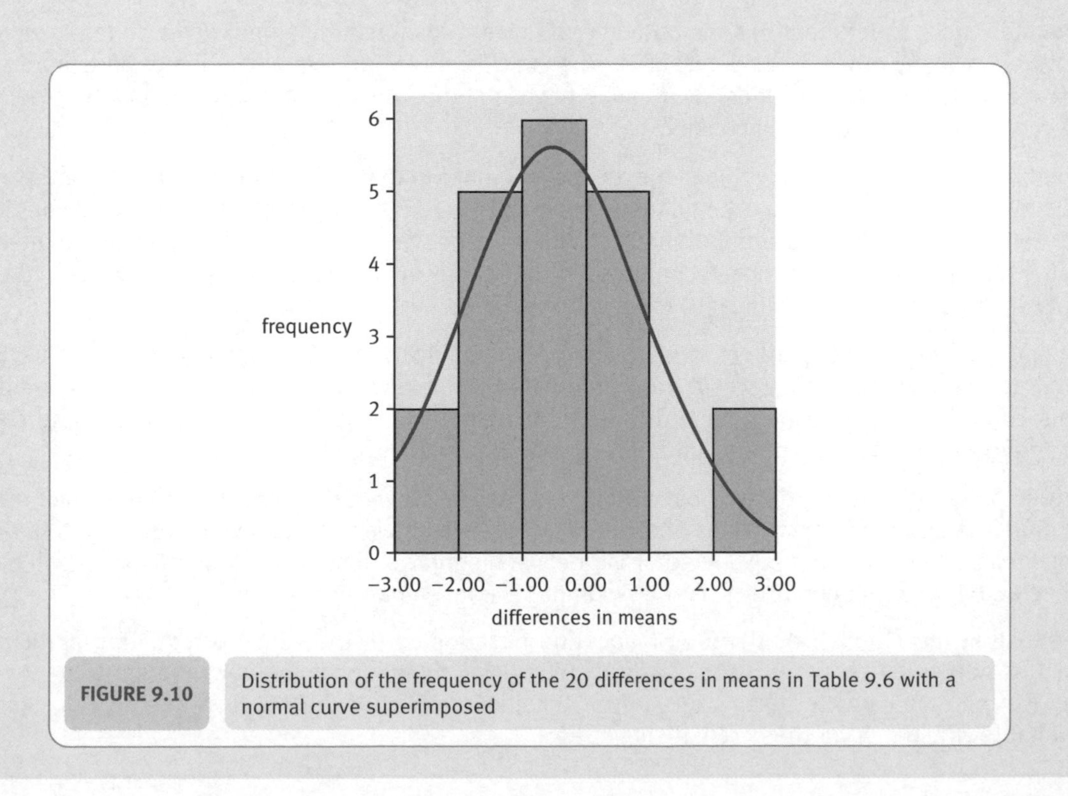

FIGURE 9.10 Distribution of the frequency of the 20 differences in means in Table 9.6 with a normal curve superimposed

9.4 Conclusion

The basics of the true or randomised experiment are simple. The major advantage of such a design is that it is easier to draw conclusions about causality since care is taken to exclude other variables as far as possible. That is, the different experimental conditions bring about differences on the dependent variable. This is achieved by randomly allocating participants to conditions or orders and standardising procedures. There are a number of problems with this. The major one is that randomisation equates groups only in the long run. For any particular experiment, it remains possible that the experimental and control groups differ initially before the experimental manipulation has been employed. The main way of dealing with this is to employ a pre-test to establish whether or not the experimental and control groups are very similar. If they are, there is no problem. If the pre-test demonstrates differences then this may bring about a different interpretation of any post-test findings. Furthermore, the more complicated the manipulation is, the more likely it is that variables other than the intended one will be manipulated. Consequently, the less easy it is to conclude that the independent variable is responsible for the differences. The less controlled the setting in which the experiment is conducted, the more likely it is that the conditions under which the experiment is run will not be the same and that other factors than the manipulation may be responsible for any observed effect.

Key points

- The laboratory experiment has the potential to reveal causal relationships with a certainty which is not true of many other styles of research. This is achieved by random allocation of participants and the manipulation of the independent variable while standardising procedures as much as possible to control other sources of variability.
- The between-subjects and within-subjects designs differ in that in the former participants take part in only one condition of the experiment whereas in the latter participants take part in all conditions (or sometimes just two or more) of the conditions. These two different types of design are analysed using rather different statistical techniques. Within-subjects designs use related or correlated tests. This enables statistical significance to be achieved with fewer participants.
- The manipulated or independent variable will consist of only two levels or conditions in the most basic laboratory experiment. The level of the manipulated variable will be higher in one of the conditions. This condition is sometimes referred to as the experimental condition as opposed to the control condition where the level of the manipulated variable will be lower.
- Within-subjects (related) designs have problems associated with the sensitisation effects of serving in more than one of the conditions of the study. There are designs that allow the researcher to detect sensitisation effects. One advantage of the between-subjects design is that participants will not be affected by the other conditions as they will not have taken part in them.
- Pre-testing to establish that random allocation has worked in the sense of equating participants on the dependent variable prior to the experimental treatment sometimes works. Nevertheless, pre-testing may cause problems due to the sensitising effect of the pre-test. Complex designs are available which test for these sensitising effects.

- The extent to which random assignment has resulted in participants being similar across either conditions or orders can be determined by a pre-test in which participants are assessed on the dependent variable before the manipulation is carried out.
- Any statistical differences between the conditions in the dependent variable at post-test are very likely to be due to the manipulated variable if the dependent variable does not differ significantly between the conditions at pre-test and if the only other difference between the conditions is the manipulated variable.

ACTIVITY

Design a basic randomised experiment to test the hypothesis that unemployment leads to crime. After thinking about this you may find it useful to see whether and how other researchers have tried to study this issue using randomised designs. How are you going to operationalise these two variables? Is it possible to manipulate unemployment and how can you do so? If you are going to carry out a laboratory experiment you may have to operationalise unemployment in a more contrived way than if you carry out an experiment in a more natural or field setting. How can you reduce the ethical problems that may arise in the operationalisation of these variables? How many participants will you have in each condition? How will you select these participants? Will you pre-test your participants? Will you use a between- or a within-subjects design? How will you analyse the results? What will you say to participants about the study before and after they take part?

CHAPTER 10

Advanced experimental design

Overview

- Few laboratory experiments consist of just an experimental and control group. The information obtained from such a study would be very limited for the time, money and effort expended. The simple experimental group–control group is often extended to include perhaps four or five different conditions. Naturally it is important that every condition of an experiment should be justifiable. A typical justification is that each group will produce a different outcome relative to the other groups.
- Most behaviours are multiply affected by a range of factors (i.e. variables). Consequently, it can be advantageous to study several factors at the same time. In this way the relative effects of the factors can be compared. The number of variables that can be manipulated in a study should be kept to an optimum. Typically no more than two or three should be used. If more are employed, the interpretation of the statistical findings becomes extremely complex and, possibly, misleading.
- The factors that are studied may include those which cannot be randomised (or manipulated) such as gender, age and intelligence. These are sometimes referred to as subject variables.
- It is common for researchers to use more than one dependent variable. This is because the independent variables may have a range of effects and also because using a number of measures of the dependent variable can be informative. Where there are several dependent variables it may be worthwhile controlling for any order effects among these variables by varying the order. This can be done systematically using a Latin square.
- Latin squares are also used to systematically vary the order of the conditions run in a within-subjects (i.e. related) design where there are a number of different conditions.

- Planned and unplanned comparisons in factorial designs have rather different requirements in terms of statistical analysis. Comparisons planned in advance of data collection have distinct advantages, for example, in terms of the ease of making multiple comparisons.
- Quite distinct from any of the above, the advanced experimenter should consider including means of controlling for potential nuisances in the research design. The use of placebos, double-blind procedures and the like help make the methodology more convincing. Quasi-controls to investigate the experience of participants in the research might be regarded as good practice as, in part, they involve discussions after the study between participant and researcher as part of a process of understanding the findings of the research. They are essentially a variant of the post-experimental debriefing interviews discussed in Chapter 9 but with a more focused and less exploratory objective.

10.1 Introduction

In this chapter we will extend our understanding of experimental design in three ways. First of all, we will look at increasing the number of levels of the independent variable so that there are three or more groups. Then we will consider more advanced designs for experiments including those in which there are two or more independent variables. This leads to extra efficiency in terms of the amount of information which can be obtained from a single study. Experimental designs where more than one dependent variable is used are also considered. In addition, we will look at aspects of experimental design which, unless carefully considered and acted upon, may result in problems in the interpretation of the findings of the research. Some of these are conventionally termed experimenter effects and come under the general rubric of the social psychology of the laboratory experiment.

The simple two-groups design – experimental and control group – provides the researcher with relatively little information for the effort involved. The design may be extended by using a greater variety of conditions but, perhaps more likely, the single independent variable is extended to two or three independent variables, perhaps more:

- Just having two conditions or levels of an independent variable, such as the amount of alcohol consumed (as in the example used in the previous chapter), tells us little about the shape of the relationship between the independent and the dependent variable. Is the relationship a linear one or curvilinear? What kind of curvilinear relationship is it? Having a number of levels of the dependent variable helps us to identify the nature of the trends because of the extra information from the additional conditions.
- If we have several independent variables then we can answer the question does the independent variable interact with other independent or subject variables? For example, is the effect of alcohol on the number of errors made similar for both males and females or is it different in the two genders?
- Does the independent variable affect more than one dependent variable? For example, does alcohol affect the number of errors made on a number of different tasks?

These three main ways of extending the basic two-group design are discussed below and highlighted in Figure 10.1.

Multiple levels of an independent variable	Multiple dependent variables	Multiple independent variables
Rather than an experimental and a control condition, there are three or more conditions	Uses several dependent variables essentially measuring facets of much the same thing	Uses two or more independent variables. They are known as two-way, three-way, etc. factorial designs
Thus there may be several experimental conditions and/or several control conditions	The problem is to combine the dependent variables in a statistically meaningful way	Allows assessment of interactions between different combinations of levels of the independent variables
Statistically these are analysed with the one-way analysis of variance and multiple comparisons tests	Multivariate analysis of variance (MANOVA) is the usual statistical approach	The two-way, three-way, etc. analysis of variance is the usual statistical approach

FIGURE 10.1 The three main ways of building more complex experimental designs

10.2 Multiple levels of the independent variable

Multiple levels of the independent variable occur when there are three or more different levels of that variable. This is sometimes described as having several levels of the treatment. An example of this is to be found in Figure 10.2. The independent variable may vary in one of two different ways – quantitative or qualitative:

- Quantitative would be, for example, when the amount of alcohol consumed in the different conditions can be arranged in order of numerical size. In general the order is from smaller quantities to larger ones ordered either across or down a table or across a graph such as Figure 10.2.
- Qualitative would be, for example, when we study the effects of the kind of music being played in the different conditions. There is generally no one way in which the levels or categories can be ordered in terms of amount. The categories will reflect a number of different characteristics such as the date when the music was recorded, the number of instruments being played and so forth. In other words, qualitative is the equivalent of a nominal, category or categorical variable. When studying the effect of a qualitative variable which varies in numerous ways, it is not possible to know what particular features of the qualitative variable produce any differences that are obtained in the study.

Multiple comparisons

The analysis of designs with more than two conditions is more complicated than one with only two conditions. For one reason, there are more comparisons to make the more conditions there are. If there are three levels or conditions we can compare:

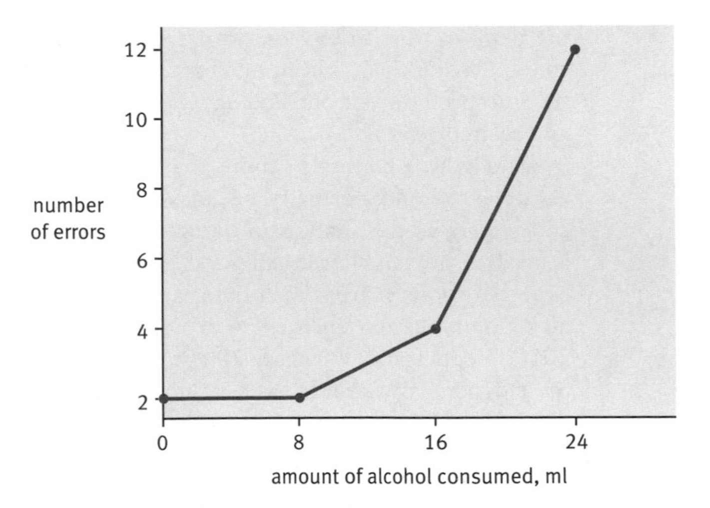

FIGURE 10.2 Number of errors as a function of amount of alcohol consumed

- condition 1 with condition 2,
- condition 1 with condition 3, *and*
- condition 2 with condition 3.

That is, a total of up to three *different* comparisons. With four conditions, there are six different comparisons we can make. With five conditions there are ten different comparisons and so on. (The number of comparisons is the sum of numbers from 1 to the number of conditions minus 1. That is, for four conditions, one less than the number of conditions is 4 – 1 or 3. So we add 1 + 2 + 3 which gives us 6 comparisons.) Whether or not one makes all possible comparisons depends on the purpose of one's research. The temptation is, of course, to do all possible comparisons. For example, you may have a study in which there are two experimental groups and two control groups. You may, for this particular study, only be interested in the differences between the experimental groups and the control groups. You may not be interested in whether the control groups differ from each other or even whether the experimental groups differ from each other. If a comparison does not matter for your purposes, then there is no necessity to include it in your analysis. In research, however, the justification for whatever number of comparisons are to be made should be presented clearly.

The more comparisons we make, so the likelihood increases of finding some of these comparisons to be statistically significant by chance. If these comparisons are completely independent of each other, the probability of finding one or more of these comparisons statistically significant can be calculated with the following formula:

$$\text{probability of statistically significant comparison} = 1 - (1 - .05)^{\text{number of comparisons}}$$

where .05 represents the 5 per cent or .05 significance level. With three comparisons, this probability or significance level is about 14 per cent or .14 [$1 - (1 - .05)^3 = .14$]. With four comparisons it is 19 per cent or .19 [$1 - (1 - .05)^4 = .19$]. With five comparisons it is 23 per cent or .23 [$1 - (1 - .05)^5 = .23$] and so on. This probability is known as the *family-wise* or *experiment-wise* error rate because we are making a number, or family, of comparisons from the same study or experiment. The point is that by making a lot of comparisons we increase the risk that some of our findings from the data are due to chance. So if some comparisons serve no purpose in a particular study, leaving them out means that there is less chance of making this sort of 'error' in the interpretation of our findings. Box 10.1 discusses this sort of interpretation 'error' and related issues.

Data which consist of more than two conditions or more than two independent variables, would normally be analysed by a group of tests of significance known collectively as the analysis of variance (ANOVA). However, there may not always be a need to do this. If we had good grounds for predicting which conditions would be expected to differ from each other and the direction of those differences, then an overall or omnibus test such as an analysis of variance may be unnecessary (e.g. Howell, 2010; Keppel and Wickens, 2004). An omnibus test simply tells us whether overall the independent variable has a significant effect but it does not tell us which conditions are actually different from each other. Regardless of whether we carry out an omnibus

Box 10.1 Key Ideas

The risks in interpreting trends in our data

Since research data in psychology are based on samples of data rather than *all* of the data then there is always a risk that the characteristics of our particular sample of data do not represent reality accurately. Some outcomes of sampling are very likely and some are very unlikely to occur in any particular study. Most randomly drawn samples will show similar characteristics to the population from which they were drawn. In statistical analysis, the working assumption usually is the null hypothesis which is that there is no difference between the different conditions on the dependent variable or that there is no relationship between two variables. In other words, the null hypothesis says that there is no trend in reality. Essentially in statistical analysis we assess the probability that the null hypothesis is true. A statistically significant statistical analysis means that it is unlikely that the trend would have occurred by chance if the null hypothesis of no trend in reality were true. The criterion which we impose to make the decision is that a trend which is likely to occur in 95 per cent of random samples drawn from a population where there is *no* trend is not statistically significant. However, trends which are so strong that they fall into the 5 per cent of outcomes are said to be statistically significant and we accept the hypothesis that there is in reality a trend between two variables – a difference or a relationship. The upshot of all of this, though, is that no matter what we decide there is a risk that we will be wrong.

Type I error refers to the situation in which we decide on the basis of our data that there is a trend but in actuality there is really no trend (see Figure 10.3). We have set the risk of this at 5 per cent. It is a small risk but nevertheless there is a risk. Psychologists tend to be very concerned about Type I errors.

Type II error refers to a different situation. This is the situation in which in reality there is a trend involving two variables but our statistical analysis fails to detect this trend at the required 5 per cent level of significance. Psychologists seem to be less worried about this in general. However, what it could mean is that really important trends are overlooked. Researchers who studied a treatment for dementia but their findings did not reach statistical significance – maybe because the sample size was too small – would be making a Type II error. Furthermore, by other criteria this error would be a serious one if the consequence was that this treatment were abandoned as a result. This stresses the importance of using other criteria relevant to decision-making in psychological research. Statistical significance is one criterion but it is most certainly not the only criterion when reaching conclusions based on research. Many professional researchers go to quite considerable lengths to avoid Type II errors by considering carefully such factors as the level of statistical significance to be used, the size of the effect (or trend) which would minimally be of interest to the researchers, and the sample size required to achieve these ends. This is known as power analysis. It is covered in detail in our companion book *An Introduction to Statistics in Psychology* (Howitt and Cramer, 2011a). Figure 10.3 gives some of the possible decision outcomes from a statistical analysis.

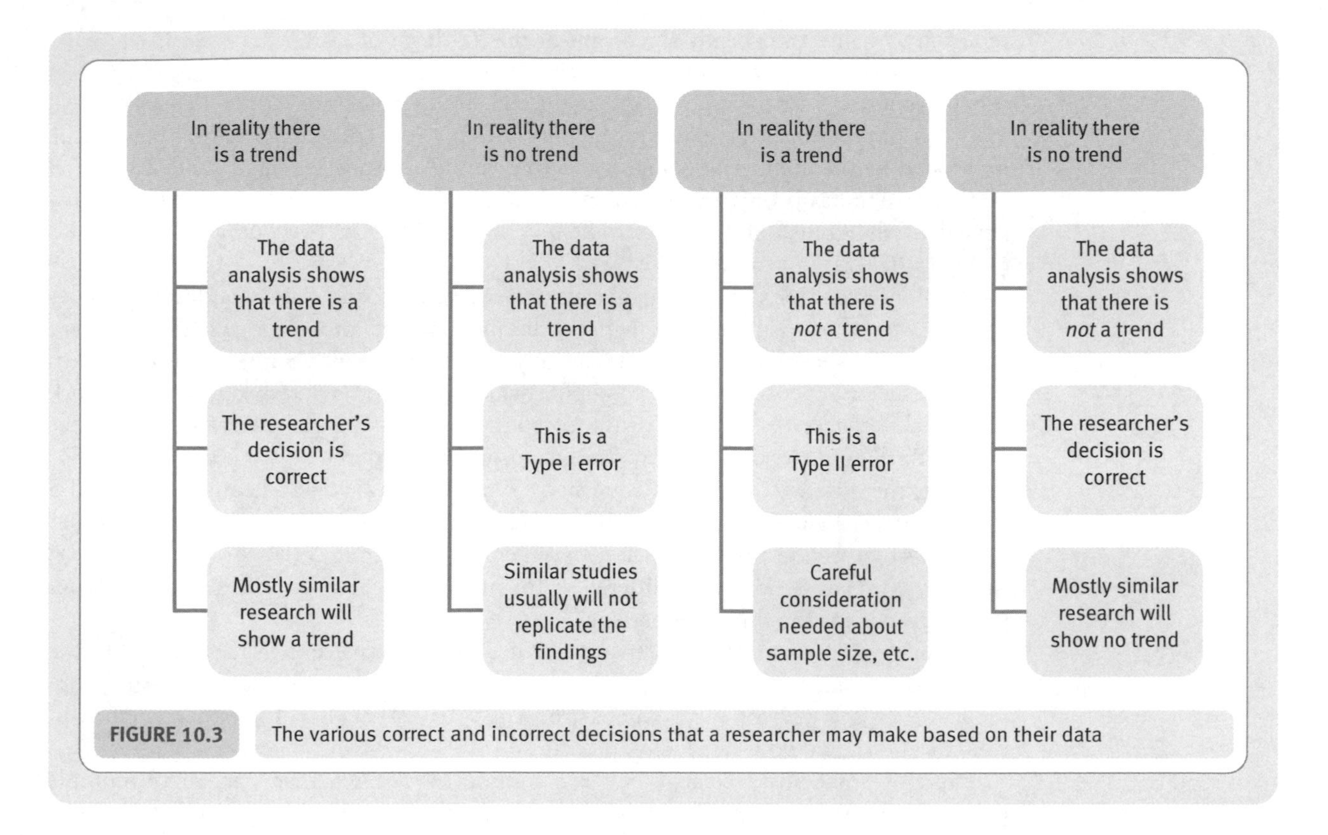

FIGURE 10.3 The various correct and incorrect decisions that a researcher may make based on their data

test, the conditions we expect to differ still have to be compared. These comparisons have been called planned or *a priori* comparisons. (*A priori* is Latin for 'from what is before'.) We can use a test such as a simple *t*-test to determine which conditions differ from each other provided that we test a limited number of comparisons (for example, Howell, 2010; Keppel and Wickens, 2004). If we make a large number of comparisons, then we should make an adjustment for the family-wise error rate. The point is that if we plan a few comparisons we have effectively pinpointed key features of the situation. The more comparisons we make then the less precision is involved in our planning. Hence the need to make adjustments when we have a high proportion of all of the possible comparisons to make. Very little student research, in our experience, involves this degree of pre-planning of the analysis. It is hard enough coming up with research questions, hypotheses and research designs to add to the burden by meticulously planning in advance on a theoretical, empirical, or conceptual basis just what comparisons we will make during the actual data analysis.

The more common situation, however, is when we lack good reasons for expecting a particular difference or for predicting the direction of a difference. The procedure in these circumstances is to employ an omnibus statistical test such as an analysis of variance. If this analysis is significant overall, unplanned, *post hoc* or *a posteriori* comparisons can be carried out. These comparisons determine which conditions differ significantly from each other. *Post hoc* is Latin for 'after this' and *a posteriori* is Latin for 'what comes after'. With the *post hoc* test, it is necessary to adjust the significance of the statistical test to take into account the number of comparisons being made. The simplest way to do this is to divide the .05 level of statistical significance by the number of comparisons to be made. This is known as a Bonferroni adjustment or test. So with three comparisons, the adjusted level is about .0167 (.05/3 = .0167). With four comparisons it is .0125 (.05/4 = .0125) and so on. Be careful! What this means is that a comparison has to be statistically significant at this adjusted level to be reported as being statistically significant at the .05 level of significance. So if we make four comparisons, only differences which are statistically significant at the .0125 level can be reported as being significant at the .05 level. It is easier to do this with SPSS Statistics output since the exact probability found for a comparison simply has to be multiplied by the number of comparisons to give the appropriate significance level. The finding is significant only if the *multiplied* exact significance is below .05. This is discussed in the companion text *Introduction to Statistics in Psychology* (Howitt and Cramer, 2011a).

The Bonferroni test is a conservative test if we are comparing all the conditions because at least one of the comparisons will not be independent of the others. Conservative basically means less likely to give statistically significant results. Suppose we wanted to compare the mean scores for three conditions which are 2, 4 and 8, respectively. If we work out the differences for any two of the comparisons, then we can derive the difference for the third comparison by subtracting the other two differences from each other. For instance, the differences between conditions 1 and 2 (4 – 2 = 2) and conditions 1 and 3 (8 – 2 = 6) are 2 and 6, respectively. If we subtract these two differences from each other (6 – 2 = 4) we obtain the difference between conditions 2 and 3 (8 – 4 = 4), which is 4. In other words, if we know the differences for two of the comparisons, we can work out the difference for the third comparison.

So, in this situation, the Bonferroni test is a conservative test in the sense that the test assumes three rather than two independent comparisons are being made. The probability level is lower for three comparisons (.05/3 = .017) than for two comparisons (.05/2 = .025), and so is less likely to occur.

There is some disagreement between authors about whether particular multiple comparison tests such as the Bonferroni test should be used for *a priori* or *post hoc* comparisons. For example, Howell (2010) suggests that the Bonferroni test should be used for making planned or *a priori* comparisons while Keppel and Wickens (2004) recommend that this test be used for making a small number of unplanned or *post hoc* comparisons! The widely used statistical package SPSS Statistics also lists the Bonferroni test as a *post hoc* test. There are also tests for determining the shape of the relationship between the levels of a quantitative independent variable and the dependent variable, which are known as trend tests (for example, Kirk, 1995).

In many instances such disagreements in the advice of experts will make little or no difference to the interpretation of the statistical analysis – that is the findings will be unaffected – even though the numbers in the statistical analyses differ to a degree. Since multiple comparison tests are quickly computed using SPSS Statistics and other statistical packages, it is easy to try out a number of multiple comparison tests. Only in circumstances in which they lead to radically different conclusions do you really have a problem. These circumstances are probably rare and equally probably mean that some comparisons which are significant with one test are marginally non-significant with another. In these circumstances, it would be appropriate to highlight the problem in your report.

10.3 Multiple dependent variables

Sometimes researchers may wish to use a number of different measures of the dependent variable within a single study. For example, we can assess the effects of alcohol on task performance in terms of both the number of errors made and the speed with which the task is carried out. Performance on a number of tasks such as simple reaction time, complex reaction time, attention span and distance estimation could be studied.

One could, of course, carry out separate studies for each of these different measures of performance. However, it would be more efficient to examine them in the same study. In these circumstances, it may be important to control for potential order effects in the measurement of the various dependent variables by randomising the order of presentation. Univariate analysis of variance (ANOVA) is not appropriate for analysing these data since it deals with only one dependent variable. Multivariate analysis of variance (abbreviated to MANOVA) is used instead since it deals with multiple dependent variables. A description of MANOVA can be found in Chapter 26 in the companion text *Introduction to Statistics in Psychology* (Howitt and Cramer, 2011a) together with instructions about how to carry one out.

There are circumstances where one would not use MANOVA. Sometimes a researcher will measure tasks such as reaction time several times in order to sample the participant's performance on this task. These data are better dealt with by simply averaging to give a mean score over the several trials of what is the same task. There would be nothing to gain from treating these several measures of the same thing as multiple dependent variables.

10.4 Factorial designs

A study which investigates more than one independent variable is known as a factorial design – see Table 10.1. Variables such as gender may be referred to as subject variables. These are characteristics of the participants which cannot be independently manipulated (and randomly assigned). Gender, age and intelligence are good examples of subject variables. They are also referred to as independent variables if they are seen as potential causes of variations in the dependent variable.

Terms such as two-way, three-way and four-way are frequently mentioned in connection with factorial research designs. The word 'way' really means factor or independent variable. Thus a one-way design means one independent variable, a two-way design means two independent variables, a three-way design means three independent variables and so forth. The phrase is also used in connection with the analysis of variance. So a two-way analysis of variance is an appropriate way of analysing a two-way research design. The number of factors and the number of levels within the factors may be indicated by stating the number of levels in each factor and by separating each of these numbers by an '×' which is referred to as 'by'. So a design having two factors with two levels and a third factor with three levels may be called a $2 \times 2 \times 3$ factorial design. The analysis of factorial designs is usually through the use of ANOVA. There are versions of ANOVA that cope with virtually any variation of the factorial design. For example, it is possible to have related variables and unrelated variables as independent variables in the same design (i.e. a mixed ANOVA).

Table 10.1 A simple factorial design investigating the effects of alcohol and gender on performance

	Females	Males
8 ml alcohol		
16 ml alcohol		

When quantitative subject variables such as age, intelligence or anxiety are to be used in a factorial design and analysed using ANOVA, they can be categorised into ranges or groups of scores. Age, for example, may be categorised as ages between 18 and 22, 23 and 30, 31 and 40, and so on. The choice of ranges to use will depend on the nature of the study. Some researchers may prefer to use the subject variable as a covariate in an analysis of covariance design (ANCOVA) which essentially adjusts the data for subject differences before carrying out a more or less standard ANOVA analysis. The use of ranges is particularly helpful if there is a non-linear relationship between the subject variable and the dependent variable. This would be assessed by drawing a scatterplot of the relationship between the subject variable and the dependent variable. If the relationship between the two seems to be a curved line then there is a non-linear relationship between the two.

The use of subject variables in factorial designs can result in a situation in which the different conditions (cells) in the analysis contain very different numbers of participants. This can happen in all sorts of circumstances but, for example, it may be easier to recruit female participants than male participants. The computer program will calculate statistics for an analysis of variance which has different numbers of participants in each condition. Unfortunately, the way that it makes allowance for these differences is less than ideal. So, if possible, it is better to have equal numbers of participants in each condition. Not using subject variables, this is generally achieved quite easily. But if there is no choice, then stick with the unequal cell sizes.

There is an alternative way of analysing complex factorial designs which is to use multiple regression. This statistical technique identifies the pattern of independent variables which best account for the variation in a dependent variable. This readily translates to an experimental design which also has independent and dependent variables. If there are any subject variables in the form of scores then they may be left as scores. (Though this requires that the relationship between the subject variable and the dependent variable is linear.) Qualitative variables (i.e. nominal, category or categorical variables) may also be included as predictors. They may need to be converted into *dummy variables* if the qualitative variable has more than two categories. A good description of dummy variables is provided by Cohen, Cohen, West and Aiken (2003), and they are explained also in the companion text *Introduction to Statistics in Psychology* (Howitt and Cramer, 2011a). Basically, a dummy variable involves taking each category of a nominal variable and making it into a new variable. Participants are simply coded as having the characteristic to which the category refers or not. For example, if the category variable is cat, dog and other (the question is 'What is the participant's favourite animal?') then this can be turned into two dummy variables such as cat and dog. Participants are coded as choosing cat *or not* and choosing dog *or not*. There is always one fewer dummy variable than the number of categories. Participants choosing 'other' will be those who have not chosen cat and dog.

The advantages of using multiple regression to analyse multifactorial designs include:

- the subject variables are not placed into fairly arbitrary categories;
- the variation (and information) contained in the subject variable is not reduced by turning the subject variable into a small number of categories or ranges of scores.

There are three main advantages to using a factorial design:

- It is more efficient, or economical, in that it requires fewer cases or observations for approximately the same degree of precision or power. For example, a two-factor factorial design might use just 30 participants. To achieve the same power running the two-factor factorial design as two separate one-factor designs, twice as many participants would be needed. That is, each one-factor design would require 30 participants. In the multifactorial design, the values of one factor are averaged across the values of the other factors. That is to say, the factorial design essentially can be considered as several non-factorial designs – hence, the economy of numbers.

- Factorial designs enable greater generalisability of the results in that a factor is investigated over a wider range of conditions. So, for example, we can look at the effects of alcohol under two levels of noise rather than, say, a single one, and in females and males rather than in just one of these two groups.
- A third advantage is that a factorial design allows us to determine whether there is an interaction between two or more factors in that the effect of one factor depends on the effect of one or more other factors. Box 10.2 deals with interactions.

Box 10.2 Key Ideas

The nature of interactions

One of the consequences of employing multifactorial designs is that the combined influences of the variables on the dependent variable may be identified. An interaction is basically a combination of levels of two or more variables which produces effects on the dependent variable which cannot be accounted for by the separate effects of the variables in question. Interactions must be distinguished from main effects. A main effect is the influence of a variable acting on its own – not in combination with any other variable. Interactions can only occur when there are two or more independent variables.

Interactions may be most easily grasped in terms of a graph such as Figures 9.4 and 9.5 in the previous chapter where the vertical axis represents the dependent variable, the horizontal axis represents one of the independent variables and the lines connecting the points in the graph represent one or more other independent variables. Thus the vertical axis shows the number of errors made, the horizontal axis represents the time of testing (pre-test and post-test) and the two lines the two alcohol conditions, 8 and 16 ml. An interaction effect occurs if the lines in the graph are substantially out of parallel, such as when the lines diverge or converge (or both). In Figures 9.2 and 9.3 the effect of differing amounts of alcohol appears to depend on the time of testing. In other words there seems to be an interaction between the amount of alcohol consumed and the time of testing. In both figures the difference in errors between the two amounts of alcohol is greater at the pre-test than the post-test. Of course, in terms of a true or randomised pre-test–post-test experimental design we would hope that the pre-test scores were similar, as illustrated in Figure 10.4, as the main purpose of randomisation is to equate groups at the pre-test. But randomisation is randomisation and what the researcher hopes for does not always happen.

In Figure 10.4 there still appears to be an interaction but the difference between the two amounts of alcohol is greater at post-test than at pre-test. Drinking 16 ml of alcohol has a greater effect on the number of errors made than 8 ml of alcohol which is what we would anticipate. In pre-test–post-test experimental designs this is the kind of interaction effect we would expect if our independent variable had an effect.

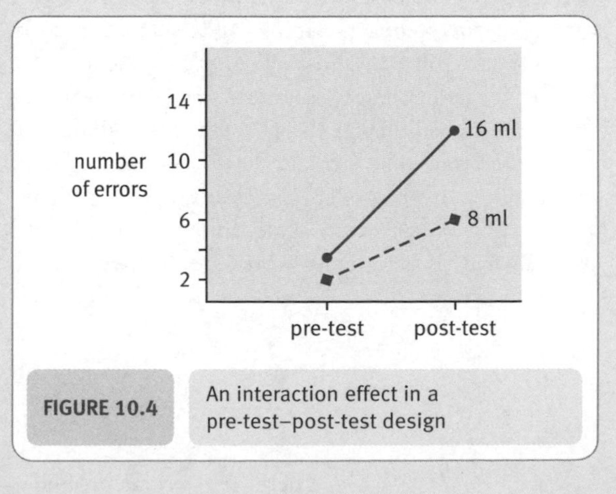

FIGURE 10.4 An interaction effect in a pre-test–post-test design

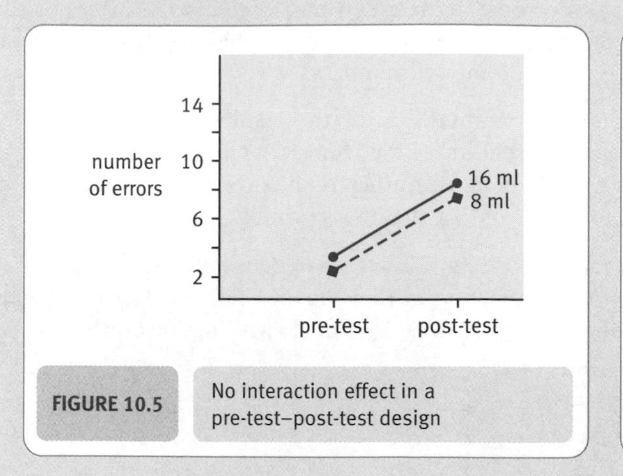

FIGURE 10.5 No interaction effect in a pre-test–post-test design

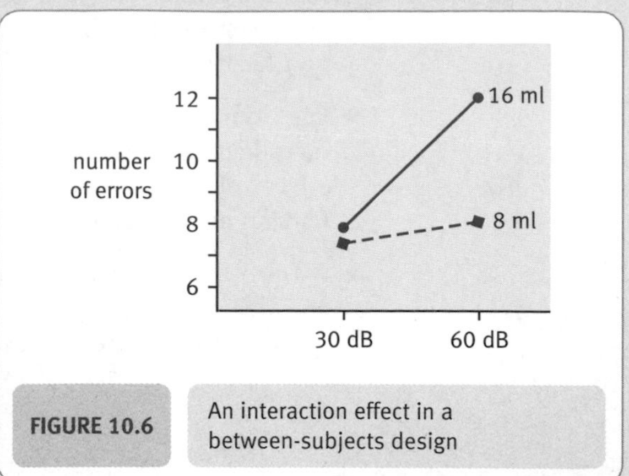

FIGURE 10.6 An interaction effect in a between-subjects design

The absence of an interaction between time of testing and alcohol is illustrated in Figure 10.5 as the two lines representing the two alcohol conditions are more or less parallel. There also appears to be a main effect for the time of testing in that the number of errors made at post-test seem to be greater than the number made at pre-test, but there does not seem to be a difference between the two alcohol conditions.

Figure 10.6 shows an apparent interaction effect for a between-subjects factorial design which consists of the two factors of amount of alcohol and level of noise. The difference in the number of errors made between the two alcohol conditions is greater for the 60 dB condition than the 30 dB condition.

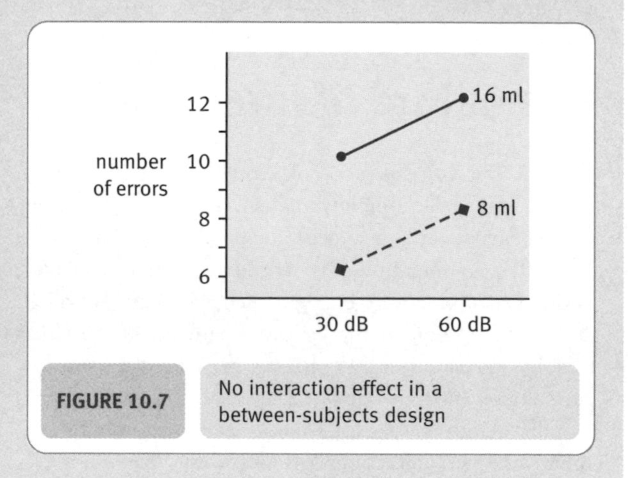

FIGURE 10.7 No interaction effect in a between-subjects design

Figure 10.7 illustrates the lack of an interaction effect for these two factors. The difference in performance between the two alcohol conditions appears to be similar for the two noise conditions.

There are circumstances where one should be very careful in interpreting the results of a study. These are circumstances such as those illustrated in Figure 10.6. In this diagram we can see that the only difference between the conditions is the 16 ml/60 dB condition. All other three conditions actually have a similar mean on the numbers of errors. This is clearly purely an interaction with no main effects at all. The problem is that the way the analysis of variance works means it will tend to identify main effects which simply do not exist. That is because to get the main effects, two groups will be combined (the two 30 dB groups and the two 60 dB groups for instance). In other words, at least part of the interaction will be subsumed under the main effects.

It is not possible to determine whether there is an interaction between two or more factors simply by looking at the plot of the scores on a graph. It is necessary to establish that this is a statistically significant interaction by carrying out a test such as an analysis of variance.

There are special problems facing a researcher designing a related design study. If participants are to be studied in every possible condition of the study then the order ought to be counterbalanced such that no order is more common than any other order. These designs are known as Latin square designs. These are discussed in Box 10.3.

Box 10.3 Key Ideas

Latin squares to control order effects

In a within-subject design you need to control for order effects by running the levels or conditions in different orders. The more conditions you have, the more orders there are for running those conditions. With three conditions called A, B and C, there are six different orders: ABC; ACB; BAC; BCA; CAB; and CBA. With four conditions there are 24 possible orders. With five conditions there are 120 possible orders and so on. We can work out the number of potential orders by multiplying the number of conditions by each of the numbers that fall below that number. For three conditions this is $3 \times 2 \times 1$ which gives 6. For five conditions it is $5 \times 4 \times 3 \times 2 \times 1$ which gives 120. Often there are more possible orders than actual participants. Suppose, for example, you only required 12 participants in a within-subjects design which has four conditions. In this situation, it is not possible to run all 24 possible orders. To determine which orders to run, one could randomly select 12 out of the 24 possible orders. However, if you do this, you could not guarantee that each condition would be run in the same ordinal position (for example, the first position) the same number of times and that each condition precedes and follows each condition once. In other words, you could not control for these order effects. The way to control for these order effects is to construct a Latin square.

A Latin square has as many orders as conditions. So a Latin square with four conditions will have four orders. To make a Latin square, perform the following steps:

- Create a random order of the conditions. This random order will be used to generate the other orders in the Latin square. There are several ways to create this initial random order. Suppose there are four conditions labelled A, B, C and D. One way is to write each letter on a separate slip of paper or index card, thoroughly shuffle the papers or cards and choose a sequence. (Randomisation is dealt with on p. 171.)
- Suppose the starting random order from step 1 is BACD. Sequentially number the conditions in this random order starting with 1. For this example, B = 1, A = 2, C = 3 and D = 4.
- To create the first order in the Latin square, put the last number or condition (N) in the third position as follows:

  1, 2, 4, 3

  which corresponds to the conditions as initially lettered B, A, D, C. If we had more than four conditions, then every subsequent unevenly numbered position (e.g. 5, 7 and so on) would have one less than the previous unevenly numbered position as shown in Table 10.2.
- To create the second order in the Latin square add 1 to each number apart from the last number N, which now becomes 1. So, in terms of our example with four conditions, the order is:

  2 (1 + 1), 3 (2 + 1), 1 (N), 4 (3 + 1)

  which corresponds to the conditions as first lettered A, C, B, D.
- To create further orders we simply proceed in the same way by adding 1 to the previous numbers except that the last number (4) becomes the first number (1). So, the order of the third row in our example becomes:

  3 (2 + 1), 4 (3 + 1), 2 (1 + 1), 1

  which corresponds to the conditions as originally lettered C, D, A, B.

  Our Latin square will look as follows:

  B, A, D, C
  A, C, B, D
  C, D, A, B
  D, B, C, A

Table 10.2 Position of unevenly numbered conditions in the first order of a Latin square

Order number	1	2	3	4	5	6	7	8	9	10
Condition number	1	2	N	3	$N-1$	4	$N-2$	5	$N-3$	6

We can see that each of the four letters occurs only once in each of the four orders or four columns of the square. Each letter is preceded and followed once by every other letter. For example, B is preceded once by C, A and D in the second, third and fourth rows, respectively. It is also followed once by A, D and C in the first, second and last rows, respectively.

- If there is an odd number of conditions, then two Latin squares are constructed. The first square is created as just described. The second square is produced by reversing the order of each of the rows in the first square so that the first condition becomes the last and the second condition becomes the second to last and so on. So, if we have five conditions and the first row of our initial Latin square is:

  C, A, D, E, B

  the first row of our reversed Latin square becomes:

  B, E, D, A, C

  With five conditions we would have 10 rows or orders.

- Participants are randomly assigned to each order. The number of participants for each order should be the same.

The Latin square could be used for controlling the order in which we measure the dependent variable when there are several of these being measured in a particular study.

10.5 The psychology and social psychology of the laboratory experiment

There is no doubt that the design of effective experiments is problematic. Some of the most troublesome issues in psychology experiments are not about the detail of design or the statistical analysis, but a consequence of the psychology laboratory being a social setting in which people interact and, it has to be said, in less than normal circumstances. Generically this can be referred to as the psychology and social psychology of the laboratory experiment. These issues are largely about the interaction between the participants and the experimenter, and the experimental procedures. Their consequence is to somewhat muddy the interpretation and validity of laboratory experiments. These are not recent ideas. Some stretch back into the history of psychology and most can be traced back 40 or 50 years. Some see these features as making experimentation untenable as the fundamental method of psychological research, others regard them as relatively trivial, but interesting, features of experimental research.

Placebo effect

Placebo effects have long been recognised in the evaluation of drugs and clinical treatments (Rivers, 1908). Careful attempts are made to control placebo effects in clinical trials though somewhat similar effects in other fields of psychological research may be left uncontrolled (Orne, 1959, 1969). A drug has two aspects: what the medication looks like and the active ingredients that it contains. Long ago, the medical researcher Beecher (1955) noted that participants who believed they were receiving the active medication but were in fact receiving bogus medication (which lacked the active ingredient) nevertheless showed similar improvement (treatment effects) to those who received the active ingredient. One possible explanation is that their expectations about the effectiveness of the drug bring about the apparent therapeutic change. The treatment that does not contain the component whose effectiveness is being evaluated is called a placebo or placebo treatment. *Placebo* is Latin for 'I shall be pleasing or acceptable'. The treatment is called a placebo because it is given to patients to please them into thinking that they are being treated.

In clinical trials of the effectiveness of drugs or clinical treatments, participants are told that either (a) they are being given the treatment (even though they are not) or (b) that they may receive either the placebo or the treatment but they are not told which they are receiving. In other words, they may not be aware of (that is they may be 'blind' to) which they are receiving. Furthermore, the person administrating the study should, ideally, be ignorant of what is actually happening to the participant. It is believed that the person responsible for, say, giving out the treatment may unconsciously convey to the participant information about which sort of treatment they are receiving. If the administrator does not know what treatment is being given, then no subtle cues may be communicated about whether or not the placebo was given, for example. In other words, it may be ideal if the administrator is not aware of, or is blind to, the treatment they are giving. This is known as a 'double-blind' – which means both the participants in the research and the administrator of the research are ignorant of whether the active or the placebo treatment has been given.

Sheldrake (1998) surveyed experimental papers that were published in important journals in a number of fields. One feature of this survey was the low frequency of the use of blind experimental procedures. Blind procedures were by far the most common in the field of parapsychology, where over four out of five studies used blind procedures. This is important since parapsychological research (i.e. into supernatural phenomena) is one type of research about which there is a great deal of scepticism. Hence the need for researchers in this field to employ the most exacting research methods since critics are almost certain to identify methodological faults. Equally clearly, some fields of research do not equally fear similar criticisms – or else blind procedures would be rather more common. So there may be an important lesson – that the more likely one's findings are to be controversial, the greater the need for methodological rigour in order to avoid public criticism.

Experimenter effects

In all of the concern about experimental design and statistical analysis, it is easy to overlook some important parameters of the experimental situation. If we concentrate on what participants do in an experiment we may ignore what effect the researcher is having. There is some evidence that the role of the researcher is not that of a neutral and unbiased collector of scientific data. Instead there is evidence that different characteristics of the experimenter may affect the outcome of the research. Some of these characteristics would include factors such as the race and gender of the researcher. But there are other features of experimenters which are shared by experimenters in general. An important one of these is that experimenters generally have a commitment to their research and the outcomes of their research. Rosnow (2002) indicates something of the extent to which the experimenter can influence the accuracy of the observations they record. For example, in an overview of a sizeable number of studies involving the observations of several hundreds of researchers, something like one in every hundred observations are incorrect when measured against objective standards. Some of the inaccuracy appears to be just that, but given that about two-thirds of the errors tended to support the experimenters' hypotheses then it would appear fair to accept that there is a small trend for researchers to make errors which favour their position on a topic (Rosenthal, 1978). Whether or not these 'consistent' errors are sufficient to determine the findings of studies overall is difficult to assess. However, even if they are insufficient acting alone, there is a range of other influences of the researcher which may be compounded with recording errors.

Of course, here we are talking about non-intentional errors of which the researchers would probably be unaware. Furthermore, we should not assume that biases solely exist at the level of data collection. There are clearly possibilities of the literature review or the conclusions including some form of systematic bias. One of the best ways of dealing with these is to be always sceptical of the claims of other researchers and check out key elements of their arguments, including those of Rosenthal. That is the essence of the notion of the scientific method anyway.

Experimenter expectancy effect

Rosenthal (1963, 1969) raised another potential problem in experimental studies. The idea is that experimenters may unintentionally influence participants into behaving in the way that the experimenter wants or desires. Barber (1973, 1976) described this as the experimenter unintentional expectancy effect. A typical way of investigating this effect involves using a number of student experimenters. Participants in the study are asked to rate the degree of success shown in photographs of several different people (Rosenthal and Rubin, 1978). Actually the photographs had been chosen because previous research had shown them to be rated at the neutral midpoint of the success scale. In this study, the student experimenters were deceived into believing either that previous research had shown the photographs to be rated as showing success or that the previous research had shown the photographs to be showing failure. However, of the 119 studies using this experimental paradigm (procedure), only 27 per cent found that student experimenters were affected by their expectations based on the putative previous research. So in most cases there was no evidence for expectancy effects. Nevertheless, if over a quarter of studies found evidence of an effect, then it should always be considered a possibility when designing research. Very few studies have included control conditions to determine to what extent expectancy effects may occur when the focus of the study is not the examination of expectancy effects.

Demand characteristics

Orne (1962, 1969) suggested that when participating in an experiment, participants are influenced by the totality of the situation which provides cues that essentially convey a hypothesis for the situation and perhaps indications of how they should behave. In many ways, the concept of demand characteristics cannot be separated from the notion of helpful and cooperative participants. Orne, for example, had noticed that participants in research when interviewed afterwards make statements indicating that they are aware that there is some sort of experimental hypothesis and that they, if acting reasonably, should seek to support the researcher in the endeavour. So, for example, a participant might say 'I hope that was the sort of thing you wanted' or 'I hope that I didn't mess up your study'. The demand characteristics explanation takes account of the totality of cues in the situation – it is not specifically about the experimenter's behaviour. The prime focus is on the participants and the influence of the totality of the situation on them. He proposed that certain features or cues of the experimental situation, including the way that the experimenter behaves, may lead the participant to behave in certain ways. These cues are called the demand characteristics of the experiment. This effect is thought to be largely unconscious in that participants are not aware of being affected in this way. Of course, some might question this as the cognitive processes involved seem quite complex. Orne only gives a few examples of studies where demand characteristics may be operating but these examples do not seem to quite clinch the matter.

One study examined sensory deprivation effects. The question was whether the apparent effects which researchers had found for the apparent effects of the deprivation of sensory stimulation for several hours could be the result of something else. Could it be that sensory deprivation effects were simply the result of the participants' *expectations* that they should be adversely affected (Orne and Scheibe, 1964)? To test this, a study was designed. In one condition participants underwent various procedures which indicated that they may be taking part in a study which had deleterious effects. For example, they were given a physical examination and were told that if they could not stand being in the sensory deprivation condition any longer then they could press a red 'panic' button and they would be released from the situation. In the other condition participants were put in exactly the same physical situation but were simply told that they were acting as the controls in a sensory deprivation study. The effects of these two conditions were examined in terms of 14 different measures (an example of a study using multiple dependent variables). However, there were significant differences in only 3 of the 13 measures where this difference was in the predicted direction.

It has to be stressed that Orne did not regard demand characteristics as just another nuisance source of variation for the experimenter to control. Indeed, the demand characteristics could not be controlled for, for example, by using sophisticated control conditions. Instead the demand characteristics needed to be understood using one major resource – the participants in the study themselves. Rosnow (2002) likens the role of demand characteristics to the greengrocer whose thumb is always on the scale. The bias may be small but it is consistent and in no sense random.

Orne's solution was to seek out information which would put the researcher on a better track to understanding the meaning of their data. Quasi-control strategies were offered which essentially change the status of the participants in the research from that of the 'subject' to a role which might be described as co-investigators. In the post-experiment interview, once the participant has been effectively convinced that the experiment is over, the experimenter and participant are free to discuss all aspects of the study. Things such as the meaning of the study as experienced by the participant could be explored. Of course, the participant needs to understand that the experimenter has concerns about the possibility that demand characteristics influenced behaviour in the experiment.

An alternative to this is to carry out a pre-inquiry. This is a mind game really in which the participants are asked to imagine that they are participating in the actual study. The experimental procedures are described to the participants in the mind experiment in a great deal of detail. The participant is then asked to describe how they believe that they would behave in these circumstances. Eventually, the experimenter is in a position to compare the conjectures about behaviour in the study with what actually happens in the study. An assessment may be made of the extent to which demand characteristics may explain the participants' actual behaviours. The problem is, of course, that the behaviours cannot be decisively identified as the consequence of demand characteristics.

Imagine an experiment (disregarding everything you learnt in the ethics chapter) in which the experimental group has a lighted cigarette placed against their skin whereas the control group has an unlighted cigarette placed against their skin. In a quasi-control pre-inquiry, participants will probably anticipate that the real participants will show some sort of pain reaction. Would we contemplate explaining such responses in the real experiment as the result of demand characteristics? Probably not. But what, for example, if in the actual experiment the participants in the lighted cigarette condition actually showed no signs of pain? In these circumstances the demand characteristics explanation simply is untenable. What, though, if the pre-inquiry study found that participants expected that they would remain stoical and stifle the expression of pain? Would we not accept the demand characteristics explanation in this case? This would surely clarify the meaning of the findings of a study in which participants know that they are participants of a study.

There are a number of demonstrations by Orne and others that participants in experiments tend to play the role of good participants. That is, they seem especially willing to carry out tasks which, ordinarily away from the laboratory, they would refuse to do or question. So it has been shown that participants in a laboratory will do endless body press-ups simply at the request of the experimenter. This situation of 'good faith' in which the participant is keen to serve the needs of the experiment may not always exist and it is a very different world now from when these studies were originally carried out in the middle of the twentieth century. But this too could be accommodated by the notion of demand characteristics.

Not surprisingly, researchers have investigated demand characteristics experimentally, sometimes using aspects of Orne's ideas. Demand characteristics have been most commonly investigated in studies manipulating feelings of elation and depression (Westermann, Spies, Stahl and Hesse, 1996).

Velten (1968) examined the effect of demand characteristics by having participants in the control conditions read various information about the corresponding experimental condition – for example, by describing the procedure used in this condition, by asking participants to behave in the way they think that participants in that condition would behave, and by asking them to act as if they were in the same mood as that condition was designed to produce. These are known as quasi-control studies. Participants in an elation condition rated their mood as significantly less depressed than those in the elation demand characteristics condition and participants in the depression condition rated their mood as significantly more depressed than those in the depression demand characteristics condition.

What to conclude? Orne-style quasi-control studies of the sort described above have one feature that would be valuable in any study, that participants and researcher get together as equals to try to understand the experience of participants in the research. Out of such interviews, information may emerge which can help the researcher understand their data. Not to talk to research participants is a bit like burying one's head in the sand to avoid exposure to problems. Researchers should want to know about every aspect of their research – whether or not this knowledge is comfortable. On the other hand, studies into the effects of demand characteristics often produce at best only partial evidence of their effects as we saw above. That is, quasi-participants who simply experience descriptions of the experimental procedures rarely if ever seem to reproduce the research findings in full. Whether such 'partial replications' of the findings of the original study are sufficient to either accept or reject the notion of demand characteristics is difficult to arbitrate on. Furthermore, it is not clear to what extent interviews with participants may themselves be subject to demand characteristics where participants tend to give experimenters the kinds of answers that participants think the experimenter wants to hear.

The important lesson learnt from the studies of the social psychology of the laboratory experiment is the futility of regarding participants in research as passive recipients of stimuli which affect their behaviour. The old-fashioned term subject seems to encapsulate this view better than anything. The modern term participants describes the situation more accurately.

10.6 Conclusion

Most true or randomised experimental designs include more than two conditions and measure more than one dependent variable, which are more often than not treated separately. Where the design consists of a single factor, the number of conditions is limited and may generally consist of no more than four or five conditions. The number of factors that are manipulated in a true or randomised design should also be restricted and may usually consist of no more than two or three manipulated variables. The reason for this advice partly rests on the difficulty of carrying out studies with many independent variables – the planning of them introduces many technical difficulties. Furthermore, the statistical analysis of very complex factorial designs is not easy and may stretch the statistical understanding of many researchers to the limit. For example, numerous complex interactions may emerge which are fraught with difficulties of interpretation. A computer program may do the number crunching for you but there its responsibility ends. It is for the researcher to make the best possible sense of the numbers, which is difficult when there are too many layers of complexity.

Hypotheses often omit consideration of effects due to basic demographic factors such as gender and age. Nevertheless, factorial designs can easily and usefully include such factors when numbers of participants permit. Of course, where the participants are, for example, students, age variation may be too small to be worthy of inclusion. Alternatively, where the numbers of females and males are very disproportionate it may also be difficult to justify looking for gender differences.

It is wise to make adequate provision in terms of participant numbers for trends to be statistically significant. Otherwise a great deal of effort is wasted. A simple but good way of doing this is to examine similar studies to inform yourself about what may be the minimum appropriate sample size. Alternatively, by running a pilot study a more direct estimate of the likely size of the experimental effect can be made and a sample size chosen which gives that size of effect the chance of being statistically significant. The more sophisticated way of doing this is to use power analysis. This is discussed in the companion book *Introduction to Statistics in Psychology* (Howitt and Cramer, 2011a). The way in which the results of factorial randomised designs are analysed can also be applied to the analysis of qualitative variables in non-randomised designs such as surveys, as we shall see in the next chapter.

A pilot study is also an opportunity for exploring the social psychological characteristics of the experiment being planned. In particular, it can be regarded as an opportunity to interview participants about their experience of the study. What they have to say may confirm the appropriateness of your chosen method but it, equally, may provide food for thought and a stimulus to reconsider some of the detail of the planned experiment.

Key points

- Advanced experimental designs extend the basic experimental group–control group paradigm in a number of ways. Several experimental and control groups may be used. More than one independent variable may be employed and several measures of the dependent variable.
- Because considerable resources may be required to conduct a study, the number of conditions run must be limited to those which are considered important. It is not advisable simply to extend the number of independent variables since this can lead to problems in interpreting the complexity of the output. Further studies may be needed when the interpretation of the data is hampered by a lack of sufficient information.
- Multifactorial designs are important since they are not only efficient in terms of the numbers of participants needed, but they can help identify interactions between the independent variables in the study. Furthermore, the relative influences of the various factors is revealed in a factorial design.
- Research which has insufficient cases to detect the effect under investigation is a waste of effort. The numbers in the cells of an experimental design need to be sufficient to determine that an effect is statistically significant. Previous similar studies may provide indications of appropriate sample sizes or a pilot study may be required to estimate the likely size of the effect of the factors. From this, the minimum sample size to achieve statistical significance may be assessed.

➔

- As with most designs, it is advantageous if the cells of a design have the same number of cases. This, for factorial designs, ensures that the effects of the factors are independent of one another. The extent to which factors are not independent can be determined by multiple regression. Many statistical tests work optimally with equal group sizes.
- Placebos and double-blind procedural controls should be routinely used in the evaluation of the effects of drugs to control for the expectations that participants and experimenters may have about the effects of the drugs being tested. Some of these procedures are appropriate in a variety of psychological studies.
- In general, it would seem that few researchers incorporate checks on demand characteristics and other social psychological aspects of the laboratory experiment. This may partly be explained by the relative lack of research into such effects in many areas of psychological research. However, since it is beneficial to interview participants about their experiences of the experimental situation, it is possible to discuss factors such as demand characteristics and expectancy effects with participants as part of the joint evaluation of the research by experimenter and participants.

ACTIVITIES

1. Answer the following questions in terms of the basic design that you produced for the exercise in Chapter 9 to investigate the effect of unemployment on crime. Can you think of reasons for breaking down the independent variable of unemployment into more than two conditions? If you can, what would these other conditions be? Are there ways in which you think participants may be affected by the manipulation of unemployment which are not part of unemployment itself? In other words, are there demand-type characteristics which may affect how participants behave? If there are, how would you test or control these? Which conditions would you compare and what would your predictions be about the differences between them? Is there more than one way in which you can operationalise crime? If there is, would you want to include these as additional measures of crime? Are there any subject or other independent variables that you think are worth including? If there are, would you expect any of these to interact with the independent variable of unemployment?

2. What is your nomination for the worst experiment of all time for this year's Psycho Awards? Explain your choice. Who would you nominate for the experimenter's hall of fame and why?

CHAPTER 11

Cross-sectional or correlational research

Non-manipulation studies

Overview

- Various terms describe research that, unlike the true or randomised experiment, does not involve the deliberate manipulation of variables. These terms include 'correlational study', 'survey study', 'observational study' and 'non-experiment'.
- Non-manipulation study is seen as the most accurate and most generic term to describe this type of study as there are problems with the others.
- There are many reasons why laboratory/experimental research cannot fulfil all of the research needs of psychology. Sometimes important variables simply cannot be manipulated effectively. Laboratory experiments can handle only a small number of variables at any one time, which makes it difficult to compare variables in terms of their relative influence. One cannot use experiments to investigate patterns or relationships among a large number of variables.
- Cross-sectional designs are typical of most psychological research. In cross-sectional designs, the same variable is measured on only one occasion for each participant. The question of causality cannot be tested definitively in cross-sectional designs though the relationships obtained are often used to support potential causal interpretations. These designs, however, help determine the direction and the strength of the association between two or more variables. Furthermore, the extent to which this association is affected by controlling other variables can also be assessed.

11.1 Introduction

The most common alternative to the true or randomised experiment is variously referred to as a non-experimental, correlational, passive observational, survey or observational study. There are inadequacies with each of these terms. An experiment has the implication of some sort of intervention in a situation in order to assess the consequences of this intervention. This was more or less its meaning in the early years of psychological research. Gradually the experiment in psychology took on the more formal characteristics of randomisation, experimental and control groups, and control of potentially confounding sources of variation. However, in more general terms, an experiment is generally defined along the lines of being a test or trial (Allen, 1992) which does not necessarily involve all of the formal expectations of the randomised psychology experiment. In this more general context, we could be interested in testing whether one variable, such as academic achievement at school, is related to another variable, such as parental income. However, this would be a very loose use of language in psychology and the key requirement of the manipulation of a variable is the defining feature. This is clear, for example, when Campbell and Stanley (1963, p. 1) state that an experiment is taken to refer to 'research in which variables are manipulated and their effects upon other variables observed'. It should be clear from this that a non-experiment in psychology refers to any research which does not involve the manipulation of a variable. However, it may be better to state this directly by referring to a non-manipulation rather than a non-experimental study.

Campbell and Stanley (1963, p. 64) use the term 'correlational' to describe designs which do not entail the manipulation of variables. Today it is a very common term to describe this sort of research. Later, however, Cook and Campbell (1979, p. 295) point out that the term 'correlational' describes a statistical technique, not a research design. So correlational methods can be used to analyse the data from an experiment just as they can be used in many other sorts of quantitative research. A particular set of statistical tests have traditionally been applied to the data from experiments (e.g. *t*-tests, analyses of variance). However, it is perfectly feasible to analyse the same experiment with a correlation coefficient or multiple regression or other techniques. It is important to appreciate that the common distinction between correlation and differences between means is more apparent than real (see Box 11.1). In the same way, data from non-experimental studies can frequently be analysed using the statistical techniques common in reports of experiments – *t*-tests and analyses of variance, for example. We would most probably apply a two-way analysis of variance to determine whether the scores on a measure of depression varied according to the gender and the marital status of participants. Although researchers would not normally do this, the same data could be analysed using multiple regression techniques. In other words, analysis of variance and multiple regression are closely linked. This is quite a sophisticated matter.

Although it never gained popularity, Cook and Campbell (1979, p. 296) suggested the term *passive observational* to describe a non-manipulation study. The adjective passive implies that the study does not involve a manipulation in this context. However, to refer to most research procedures as being passive reflects the situation very poorly. For instance, observation itself is often thought of as active rather than passive. However, like other forms of human perception, there is a degree of selectivity in terms of what is being observed (Pedhazur and Schmelkin, 1991, p. 142). Furthermore, even the value of the term 'observational' in this context is problematic since it can be equally applied to the data collection methods in experiments and other types of research. As observation can be used in a true or randomised experiment, this term does not exclude randomised

Box 11.1 Key Ideas

Tests of correlation versus tests of difference

Although at first there may seem to be a confusing mass of different statistical techniques, many of them are very closely related as they are based on the same general statistical model. For example, both a Pearson's product moment correlation (r) and an unrelated t-test for data with similar variances can be used to determine the relationship between a dichotomous variable such as gender and a continuous variable such as scores on a measure of depression. Both these tests will give you the same significance level when applied to the same data. The relationship between the two tests is

$$r = \sqrt{\frac{t^2}{t^2 + \text{df}}}$$

where df stands for the degrees of freedom. The degrees of freedom are two fewer than the total number of cases.

The dichotomous variable could equally be the experimental condition versus the control condition. Hence the applicability of both tests to simple experiments. The dichotomous variable is coded 1 and 2 for the two different values whether the variable being considered is gender or the independent variable of an experiment (this is an arbitrary coding and could be reversed if one wished).

studies. Cook and Campbell (1979, p. 296) themselves objected to using the term 'observational study' because it would apply to what they called quasi-experiments which did not involve randomisation.

The term 'survey' is also not particularly appropriate either. It too refers to a method of data collection which typically involves asking people questions. It also has the connotation of drawing a precise sample from a population such as in stratified random sampling (see p. 233). Many studies in psychology have neither of these features and yet are not randomised experiments.

In view of the lack of a satisfactory term, we have given in to the temptation to use non-manipulation study to refer to this kind of study. This term is less general than non-experimental, refers to the essential characteristic of an experiment and does not describe a method of data collection. However, we realise we are even less likely than Cook and Campbell (1979, p. 295) 'to change well-established usage'. In the majority of the social sciences, the distinction would not be very important. In psychology it is important for the simple reason that psychology alone among the social sciences has a strong commitment to laboratory experiment. Of course, medicine and biology do have a tradition of strict experimentation and may have similar problems over terminology. Psychology has its feet in both the social and the biological sciences.

11.2 Cross-sectional designs

The most basic design for a cross-sectional study involves just two variables. These variables may both be scores, may both be nominal categories, or there may be a mixture of nominal and score variables. For example, we could examine the relationship between gender and a diagnosis of depression. In this case both variables would consist of two binary values – male versus female, diagnosed as depressed versus not diagnosed as depressed. Such basic designs would provide very limited information which can restrict their interest to researchers and consequently their use in professional research. They are probably too simple to warrant the time and effort expended when they could benefit from the collection of a wider variety of data with possibly little or no more effort on the part of an experienced researcher. This would be the more usual approach and it is one that should be naturally adopted by student researchers.

So, ideally, you should think in terms of a minimum of three variables for a cross-sectional study but realistically there are advantages in extending this further. The reason for considering a minimum of three variables is that the third variable introduces the possibility of including controls for potentially confounding variables or investigating possible intervening variables. There is often every advantage of introducing more variables and more than one measure of the same variable. This is *not* an invitation to throw into a study every variable that you can think of and have a means of measuring. The reasons for adding in more than the minimum number of variables is that the additional information they yield has the potential to clarify the meaning of the relationship between your primary variables of interest. Ideally this is a careful and considered process in which the researcher anticipates the possible outcomes of the research and adds in additional variables which may contribute positively to assessing just what the outcome means. Merely throwing in everything is likely to lead to more confusion rather than clarification. So don't do it.

The cross-sectional design is as difficult to execute as any other form of study, including the laboratory experiment. The skills required to effectively carry out field work are not always the same as those for doing experiments, but they are in no sense less demanding. Indeed, when it comes to the effective statistical analysis of cross-sectional data, this may be more complex than that required for some laboratory experiments. The reason for this is that non-manipulation studies employ statistical controls for unwanted influences whereas experimental studies employ procedural controls to a similar end. Furthermore, the cross-sectional study may be more demanding in terms of numbers of participants simply because the relationship between the variables of interest is generally smaller than would be expected in a laboratory experiment. In the laboratory, it is possible to maximise the obtained relationships by controlling for the 'noise' of other variables – that is by standardising and controlling as much as possible. In a cross-sectional design, we would expect the relationships between variables to be small and a correlation of about .30 would be considered quite a promising trend by many researchers. For a correlation of this size to be statistically significant at the two-tailed 5 per cent or .05 level would require a minimum sample size of over 40.

The need for statistical controls for the influence of third variables in cross-sectional and all non-manipulation studies makes considerable demands on the statistical knowledge of the researcher. Many of the appropriate statistics are not discussed in many introductory statistics texts in psychology. One exception to this is *Introduction to Statistics in Psychology* (Howitt and Cramer, 2011a). Given the complexity of some of the statistical techniques together with the substantial numbers of variables that can be involved means that the researcher really ought to use a computer program capable of analysing these sorts of data well. The companion computing text *Introduction to SPSS Statistics in Psychology* (Howitt and Cramer, 2011b) will help you make light work of this task.

Because many variables which are of interest tend to be correlated with each other, samples have to be larger when the relationship between three or more variables are investigated together. It is difficult to give an exact indication of how big a sample should be because this depends on the size of the associations that are expected, but in general the size of the sample should be more than 60.

11.3 The case for non-manipulation studies

There are a number of circumstances which encourage the use of non-manipulation studies just as there are other circumstances in which the randomised laboratory experiment may be employed to better effect (see Figure 11.1):

- *Naturalistic research settings* Generally speaking, randomised experiments have a degree of artificiality which varies but is probably mostly present. Although there have been a number of successful attempts to employ randomised experiments in the 'field' (natural settings), these have been relatively few and risked losing the advantages of the laboratory experiment. Consequently, given that research is a matter of choices, many psychologists prefer not to do randomised experiments at all. There are arguments on all sides, but research is a matter of balancing a variety of considerations and that balance will vary between researchers and across circumstances. So non-manipulation studies can seem to be much more naturalistic.
- *Manipulation not possible* It is not always possible, practical or ethical to manipulate the variable of interest. This would be the case, for example, if you were interested in looking at the effects of divorce on children. In this situation you could compare

Non-manipulative studies can be much more naturalistic

In much research it is not possible to manipulate variables

To establish that there is an association before carrying out an experiment to further examine this

To understand what the strength of the real-life relationship is between two variables

To determine what variables might be potentially the most important influences on other variables

To predict values of one variable from another, e.g. when making selection choices

Developing explanatory models in real life prior to testing aspects of the model in the laboratory

To understand the structural features of things such as intelligence

To study how something changes over time

To understand the temporal direction of associations – what changes come before other changes

FIGURE 11.1 A summary of the various uses of non-manipulative studies

children from parents who were together with children from parents who were divorced. However, divorce cannot be assigned at random by the researcher.

- *Establishing an association* You may wish to see whether there is a relationship between two or more variables before committing resources to complex experiments designed to identify causal relationships between those variables. For example, you may wish to determine whether there is an association between how much conflict there is in a relationship and how satisfied each partner is with the relationship before seeing whether reducing conflict increases satisfaction with the relationship. Finding an association does not mean that there is a causal relationship between two variables. This could be an example of where a third variable is confusing things. For example, low income may make for stressful circumstances in which couples are in conflict more often and are less satisfied with their relationship because shortage of cash makes them less positive about life in general. In this example, this confounding factor of income makes it appear that conflict is causally related to dissatisfaction when it is not. Conversely, the failure to find an association between two variables does not necessarily mean that those variables are not related. The link between the variables may be suppressed by other variables. For instance, there may be an association between conflict and dissatisfaction, but this association may be suppressed by the presence of children. Having children may create more conflict between partners, but may also cause them to be more fulfilled as a couple.
- *Natural variation* In experiments, every effort is made to control for variables which may influence the association between the independent and dependent variables. In a sense, by getting rid of nuisance sources of variation, the key relationship will be revealed at its strongest. But what if your desire is to understand what the relationship is when these other factors are present, as they normally would be in real life? For example, when manipulating a variable you may find that there is a very strong association between the dependent variable and the independent variable but this association may be weaker when it is examined in a natural setting.
- *Comparing the sizes of associations* You may want to find out which of a number of variables are most strongly associated with a particular variable. This may help decide which ones would be the most promising to investigate further or which ones need controlling in a true experiment. For example, if you wanted to develop a programme for improving academic achievement at school, it would be best to look at those variables which were most strongly related to academic achievement rather than those which were weakly related to it.
- *Prediction and selection* You may be interested in determining which variables best predict an outcome. These variables may then be used for selecting the most promising candidates. For example, you may be interested in finding out which criteria best predict which prospective students are likely to be awarded the highest degree marks in psychology and use these criteria to select applicants.
- *Explanatory models* You may want to develop what you consider to be an explanatory model for some behaviour and to see whether your data fit that model before checking in detail whether your assumptions about causes are correct. For example, you may think that children with wealthier parents perform better academically than children with poorer parents because of differences in the parents' interest in how well their children do academically. It may be that children with wealthier parents have parents who show more interest in their academic progress than children with poorer parents. As a consequence, children of wealthier parents may try harder and so do better. If this is the case, parental interest would be a mediating or intervening variable which mediates or intervenes between parental wealth and academic achievement.

- *Structure* You may be interested in determining what the structure is of some characteristic such as intelligence, personality, political attitudes or love. For example, you may be interested in seeing whether there is a general factor of intelligence or whether there are separate factors of intelligence such as memory, verbal ability, spatial ability and so on.
- *Developing or refining measures* You may want to develop or refine a measure in which you compare your new or refined measure with some criterion. For instance, you may want to refine a measure of social support. More social support has been found to be related to less depression so you may wish to see whether your refined measure of social support correlates more strongly with depression than the original measure.
- *Temporal change* You may wish to see whether a particular behaviour changes over time and, if it does, to what variables those changes are related. For example, has the incidence of divorce increased over the last 50 years and, if so, with what factors is that increase associated?
- *Temporal direction of associations* You may wish to determine what the temporal direction of the association is between two variables. For example, does parental interest in a child's academic achievement at school affect the child's achievement or is the causal direction of this association the other way round with the child's academic achievement influencing the parents' interest in how well their child is doing academically? Of course, both these casual sequences may be possible. An association where both variables affect each other is variously known as a bi-directional, bilateral, non-recursive, reciprocal or two-way association.

Chapter 12 discusses methods of researching changes over time.

11.4 Key concepts in the analysis of cross-sectional studies

There are a number of conceptual issues in the analysis of cross-sectional studies which need to be understood as a prelude to the more purely statistical matters (see Figure 11.2).

Varying reliability of measures

The concept of reliability is quite complex and is dealt with in detail in Chapter 15. There are two broad types of reliability. The first type is the internal consistency of a psychological scale or measure (i.e. how well the items correlate with the other items ostensibly measuring the same thing). This may be measured as the split-half reliability but is most often measured as Cronbach's (1951) alpha which is a more comprehensive index than that of split-half reliability. Generally speaking a reliability of about .70 would be regarded as satisfactory (Nunnally, 1978). The other type of reliability is the stability of the measure over time. This is commonly measured using test–retest reliability. All of these measures vary from 0 to 1.00 just like a correlation coefficient.

The crucial fact about reliability is that it limits the maximum correlation a variable may have with another variable. The *maximum* value of the correlation of a variable with a reliability of .80 with any other variable is .80, that is, the figure for reliability. The *maximum* value that the correlation between two variables may have is the square root of the product of the two reliabilities. That is, if one reliability is .80 and the reliability of the other measure is .50 then the maximum correlation between these two variables

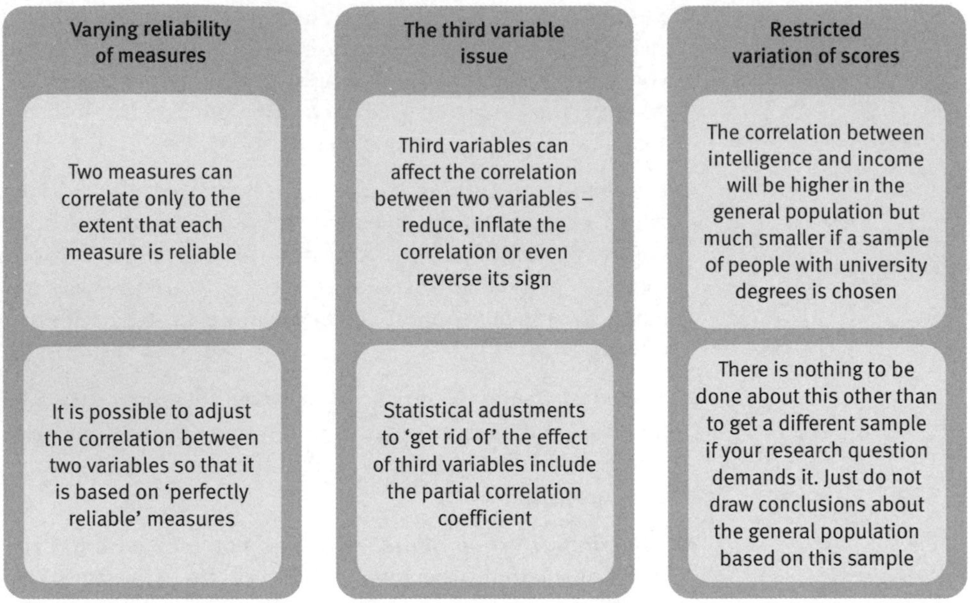

Varying reliability of measures	The third variable issue	Restricted variation of scores
Two measures can correlate only to the extent that each measure is reliable	Third variables can affect the correlation between two variables – reduce, inflate the correlation or even reverse its sign	The correlation between intelligence and income will be higher in the general population but much smaller if a sample of people with university degrees is chosen
It is possible to adjust the correlation between two variables so that it is based on 'perfectly reliable' measures	Statistical adustments to 'get rid of' the effect of third variables include the partial correlation coefficient	There is nothing to be done about this other than to get a different sample if your research question demands it. Just do not draw conclusions about the general population based on this sample

FIGURE 11.2 Factors which alter the apparent relationships between two variables

is .63 ($\sqrt{(.80 \times .50)} = \sqrt{.40} = .632$). Remember, that is the maximum and that we suggested that quite a good correlation between two variables in a cross-sectional study might be .30. So that correlation might be stronger than it appears if it were not for the influence of the lack of reliability of one or both of the variables.

If one knows the reliabilities (or even one reliability) then it is quite easy to correct the obtained correlation for the unreliability of the measures. One simply divides the correlation coefficient by the square root of the product of the two variables. So in our example, the correlation is .30 divided by the square root of $.80 \times .50$. This gives $.30/\sqrt{.40} = .30/.63 = .48$. This is clearly indicative of a stronger relationship than originally found, as might be expected.

While such statistical adjustments are a possibility and very useful if one has the reliabilities, this is not always the case. The other approach is to ensure that one's measures have the best possible opportunity for being reliable. This might be achieved, for example, by standardising one's procedures to eliminate unnecessary sources of variability. So, for example, if different interviewers ask about a participant's age in different ways then this will be a source of unnecessary variation (and consequently unreliability). For example, 'What age are you now?', 'About what age are you?' and 'Do you mind telling me your age?' might produce a variety of answers simply because of the variation of wording the question. For example, 'Do you mind telling me your age?' might encourage the participant to claim to be younger than they are simply because the question implies the possibility that the participant might be embarrassed to reveal their age.

Another problem for student researchers is the ever-increasing sophistication of the statistics used by professional researchers when reporting their findings. For example, the statistical technique of structural equation modelling is quite commonly used to correct for reliability. Structural equation modelling is also known as analysis of covariance structures, causal modelling, path analysis and simultaneous equation modelling. In their survey of statistical tests reported in a random sample of papers published in the *Journal of Personality and Social Psychology*, Sherman and his colleagues (1999) found that 14 per cent of the papers used this technique in 1996 compared with 4 per cent in 1988 and 3 per cent in 1978. A brief introduction to structural equation modelling may be found in our companion book, *Introduction to SPSS Statistics in Psychology* (Howitt and Cramer, 2011b). For the example above, structural equation modelling gives a standardised coefficient which has the same value as the correlation corrected for unreliability. We cannot imagine that undergraduate students would use this technique except when closely supervised by an academic but postgraduate students may well be expected to use it.

The third variable issue

The confounding variable problem is the classic stumbling block to claiming causal relationships in non-experimental research. The problem is really two fold. The first aspect is that we cannot be sure that the relationship between two variables cannot be explained by the fact that both of them are to a degree correlated with a third variable; these relationships may bring about the original correlation. This, remember, is in the context of trying to establish whether variable A is the *cause* of variable B. The other problem is that it is very difficult to anticipate quite what the effect of a third variable is – it actually can increase correlations as well as decrease them. Either way it confuses the meaning of the correlation between variables A and B.

Suppose we find that the amount of support in a relationship is positively related to how satisfied partners are. The correlation is .50 as shown in Figure 11.3. Further suppose that the couple's income is also positively related to both how supportive the partners are and how satisfied they are with the relationship. Income is correlated .60 with support and .40 with satisfaction. Because income is also positively related to both support and satisfaction it is possible that part or all of the association between support and satisfaction is accounted for by income. To determine if this is the case, we can partial out the influence of income. In other words, we can remove the influence of income. One way of doing this is to use partial correlation. This is the statistical terminology for what psychologists would normally call controlling for a third variable. That is, partialling = controlling.

Partialling is quite straightforward computationally. The basic formula is relatively easy to compute. Partialling is discussed in detail in the companion book *Introduction to Statistics in Psychology* (Howitt and Cramer, 2011a). However, we recommend using SPSS Statistics or some other package since this considerably eases the burden of calculation and risk of error when controlling for several variables at the same time. It is also infinitely less tedious.

The (partial or first-order) correlation between support and satisfaction is .35 once income has been partialled out. This is a smaller value than the original (zero order) correlation of .50. Consequently, income explains part of the association between support and satisfaction. How is this calculated?

The following formula is used to partial out one variable where A refers to the first variable of support, B to the second variable of satisfaction and C to the third or confounding variable of income:

$$r_{AB.C} = \frac{r_{AB} - (r_{AC} \times r_{BC})}{\sqrt{(1 - r^2_{AC}) \times (1 - r^2_{BC})}}$$

support .50 satisfaction
.60 .40
income

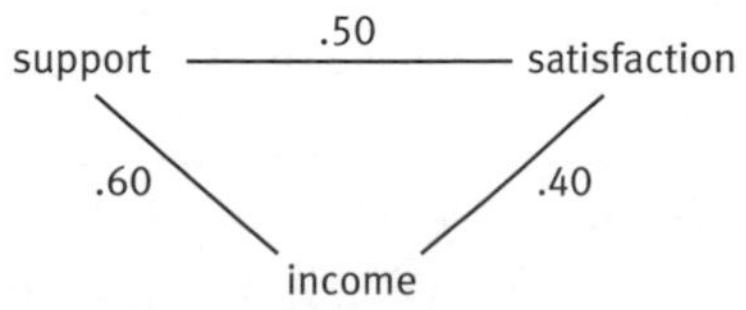

FIGURE 11.3 Correlations between support, satisfaction and income

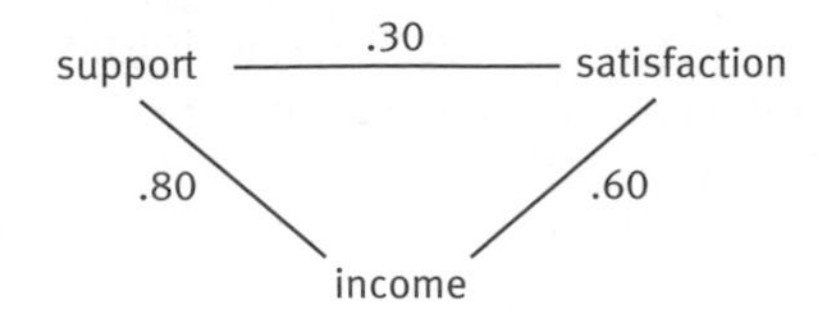

FIGURE 11.4 Partialling leading to a change in sign of the association between support and satisfaction

We can substitute the correlations in Figure 11.3 in this formula:

$$r_{\text{AB.C}} = \frac{.50 - (.60 \times .40)}{\sqrt{(1 - .60^2) \times (1 - .40^2)}} = \frac{.260}{.733} = .35$$

Sometimes the influence of the third variable is to make the association between the two main variables bigger than it actually is. This kind of variable is known as a *suppressor* variable. The sign of the partial correlation can also be opposite to that of the original correlation (Cramer, 2003). This radical change occurs when the correlation has the same sign as, and is smaller than, the product of the other two correlations. An example of this is shown in Figure 11.4. The correlation between support and satisfaction is .30. The signs of all three correlations are positive. The product of the other two correlations is .48 (.60 × .80 = .48) which is larger than the correlation of .30 between support and satisfaction. When income is partialled out, the correlation between support and satisfaction becomes −.38. In other words, when income is removed, more support is associated with less rather than greater satisfaction. Rosenberg (1968) refers to variables which when partialled out change the direction of the sign between two other variables as *distorter* variables. He discusses these in terms of contingency tables of frequencies rather than correlation as is done here. Cohen and Cohen (1983), on the other hand, include change of sign as a suppressor effect. Different types of third variables are discussed in detail in Chapter 12.

Restricted variation of scores

This is probably the most technical of the considerations which lower the correlation between two variables. Equally, it is probably the least well recognised by researchers. A good example of a reduced correlation involves the relationship between intelligence and creativity in university students. University students have a smaller range of intelligence than the general population because they have been selected for university as they are more intelligent. The correlation between intelligence and creativity is greater in samples where the range of intelligence is less restricted.

To be formal about this, the size of the correlation between two variables is reduced when:

- the range or variation of scores on one variable is restricted *and*
- when the scatter of the scores of two variables about the correlation line is fairly constant over the entire length of that line (e.g. Pedhazur and Schmelkin, 1991, pp. 44–5).

A correlation line is the straight line which we would draw through the points of a scattergram (i.e. it is a regression line) but the scores on the two variables have been turned into *z*-scores or standard scores. This is simply done by subtracting the mean of the scores and then dividing it by the standard deviation. This straight line

Table 11.1 Scores on two variables for ten cases

Case number	A	B	Case number	A	B
1	1	1	6	6	8
2	1	3	7	7	4
3	2	4	8	8	6
4	3	6	9	9	7
5	4	2	10	9	9

best describes the linear relationship between these two variables (see, for example, Figure 11.5). If the scatter of scores around the correlation line is not consistent, then it is not possible to know what the size of the correlation is as this will vary according to the scatter of the scores.

The effects of restricting range can be demonstrated quite easily with the small set of ten scores shown in Table 11.1. The two variables are, respectively, called A and B. The scores of these two variables are plotted in the scattergram in Figure 11.5 which also shows the correlation line through them. Although the set of scores is small we can see that they are scattered in a consistent way around the correlation line. The correlation for the ten scores is about .74. If we reduce the variation of scores on B by selecting the five cases with scores either above or below 5, the correlation is smaller at about .45. The correlation is the same in these two smaller groups because the scatter of scores around the correlation line is the same in both of them.

Of course, you need to know whether you have a potential problem due to the restricted range of scores. You can gain some idea of this if you know what the mean or the standard deviation of the unrestricted scores is:

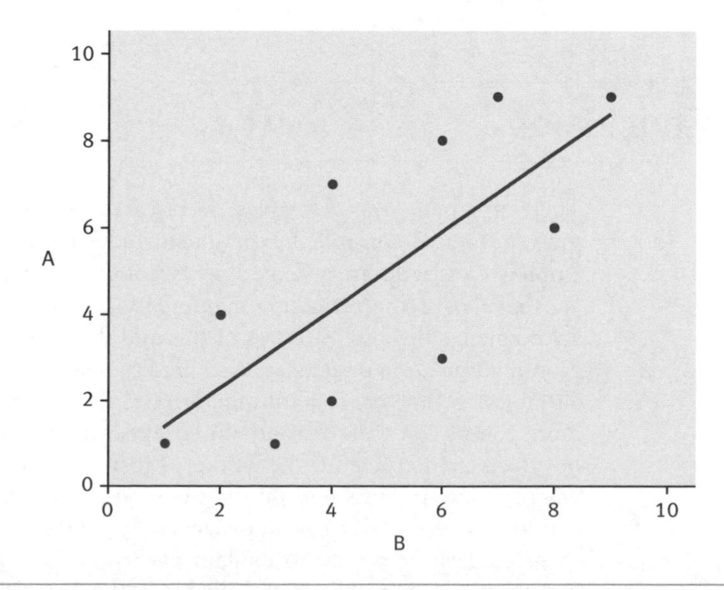

FIGURE 11.5 Scattergram with a correlation line

- If the mean score is much higher (or lower) than the mean score of unrestricted scores, then the variation in scores is more likely to be restricted as the range for scores to be higher (or lower) than the mean is reduced. For example, the mean score for the two variables in Table 11.1 is 5.00. The mean score for the five cases scoring higher than 5 on variable B is 7.20 [(6 + 8 + 6 + 7 + 9)/5 = 36/5 = 7.20]. As the mean for these five scores is higher than the mean of 5.00 for the ten cases, the potential range for these five scores to be higher is less than that for the ten scores.
- The standard deviation (or variance) is a direct measure of variance. If the standard deviation of a set of scores is less than that for the unrestricted scores, then the variance is reduced. The standard deviation for variable B of the ten scores is about 2.63, whereas it is about 1.30 for the five scores both above and below the mean score of 5.

It is important to understand that generally in your research the effects of the range of scores may be of no consequence. For example, if one is interested in the relationship between creativity and intelligence in university students then there is no problem. However, it would be a problem if one were interested in the general relationship between intelligence and creativity. In this case, the restriction on the range of intelligence in the university sample might be so great that no significant relationship emerges. It is misleading to conclude from this that creativity is not related to intelligence since it may well be in the general population. Equally, you may see that studies, ostensibly on the same topic, may appear to yield seemingly incompatible findings simply because of the differences in the samples employed.

Another implication, of course, is of the undesirability of using restricted samples when exploring general psychological processes. The study of university students as the primary source of psychological data is not bad simply because of the restrictions of the sampling but also because of the restrictions likely on the distributions of the key variables.

There is more information on statistics appropriate to the analysis of cross-sectional designs in Chapter 12. Multiple regression and path analysis are discussed there and can help the researcher take full advantage of the fullness of the data which cross-sectional and other non-manipulation designs can provide.

11.5 Conclusion

There are numerous examples of research which cannot meet the requirements of the randomised controlled experiment. Indeed, psychology is somewhat unusual in its emphasis on laboratory experiments compared with other social science disciplines. We have used the term non-manipulation design for this sort of study though we acknowledge the awkwardness of this and the many other terms for this type of design. Non-manipulation designs are also used to determine the size of the association between variables as they occur naturally. Studies using these designs generally involve testing more cases than true or randomised experiments because the size of the associations or effects are expected to be weaker. Most of these studies use a cross-sectional design where cases are measured on only one occasion (Sherman *et al.*, 1999). Although the causal order of variables cannot generally be determined from cross-sectional designs, these studies often seek to explain one variable in terms of other variables. In other words, they assume that one variable is the criterion or dependent variable while the other variables are predictor or independent variables.

Key points

- The main alternative to controlled and randomised experiments is the cross-sectional or non-manipulation study. There are numerous problems with the available terminology. We have used the term non-manipulation study.
- Non-manipulation studies enable a large number of variables to be measured relatively easily under more natural conditions than a true or randomised study. This allows the relationships between these variables to be investigated. The cost is that these studies can be complex to analyse especially when questions of causality need to be raised.
- The more unreliable measures are, the lower the association will be between those measures. Adjustments are possible to allow for this.
- Partial correlation coefficients which control for third variable effects are easily computed though a computer is probably essential if one wishes to control for several third variables.
- Restricting the range of scores on one of the variables will reduce the correlation between two variables.

ACTIVITIES

1. Design a non-manipulation study that investigates the hypothesis that unemployment may lead to crime. How would you measure these two variables? What other variables would you investigate? How would you measure these other variables? How would you select your participants and how many would you have? What would you tell participants the study was about? How would you analyse the results?

2. What would be the difficulties of studying in the psychology laboratory the hypothesis that unemployment may lead to crime?

CHAPTER 12

Longitudinal studies

Overview

- Longitudinal studies examine phenomena at different points in time.
- A panel or prospective study involves looking at the same group of participants on two or more distinct occasions over time.
- Ideally exactly the same variables are measured on all occasions, although you will find studies where this is not achieved.
- Longitudinal studies may be used to explore the temporal ordering or sequence of these variables (i.e. patterns of variation over time). This is useful in determining whether the association is two-way rather than one-way, that is, both variables mutually affecting each other although possibly to different degrees.
- The concepts of internal and external validity apply particularly to longitudinal studies. The researcher needs to understand how various factors such as history and changes in instrumentation may threaten the value of a study.
- Cross-lagged correlations are the correlations between variable *X* and variable *Y* when these variables are measured at different points in time. A lagged correlation is the correlation between variable *X* measured at time 1 and variable *X* measured at time 2.
- Multiple regression and path analysis are important statistical techniques in the analysis of complex non-manipulation studies. The extra information from a longitudinal study adds considerably to their power.

12.1 Introduction

Longitudinal studies offer the important advantage that they assess patterns of change over time. This enables a fuller interpretation of data than is possible with cross-sectional designs of the sort discussed in Chapter 11. It would appear self-evident that the study of change in any psychological phenomenon at different points in the life cycle is important in its own right. So, for example, it is clearly important to understand how human memory changes (or does not change) at different stages in life. There are numerous studies which have attempted to do this for all sorts of different psychological processes. Despite this, there is a quite distinct rationale for studying change over time which has much less to do with life cycle and other developmental changes.

Remember that one of the criteria by which cause and effect sequences may be studied is that the cause must precede the effect and the effect must follow the cause. Longitudinal studies by their nature allow the assessment of the relationship between two variables over a time period. In other words, one of the attractions of longitudinal studies is that they may help to sort out issues of causality. Hence, you may find causality to be the central theme of many longitudinal studies to the virtual exclusion of the process of actually studying change over time for its own sake. Most frequently, variables are measured only once in the majority of studies. These are referred to as cross-sectional studies as the variables are measured across a section of time. These designs were discussed in detail in Chapter 11. Studies in which variables are measured several times at distinct intervals have been variously called longitudinal, panel or prospective studies. However, each of these terms implies a somewhat different type of study (see Figure 12.1). For example, a panel study involves a group of participants (a panel) which is studied at different points in time. On the other hand, a longitudinal study merely requires that data be collected at different points in time.

So there are various kinds of longitudinal designs depending on the purpose of the study. Designs where the same people are tested on two or more occasions are sometimes referred to as *prospective studies* (Engstrom, Geijerstam, Holmbery and Uhrus, 1963) or *panel studies* (Lazarsfeld, 1948) as we have indicated. This type of design was used to study American presidential elections, for example. As you can imagine, because American

Panel study	Longitudinal study	Retrospective study
This involves a group of participants (the panel) who are studied at several points in time using similar measures	A study in which samples are studied at different times – there is no implication that it is the same group each time	In part, these studies ask participants to answer questions, etc. from a past perspective, i.e. retrospectively
This sort of study has great potential to help reveal causal pathways	Although capable of showing general changes over time, it cannot make strong claims about causality	Perceptions of the past may be affected by the present so this design is weak in terms of establishing causality

FIGURE 12.1 Types of study to investigate changes over time and the causal sequences involved

elections take place over a number of months every four years, voters are subject to a great deal of media and other pressure. Some change their minds about the candidates, some change their minds about the parties they were intending to vote for. Some change their minds again later still. So there are enormous advantages in being able to interview and re-interview the same group of participants at different points during the election. The alternative would be to study the electorate at several different points during the election but using different samples of the electorate each time. This causes difficulties since although it is possible to see what changes over time, it is not possible to relate these changes to what went before easily. So such a design might fail to provide the researcher with information about what sort of person changed their minds under the influence of, say, the media and their peers.

There would be enormous benefit in being able to study criminals over the long term. Some such studies have been done. For example, Farrington (1996) has studied the same

Box 12.1 Key Ideas

Threats to internal and external validity

The concepts of internal and external validity originate in the work of Campbell and Stanley (1963). They are particularly important and salient to longitudinal studies.

Internal validity is concerned with the question of whether or not the relationship between two variables is causal. That is, does the study help the researcher identify the cause and effect sequence between the two variables? It also refers to the situation where there is no empirical relationship between two variables. The question is, then, whether this means that there is no causal relationship or whether there is a relationship which is being hidden due to the masking influence of other variables. Cook and Campbell (1979) list a whole range of what they refer to as 'threats to internal validity'. Some of these are listed below and described briefly:

- *History* Changes may occur between a pre-test and a post-test which are nothing to do with the effect of the variable of interest to the researcher. In laboratory experiments participants are usually protected from these factors. Greene (1990) was investigating the influence of eyewitness evidence on 'juries' under laboratory conditions. She found that a spate of news coverage of a notorious case where a man had been shown to be unjustly convicted on the basis of eyewitness evidence affected things in the laboratory. Her 'juries' were, for a period of time, much less likely to convict on the basis of eyewitness testimony.
- *Instrumentation* A change over time may be due to changes in the measuring instrument over time. In the simplest cases, it is not unknown for researchers to use different versions of a measuring instrument at different points in time. But the instrumentation may change for other reasons. For example, a question asking how 'gay' someone felt would have had a very different meaning 50 years ago from today.
- *Maturation* During the course of a longitudinal study a variety of maturation changes may occur. Participants become more experienced, more knowledgeable, less energetic and so forth.
- *Mortality* People may drop out of the study. This may be systematically related to the experimental condition or some other characteristics. This will not be at random and may result in apparent changes when none has occurred.
- *Statistical regression* If groups of people are selected to be, say, extremely high and extremely low on aggression at point 1 in time, then their scores on aggression at point 2 in time will tend to converge. That is, the high scorers get lower scores than before and the low scorers get higher scores than before. This is purely a statistical artefact known as regression to the mean.
- *Testing* People who are tested on a measure may be better on that measure when they are retested later simply because they are more familiar with its contents or because they have had practice.

External validity is closely related to the issue of generalisation discussed in detail in Chapter 4. It has to do with generalising findings to other groups of individuals, other geographic settings and other periods of time.

delinquent children from early childhood through adulthood. As can be imagined, the logistical difficulties are enormous. Another example is the study of criminals who start their criminal careers late on in life. This is much more problematic to study and is, as a consequence, little investigated. If you wished to study the criminal careers of late-onset criminals then an enormous sample of children would be required. Some may turn out to be late-onset criminals but the vast majority would not. It is obviously easier to start with a sample of delinquents and study their progress than to try to obtain a sample of children, some of whom will turn criminal late in life. Hence the rarity of such studies.

Retrospective studies are ones in which information is sought from participants about events that happened prior to the time that they were interviewed. Usually this also involves the collection of information about the current situation. Of course, it is perfectly possible to have a study which combines the retrospective design and the prospective design. The sheer logistical requirements of longitudinal studies cannot be overestimated: following a sample of delinquent youth from childhood into middle age has obvious organisational difficulties. Furthermore, the timescale is very long – possibly as long as a typical academic career – so alternatives may have to be contemplated such as using retrospective studies in which the timescale can be truncated in real time by carrying out retrospective interviews with the offenders as adults to find information about their childhood. These adults can then be studied into middle age within a more practical timescale. However, their recollections may not be accurate. Not surprisingly, longitudinal research of all sorts is uncommon though there are good examples available.

12.2 Panel designs

Panel or prospective studies are used to determine the changes that take place in people over time. For example, in the late 1920s in the United States there were a number of growth or developmental studies of children such as the Berkeley Growth Study (Jones, Bayley, MacFarlune and Honzik, 1971). This study was started in 1928 and was designed to investigate the mental, motor and physical development in the first 15 months of life of 61 children. It was gradually expanded to monitor changes up to 54 years of age. In the UK the National Child Development Study was begun in 1958 when data were collected on 17 000 children born in the week of 3–9 March (Ferri, 1993). This cohort has been surveyed on six subsequent occasions at ages 7, 11, 16, 23, 33 and 41/42. Most panel studies are of much shorter duration.

Figure 12.2 gives a simple example of a panel design with three types of correlation (see Figure 12.1). Data are collected at two different points in time on the same sample of individuals. One variable is the supportiveness of one's partner and the other variable is

Synchronous correlation	Auto-, test-rated or lagged correlation	Cross-lagged correlation
Cross-sectional correlations – the usual correlation between variables *X* and *Y* measured at the same point in time	The correlation between, e.g., variable *X* measured at time 1 and variable *X* measured at time 2	The correlation between, e.g., variable *X* measured at time 1 and variable *Y* measured at time 2

FIGURE 12.2 Types of correlation coefficients in longitudinal analyses

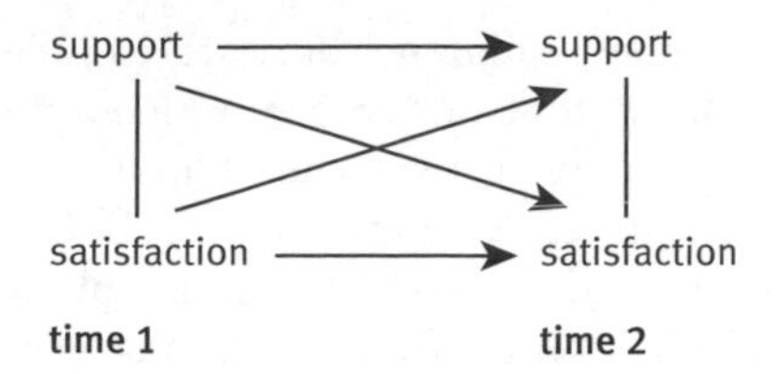

FIGURE 12.3 A two-wave panel design

relationship satisfaction. The question is: does supportiveness lead to relationship satisfaction? In Figure 12.3 there are essentially *two* measures of that relationship measured, but at different points in time. That is, there is a relationship between supportiveness and satisfaction measured at Time 1 and another measured at Time 2. These relationships assessed at the same time are known as cross-sectional or *synchronous correlations*. Generally speaking, they are just as problematic as any other cross-sectional correlations to interpret and there is no real advantage to having the two synchronous correlations available in itself.

There is another sort of relationship to be found in Figure 12.3. This is known as a *cross-lagged* relationship. A lag is, of course, a delay. The cross-lagged relationships in this case are the correlation of supportiveness at Time 1 with satisfaction at Time 2, and the correlation of satisfaction at Time 1 with supportiveness at Time 2.

So perhaps we find that $\text{supportiveness}_{\text{Time1}}$ is correlated with $\text{satisfaction}_{\text{Time2}}$. Does this correlation mean that supportiveness causes satisfaction? It would be congruent with that idea but there is something quite simple that we can do to lend the idea stronger support. That is, we can partial out (control for) $\text{satisfaction}_{\text{Time1}}$. If we find that by doing so, the correlation between $\text{supportiveness}_{\text{Time1}}$ and $\text{satisfaction}_{\text{Time2}}$ reduces to zero then we have an interesting outcome. That is, $\text{satisfaction}_{\text{Time1}}$ is sufficient to account for $\text{satisfaction}_{\text{Time2}}$.

Of course, there is another possibility which has not been eliminated. That is, there is another causal sequence in which satisfaction may be the cause of supportiveness. At first this may seem less plausible, but if one is satisfied with one's partner then they are probably seen as more perfect in many respects than if one is dissatisfied. Anyway, this relationship could also be tested using cross-lagged correlations. A correlation between $\text{satisfaction}_{\text{Time1}}$ with $\text{supportiveness}_{\text{Time2}}$ would help establish the plausibility of this causal link. However, if we control for $\text{supportiveness}_{\text{Time1}}$ and find that the correlation declines markedly or becomes zero, then this undermines the causal explanation in that supportiveness at Time 1 is related to supportiveness at Time 2.

The cross-lagged correlations should generally be weaker than the cross-sectional or synchronous correlations at the two times of measurement because changes are more likely to have taken place during the intervening period. The longer this period, the more probable it is that changes will have occurred and so the weaker the association should be. This also occurs when the same variable is measured on two or more occasions. The longer the interval, the lower the test–retest or auto-correlation is likely to be. If both cross-lagged correlations have the same sign (in terms of being positive or negative) but one is significantly stronger than the other, then the stronger correlation indicates the temporal direction of the association. For example, if the association between support at Time 1 and satisfaction at Time 2 is more positive than the association between satisfaction at Time 1 and support at Time 2, then this difference implies that support leads to satisfaction.

There are several problems with this sort of analysis:

- The size of a correlation is affected by the reliability of the measures. Less reliable measures produce weaker correlations as we saw in Chapter 11. Consequently, the reliability of the measures needs to be taken into account when comparing correlations.

- The difference between the two cross-lagged correlations does not give an indication of the size of the possible causal association between the two variables. For example, the cross-lagged correlation between support at Time 1 and satisfaction at Time 2 will most probably be affected by satisfaction at Time 1 and support at Time 2. To determine the size of the cross-lagged association between support at Time 1 and satisfaction at Time 2 controlling for satisfaction at Time 1 and support at Time 2 we would have to partial out satisfaction at Time 1 and support at Time 2.
- This method does not indicate whether both cross-lagged associations are necessary in order to provide a more satisfactory explanation of the relationship. It is possible that the relationship is reciprocal but that one variable is more influential than the other.

The solution to the above problems may lie in using *structural equation modelling* which is generally the preferred method for this kind of analysis. It takes into account the unreliability of the measures. It provides an indication of the strength of a pathway taking into account its association with other variables. It also offers an index of the extent to which the model fits the data and is a more satisfactory fit than models which are simpler subsets of it. There are various examples of such studies (e.g. Cramer, Henderson and Scolt, 1996; Fincham, Beach, Harold and Osborne, 1997; Krause, Liang and Yatomi, 1989). However, one of the problems with structural equation modelling is that once the test–retest correlations are taken into account, the size of the cross-lagged coefficients may become non-significant. In other words, the variable measured at the later point in time seems to be completely explained by the same variable measured at the earlier point in time (for example, Cramer, 1994, 1995).

12.3 Different types of third variable

The general third-variable issue was discussed in the previous chapter. Conceptually there is a range of different types of third variable (see Figure 12.4). They are distinguishable only in terms of their effect and, even then, this is not sufficient. We will illustrate these different types by reference to the issue of whether the supportiveness of one's partner in relationships leads to greater satisfaction with that partner.

Mediator (or intervening or mediating) variables

A variable which reduces the size of the correlation between two other variables may act as an explanatory link between the other two variables. In these circumstances it is described as a mediating or intervening variable. Making the distinction between a confounding and an intervening variable is not usually easy to do theoretically. With cross-sectional data it is not possible to establish the causal or the temporal direction between two variables. Nonetheless, researchers may suggest a direction even though they cannot determine this with cross-sectional data. For example, they may suggest that having a supportive relationship may lead to greater satisfaction with that relationship when it is equally plausible that the direction of the association may be the other way round or that the direction may be both ways rather than one way.

Suppose that we think that the direction of the association between supportiveness and satisfaction goes from support to satisfaction. With a variable which can change, like income, it is possible to argue that it may act as an intervening variable. For example, having a supportive partner may enable one to earn more which, in turn, leads to greater satisfaction with the relationship. It is easier to argue that a variable is a confounding one when it is a variable which cannot change like age or gender. Support cannot affect

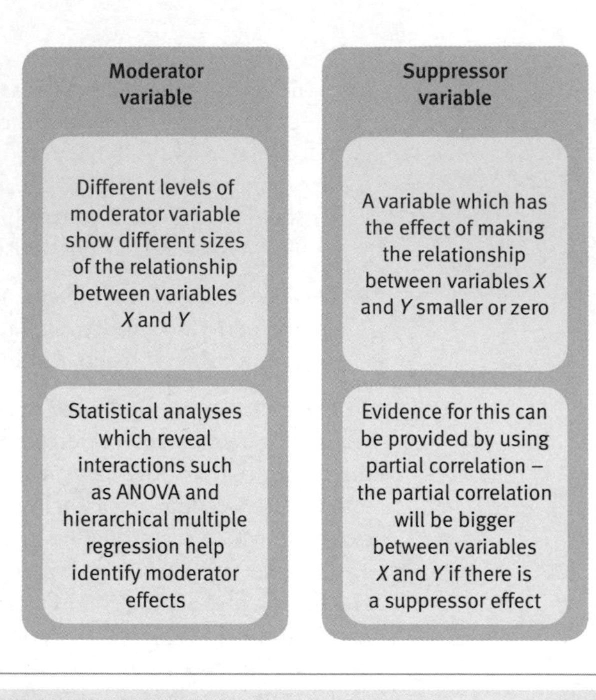

Mediator variable
A variable which 'accounts for' the relationship between variables X and Y
Longitudinal designs can help provide evidence about the causal directions involved and thus 'establish' the role of the causal variable
Moderator variable
Different levels of moderator variable show different sizes of the relationship between variables X and Y
Statistical analyses which reveal interactions such as ANOVA and hierarchical multiple regression help identify moderator effects
Suppressor variable
A variable which has the effect of making the relationship between variables X and Y smaller or zero
Evidence for this can be provided by using partial correlation – the partial correlation will be bigger between variables X and Y if there is a suppressor effect

FIGURE 12.4 Three important types of 'third variable'

age or gender, and so this kind of variable cannot be an intervening one. However, many variables in psychology can change and so are potential intervening variables.

Moderator (moderating) variables

The size or the sign of the association between two variables may vary according to the values of a third variable, in which case this third variable is known as a moderator or moderating variable. For example, the size of the association between support and satisfaction may vary according to the gender of the partner. It may be stronger in men than in women. For example, the correlation between support and satisfaction may be .50 in men and .30 in women. If this difference in the size of the correlations is statistically significant, we would say that gender moderates the association between support and satisfaction.

If we treated one of these variables, say support, as a dichotomous variable and the other as a continuous variable, we could display these relationships in the form of a graph, as shown in Figure 12.5. Satisfaction is represented by the vertical axis, support by the horizontal axis and gender by the two lines. The difference in satisfaction between women and men is greater for those with more support than those with less support. In other words, we have an interaction between support and gender like the interactions described for experimental designs. A moderating effect is an interaction effect.

If the moderating variable is a continuous rather than a dichotomous one, then the cut-off point used for dividing the sample into two groups may be arbitrary. Ideally the two groups should be of a similar size, and so the median score which does this can be used. Furthermore, the natural variation in the scores of a continuous variable is lost when it is converted into a dichotomous variable. Consequently, it is better to treat a continuous variable as such rather than to change it into a dichotomous variable.

The recommended method for determining the statistical significance of an interaction is to conduct a hierarchical multiple regression (Baron and Kenny, 1986). The two main variables or predictors, which in this example are support and gender, are standardised and entered in the first step of the regression to control for any effects they may have. The interaction is entered in the second step. The interaction is created by multiplying the two standardised predictors together provided that neither of these are a categorical variable

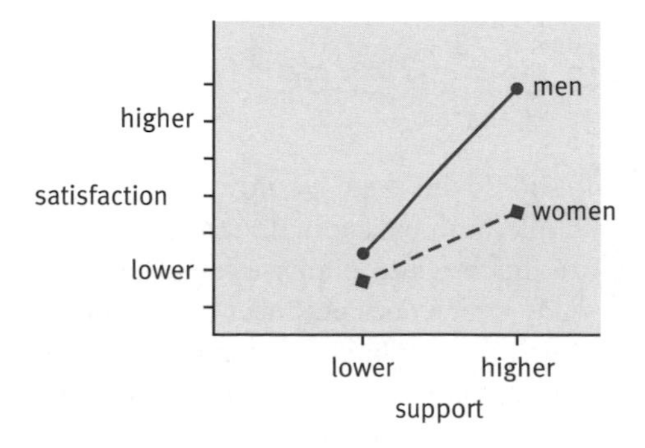

FIGURE 12.5 The association between support and satisfaction moderated by gender

with more than two categories. If this interaction is significant, there is a moderating effect as this means that the interaction accounts for a significant proportion of the variance in the criterion, which in this example is satisfaction. The nature of the interaction effect needs to be examined. One way of doing this is to divide the sample into two based on the median of the moderating variable, produce a scatterplot of the other two variables for the two samples separately and examine the direction of the relationship in the scatter of the two variables. The calculation steps to assess for moderator variables can be found in the companion statistics text, *Introduction to Statistics in Psychology* (Howitt and Cramer, 2011a).

Suppressor variables

Another kind of confounding variable is one that appears to suppress or hide the association between two variables so that the two variables seem to be unrelated. This type of variable is known as a suppressor variable. When its effect is partialled out, the two variables are found to be related. This occurs when the partial correlation is of the opposite sign to the product of the other two correlations and the other two correlations are moderately large. When one of the other correlations is large, the partial correlation is large (Cramer, 2003). Typically the highest correlations are generally those in which the same variable is tested on two occasions that are not widely separated in time. When both the other correlations are large, the partial correlation is greater than 1.00! These results are due to the formula for partialling out variables and arise when correlations are large, which is unusual.

There appear to be relatively few examples of suppressor effects. We will make up an example to illustrate one in which we suppose that support is not correlated with the satisfaction with the relationship (i.e. $r = .00$). Both support and satisfaction are positively related to how loving the relationship is, as shown in Figure 12.6. If we partial out love, the partial correlation between support and satisfaction changes to −.43. In other words we now have a moderately large correlation between support and satisfaction whereas the original or zero-order correlation was zero. This partial correlation is negative because the product of the other two correlations is positive.

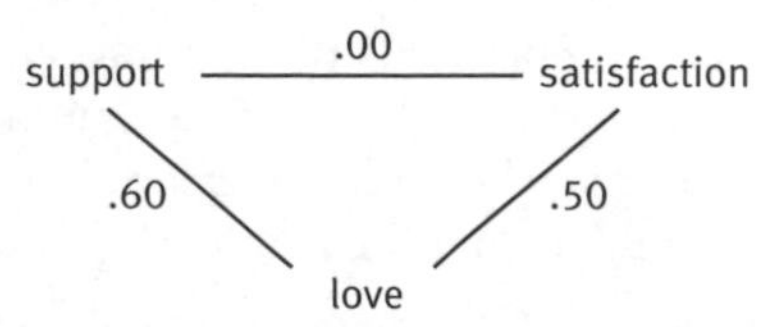

FIGURE 12.6 Example of a suppressed association

12.4 Analysis of non-experimental designs

It should be clear by now that the analysis of non-experimental designs is far from simple both conceptually and statistically. Furthermore, the range and scope of studies are much wider than we have suggested so far. Subsumed under this heading is every study which does not meet the requirements of a randomised experimental design. Quite clearly it is unlikely that any single chapter can cover every contingency. So you will find in many of the remaining chapters of this book a whole range of different styles of non-experimental data collection and analysis methods. To the extent that they are quantitative studies, they share a number of characteristics in terms of analysis strategies. In this section, we will briefly review two of these as examples. They are both dealt with in detail in the companion statistics text, *Introduction to Statistics in Psychology* (Howitt and Cramer, 2011a). Although some such designs may be analysed using the related *t*-test and the related analysis of variance, for example, the variety of measures usually included in such studies generally necessitates the use of more complex statistics designed to handle the multiplicity of measures.

Multiple regression

Multiple regression refers to a variety of methods which identify the best pattern of variables to distinguish between higher and lower scorers on a key variable of interest. For example, multiple regression would help us identify the pattern of variables which differentiates between different levels of relationship satisfaction. Using this optimum pattern of variables, it is possible to estimate with a degree of precision just how much satisfaction a person would feel given their precise pattern on the other variables. This could be referred to as a model of relationship satisfaction. (A model is a set of variables or concepts which account for another variable or concept.)

Another way of looking at it is to regard it as being somewhat like partial correlation. The difference is that multiple regression aims to understand the components which go to make up the scores on the key variable – the criterion or dependent variable. In this case, the key variable is relationship satisfaction. There is a sense in which multiple regression proceeds simply by partialling out variables one at a time from the scores on relationship satisfaction. How much effect on the scores does removing income, then social class, then supportiveness have? If we know the sizes of these effects, we can evaluate the possible importance of different variables on relationship satisfaction.

Technically, rather than use the partial correlation coefficient, multiple regression uses the part correlation coefficient or semi-partial correlation coefficient. The proportion of variance attributable to or explained by income or supportiveness is easily calculated. The proportion is simply the square of the part or semi-partial correlation. (This relationship is true for many correlation coefficients too.) We use the part or semi-partial correlation because it is only one variable that we are adjusting (relationship satisfaction). Partial correlation actually adjusts two variables, which is not what we need.

The part correlation between support and satisfaction partialling out income for the correlations shown in Figure 11.3 is .33 which squared is about .11. What this means is that support explains an additional 11 per cent of the variance in satisfaction to the 16 per cent already explained by income. The following formula is used to calculate the part correlation in which one variable is partialled out, where B refers to satisfaction, A to support and C to income:

$$r_{BA.C} = \frac{r_{BA} - (r_{BC} \times r_{AC})}{\sqrt{(1 - r^2_{AC})}}$$

If we insert the correlations of Figure 11.3 into this formula, we find that the part correlation is .33:

$$\frac{.50 - (.40 \times .60)}{\sqrt{(1 - .60^2)}} = \frac{.26}{.80} = .33$$

Multiple regression has three main uses:

- To predict what the likely outcome is for a particular case or group of cases. For example, we may be interested in predicting whether a convicted prisoner is likely to re-offend on the basis of information that we have about them.
- To determine what the size, sign and significance of particular associations or paths are in a model which has been put forward to explain some aspect of behaviour. For example, we may wish to test a particular model which seeks to explain how people become involved in criminal activity. This use is being increasingly taken over by the more sophisticated statistical technique of structural equation modelling.
- To find out which predictors explain a significant proportion of the variance in the criterion variable such as criminal activity. This third use differs from the second in that generally a model is not being tested.

There are three main types of multiple regression for determining which predictors explain a significant proportion of the variance in a criterion:

- *Hierarchical or sequential multiple regression* In this, the group of predictors is entered in a particular sequence. We may wish to control for particular predictors or sets of predictors by putting them in a certain order. For example, we may want to control for basic socio-demographic variables such as age, gender and socio-economic status before examining the influence of other variables such as personality or attitudinal factors.
- *Standard or simultaneous multiple regression* All of the predictors are entered at the same time in a single step or block. This enables one to determine what the proportion of variance is that is uniquely explained by each predictor in the sense that it is not explained by any other predictor.
- *Stepwise multiple regression* In this, statistical criteria are used to select the order of the predictors. The predictor that is entered first is the one which has a significant and the largest correlation with the key variable (the criterion or dependent variable). This variable explains the biggest proportion of the variation of the criterion variable because it has the largest correlation. The predictor that is entered second is the one which has a significant and the largest part correlation with the criterion. This, therefore, explains the next biggest proportion of the variance in the criterion. This part correlation partials out the influence of the first predictor on the criterion. In this way, the predictors are made to contribute independently to the prediction. The predictor that is entered next is the one that has a significant and the next highest part correlation with the criterion. This predictor partials out the first two predictors. If a predictor that was previously entered no longer explains a significant proportion of the variance, it is dropped from the analysis. This process continues until there is no predictor that explains a further significant proportion of the variance in the criterion.

One important feature of multiple regression needs to be understood otherwise we may fail to appreciate quite what the outcome of an analysis means. Two or more predictors may have very similar correlations or part correlations with the criterion, but the one which has the highest correlation will be entered even though the difference in the size of the correlations is tiny. If the predictors themselves are highly related then those with the slightly smaller correlation may not be entered into the analysis at all. Their absence may give the impression that these variables do not predict the criterion when they do. Their correlation with the criterion may have been slightly weaker because the measures of these predictors may have been slightly less reliable. Consequently, when interpreting the results of a stepwise multiple regression, it is important to look at the size of the correlation between the predictor and the criterion. If two or more predictors are similarly correlated with the criterion, it is necessary to check whether these predictors are measuring the same rather than different characteristics.

An understanding of multiple regression is very useful as it is commonly and increasingly used. Sherman and his colleagues (1999) found that 41 per cent of the papers they randomly sampled in the *Journal of Personality and Social Psychology* used it in 1996 compared with 21 per cent in 1988 and 9 per cent in 1978.

Path analysis

A path is little more than a route between two variables. It may be direct but it can be indirect. It can also be reciprocal in that two variables mutually affect each other. For a set of variables, there may be a complex structure of paths, of course. Figure 12.7 has examples of such paths and various degrees of directness. Multiple regression can be used to estimate the correlations between the paths (these are known as path coefficients). However, structural equation modelling is increasingly used instead. This has three main advantages over multiple regression:

- The reliabilities of measures are not taken into account in multiple regression but they are in structural equation modelling. As explained earlier, reliability places a strict upper limit on the maximum correlation between any two variables.
- Structural equation modelling gives an index of the extent to which the model provides a satisfactory fit to the data. This allows the fit of a simpler subset of the model to be compared with the original model to see if this simpler model provides as adequate a fit as the original model. Simpler models are generally preferred to more complicated ones as they are easier to understand and use.
- Structural equation modelling can explain more than one outcome variable at the same time, like the two presented in the path diagram of Figure 12.7.

The path diagram or model in Figure 12.7 seeks to explain the association between depression and satisfaction with a romantic relationship in terms of the four variables of attitude similarity, interest similarity, love and negative life events. The path diagram shows the assumed relationship between these six variables. The temporal or causal sequence moves from left to right. The direction of the sequence is indicated by the arrow of the line. So interest similarity leads to love which in turn leads to satisfaction. There is a direct association between interest similarity and satisfaction, and an indirect association which is mediated through love. In other words, interest similarity has, or is assumed to have, both a direct and an indirect effect. The association between satisfaction and depression

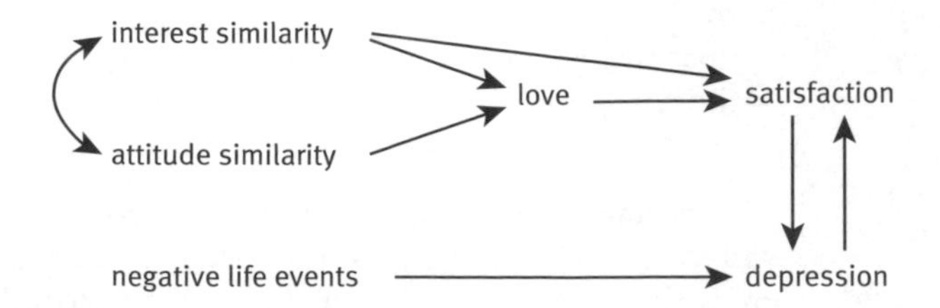

FIGURE 12.7 A path diagram with six variables

is a reciprocal one as the arrows go in both directions. Being satisfied results in less depression and being less depressed brings about greater satisfaction. The lack of a line between two variables indicates that they are not related. So attitude similarity is not related to negative life events, depression or satisfaction. The curved line with arrows at either end shows that interest similarity and attitude similarity are related but that they are not thought to affect each other. Ideally we should try to develop a model such as we have done here to explain the relationships between the variables we have measured in our study.

12.5 Conclusion

Where the primary aim is to determine the temporal ordering of variables a panel or prospective study is required. In these, the same participants are studied on two or more occasions. The main variables of interest should be measured on each of these occasions so that the size of the temporal associations can be compared. Statistical analysis for non-manipulation studies is generally more complicated than that of true or randomised studies. This is especially the case when causal or explanatory models are being tested. Familiarity with statistical techniques such as multiple regression is advantageous for a researcher working in this field.

Key points

- Panel or prospective designs measure the same variables in the same cases on two or more occasions. It is possible to assess whether variables may be mutually related.
- Longitudinal studies may be especially influenced by a number of threats to their internal validity and external validity. For example, because of the time dimension involved the participants may simply change because they have got a little older.
- There are a number of types of variable which may play a role in the relationship between two variables. These include intervening variables and suppressor variables. Conceptually it is important to distinguish between these different sorts of third variables although they are very similar in practice.
- The complexity of these designs encourages the use of complex statistical techniques such as multiple regression and path analysis.

ACTIVITIES

1. Draw a path diagram for the following:
   - Couples who are similar fall in love more intensely.
   - They marry but tend to grow apart and stop loving each other.
2. Couples who love each other tend to have better sexual relationships with each other. It is found that once couples have a baby, the physical side of their relationship declines. What sort of variable is the baby?

CHAPTER 13

Sampling and population surveys

Overview

- When we want to make inferences about a finite population such as the people of Britain, ideally we should obtain a representative sample of that population.
- The size of the sample to use is determined by various factors. How confident we are that the results of the sample represent those in the population is usually set at the 95 per cent or .95 level although it might be set higher. The bigger the variation is in the characteristic that we are interested in estimating, the bigger the sample has to be. The smaller we want the sampling error or margin of error to be, the larger the sample has to be.
- Probability sampling is where every unit or element in the population has an equal and a known probability of being selected.
- Where the population is spread widely, multi-stage sampling may be used where the first stage is to select a limited number of areas from which further selections will be made.

13.1 Introduction

As we have seen, when psychologists test a generalisation or hypothesis they always do so on a relatively limited sample of people, usually those who are convenient to recruit. Although these people are typically no longer students (Bodner, 2006), they are generally not a representative or random sample of people. The time, effort and expense of recruiting such a sample are deterrents. Research psychologists are usually under some pressure to conduct their research as quickly and economically as possible. Obtaining a more representative sample of participants would hinder their work by introducing more constraints in terms of time and money. To the extent that one believes that the idea being tested applies widely, then one would be disinclined to test a more representative sample – it wouldn't really be necessary. If you thought that different types of people were likely to produce rather different responses in the study, then you might include these groups in your study to see if this were the case – that is, you would seek to improve the degree to which your sampling of participants was representative. However, even in these circumstances a sample with very different characteristics to the ones already studied might be just as informative as a more representative sample.

The issue of whether the assumption that findings apply widely (i.e. are generalisable) ought to be more controversial than it is in psychology. Just to indicate something of the problem, there have been very carefully controlled laboratory studies which have produced diametrically opposite findings from each other. For example, using more sophisticated students from the later years of their degree has on occasion produced findings very different from the findings of a study using first year students (Page and Scheidt, 1971). So what is true of one type of participant is not true for another type. The variability of findings in research, although partly the consequence of sampling variation, is also due to other sources of variability such as the characteristics of the sample. So possibly the question of whether to use convenience samples rather than representative samples is best addressed by considering what is known about the behaviour of different groups of people in relation to the topic in question. Past research on a particular topic, for example, may indicate little or no evidence that different samples produce very different findings. In that case, the researcher may feel confident enough to use a convenience sample for their research.

Researchers should know how representative samples may be obtained – if only as a possible ideal sampling scenario for the quantitative researcher. Furthermore, some studies have as their aim to make statements about a representative sample of people. Researchers who are interested in how the general public behave may be less inclined to pay much attention to the results of a study which is solely based on students. Sociologists, for example, have lampooned psychology, sometimes unfairly, for its dependency on university students. Studies which have found similar results in a more representative sample lend extra credibility to their findings and allow generalisation. Figure 13.1 shows different types of sampling.

13.2 Types of probability sampling

The characteristics of very large populations can be estimated from fairly small samples as is argued towards the end of this chapter. The problem is ensuring that the sample is representative of the population if you want to generalise your results to that population. A distinction is made between *probability* and *non-probability sampling*:

Random sample	Stratified random sample	Non-random sample
Simple random sample – straightforward random sampling from a list of names	Stratified sample – structured e.g. by gender so that random sample includes both genders proportionately	Quota sample – participants recruited because they fit one of various researcher-specified categories
		Convenience sample – the selection of research participants because they are easily recruited
Systematic sample – every *n*th case in a list after a random starting point	Disproportionate stratified sample – involves oversampling from some groups of interest which are relatively uncommon	Snowball sample – participants are asked to suggest other similar people to participate
		Purposive sample – selecting participants because they are of theoretical interest to the research
Multi-stage sample – where, say, universities are selected at random and then students selected at random from each of these universities	Cluster sample – geographical areas selected and then random sample drawn within each cluster	Theoretical sampling – describes selecting further participants to 'test' as ideas develop (grounded theory)

FIGURE 13.1 The different types of sample

- Probability sampling is typically used when we have a clearly defined and accessible population which we want to make inferences about or when we want to know how characteristic a behaviour is of a particular population, such as the people of Britain.
- Non-probability sampling is normally used in psychological research. This is because we are generally not interested in getting precise population estimates of a particular feature or characteristic in psychology. In psychology, the emphasis in research tends to be on relationships between variables and whether or not this relationship differs significantly from a zero relationship.

The main advantage of probability sampling is that every person or element in the population has an equal and known probability of being selected. Suppose we want to use probability sampling to select 10 per cent or 100 people out of a total population of 1000 people. In accordance with the concept of random sampling, everyone in that sample should have an equal probability of .10 of being selected (100/1000 = .10). The simplest procedure is to give each member of the population a number from 1 to 1000 and draw a hundred of these numbers at random.

There are various ways of drawing a sample of 100 numbers representing the 100 people we need for the probability sampling:

- We could use a statistical package such as SPSS Statistics. We would enter the numbers 1 to 1000 in one of the columns. We would then select *Data*, *Select cases . . .*, *Random sample of cases*, *Sample . . .*, *Exactly*, and then enter 100 cases from the first 1000 cases. The 100 numbers that were selected would be the 100 people in the sample. Alternatively, you will find applets on the Web which will generate a random sample of cases for you.
- We could write the numbers 1 to 1000 on 1000 index cards or slips of paper, shuffle them and then select 100 cards or slips of paper. These 100 numbers would represent the people in our sample.
- We could use a table of random numbers which can be found in the back of many introductory textbooks on statistics. These tables usually consist of rows and columns of pairs of numbers such as 87 46 and so on. The 1000th person is represented by the number 000. Each person in the population has a distinct three digit number. So we need a way of selecting three-digit numbers from the table of random numbers. There are no rules for this and you can decide on your own system. One could simply choose a random starting point in the table (eyes shut, using a pin) and record the first three digits after this as the first random selection. The person corresponding to this number is the first participant selected to be in the sample. The next three digits would give the number of the second member of the sample and so forth. Of course, one could go backwards through the table if one chose. If we select the same number more than once, we ignore it as we have already selected the individual represented by that number. Our sample of 100 individuals would be complete once we had selected 100 sets of three numbers from the table. As long as one is consistent, one can more or less decide on whatever rule one wishes for selecting numbers.

This form of probability sampling is called *simple random sampling*. An alternative procedure is to select every 100th person on the list. We have to decide what our starting point is going to be which can be any number from 1 to 100 and which we can choose using a random procedure. Let us suppose it is 67. This would be the first person we select. We then select every 100th person after it, such as 167, 267, 367 and so on. This procedure is known as *systematic sampling*. The advantage of systematic sampling is that it is simpler to use with a printed list such as a register of electors. It is quicker than random sampling and has the advantage that people close together on the list (for example, couples) will not be selected. Its disadvantage is that it is not completely random. The list may not be arranged in what is effectively a random order for some reason. Generally speaking, though, so long as the complete list is sampled from (as the above method will ensure), there are unlikely to be problems with systematic sampling. If the entire list is not sampled, then this method may introduce biases. For example, if the researcher simply took every, say, 75th case then those at the end of the list could not be included in the sample.

Neither simple random sampling nor systematic sampling ensure that the sample will be representative of the population from which the sample was taken. For example, if the population contained equal numbers of females and males, say 500 of each, it is possible that the sample will contain either all, or a disproportionate number of, females or males. It may be important that the sample is representative of the population in respect of one or more characteristics such as gender. This is achieved by dividing the population into groups or strata representing that characteristic, such as females and males. Then the random sampling is essentially done separately for each of these two groups. In terms of our example of selecting a sample of 100 people, we would select 50 from the 500 females and 50 from the 500 males. This form of sampling is known as *stratified random sampling* or just *stratified sampling*.

As the proportion of females in the sample (.50) is roughly the same as the proportion of females in the population (approximately .50), this kind of stratified sampling may be called proportionate stratified sampling. It may be distinguished from *disproportionate stratified sampling* in which the sampling is not proportionate to the size of the group in the population. Disproportionate stratified sampling is used when we want to ensure that a sufficient number of people are sampled of whom there are relatively few in the population. For example, we may be keen to determine the behaviour of unemployed people in our population of 1000 people. Suppose there are only 50 unemployed people in our population. If we used proportionate stratified sampling to select 10 per cent or 100 people from our population, then our sample of unemployed people is 5 (10/100 × 50 = 5) which is too few to base any generalisations on. Consequently, we may use disproportionate stratified sampling to obtain a bigger sample of unemployed people. Because the number of unemployed people is small, we may wish to have a sample of 25 of them, in which case the proportion of unemployed people is .50 (25/50 = .50) instead of .10 (5/50 = .10). As our overall sample may be still limited to 100 people, the number of people in our sample who are not unemployed is now 75 instead of 95. So, the proportion of people who are not unemployed is smaller than (95/950 = .10) and is about .08 (75/950 = .0789).

One of the problems with stratified sampling is that relevant information about the characteristic in question is needed. For a characteristic such as gender this is easily obtained from a person's title (Mr, Miss, Ms or Mrs) but this is the exception rather than the rule. Otherwise, more work is involved in obtaining information about that characteristic prior to sampling.

If our population is dispersed over a wide geographical area we may use what is called *cluster sampling* in order to restrict the amount of time taken to draw up the sampling list or for interviewers to contact individuals. For example, if we wanted to carry out a probability survey of all British students it would be difficult and time-consuming to draw up a list of all students from which to select a sample. What we might do instead is to select a few universities from around the country and sample all the students within those universities. The universities would be the group or cluster of students which we would sample. The clusters need not be already existing ones. They may be created artificially. For example, we may impose a grid over an area and select a number of squares or cells within that area. The advantage of cluster sampling is that it saves time and money. Its disadvantage is that it is likely to be less representative of the population because the people within a cluster are likely to be more similar to one another. For example, students at one university may be more inclined to come from fee-paying rather than state schools or to be female rather than male.

Another form of probability sampling is called *multi-stage sampling* in which sampling is done in a number of stages. For example, we could have a two-stage sample of university students in which the first stage consists of sampling universities and the second stage of sampling students within those universities.

The *representativeness of a sample* can be evaluated against other information about the population from which it was drawn. For example, in trying to determine whether our sample is representative of the British population we can compare it with census or other national data on characteristics such as gender, age, marital status and employment status. It should be noted that these other sources of information will not be perfectly accurate and will contain some degree of error themselves.

13.3 Non-probability sampling

The cost, effort and time involved in drawing up a representative or probability sample are clearly great. A researcher without these resources may decide to use a *quota sample* instead. In a quota sample an attempt is made to ensure different groups are represented in the proportion in which they occur within that society. So, for example, if we know that 5 per cent of the population are unemployed, then we may endeavour to ensure that 5 per cent of the sample are unemployed. If our sample consists of 100 people, we will look for 5 people who are unemployed. Because we have not used probability sampling then we may have a systematically biased sample of the unemployed. The numbers of people in the different groups making up our sample do not have to be proportionate to their numbers in society. For example, if we were interested in looking at how age is related to social attitudes, we may choose to use equal numbers of each age group no matter their actual frequencies in the population.

Where we need to collect a sample of very specific types of people then we may use *snowball sampling*. So this would be an appropriate way of collecting a sample of drug addicts, banjo players or social workers experienced in highly publicised child abuse cases. Once we have found an individual with the necessary characteristic we ask them whether they know of anyone else with that characteristic who may be willing to take part in our research. If that person names two other people and those two people name two further individuals then our sample has snowballed from one individual to seven.

There are a number of other versions of non-probability sampling: (a) quota sampling is used in marketing research, etc. and requires that the interviewer approaches people who are likely to fill various categories of respondent required by the researcher (e.g. females in professional careers, males in manual jobs, etc.); (b) convenience sampling is used in much quantitative research in psychology and simply uses any group of participants readily accessible to the researcher; (c) purposive sampling is recruiting specified types of people because they have characteristics of interest to the theoretical concerns of the researcher; and (d) theoretical sampling comes from grounded theory (see Chapter 21) and occurs after some data are collected and an analysis formulated such that further recruits to the study inform or may challenge the developing theory in some way.

In general, psychologists would assume a sample to be a non-random one unless it is specifically indicated. If it is a random sample, then it is necessary to describe in some detail the particular random procedure used to generate that sample. In most psychological research the sample will be a convenience one and it may be sufficient to refer to it as such.

13.4 National surveys

Most of us are familiar with the results of the national opinion polls which are frequently reported in the media. However, national studies are extremely uncommon in psychological research. Using them would be an advantage but not always a big one. Generally, there is no great need to collect data from a region or country unless you are interested in how people in that region or country generally behave. Researchers from other social sciences, medical science and similar disciplines are more likely to carry out national surveys than psychologists. This is partly because these researchers are more interested in trends at the national and regional levels. Nonetheless the results of these surveys are often of relevance to psychologists.

It is not uncommon for such surveys to be placed in an archive accessible to other researchers. For example, in Britain many social science surveys are archived at the Economic and Social Research Council (ESRC). The datasets that are available there for further or secondary analysis are listed at the following website: http://www.data-archive.ac.uk/

Major surveys include the British Crime Survey, British Social Attitudes and the National Child Development Study. International datasets are also available. Students are not allowed direct access to these datasets but they may be obtained via a lecturer who has an interest in them and the expertise to analyse them. This expertise includes an ability to use a statistical computer package such as SPSS Statistics which is widely used and taught to students. There are books which show you how to use SPSS such as the companion computing text, *Introduction to SPSS Statistics in Psychology* (Howitt and Cramer, 2011b).

Box 13.1 Research Example

A national representative survey

The British Social Attitudes Survey is a good example of the use of a representative sample in research which may be of relevance to psychologists. The detail is less important to absorb than the overall picture of the meticulous nature of the process and the need for the researcher to make a number of fairly arbitrary decisions. This survey has been carried out more or less annually since 1983. The target sample for the 2008 survey was 9060 adults aged 18 or over living in private households (Park *et al.*, 2010, p. 270). The *sampling list* or *frame* was the Postcode Address File. This is a list of addresses (or postal delivery points) which is compiled by, and which can be bought from, the Post Office. The multi-stage sampling design consisted of three stages of selection:

- selection of postcode sectors;
- selection of addresses within those postcode sectors;
- selection of an adult living at an address.

The postcode sector is identified by the first part of the postcode. It is LE11 3 for Loughborough, for example. Any sector with fewer than 1000 addresses was combined with an adjacent sector. Sectors north of the Caledonian Canal in Scotland were excluded due to the high cost of interviewing there. The sectors were stratified into:

- 37 sub-regions;
- three equal-sized groups within each sub-region varying in population density; and
- ranking by the percentage of homes that were owner-occupied.

The sampling frame may look something like that shown in Table 13.1 (Hoinville and Jowell, 1978, p. 74).

The total number of postcode sectors in the United Kingdom in 2005 was then 11 598 (Postal Geography, n.d.) from which 302 sectors were selected. The probability of selection was made proportional to the number of addresses in each sector. The reason for using this procedure is that the number of addresses varies considerably among postcode sectors. If postcode sectors have an equal probability of being selected, the more addresses a postcode sector has, the smaller the chance or probability is that an address within that sector will be chosen. So not every address has an equal probability of being selected for the national sample.

To ensure that an address within a sector has an equal probability of being chosen, the following procedure was used (Hoinville and Jowell, 1978, p. 67). Suppose we have six postcode sectors and we have to select three of these sectors. The number of addresses in each sector is shown in Table 13.2. Altogether we have 21 000 addresses. If we use systematic sampling to select these three sectors, then we need to choose the random starting point or address from this number. We could do this using a five-digit sequence in a table of random numbers. Suppose this number was 09334. If we add the number of addresses cumulatively as shown in the third column of Table 13.2, then the random starting point is in the second postcode sector. So this would be the first postcode sector chosen randomly.

In systematic sampling we need to know the sampling interval between the addresses on the list. This is simply the total number of addresses divided by the number of samples (21 000/3 = 7000). As the random starting point is greater than 7000, a second point is 7000 below 9334 which is 2334 (9334 − 7000 = 2334). This point falls within the first postcode sector which is the second postcode sector to be selected. A third point is 7000 above 9334 which is 16 334 (9334 + 7000 = 16 334). This point falls within the fourth postcode sector which is the third postcode sector to be chosen. So these would be our three sectors.

Thirty addresses were systematically selected in each of the 302 postcode sectors chosen. This gives a total of 9060 addresses (30 × 302 = 6200). A random start point was chosen in each sector and 30 addresses selected at equal fixed intervals from that starting point. The numbers of adults aged 18 and over varies at the different addresses. A person was selected at random at each address using a computerised random selection procedure.

Response rates were affected by a number of factors:

- About 10 per cent of the addresses were out of the scope of the survey (for example, they were empty, derelict or otherwise not suitable).
- About 30 per cent of the 9060 refused to take part when approached by the interviewer.
- About 4 per cent of the 9060 could not be contacted.

Table 13.1 Part of sampling frame for the British Social Attitudes Survey

Region 01	Percentage owner or non-manual occupiers
Highest density group	
Postcode sector	65%
Postcode sector	60%
.	
.	
.	
Postcode sector	40%
Intermediate density group	
Postcode sector	78%
Postcode sector	75%
.	
.	
.	
Postcode sector	60%
Lowest density group	
Postcode sector	79%
Postcode sector	74%
.	
.	
.	
Postcode sector	55%
.	
.	
.	
Region 37	
.	
.	
.	

Source: adapted from Hoinville and Jowell (1978)

➔

Table 13.2 Example of sampling sectors with probability proportional to size

Sector	Size	Cumulative size	Points
1	4 000	0–4 000	2 334 2nd point
2	6 000	4 001–10 000	9 334 random start
3	5 000	10 001–15 000	
4	2 000	15 001–17 000	16 334 3rd point
5	3 000	17 001–20 000	
6	1 000	20 001–21 000	

Source: adapted from Hoinville and Jowell (1978)

- About 5 per cent of the original 9060 did not respond for some other reason.

This means that the response rate for the final survey was about 50 per cent of the original sample of 9060. This is quite a respectable figure and many surveys obtain much lower return rates. Of course, the non-participation rate may have a considerable impact on the value of the data obtained. There is no reason to believe that non-participants are similar to participants in their attitudes.

13.5 Socio-demographic characteristics of samples

National samples usually gather socio-demographic information about the nature of the sample studied. Which characteristics are described depends on the kind of participants and the purpose of the study. If the participants are university students, then it may suffice to describe the number of female and male students and the mean and standard deviation of their age either together or separately. If a substantial number or all of the participants are not students, then it is generally necessary to provide further socio-demographic information on them such as how well educated they are, whether they are working and what their social status is. These socio-demographic characteristics are not always easy to categorise, and the most appropriate categories to use may vary over time as society changes and according to the particular sample being studied. When deciding on which characteristics and categories to use, it is useful to look at recent studies on the topic and see what characteristics were used. Two socio-demographic characteristics which are problematic to define and measure are social status and race or ethnicity.

In the United Kingdom one measure of social status is the current or the last job or occupation of the participant (e.g. Park *et al.*, 2010, pp. 276–8). The latest government scheme for coding occupations is the Standard Occupational Classification 2000 (Great Britain Office for National Statistics, 2000; http://www.ons.gov.uk/about-statistics/classifications/current/ns-sec/cats-and-classes/analytic-classes/index.html). The previous version was the Standard Occupational Classification 1990 (OPCS, 1991). The main socio-economic grouping based on the latest scheme is the National Statistics Socio-Economic Classification which consists of the following eight categories:

- Employers in large organisations and higher managerial and professional occupations.
- Lower professional and managerial and higher technical and supervisory occupations.
- Intermediate occupations.

- Employers in small organisations and own account workers.
- Lower supervisory and technical occupations.
- Semi-routine occupations.
- Routine occupations.
- Never worked and long-term unemployed.

Previous schemes include the Registrar General's Social Class, the Socio-Economic Group and the Goldthorpe (1987) schema. There is a computer program for coding occupations based on the Standard Occupational Classification 2000 (Great Britain Office for National Statistics, 2000; http://www.ons.gov.uk/about-statistics/classifications/current/SOC2000/about-soc2000/index.html) called Computer-Assisted Structured COding Tool (CASCOT). CASCOT is available free online (http://www2.warwick.ac.uk/fac/soc/ier/publications/software/cascot/choose_classificatio/).

A measure of race may be included to determine how inclusive the sample is and whether the behaviour you are interested in differs according to this variable. Since 1996 the British Social Attitudes Survey measured race with the following question and response options (http://www.britsocat.com/). The percentage of people choosing these categories in the 2007 survey is shown after each option.

To which one of these groups do you consider you belong?	
Black: of African origin	1.21%
Black: of Caribbean origin	1.65%
Black: of other origin (please state)	0.13%
Asian: of Indian origin	2.04%
Asian: of Pakistani origin	1.43%
Asian: of Bangladeshi origin	0.62%
Asian: of Chinese origin	0.43%
Asian: of other origin (please state)	1.18%
White: of any European origin	88.09%
White: of other origin (please state)	1.10%
Mixed origin (please state)	1.07%
Other (please state)	0.69%

None of the 4123 people in the sample did not know which category they fell in and only 0.39 per cent or 16 people did not answer this question.

13.6 Sample size and population surveys

When carrying out research, an important consideration is to estimate how big a sample is required. For surveys, this depends on a number of factors:

- How big the population is.
- How many people you will be able to contact and what proportion of them are likely to agree to participate.

- How variable their responses are.
- How confident you want to be about the results.
- How accurate you want your estimate to be compared with the actual population figure.

Not everyone who is sampled will take part. Usually in national sampling some sort of list of members of the population is used. This is known as the *sampling frame*. Lists of the electorate or telephone directories are examples of such lists though they both have obvious inadequacies. Some of the sample will have moved from their address to an unknown location. Others may not be in when the interviewer calls even if they are visited on a number of occasions. Others refuse to take part. It is useful to make a note of why there was no response from those chosen to be part of the sample. The *response rate* will differ depending on various factors such as the method of contact and the topic of the research. The response rate is likely to be higher if the interviewer visits the potential participant than if they simply post a questionnaire. The most probable response rate may be estimated from similar studies or from a pilot or exploratory study.

How variable the responses of participants are likely to be can be obtained in similar ways. It is usually expressed in terms of the *standard* deviation of values which is similar to the average extent to which the values deviate from the mean of the sample.

Confidence interval

When one reads about the findings of national polls in the newspapers, statements like this appear: 'The poll found that 55 per cent of the population trust the government. The margin of error was plus or minus 2 per cent.' Of course, since the finding is based on a sample then we can never be completely confident in the figure obtained. Usually the confidence level is set at 95 per cent or .95. The interval is an estimate based on the value obtained in the survey (55 per cent of the population) and the variability in the data. The variability is used to estimate the range of the 95 per cent of samples that are most likely to be obtained if our data were precisely the same as the values in the entire population. The single figure of 55 per cent trusting the government is known as a *point estimate* since it gives a single value. Clearly the confidence interval approach is more useful since it gives some indication of the imprecision we expect in our data. This is expressed as the margin of error.

One could think of the confidence interval being the range of the most common sample values we are likely to obtain if we repeated our survey many times. That is, the 95 most common sample values if we repeated the study 100 times. If this helps you to appreciate the meaning of confidence intervals then all well and good. Actually it is not quite accurate since it is true only if our original sample data are totally representative of the population. This is not likely to be the case of course, but in statistics we operate with best guesses, not certainties. If the confidence interval is set at 95 per cent or .95 it means that the population value is likely to be in the middle 95 per cent of possible sample means given by random sampling.

The confidence level is related to the notion of statistical significance that was introduced in Chapter 4. A detailed discussion of confidence intervals may be found in Chapter 37 of the companion text *Introduction to Statistics in Psychology* (Howitt and Cramer, 2011a). Confidence intervals apply to any estimate based on a sample. Hence, there are confidence intervals for virtually all statistics based on samples. Both statistical significance and the confidence level are concerned with how likely it is that a result will occur by chance. Statistical significance is normally fixed at 5 per cent or .05. This means that the result will be obtained by chance on 5 times out of 100 or less. If we find that a result is statistically significant it means that the result is so extreme that it is unlikely to occur by chance. A statistically significant finding is one which is outside of the middle 95 per cent of samples defined by the confidence interval.

Sampling error (margin of error) and sample size

How accurately one's data reflect the true population value is dependent on something known as sampling error. Samples taken at random from a population vary in terms of their characteristics. The difference between the mean of your sample and the mean of the population of the sample is known as the *sampling error*. If several samples are taken from the same population their means will vary by different amounts from the value in the population. Some samples will have means that are identical to that of the population. Other samples will have means which differ by a certain amount from the population value. The variability in the means of samples taken from a population is expressed in terms of a statistical index known as the *standard error*. This is a theoretical exercise really as we never actually know what the population mean is – unless we do research on the entire population. Instead we estimate the mean of the population as being the same as the mean for our sample of data. This estimate may differ from the population mean, of course, but it is the best estimate we have. It is possible to calculate how likely the sample mean is to differ from the population mean by taking into account the variability within the sample (the measure of variability used is the standard deviation of the sample). The variability within the sample is used to estimate the variability in the population which is then used to estimate the variability of sample means taken from that population.

If we want to be 95 per cent or .95 confident of the population mean, then we can work out what the sampling error is using the following formula, where t is the value for this confidence level taking into account the size of the sample used (Cramer, 1998, pp. 107–8):

$$\text{sampling error} = t \times \frac{\text{standard deviation}}{\sqrt{\text{sample size}}}$$

The standard deviation is calculated using the data in our sample. The sample size we are considering is either known or can be decided upon. The appropriate t value can be found in the tables of most introductory statistics textbooks such as the companion text *Introduction to Statistics in Psychology* (Howitt and Cramer, 2011a).

If we substitute values for the standard deviation and the sample size, we can see that the sampling error becomes progressively smaller the larger the sample size. Say, for example, that the standard deviation is about 3 for scores of how extroverted people are (Cramer, 1991). For a sample size of 100 people, the t value for the 95 per cent confidence level is 1.984 and so the sampling error is about 0.60:

$$\text{sampling error} = 1.984 \times \frac{3}{\sqrt{100}} = 1.984 \times \frac{3}{10} = \frac{5.952}{10} = 0.5952 = 0.60$$

If the mean score for extroversion for the sample was about 16, then the sample mean would lie between plus or minus 0.60 on either side of 16 about 95 per cent of the time for samples of this size. So the mean would lie between 15.40 (16 – 0.60 = 15.40) and 16.60 (16 + 0.60 = 16.60). These values would be the 95 per cent or .95 confidence limits. The confidence interval is the range between these confidence limits which is 1.20 (16.60 – 15.40 = 1.20). The confidence interval is simply twice the size of the sampling error (0.60 × 2 = 1.20). It is usually expressed as the mean plus or minus the appropriate interval. So in this case the confidence interval is 16.00 ± 0.60.

If the sample size is 400 people instead of 100, the t value for the 95 per cent confidence level is slightly smaller and is 1.966. The sampling error for the same standard deviation is also slightly smaller and is about 0.29 instead of about 0.60:

$$1.966 \times \frac{3}{\sqrt{400}} = 1.966 \times \frac{3}{20} = \frac{5.898}{20} = 0.2949 = 0.29$$

In other words, the sampling error in this case is about half as small for a sample of 400 as for a sample of 100.

We can also see that if the variation or standard deviation of the variable is greater, then the sampling error will be greater. If the standard deviation was 6 instead of 3 with this sample and confidence level, then the sampling error would be about 0.59 instead of about 0.29:

$$1.966 \times \frac{6}{\sqrt{400}} = 1.966 \times \frac{9}{20} = \frac{11.796}{20} = 0.5898$$

The sampling error is sometimes known as the margin of error and may be expressed as a percentage of the mean. If the mean of the extroversion scores is 16 and the sampling error is about 0.60, then the sampling error expressed as a percentage of this mean is 3.75 per cent (0.60/16 × 100 = 3.75). If the sampling error is about 0.29, then the sampling error given as a percentage of this mean is about 1.81 per cent (0.29/16 × 100 = 1.8125). A margin of error of 2 per cent for extroversion means that the mean of the population will vary between 0.32 (2/100 × 16 = 0.32) on either side of 16 at the 95 per cent confidence level. In other words, it will vary between 15.68 (16 – 0.32 = 15.68) and 16.32 (16 + 0.32 = 16.32).

Suppose that we want to estimate what sample size is needed to determine the population mean of extroversion for a population of infinite size at the 95 per cent confidence level with a margin of error of 2 per cent. We apply the following formula, where 1.96 is the z value for the 95 per cent confidence level for an infinite population:

$$\text{sample size} = \frac{1.96^2 \times \text{sample standard deviation}^2}{\text{sampling error}^2}$$

If we substitute the appropriate figures in this formula, we can see that we need a sample of 346 to determine this:

$$\frac{1.96^2 \times 3^2}{0.32^2} = \frac{3.84 \times 9}{0.10} = \frac{34.56}{0.10} = 345.60$$

If the margin of error was set at a higher level, then the sample size needed to estimate the population characteristic would be smaller. If we set the margin of error at, say, 5 per cent rather than 2 per cent, the sampling error would be 0.80 (5/100 × 16 = 0.80) instead of 0.32 and the sample required would be 54 instead of 346.

$$\frac{1.96^2 \times 3^2}{0.80^2} = \frac{3.84 \times 9}{0.64} = \frac{34.56}{0.64} = 54.00$$

Remember that the above formula only deals with a situation in which we have specified a particular margin of error. It has very little to do with the typical situation in psychology in which the researcher tests to see whether or not a relationship differs significantly from no relationship at all.

It should be noted that the formula for calculating sampling error for proportionate stratified sampling and cluster sampling differs somewhat from that given above which was for simple random sampling (Moser and Kalton, 1971, pp. 87, 103). Compared with simple random sampling, the sampling error is likely to be smaller for proportionate stratified sampling and larger for cluster sampling. This means that the sample can be somewhat smaller for proportionate stratified sampling but somewhat larger for cluster sampling than for simple random sampling.

Table 13.3 Sample size for varying finite populations with 95 per cent confidence level, 2 per cent sampling error and standard deviation of 3

Population size	Sample size
1 000	257
5 000	324
10 000	334
100 000	345
250 000	346
infinite	346

Sample size for a finite population

The previous formula assumes that we are dealing with an infinitely large population. When dealing with big populations, this formula is sufficient for calculating the size of the sample to be used. When the populations are fairly small, we do not need as many people as this formula indicates. The following formula is used for calculating the precise number of people needed for a finite rather than an infinite population where n is the size of the sample and N is the size of the finite population (Berenson, Levine and Krehbiel, 2009):

$$\text{adjusted } n = \frac{n \times N}{n + (N - 1)}$$

We can see this if we substitute increasingly large finite populations in this formula while the sample size remains at 346. This has been done in Table 13.3. The first column shows the size of the population and the second column the size of the sample needed to estimate a characteristic of this population. The sample size can be less than 346 with finite populations of less than about 250 000.

When carrying out a study we also need to take account of the response rate or the number of people who will take part in the study. It is unlikely that we will be able to contact everyone or that everyone we contact will agree to participate. If the response rate is, say, 70 per cent, then 30 per cent will not take part in the study. Thus, we have to increase our sample size to 495 people (346/.70 = 494.29). A 70 per cent response rate for a sample of 495 is 346 (.70 × 495 = 346.50). Often response rates are much lower than this.

13.7 Conclusion

Most psychological research is based on convenience samples which are not selected randomly and which often consist of students. The aim of this type of research is often to determine whether the support for an observed relationship is statistically significant. It is generally not considered necessary to ascertain to what extent this finding is characteristic of a particular population. Nonetheless where this is possible, it is useful to know the degree to which our findings may be typical of a particular population.

Consequently, it is important to understand what the basis is for selecting a sample which is designed to be representative of a population. Furthermore, in some cases, the population will be limited in size so that it is possible with relatively little effort to select a sample randomly. For example, if we are interested in examining the content of, say, recorded interactions or published articles, and we do not have the time or the resources to analyse the whole content, then it is usually appropriate to select a sample of that content using probability sampling. The great advantage of probability sampling is that the sample is likely to be more representative of the population and that the sampling will not be affected by any biases we have of which we may not even be aware.

Key points

- A random or probability sample is used to estimate the characteristics of a particular finite population. The probability of any unit or element being selected is equal and known.
- The population does not necessarily consist of people. It may comprise any unit or element such as the population of articles in a particular journal for a certain year.
- The size of the sample to be chosen depends on various factors such as how confident we want to be that the results represent the population, how small we want the sampling error to be, how variable the behaviour is and how small the population is. Bigger samples are required for higher confidence levels, smaller sampling errors, more variable behaviour and bigger populations.
- Fairly small samples can be used to estimate the characteristics of very large populations. The sample size does not increase directly with the population size.
- Where the population is widely dispersed, cluster sampling and multi-stage sampling may be used. In the first stage of sampling a number of clusters such as geographical areas (for example, postcode sectors) may be chosen from which further selections are subsequently made.
- Where possible the representativeness of the sample needs to be checked against other available data about the population.
- Where the sample is a convenience one of undergraduate students, it may suffice to describe the number of females and males, and the mean and standard deviation of their ages. Where the sample consists of a more varied group of adults, it may be necessary to describe them in terms of other socio-demographic characteristics such as whether or not they are employed, the social standing of their occupation and their racial origin.

ACTIVITIES

1. How would you randomly select ten programmes from the day's listing of a major TV channel to which you have ready access?
2. How would you randomly select three 3-minute segments from a 50-minute TV programme?
3. How would you randomly select ten editions of a major Sunday newspaper from last year?

PART 3

Fundamentals of testing and measurement

CHAPTER 14

Psychological tests

Their use and construction

Overview

- Psychological tests and measures are commercially available, can be sometimes found in the research literature or may be created by the researcher. The construction of a psychological test is relatively easy using statistical packages.
- Tests used for clinical and other forms of assessment of individuals need to be well standardised and carefully administered. Measures used for research purposes only do not need the same degree of precision to be useful.
- Psychologists tend to prefer 'unidimensional' scales which are single dimensional 'pure' measures of the variable in question. However, multidimensional scales may be more useful for practical rather than research applications.
- Item analysis is the process of 'purifying' the measure. Item–total correlations simply correlate each individual item with the score based on the other items. Those items with high correlations with the total are retained. An alternative is to use factor analysis, which identifies clusters of items that measure the same thing.

14.1 Introduction

Standardised tests and measures are the major tools used extensively in psychological work with clients (for example, clinical psychology, educational psychology, occupational psychology). They are also frequently used in research. In many ways standardised tests and measures are very characteristic of psychology. The term *standardised* can mean several things:

- That consistency of results is achieved by the use of identical materials, prescribed administration procedures and prescribed scoring procedures. That is to say, variability in the ways in which different psychologists administer the test or measure is minimised. Way back in 1905 when Alfred Binet and Theodore Simon presented the world's first psychological scale – one to essentially measure intelligence – he was adamant about the detail of the assessment setting. For example, he suggested an isolated, quiet room in which the child was alone with the test administrator and, ideally, an adult familiar to the child to help reassure the child. However, the familiar adult should be 'passive and mute' and not intervene in any way (Binet and Simon, 1904, 1916).

- That the consistency of interpretation of the test is maximised by providing normative or standardisation data for the test or measure. This means that the test or measure has been administered to a large, relevant sample of participants. In this way, it is possible to provide statistical data on the range and variability of scores in such a sample. As a consequence, the psychologist is able to compare the scores of their participants with those of this large sample. These statistical data are usually referred to as the norms (or normative data) but they are really just the standard by which individual clients are judged. Often tables of *percentiles* are provided which indicate for any given score on the test or measure, the percentage of individuals with that score or a lower score (see the companion book *Introduction to Statistics in Psychology*, Howitt and Cramer, 2011a). Norms may be provided for different genders and/or age groups, and so forth.

Standardised tests and measures are available for many psychological characteristics including attitudes, intelligence, aptitude, ability, self-esteem, musicality, personality and so forth. Catalogues of commercially available measures are published by a number of suppliers. These may be quite expensive, elaborate products. Their commercial potential and practical application partly explains the cost. For example, there is a big market for tests and measures for the recruitment and selection of employees by businesses, especially in the USA. Selection interviews are not necessarily effective ways of assessing the abilities and potential of job applicants. Standardised selection tests assessing aptitude for various types of employment may help improve the selection process. By helping to choose the best employee, the costs of training staff and replacing those who are unsuited to the work are minimised.

Similarly there are commercially available tests and measures designed for work with clinical patients or schoolchildren. In these contexts, tests and measures may be used as screening instruments in order to identify potential difficulties in individuals. For example, if there were a simple effective test for dyslexia then it could be given to classes of children en masse in order to identify individuals who may require further assessment and treatment/support for dyslexia.

Although many of these commercially available tests and measures are employed in research, they are often designed primarily with the needs of practitioners in mind. They may not always be the ideal choice for research:

- They are often expensive to buy. Given that research tends to use large samples, the cost may be prohibitive.
- They are often expensive to administer. Many commercial tests and measures are administered on a one-to-one basis by the psychologist. Psychologists may also require training in their use which is a further cost. Some tests may take as much as two hours or more to administer, and this is not only a further cost, but may also be a deterrent to individuals from participating in the research.
- They are often intended for use with special populations. The tests and measures used by clinical psychologists, for example, may be helpful in identifying schizoid thought tendencies in psychiatric settings but have no value when applied to non-clinical populations.
- Some of the tests and measures are restricted in their circulation, such as to qualified clinical psychologists. Students, especially, may have no access to them. University departments, though, often have a variety of tests for use by students under the supervision of a member of staff.

There is no guarantee that there is a test or measure available for the variables that the researcher needs to measure.

As a consequence, researchers may need to consider constructing new tests or measures rather than relying on commercially available ones. There are many *research instruments* which have been developed which are not available through commercial sources. These can often be found in relevant journal articles, books, websites or directly from their author. Locating these tests and measures will entail a review of the literature in the research field in question. Research studies in your chosen field will often describe or make use of these research instruments. One advantage of using the same measures as other researchers is that they are recognised by the research community as effective measures. Care needs to be taken, however, since the purposes of your research may not be exactly the same as that of previous researchers or the instrument may be unsuitable for other reasons. For example, the research instrument may have been designed for a different culture or a different age group. Hence it may need some modification to make it suitable for the particular group on which you wish to use it. There are circumstances in which the research instrument appears so unsatisfactory that the researcher decides to create an entirely new instrument.

The mechanics of test construction are fairly straightforward and, with the availability of SPSS Statistics and other computer packages, it is feasible to produce bespoke measuring instruments even as part of student research.

14.2 The concept of a scale

Psychologists frequently refer to scales in relation to psychological tests and measures. In general English dictionaries, the term 'scale' is defined as a graded classification system. This will probably suffice to understand the use of the concept in test construction. That is, individuals are numerically graded in terms of their scores on the measure. There are two important ways of creating such graded scales:

- Providing a series of test or measurement items which span the range from lowest to highest. So, if a measure of intelligence is required, a whole series of questions is provided which vary in terms of their difficulty. The most difficult question that the participant can answer is an indicator of their level of intelligence. The difficulty of an item is assessed simply by working out the percentage of a relevant sample who answer the question correctly. This approach was applied to the assessment of attitudes using the Thurstone Scale. For example, in order to measure racial attitudes a series of statements is prepared from the least racist to the most racist. The items are judged by a panel of judges in terms of the extent of the racism in the statement. The most hostile item that a participant agrees with is an indicator of their level of racism. This is known as the 'method of equal-appearing intervals' because the test constructor endeavours to make the items cover all points of the possible range evenly.

- A much more common way of constructing psychological tests and measures operates on a quite distinct principle, although the outcomes of the two methods are often substantially the same. In the method of summated scores the researcher develops a pool of items to measure whatever variable is to be measured. The final score is based on the sum of the items. Usually, an additional criterion is introduced which is that the items should correlate with the total scores on the test or measure. We will return to this in the next section. It is the most commonly used method.

Psychological tests and measures are frequently described as unidimensional or multidimensional. A *unidimensional* scale is one in which the correlations of the items with each other are determined as a result of a single underlying dimension. This is analogous to measuring the weights of 30 people using ten different sets of bathroom scales – there will be strong intercorrelations between the weights as assessed by different sets of bathroom scales. A *multidimensional* scale has two or more underlying dimensions which result in a pattern of intercorrelations between the items in which there are distinct clusters or groups of items which tend to intercorrelate with each other but not with other items so well or not at all. This is analogous to measuring the weights of 30 people using ten different sets of bathroom scales and their heights using five different tape measures. In this case, we would expect for the sample of people:

- strong intercorrelations of their weights as measured using the different sets of bathroom scales;
- strong intercorrelations of the heights as measured with the five different tape measures;
- poor intercorrelations between the ten sets of bathroom scale measures and the five sets of tape measure measures.

This is simply because our 15 different measures (analogous to 15 different items on a questionnaire) are measuring two different things: weight and height.

Which of these is the best? The short answer is that for most purposes of *research* the ideal is a unidimensional scale since this implies a relatively 'pure' measurement dimension. That is, a unidimensional scale can be thought of as aiming to measure a single concept. However, multidimensional scales are sometimes more useful in practical situations. For example, a multidimensional measure of intelligence is likely to predict success at university better than a unidimensional one. This is because university performance is determined by a variety of factors (for example, maths ability, comprehension, motivation and so forth) and not just one. Consequently a measure based on a variety of factors is more likely to be predictive of university success.

Measurement in psychology is beset with a number of fundamental and generally unavoidable problems. Many of these are to do with the weakness or imprecision of measurement in psychology. In the physical world, a centimetre is a standard, well-established and precisely measurable amount. Psychological variables cannot be measured with the same degree of precision. Every psychological measure that we know of suffers from a degree of variability, that is to say, the measurement will vary somewhat each time it is taken – apparently in an unsystematic or random fashion. If we measure age by asking participants their age, we might expect a degree of imprecision – some participants may deliberately lie, others will have forgotten their age, we may mishear what they say and so forth. This occurs when we are measuring something as easy to define as age so one can imagine that the problem is worse when measuring a difficult to define (or unclear) concept such as self-esteem, happiness or cognitive distortions.

Since psychological concepts are often not precisely definable, psychologists tend to measure concepts using a variety of test or measurement items rather than a single item. The idea is that by using a number of imprecise measures of the concept in question, the aggregate of these measures is likely to be a better measure than any of the constituent individual items.

There is nothing wrong with using a single item to measure a psychological variable – one would not measure age by using a 20-item age scale, for example. However, we would use a long scale to measure a less clear variable such as happiness. So the use of scaling is really confined to circumstances in which you wish to get a decent measure of a variable that is difficult to measure. Thus, you would not use scaling if you wished to measure gender or age. A single question will generally produce high quality and highly valid answers to these questions.

It's a bit like finding out the cost of the tube fare to Oxford Street in London by asking lots of friends. Probably none of your friends knows the precise fare, but several would have a rough idea. By combining several rough estimates together by averaging, the probable outcome is a reasonable estimate of the train fare. Obviously it would be better to use a more accurate measure (e.g. phone London Underground) but if this is not possible the rough estimates would do. In other words, there is an objective reality (the actual fare that you will pay) but you cannot find that out directly. This is much the same as psychological variables – there may be an objective reality of happiness but we can only measure it indirectly using an aggregate of imprecise measures.

14.3 Scale construction

At this point, it is important to stress that psychological tests and measures are not created simply on the back of statistical techniques. Ideally, the psychologist constructing a measure will be familiar with the relevant theory and research concerning the thing to be measured. They may also be familiar with related concepts, the opinion of experts, and information from samples of individuals about how they understand and experience aspects of the concept. For example, just what is depression like experientially? Such information concerning the concept can contribute to a more insightful and pertinent set of items to begin the research. The following are worth emphasising:

- Every effort should be made to specify the nature of the concept we wish to measure – just what do we mean by loneliness, depression or staff burnout, for example? Often by reflecting on this we begin to realise that potentially there may be many different features of the concept which we need to incorporate into the *pool of items* from which we will develop the test or measure.

- Even after we have developed our understanding of the concept as well as we can, we may find it impossible to phrase a single question to assess it. Take loneliness; is a question such as 'How many friends do you have?' a good measure of loneliness? It depends on many things – what an individual classifies as a friend, whether loneliness is determined by the number rather than the quality of friendships, the age of the individual since an elderly person may have fewer friends simply as a consequence of bereavements, and so forth. In short, there are problems in turning a variable into a measure of that variable. This does not mean that the question is useless as a measure of the concept, merely that it is not a particularly accurate measure.

- Variables do not exist in some sort of rarefied form in the real world. They are notions which psychologists and other researchers find extremely useful in trying to understand people. So sometimes it will appear appropriate to a researcher to measure a range of things which seem closely related. For example, loneliness might be considered to involve a range of aspects – few friendships, feelings of isolation, no social support, geographical isolation and so forth.

Once a pool of items for potential inclusion has been developed, the next stage is to administer the first draft of the test to a suitable and as substantial a sample of individuals as possible. Advice on how to formulate questions is to be found in Box 14.1.

Box 14.1 Practical Advice

Writing items for questionnaires

Writing questions or items for a psychological measure requires one to focus on one key matter – trying to concoct items that are as unambiguous and clear as possible. The other main criterion has to be that they seem to measure a range of aspects of the topic. Of course, these are not simple matters to achieve and it is easy to rush the job and create an unsatisfactory measure. One needs to understand the topic at as many levels as possible. For example, what do you think the important things are likely to be? Then what do people you know regard as important aspects of the topic? Then what does a focus group or some other group of research participants talk about when they are asked to discuss the topic? How have previous researchers attempted to measure a similar topic? What does the empirical evidence indicate about the major dimensions of the topic? What does theory say about the topic?

Once again the important lesson is to research and explore the topic in a variety of ways. Only in this way can you acquire the depth of knowledge to create a good measure. To be frank, anyone can throw together a list of questions, but it requires commitment and work to write a good questionnaire. If possible, put together elements from all of the resources that you have. Finally, do not forget that once you have the questionnaire, there are a number of processes that you will need to go through to assess its adequacy. These include item analysis, reliability assessment and perhaps validity assessment. These processes contribute to the adequacy of the measure and may help you eliminate inadequate items or excess items.

Nevertheless, here are a few tips:

- Use short and simple sentence structures.
- Short, everyday words are better than long ones.
- Avoid complex or problematic grammar, such as the use of double negatives. For example, 'You ain't seen nothing yet.'
- Leading questions which suggest the expected answer should be avoided largely because of the limiting effect this will have on the variability of the answers. For example, 'Most people think it essential to vote in elections. Do you agree?'
- Choose appropriate language for the likely participants – what would be appropriate to ask a group of high court judges may be inappropriate to a group of nursery children.
- Tap as many resources for items and questions as feasible.
- Accept that you cannot rely on yourself alone as a satisfactory source of questions and ideas for questions.
- People similar to the likely participants in your research are a good starting point for ideas.
- Relax – expertise in question and item writing is a rare commodity. Most researchers mix trial and error with rigorous item analysis as a substitute.

You may wish to consult Chapter 16 on coding data in order to appreciate the variety of ways in which the researcher can structure the answers further.

Let us assume that we have gone through that process and have a list of such items. For illustrative purposes we have ten different items but the list would probably be 30 or 40 items. We have decided to attempt to measure *honesty*. Our ten items are:

Item 1	I am an honest person.
Item 2	I have frequently told little lies so as not to offend people.
Item 3	If I found money in the street I would hand it in to the police.
Item 4	I have never told even the slightest untruth.
Item 5	I would always return the money if I knew that I had been given too much change in a shop.
Item 6	I would never make private phone calls from work.
Item 7	I have shoplifted.
Item 8	It is always best to tell the truth even if it hurts.
Item 9	I usually tell the boss what I think they would like to hear even if it is not true.
Item 10	If I were to have an affair, I would never tell my partner.

Several things are readily apparent about this list:

- There is a wide range of items which seem to be measuring a variety of things. Probably all of the items are measuring something that may be regarded as honesty (or lack of it).
- Some of the items are positively worded in terms of honesty (for example, Items 1, 6 and 8). That is, agreeing with these items is indicative of honesty. Other items are negatively worded in that *disagreeing* with them is indicative of honesty (for example, Items 7, 9 and 10). Often positively and negatively worded items are both included deliberately in order to help deal with 'response sets'. Briefly, it has been established that some people tend to agree with items no matter the content of the item. Thus they have a tendency to agree with an item but also *agree* with an item worded in the opposite direction. That is they might agree with the statement that 'I am an honest person' and also agree with the statement that 'I am not an honest person'. One way of dealing with this is to use both positively and negatively worded items – mixing items for which agreement is indicative of the variable with those for which disagreement is indicative of the variable. Many questionnaires can be found which do *not* do this, however.
- You must remember to reverse score the negatively worded items – if scored in the same way as the positively worded items then the positively worded items would be cancelled out by the negatively worded ones.
- One of the items (Item 4: I have never told even the slightest untruth) seems unlikely to be true of any human. Items like this are sometimes included in order to assess faking 'good' or 'social desirability', that is, trying to give an impression of meeting social standards even unobtainable ones. On the other hand, it is possible that the researcher has simply written a bad item. That is, if everyone disagrees with an item then it cannot discriminate between people in terms of, in this case, honesty. Useful items need to demonstrate variability (variance) among participants in the research.

A careful reading through of the items seems to suggest that there are at least two different sorts of honesty being measured – one is *verbal honesty* (not lying, basically) and the other is *not stealing*. It could well be that this questionnaire is multidimensional in that it is measuring two distinct things. The usual way of assessing this is by examining empirically whether the people who are verbally honest also tend not to steal. Basically this is a matter of correlating the different items one with another.

Box 14.2 Key Ideas

Item analysis

Item analysis refers to the process of examining each item on the scale in order to identify its good features and inadequacies. The following are the main features involved:

- Items which show little variation over the sample should be dropped. This is because such items contribute little or nothing to variations in the total score on the test. Low variability may be assessed by calculating a measure of variation (for example, variance, standard error or standard deviation) or by examining a histogram of the scores on each item.
- Ideally, all items should show similar levels of variation and as much variation in response as possible. If the items do *not* have similar variability then problems may arise if one simply sums the scores on the items on the scale to get a total. If the items do not have similar variabilities, then the proper procedure would be to turn the scores on the individual items into standard scores (see the companion book *Introduction to Statistics in Psychology* (Howitt and Cramer, 2011a, Chapter 5). In some psychological tests and measures, you will find that certain items are given extra scoring weight. This is to take into account this very problem. All other things being equal, an item with large variability would be preferred over one with low variability.
- Items which are omitted (not replied to) or are commented on by a number of participants should be considered for dropping from the scale. Comments and omissions are indicative that the participants are having difficulty knowing the meaning of the item. Rephrasing the item is an option but this means that the scale should be re-administered to a new sample.
- The final stage of item analysis is to examine the consistency with which the individual items contribute to whatever is being measured by the total scale. Item–whole correlation and factor analytic approaches to doing this are discussed in the main body of the text.

Scaling – or the process of developing a psychological scale – deals with the ways in which items are combined in order to get a better measure of a concept than could be achieved using a single item. The common methods of psychological scaling employed by most modern researchers are built on one of two general principles:

- If we sum the scores on the individual items of a test or measure to give the *total score*, then each of the individual items should correlate with this total score if the items are measuring the same thing as each other. *Items which do not correlate with the total score are simply not measuring what the majority of the other items are measuring and may be eliminated from the scale.* This is also known as the *item–whole* or *item–total* approach to scale construction.
- If items on the scale are measuring the same thing then they should correlate substantially with each other (and the total score as well for that matter). Items which measure very different things will correlate with each other either very poorly or not at all. This is the basis of internal consistency approaches to scale construction as well as the *factor analytic* methods. These are discussed later in this chapter. With a multidimensional scale, sometimes you will find distinct groups of items which correlate well with each other but not with other groups of items.

We will consider each of these approaches in turn. They are known as *item analysis* techniques; see Box 14.2 and Figure 14.1.

Item–whole or item–total approach to scale construction

The purpose of item analysis is to eliminate bad items that do not measure the same thing as the scale in general. These are not laborious statistical techniques if one uses a

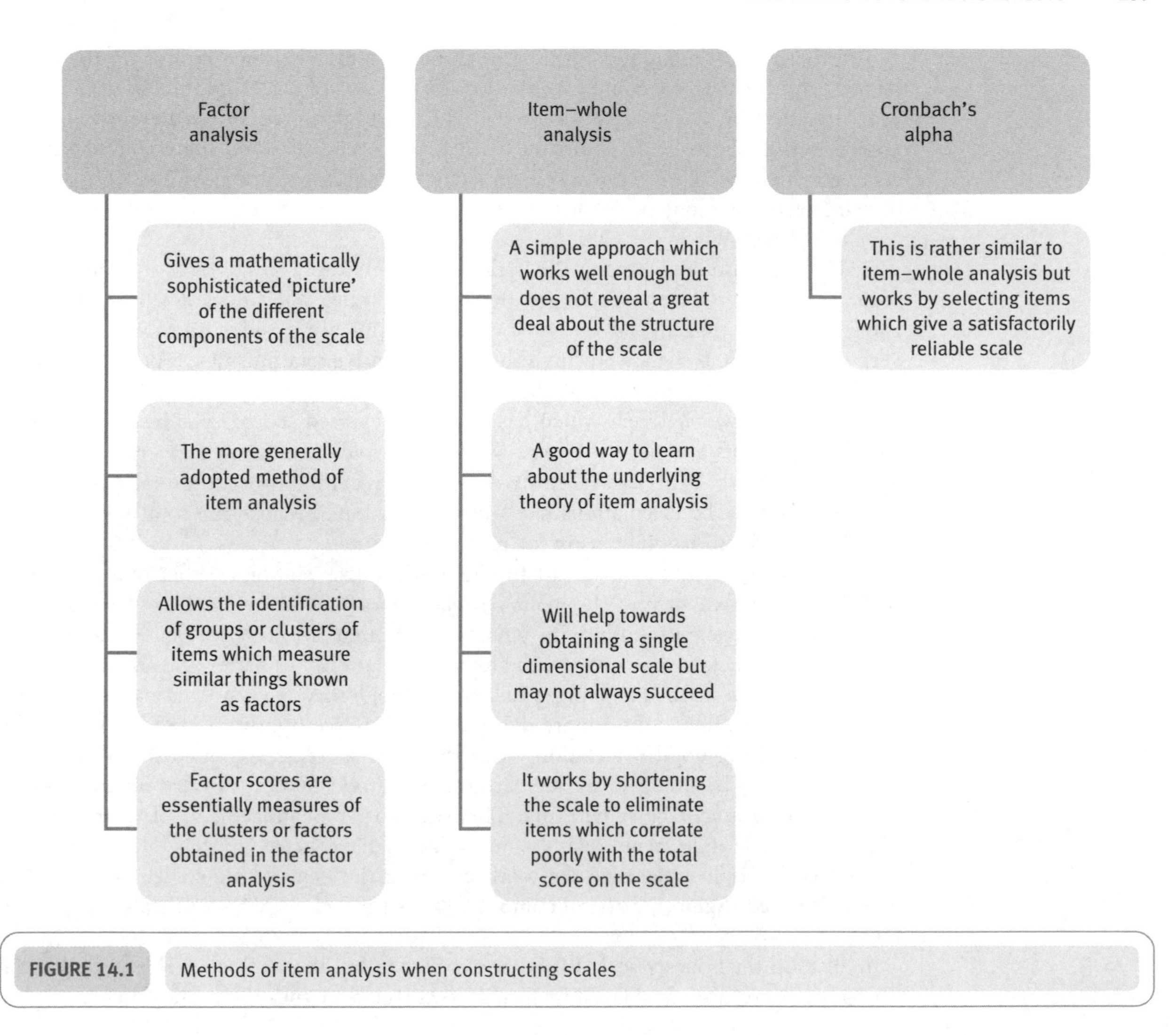

Factor analysis
Gives a mathematically sophisticated 'picture' of the different components of the scale
The more generally adopted method of item analysis
Allows the identification of groups or clusters of items which measure similar things known as factors
Factor scores are essentially measures of the clusters or factors obtained in the factor analysis
Item–whole analysis
A simple approach which works well enough but does not reveal a great deal about the structure of the scale
A good way to learn about the underlying theory of item analysis
Will help towards obtaining a single dimensional scale but may not always succeed
It works by shortening the scale to eliminate items which correlate poorly with the total score on the scale
Cronbach's alpha
This is rather similar to item–whole analysis but works by selecting items which give a satisfactorily reliable scale

FIGURE 14.1 Methods of item analysis when constructing scales

computer program, although historically much time would have been spent doing the same task. So what would we expect our data to show if we had managed to produce a good measure of, say, honesty? Remember that we are doing little more than simply adding up the answers to a range of questions about honesty to give a total score:

- The item–whole method of item analysis involves calculating a total score for honesty. The most obvious way of doing this is simply to add up (for each individual in the sample) the total of their scores on the (ten) individual items. (Don't forget to reverse score items as appropriate.) In this way, you have a total score on the scale for each participant. If the items are measuring the same thing then the total should also be measuring the same thing as the individual items. This total score is also referred to as the *whole*-scale score.
- If the total (or whole-scale) score consists of the sum of several items which individually measure the same thing as the total score (but not so well), then scores on individual items should correlate with the total score on the scale. If an item does not correlate with the whole-scale score (total score) then that item is clearly measuring something different from what the scale is measuring. It can safely be eliminated.

By dropping items, a shorter and probably more consistent scale will be obtained.

Another way of doing much the same thing is to take extreme groups on the whole or total test. That is, we could take the top 25 per cent of scores and the bottom 25 per cent of scores on the entire scale. Items which are answered very differently by high scorers on the entire scale compared with low scorers are good items and should be retained. Items which are answered similarly by high scorers and low scorers are not discriminating and may be dropped from the scale. There is no advantage of this method over using item–whole correlations.

Table 14.1 contains, among other things such as the average of scores on the full scale, the item–total correlations for our honesty scale. What does it tell us? The first thing to note is that all but one of the relationships are positive and this one case is very close to zero. If there were any substantial negative relationships, especially sizeable ones, then that item has probably been scored the wrong way round. That is, it might be a negatively worded item which has not been reversed scored. The researcher needs to check that this is indeed the case. Wrongly scored items should be rescored in the opposite direction. (This can be easily done using recoding procedures such as those in SPSS Statistics.) The calculations have all to be redone because the total score will be incorrect – one more good reason for using a computer.

The most important function of the item–whole correlation coefficients, though, is that they show us which of the items correlate poorly with the total score, that is, the items which correlate weakly with whatever it is that the scale measures. Looking at Table 14.1, it is clear that some of the items correlate rather better with the total score than others. If we wished to shorten the scale (though this one is not very long anyway), then the obvious items to drop are the ones which relate poorly to the total score. The items which have good correlations with the total score are retained for inclusion in the final scale – by doing so we increase the likelihood that all of our remaining items are measuring much the same thing. This is a matter of judgement, of course, but is easily reversible if it seems that too many items have been omitted. However, since items have been dropped then the total score and the item–whole correlations have to be recalculated. Again, statistical computer software such as SPSS Statistics make this a fairly minimal chore.

Item 8 on the honesty scale ('It is always best to tell the truth even if it hurts') is the obvious item to drop first given its near zero correlation with the total of items.

Table 14.1 Item–total correlation

	Scale mean if item deleted	Scale variance if item deleted	Corrected item–total correlation
Honest	26.00	40.33	0.38
Offend	25.38	37.76	0.43
Street	25.77	37.36	0.46
Untruth	25.69	35.23	0.74
Change	26.00	37.00	0.56
Phone	25.31	36.40	0.55
Shoplift	25.85	36.97	0.69
Hurts	25.23	44.86	−0.04
Boss	25.38	37.76	0.43
Affair	25.54	37.94	0.36

So as a reminder, just what has dropping items achieved?

- One result is that the scale becomes a more refined measure of whatever it is that it measures. That is to say, remaining items increasingly tend to measure the same thing.
- The scale is shortened – this may be very important in some research contexts because participants may be more prepared to complete a short measure than a long measure, for example. Be careful though since a short scale may not be as reliable as a longer scale (see Chapter 15 for a discussion of reliability) all other things being equal.

What constitutes a good item–whole correlation cannot be defined in absolute terms. It would be unwise to retain items which fail to meet the minimum criterion of statistical significance. For tests and measures developed solely for the purpose of research with substantial samples of participants, tests and measures with just a few items may be preferred simply because they place less demand on participants. There is a trade-off between length of the test or measure and the number of participants in the study. The greater the number of participants then the shorter the scale may be.

A small refinement

Item–total correlation analysis may be refined especially when the scale consists of relatively few items. This modification involves correlating the item with the total score *minus* the score on that particular item. Put another way, this is merely the correlation of the item with the sum of all of the *other* items. Because the item–whole correlations include the correlation of the item with itself, then this figure will always be inflated somewhat. The extent of the inflation depends on the number of items contributing to the total score on the test – the fewer the items then the greater the impact of any one item. So by dropping the item in question from the total score on the test or measure, we get a better indicator. This amount of inflation of the correlation is probably negligible when we have a lot of items; it is more influential when we have few items. The adjustment is straightforward and is recommended as the preferred approach. Computer programs such as SPSS Statistics will do both versions of the analysis so there is virtually no additional effort required.

This form of item analysis is very much a process and not a single step. By reducing the number of items one at a time, the value and influence of each variable may be assessed. The researcher simply removes items from the scale in order of their item–whole correlations. The item with the lowest item–whole correlation at any stage is normally the next candidate for omission. Box 14.3 explains another approach – how Cronbach's (1951) alpha coefficient may be used similarly to shorten scales and to increase measurement consistency.

The factoring approach

The item-analysis approach described above is important since it is the basis of many common psychological tests and measures. There is an alternative – factor analysis – which is much more feasible than in the past because of the availability of high-speed computers. Factor analysis was developed early in the history of modern psychology as a means of studying the structure of intelligence (and consequently measures of intelligence). Its primary uses are in the context of psychological test and measure construction. Once it was a specialised field but now it is readily available and calculated in seconds using statistical packages such as SPSS Statistics.

First, in factor analysis the computer calculates a matrix of correlations between all of the items on the test or measure (this is provided in Table 14.2). Then mathematical routines are calculated which detect patterns in the relationships between items on the psychological test. These patterns are presented in terms of factors. A factor is simply

Box 14.3 Practical Advice

Using Cronbach's alpha to shorten scales and increase consistency of items

There is another way of eliminating items which are not measuring what the scale measures particularly well. This is based on (Cronbach's) coefficient alpha. This is dealt with in more detail in Chapter 15. It can be regarded for now as an index of the consistency with which all of the items on the scale measure whatever the scale is measuring. It is possible (using a computer program such as SPSS Statistics) to compute the alpha coefficients of the test. There is an option which computes the alpha coefficients of the test with each of the items omitted in turn. This means that there will be as many alpha reliabilities as items in the test. Items which are not measuring the same thing as the other items may be dropped without reducing the size of the alpha reliability coefficient – simply because they are adding nothing to the consistency of the test. (This is by definition since if they added something to the consistency of the test, removing them would lower the reliability of the test.)

The researcher simply looks through the list of alpha coefficients, and the lowest alpha reliability is selected. The item with this alpha may be omitted from the scale as this item is not a good measure of what the scale itself measures. This process is repeated for the 'new' scale and an item dropped. Eventually a shortened scale will emerge which has a sufficiently high alpha coefficient. One of .70 or so is usually regarded as satisfactory.

Cronbach's alpha coefficient is also known as the alpha reliability (see Chapter 15).

Table 14.2 Correlation matrix for the ten-item honesty scale

	Honest	Offend	Street	Untruth	Change	Phone	Shoplift	Hurts	Boss	Affair
Honest	1	−0.169	0.540	0.583	0.553	0.431	0.476	0.239	−0.169	−0.283
Offend	−0.169	1	−0.037	0.196	−0.027	0.046	0.303	0.090	0.999	0.676
Street	0.540	−0.037	1	0.554	0.464	0.583	0.448	−0.004	−0.037	0.082
Untruth	0.583	0.196	0.554	1	0.771	0.720	0.703	−0.077	0.196	0.208
Change	0.553	−0.027	0.464	0.771	1	0.553	0.717	0.035	−0.027	0.078
Phone	0.431	0.046	0.583	0.720	0.553	1	0.341	−0.288	0.046	0.504
Shoplift	0.476	0.303	0.448	0.703	0.717	0.341	1	0.144	0.303	0.164
Hurts	0.239	0.090	−0.004	−0.077	0.035	−0.288	0.144	1	0.090	−0.314
Boss	−0.169	0.999	−0.037	0.196	−0.027	0.046	0.303	0.090	1	0.676
Affair	−0.283	0.676	0.082	0.208	0.078	0.504	0.164	−0.314	0.676	1

an empirically based hypothetical variable which consists of items which are strongly associated with each other. Usually, there will be several factors which emerge in a factor analysis. The precise number depends on the data and it can be that there is simply one significant or dominant factor. More practical details on factor analysis can be found in the two companion texts *Introduction to Statistics in Psychology* (Howitt and Cramer, 2011a) and *Introduction to SPSS Statistics in Psychology* (Howitt and Cramer, 2011b).

Box 14.4 Key Ideas

Phi and point–biserial correlation coefficients

Before computers, psychological test construction required numerous, time-consuming calculations. The phi and point–biserial correlation coefficients were developed as ways of speeding up the calculations by using special formulae in special circumstances. These formulae are now obsolete because computers can do the calculations quickly and easily – see the companion text *Introduction to SPSS Statistics in Psychology* (Howitt and Cramer, 2011b).

The phi coefficient is merely the Pearson correlation coefficient calculated between two binary (binomial or yes/no) variables. Many psychological tests have this form – one simply agrees or disagrees with the test item. So the phi coefficient provided a quicker way of calculating a correlation matrix between the items on a test.

The point–biserial correlation is merely the Pearson correlation calculated between a binary (yes/no) test variable and a conventional score. Thus item–whole (item–total) correlations could be calculated using the point–biserial correlation. One variable is the binomial (yes/no) item and the other variable is the total score on the test.

Each individual test item has some degree of association with each of the major patterns (i.e. the factors found through factor analysis). This degree of association ranges from a zero relationship through to a perfect relationship. In factor analysis, the relationship of a test item to the factor is expressed in terms of a correlation coefficient. These correlation coefficients are known as *factor loadings*. So a factor loading is the correlation coefficient between an item and a factor. Usually there will be more than one factor but not necessarily so. So each test item has a loading on each of several factors. This is illustrated for our honesty scale in Table 14.3. This table is a factor-loading matrix – it gives the factor loadings of each of the test items on each of the factors. Since they are correlation coefficients, factor loadings can range from –1.0 through 0.0 to +1.0. They would be interpreted as follows:

- A factor loading of 1.0 would indicate a perfect correlation of the item with the factor in question. It is unlikely that you will get such a factor loading.

Table 14.3 Factor loadings for the honesty scale

	Factor 1	Factor 2	Factor 3
Honest	0.641	−0.520	0.260
Offend	0.278	0.904	0.276
Street	0.700	−0.261	0.009
Untruth	0.919	−0.007	−0.003
Change	0.818	−0.262	0.005
Phone	0.777	−0.002	−0.492
Shoplift	0.797	0.003	0.331
Hurts	−0.002	−0.119	0.873
Boss	0.278	0.904	0.276
Affair	0.353	0.794	−0.370

- A factor loading of .8 would be a high value and you would often find such values in a factor analysis. It means that the item correlates well with the factor though less than perfectly.
- A factor loading of .5 would be a moderate value for a factor loading. Such factor loadings are of interest but you should bear in mind that a correlation of .5 actually means that only .25 of the variation of the item is accounted for by the factor. (See Chapter 4 of this book and Chapter 7 of the companion text *Introduction to Statistics in Psychology*, Howitt and Cramer, 2011a.)
- A factor loading of .2 generally speaking should be regarded as very low and indicates that the item is poorly related to the factor.
- A factor loading of .0 means that there is no relationship between that item and the factor. That is, none of the variation in the item is associated with that factor.
- Negative (–) signs in a factor loading should be interpreted just as a negative correlation coefficient would be. If the item were to be reverse scored, then the sign of its factor loadings would be reversed. So a negative factor loading may simply indicate an item which has not been reverse scored.

All of this may seem to be number crunching rather than psychological analysis. However, the end point of factor analysis is to put a psychological interpretation on the factors. This is done in a fairly straightforward manner, though it does require a degree of creativity on the researcher's part. The factor loadings refer to items which are usually presented verbally. It is possible to take the items with high factor loadings and see what the pattern is which defines the factor. This merely entails listing the items which have high loadings with factor 1, first of all. If we take our cut-off point as .5 then the items which load highly on factor 1 in descending order of size are:

Item 4 (loading = .919) 'I have never told even the slightest untruth.'

Item 5 (loading = .818) 'I would always return the money if I knew that I had been given too much change in a shop.'

Item 7 (loading = .797) 'I have shoplifted.'

Item 6 (loading = .777) 'I would never make private phone calls from work.'

Item 3 (loading = .700) 'If I found money in the street I would hand it in to the police.'

Remember that some of the items would have been reverse scored so that a high score is given to the honest end of the continuum.

The next step is to decide what is the common theme in these high loading items. This simple step may be enough for you to say what the factor is. It can be helpful to compare the high loading items on a factor with the low loading items – they should be very different. The success of this process depends as much on the insight of the researcher about psychological processes as it does on their understanding of the mechanics of factor analysis.

Looking at the items which load highly on the first factor, mainly they seem to relate to matters of theft or white-collar crime (e.g. abusing the phone at work). So we might wish to label this factor as 'financial honesty' but there may be better descriptions. Further research may cause us to revise our view but in the interim this is probably as good as we can manage.

The process is repeated for each of the factors in turn. It is conventional to identify the factors with a brief title.

Just what can be achieved with factor analysis?

- It demonstrates the number of underlying dimensions to your psychological test.
- It allows you to dispense with any items which do not load highly on the appropriate factors, that is, the ones which do not seem to be measuring what the test is designed to measure. In this way, it is possible to shorten the test.
- It is possible to compute factor scores. This is easy with SPSS Statistics – see the companion text *Introduction to SPSS Statistics in Psychology* (Howitt and Cramer, 2011b). A factor score is merely a score based on the participants' responses to the test items which load heavily on the various factors. So instead of being a score on the test, a factor score is a score on one of the factors. One advantage of using factor scores is that they are standardised scores unaffected by differences in the variance of each of the items. As an alternative, it is possible to take the items which load heavily on a factor and derive a score by totalling those items. This is not so accurate as the factor score method. A disadvantage of using factor scores is that they are likely to vary from sample to sample.

Factor analysis generates variables (factors) which are pure in the sense that the items which load highly on a factor are all measuring the same underlying thing. This is not true of psychological tests created by other means.

14.4 Item analysis or factor analysis?

We have described item analysis and factor analysis. In many ways they seem to be doing rather similar jobs. Just what is the difference between the two in application?

- Factor analysis works using the intercorrelations of all of the items with one another. Item analysis works by correlating the individual items with the total score. Factor analysis is more subtle as a consequence since the total score obtained by adding together items in item analysis might include two or more somewhat distinct sets of items (though they are treated as if they were just a single set).
- Factor analysis allows the researcher to refine their conceptualisation of what the items on the test measure. That is, the factors are fairly refined entities which may allow psychological insight into the scale. Item analysis merely provides a fairly rough way of ridding a scale of bad items which are measuring somewhat different things from those measured by the scale. In that sense it is much cruder.

It should be mentioned that extremely refined scales may not be as effective at measuring complex things as rather cruder measures. For example, we could hone our honesty scale down by factor analysis such that we have just one measure. The trouble is that honest behaviour, for example, may be multiply determined such that a refined measure does not predict honesty very well. In contrast, a cruder test that measures different aspects of honesty may do quite a good job at predicting honest behaviour simply because it is measuring more aspects of honesty. In other words, there may be a difference between a test useful for the development of psychological theory and a test which is practically useful for the purpose of, say, clinical, educational or forensic practice.

14.5 Other considerations in test construction

Of course, this chapter just outlines some of the central features of psychological test construction. There are numerous other considerations that warrant attention:

- There should be instructions for the participants about the completion of the test and, usually, instructions for the researcher to indicate the standard methods of administering the test. These instructions can be extremely detailed and for some tests fairly complex manuals are provided.
- Tests intended for administration to individuals as part of a psychological assessment may contribute significantly to decisions made about the future of that individual. In these circumstances precision is of major importance. Often the researcher will provide tables of norms which are data on how the general (or some other) population score on the test. In this way, a particular clinical client may be compared with other individuals in the test. Norms, as they are called, are often presented as percentiles which are the cut-off points for the bottom 10 per cent, bottom 20 per cent or bottom 50 per cent of scores in that population. Norms may be subdivided by features such as gender or age for greater precision in the comparison.
- Tests for research purposes do not require the same degree of precision or development as tests for practical purposes. This does not mean that the same high standards that are needed for clinical work are inappropriate, merely that for research involving a substantial number of participants sometimes circumstances will demand that a weaker or less precise test is used.

14.6 Conclusion

Writing appropriate and insightful items to measure psychological characteristics can be regarded as a skill involving a range of talents and abilities. In contrast, the creation of a worthwhile psychological scale based on these items is relatively simple once the basics are appreciated. The modern approach based on factor analysis using high-speed computers can be routinely applied to data requiring scaling. Since factor analysis identifies the major dimensions underlying the intercorrelations of the items of the test, the outcome of the process may be a unidimensional scale or a multidimensional scale according to the choices made by the researcher. It is up to the researcher whether the items selected constitute a single dimension or whether more than one dimension is retained. Scaling basically works to make the items of the scale consistent with each other and to remove any which are not consistent with the others. However, at the end of the process we will have, hopefully, a scale high on internal consistency. This does not mean that the scale is anything other than internally consistent. There is another important job to be done, that is, to assess the fitness of the measure for its purpose. This is largely a question of its validity but to some extent also one of its reliability. These are dealt with in Chapter 15.

Key points

- Standardised tests are available for assessment purposes. They may be suitable for research purposes also but not necessarily so. They may be too long, too time-consuming or in some other way not fully appropriate to the purpose of the research. Hence a researcher may find it necessary to develop new measures.
- Many psychological characteristics are difficult to measure with precision using any single item or question. Consequently, it is common to combine several items in order to obtain a more satisfactory measure. This involves selecting sets of items which empirically appear to be measuring much the same thing. This process is known as item analysis.
- The most common methods of item analysis are item–whole (or item–total) correlations and factor analysis. The item–whole method simply selects items which correlate best with the sum of the items. That is, items which measure the same thing as the total of the items are good items. Factor analysis is a set of complex mathematical procedures which identifies groups of items which empirically are highly intercorrelated with each other. A factor then is the basis of a single dimensional scale.
- There are other skills required in scale construction such as the ability to write good items, though the processes of item analysis may well get rid of badly worded items because they do not empirically relate well to other items.
- Some items need to be reverse scored if they are worded in the opposite direction to the majority of items.
- Internal consistency of items does not in itself guarantee that the scale can be vouchsafed as a useful measure of the thing it is intended to measure.

ACTIVITIES

1. We made up the data for the honesty scale. Why don't you carry out the research properly? Take our items, turn them into a questionnaire, get as many people as possible to fill it in, and once that is done analyse them using SPSS Statistics. This is quite easy if you use the companion text *Introduction to SPSS Statistics in Psychology* (Howitt and Cramer, 2011b). Were our made-up data anything like your data?
2. Try extending our honesty scale by including items which measure extra facets of honesty which were not included in our original. How satisfactory empirically is your new scale? How could you assess its validity?

CHAPTER 15

Reliability and validity

Overview

- Reliability and validity are the means by which we evaluate the value of psychological tests and measures.
- In addition, objectivity indicates the extent to which different administrators of the test would get the same outcome when testing a particular participant or client.
- Reliability is about (a) the consistency of the items within the measure and (b) the consistency of a measure over time. Validity concerns the evidence that the measure actually measures what it is intended to measure.
- Both reliability and validity are multifaceted concepts and there are a number of approaches to each. For example, validity ranges from a measure's correlation with similar measures through to a thorough empirical and theoretical assessment of how the measure performs in relation to other variables.
- Reliability and validity are *not* inherent characteristics of measures. They are affected by the context and purpose of the measurement. So, for example, a measure that is valid for one purpose may not be valid for another purpose.

15.1 Introduction

We have created our measure using the item analysis procedures described in Chapter 14. What next? Usually the answer is to assess the reliability and validity of the measure. There are several different sorts of reliability and validity which need to be differentiated. Reliability includes internal, test–retest and alternate-forms reliabilities. Validity includes face, content, concurrent and construct validity. These different types of reliability and validity are not different ways of assessing the same thing but different ways of assessing different aspects of reliability and validity. A measure produced using the item-analysis methods described in Chapter 14 may be useful for many purposes, but what these are depends partly on the reliability and validity of the measure. Many psychological measures, for example, consist of a list of questions which, at best, can only partially capture the characteristics of the things to which they refer. Depression, for instance, cannot be fully captured by the words used to measure it. Consequently, the question of just how well a test or measure captures the essence of a particular concept needs to be asked.

There are a number of criteria to consider. These apply to both the assessment of individuals and measures being used for research purposes:

- *Objectivity* The test or measure should yield similar outcomes irrespective of who is administering the measure – though this is only true with trained personnel who know just how the test should be administered. The opposite of objectivity is subjectivity, that is, the outcome of the measure will depend on who is administering the test. There are some measures which are more reliant on the judgement of the administrator (for example, Hare's psychopathology scale) than others. Training may be more intense in the use of such scales.
- *Reliability* This term has a number of distinct meanings as we will see later. One important meaning is reliability or consistency of the measure at different points in time or across different circumstances. If one realistically expects a psychological characteristic to remain relatively stable over time, the measure used for this characteristic should be relatively stable. A measure of dyslexia which one month indicates that ten children in a school class may have a problem of dyslexia but the next month picks out ten totally different children from the same class would not be reliable. Since dyslexia is a stable characteristic, then the test is patently useless. A reliable test would pick out more or less the same group of children as potentially dyslexic no matter when the test was administered. Of course, if the psychological characteristic is relatively unstable – perhaps a person's mood such as how happy they feel – then we would not expect that measure to be particularly stable. Such a measure may be stable in the short term, that is, with a similar mood on Monday morning compared with Monday afternoon but unstable from week to week. There are relatively few measures which involve unstable characteristics so generally reliability over time is regarded as important in psychology.
- *Validity* Broadly speaking, this refers to the extent to which a measure assesses what it is claimed to measure. There are a variety of ways of assessing validity – none of which is identical to any of the others. The types of question raised in judging validity range from whether the items in the measure appear to assess what they are intended to measure (face validity) to whether the measure of variable A is capable of distinguishing variable A from, say, variables C and D (discriminative validity).

The different types of reliability and validity will be dealt with in detail in subsequent sections.

The concepts of reliability and validity need to be understood in relation to reasonable expectations about the characteristics of a good measure of the psychological concepts in question. We have seen that this is not quite the same as suggesting that a good measure will maximise reliability and validity – although often they will. Reliability and validity are not in built qualities of psychological measures. Reliability and validity will vary with the context and purpose of the measurement, and among different samples of participants. Measurements designed purely for research purposes can be useful despite relatively low levels of reliability and validity. On the other hand, tests designed for the assessment of individuals in, say, clinical or educational settings of necessity have to have much higher levels of reliability and validity since they are used to assess individuals. The inadequacies of measures for research purposes can be partially compensated for by having larger sample sizes though this may be problematic in itself. Different criteria apply to research and individual assessment. In other words, there may be measuring instruments that are adequate for research purposes but unsatisfactory for assessing individuals and vice versa. The reasons for this include the following:

- Research is almost always based on a sample of individuals rather than a particular case. That means that a psychological test that discriminates between groups of people may be useful for research purposes despite being hopelessly imprecise for the assessment of individuals. A forensic psychologist, for example, may need to assess the intelligence of an offender to determine whether their intellectual functioning is so low as to render them incapable of a plea because they are incapable of understanding relevant concepts.

- It takes quite a long period of time and substantial effort to maximise the reliability and validity of a measure. Measures may be required for variables which have not yet been adequately researched. The upshot of this is that the researcher may be left with a choice between constructing a new measure for research purposes or using a poorly documented measure simply because it is available.

- Even if there appears to be a satisfactory measure already available, one should not assume that it is satisfactory without carefully examining the measure and research on it or using it. For example, depression as a clinical syndrome may be different from depression as it is experienced by people in general. To use a clinical measure, then, may be problematic if the research is on depression in non-clinical samples since the test was intended for extreme (clinical) groups, and may not discriminate among non-clinical individuals.

- Measures useful for the assessment of individuals often take a lot of time to administer: perhaps more than two hours for a major test. This amount of time may not be available to the researcher who is dealing with large samples. Hence, a less good measure might be the pragmatic choice.

Where does one find out information concerning the properties of ready-made psychological tests? The following are the main sources of information about psychological measures:

- The instruction manual for the test (if there is one). Students may find that their department has a collection of tests and measures for teaching and research purposes.
- Books or journal articles about the measure. Published research on a particular measure may be accessed through normal psychological databases (see Chapter 7). The published research may be extensive or sparse – it largely depends on the measure in question.
- Catalogues of published tests. These are obtainable from test publishers. Many university departments of psychology hold copies of such catalogues.
- The Internet is a useful source – some tests are published on it.

Of course, exercise caution – what may be seen as a perfectly satisfactory test or measure by others may be flawed from your perspective.

15.2 Reliability of measures

This section reviews the major types of reliability (see Figure 15.1). All reliability concerns the consistency of the measure but the type of consistency varies in the different types of reliability. The broad types of consistency dealt with are internal consistency and consistency over different measurements, such as different points in time. Internal consistency was discussed in Chapter 14. There are several measures of internal consistency which can be readily computed using programs such as SPSS Statistics – Cronbach's (1951) coefficient alpha and split-half reliability are examples. Stability across measures involves the stability of the measure over different versions of the test or across different points in time. Consistency over time has to be evaluated in the light of the interval between administrations of the test. Reliability over a one-week period will be greater than reliability for the same measure over a month. Reliability and validity are both essentially assessed in the form of variants of the correlation coefficient.

Internal reliability

Internal reliability indicates how consistently *all* of the items in a scale measure the concept in question. If a scale is internally reliable, any set of items from the scale could be selected and they will provide a measure that is more or less the same as any other group of items taken from that scale. One traditional way of calculating internal reliability is to calculate scores on half of the items in the test and correlate these scores with those for the same individuals on the remainder of the test. Such procedures form

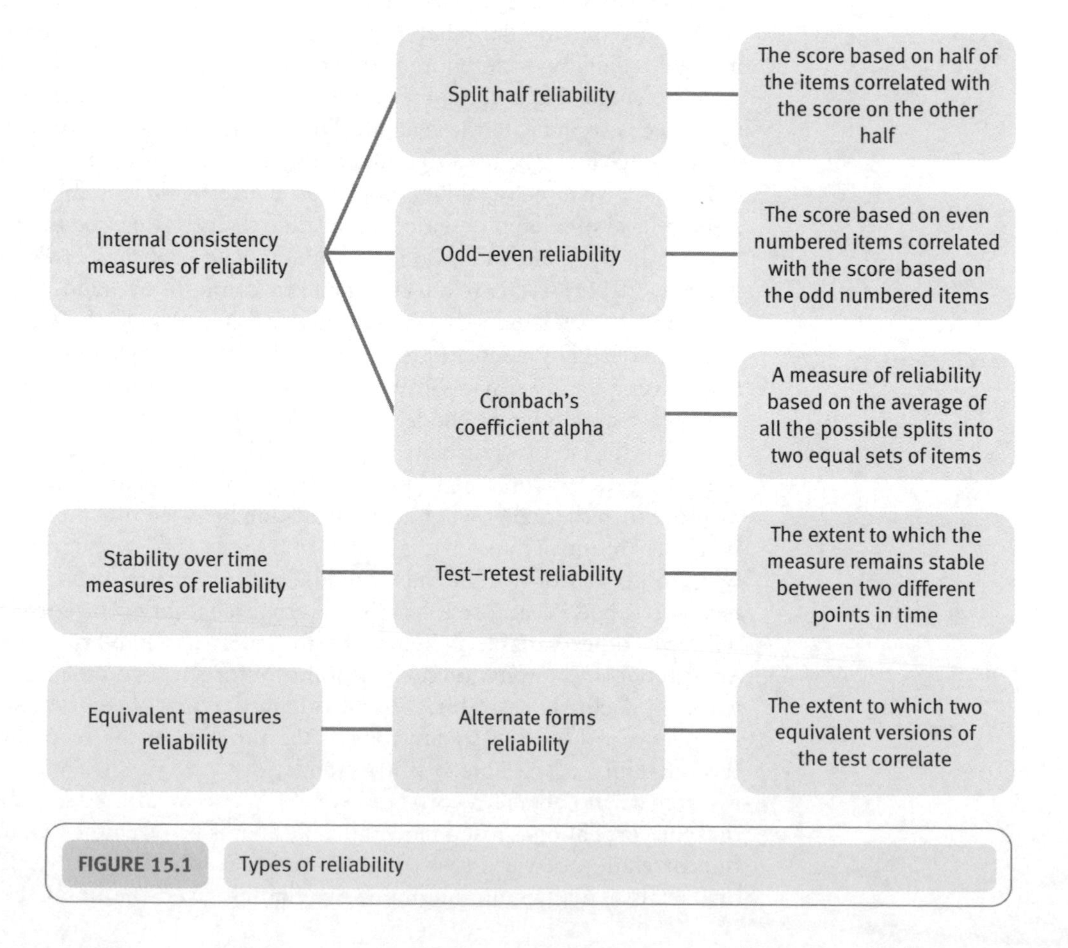

FIGURE 15.1 Types of reliability

the basis of several measures of internal reliability. Alpha reliability may be construed as just a variant on this theme:

- *Split-half reliability* The first half of the items on the test are summed then the second half of the items summed (for each participant). The Pearson correlation between the two halves is calculated – this is referred to as the split-half reliability (though a correction for length of the scale is sometimes applied such that the reliability is for the full-length scale).
- *Odd–even reliability* The two halves are created differently for this. One half is the odd-numbered items (for example, items 1, 3, 5, 7, etc.) and the other half is the even-numbered items (for example, items 2, 4, 6, 8, etc.). The correlation between these two sets of scores is the odd–even reliability. Once again, an adjustment for the length of the scale is often made.
- *Alpha reliability (Cronbach's alpha)* Split-half and odd–even reliability are dependent on which items are selected for inclusion in the two halves. Alpha reliability improves on this merely by being an average of every possible half of the items correlated with every possible other half of the items. Thus alpha reliability is the average of all possible split-half reliabilities. Since alpha reliability takes all items into account and all possible ways of splitting them, then the alpha reliability gives the best overall picture. Fortunately, the calculation of alpha reliability can be achieved more directly using an analysis of variance-based method. This is described in the companion text *Introduction to Statistics in Psychology* (Howitt and Cramer, 2011a). All of these forms of reliability are easily calculated using SPSS Statistics.

The first two measures of internal reliability or consistency have a minor drawback, that is, they are the internal reliability of halves of the items rather than a measure of the internal reliability of the entire scale. It is possible to estimate the reliability of the entire scale by employing what is known as the Spearman–Brown formula. In general, what this does is to indicate the reliability of a scale longer or shorter than the actual scale length. Thus it can be used to estimate the reliability of the full scale, or it could be used to estimate what the reliability of an even shorter scale would be. (If one can achieve the desired level of reliability with a short scale then this may be preferred.) The procedure is given in the companion book *Introduction to Statistics in Psychology* (Howitt and Cramer, 2011a) – it is relatively simple to compute by hand. SPSS Statistics does not compute the Spearman–Brown formula but it can give the Guttman reliability. The Guttman reliability coefficient is much like the split-half reliability adjusted to the full-scale length using the Spearman–Brown formula and is actually more generally applicable than the Spearman–Brown method.

Why is the internal reliability important? The better the internal reliability of a measure then the better the measure (all other things being equal). Furthermore, the better the measure then the higher will be the correlation between that measure and other variables. It is fairly intuitive that a bad measure of honesty, say, will correlate less well with, say, lying to get out of trouble than would a good measure of honesty. One reason why a measure is bad is because it has low internal reliability. The correlation of any variable with any other variable is limited by the internal reliability of the variable. So when interpreting any correlation coefficient then information on the reliability of the variables involved is clearly desirable. The maximum size that the correlation between variables A and B can be is the square root of the product of the reliability of variable A and the reliability of variable B (for example, $\sqrt{.8 \times .9} = .85$). Without knowing this, the researcher who obtains a correlation of .61 between variable A and variable B might feel that the correlation is just a moderate one. So it is between the two measures, but actually the correlation between two variables could be only .85 at most. Hence the correlation of 0.61 is actually quite an impressive finding given the unreliability of the measures.

There is a simple correction which allows one to adjust a correlation by the reliabilities of one or both of the variables. This is described in Chapter 4. Given the difference that this adjustment can make to the interpretation of the obtained relationships, it is probably too often neglected by researchers.

Finally, a note of caution is appropriate. Many textbooks stress that internal reliability is an important and essential feature of a good measure. This is true, but only up to a point. If one is trying to measure a carefully defined psychological construct (intelligence, extroversion, need for achievement) then internal reliability should be as high as can be practically achieved. On the other hand, the measurement of such refined psychological concepts may not be the main objective of the measure. Since much human behaviour is multiply determined (being the result of the influence of a number of variables though not necessarily all of them) then a measure that measures a lot of different variables may actually be better at predicting behaviour.

A good example of this is Hare's psychopathy checklist (Hare, 1991). Psychopaths are known to be especially common in criminal populations, for example. The checklist is scored by simply adding up all of the features of psychopathy that an individual possesses. These features include glibness, pathological lying, manipulativeness and grandiose estimates of self-worth. The score is simply the number of different characteristics of psychopathy manifested by that individual. For a characteristic such as psychopathy which is a syndrome of diagnostic features, internal reliability is less crucial than including all possible diagnostic features of the syndrome.

Stability over time or different measures

Psychologists also apply the consistency criterion to another aspect of measurement: how their tests and measures perform at a time interval or across similar versions of a test. There are several different types of this:

- *Test–retest reliability* This is simply the correlation between scores from a sample of participants on a test measured at one point in time with their scores on the same test given at a later time. The size of test–retest reliability is basically limited by the *internal reliability* of the test. Of course, the test–retest reliability is affected by any number of factors in addition. The longer the interval between the test and retest the more opportunity there is for the characteristics of individuals to simply change, thus affecting the test–retest reliability adversely. However, test–retest reliability may be affected by carry-over from the first administration of the test to the second. That is, participants may simply remember their answers to the first test when they complete the retest.

- *Alternate-forms reliability* A test may be affected by 'memory' contamination if it is used as a retest instrument. This may be a simple learning effect, for example. Consequently, many tests are available in two versions or two forms. Since these contain different items, some of the 'memory' contamination effects are cancelled out – though possibly not all. The relationship between these two alternate forms is known as the alternate-forms reliability. Once again, the maximum value of alternate-forms reliability is the product of the two internal reliabilities. If the alternate-forms reliability is similar to this value then it seems clear that the two forms of the test are assessing much the same things. If the alternate-forms reliability is much lower than the maximum possible, then the two forms are measuring rather different things. A correlation between two tests does not mean that they are measuring the same thing – it means that they are partially measuring the same thing. The bigger the correlation up to the maximum given the reliabilities of the tests, the more they are measuring substantially the same things. As with test–retest reliability, alternate-forms reliability is limited by the internal reliability of the tests.

The reason for the close relationship between internal reliability and other forms of reliability has already been explained. To repeat, the lower the internal reliability of a test then the lower the maximum correlation of that test with any other variable (including alternate forms of the test and retests on the same variable). The bigger the internal reliability value then, all other things being equal, the bigger the correlation of the test with any other variable it correlates with. In other words, there is a close relationship between different forms of reliability despite their superficial differences.

A measure should be reliable over time if the concept to which it refers is chronologically stable. We do not expect a thermometer to give the same readings day after day. However, we might expect that bathroom scales will give more or less stable readings of our weight over a short period of time. That is, we expect the temperature to vary but our weight should be largely constant. In the same way, reliability over time (especially test–retest reliability) should only be high for psychological characteristics which are themselves stable over time. Psychological characteristics which are not stable over time (attention, happiness, alertness, etc.) should *not* necessarily give good levels of test–retest reliability. Characteristics which we can assume to be stable over time (intelligence, honesty, religious beliefs) should show strong test–retest reliability. In other words, reliability must be carefully assessed against how it is being measured and what is being measured. That accepted, psychologists tend to want to measure stable and enduring psychological characteristics, hence test–retest reliability is generally expected to be good for most psychological tests and measures.

15.3 Validity

Validity is usually defined as 'whether a test measures what it is intended to measure'. This fits well with the dictionary definition of the term *valid* as meaning whether something is well founded, sound or defensible. The following should be considered when examining the validity of a test:

- Validity is not a property of a test itself but a complex matter of the test, the sample on which a test is used, the social context of its use and other factors. A test which is good at measuring religious commitment in the general population may be hopelessly inadequate when applied to a sample of priests. A test which is a good measure in research may prove to be flawed as a part of job selection in which applicants will put their best face forward (that is, maybe not tell the truth). Famously this issue is often put as a question 'Valid for what?' The implication of this is that validity is not an inherent feature of a test or measure but something that should be expected to vary according to the purpose to which the test or measure is being put.

- Reliability and validity are, conceptually, quite distinct and there need not be any necessary relationship between the two. So be very wary about statements which imply that a valid test or measure has to be reliable. We have already seen that in psychology the emphasis in measurement is generally on relatively stable and enduring characteristics of people (for example, their creativity). Such a measure should be consistent over time (reliable). It also ought to distinguish between inventors and the rest of us if it is a valid measure of creativity. A measure of a characteristic which varies quite rapidly over time will not be reliable over time – if it is then we might doubt its validity. For example, a valid measure of suicide intention may not be particularly stable (reliable) over time though good at identifying those at risk of suicide. How reliable it is will depend on the interval between the test and retest.

- Since validity is often expressed as a correlation between the test or measure and some other criterion, the validity coefficient (as it is called) will be limited by the reliability of the test or measure. Once again, the maximum correlation of the test or measure with any other variable has an upper limit determined by the internal reliability.

15.4 Types of validity

There are a number of generally recognised types of validity – face, content, criterion; that is, concurrent and predictive validity, construct, known-group and convergent validity (see Figure 15.2). (Other types of validity such as internal validity and external validity concern research design rather than measurement as such. These are dealt with in Chapter 12.) Over the years, the distinction between these different types of validity has become a little blurred in textbooks. We will try to reinstate the distinctive features of each.

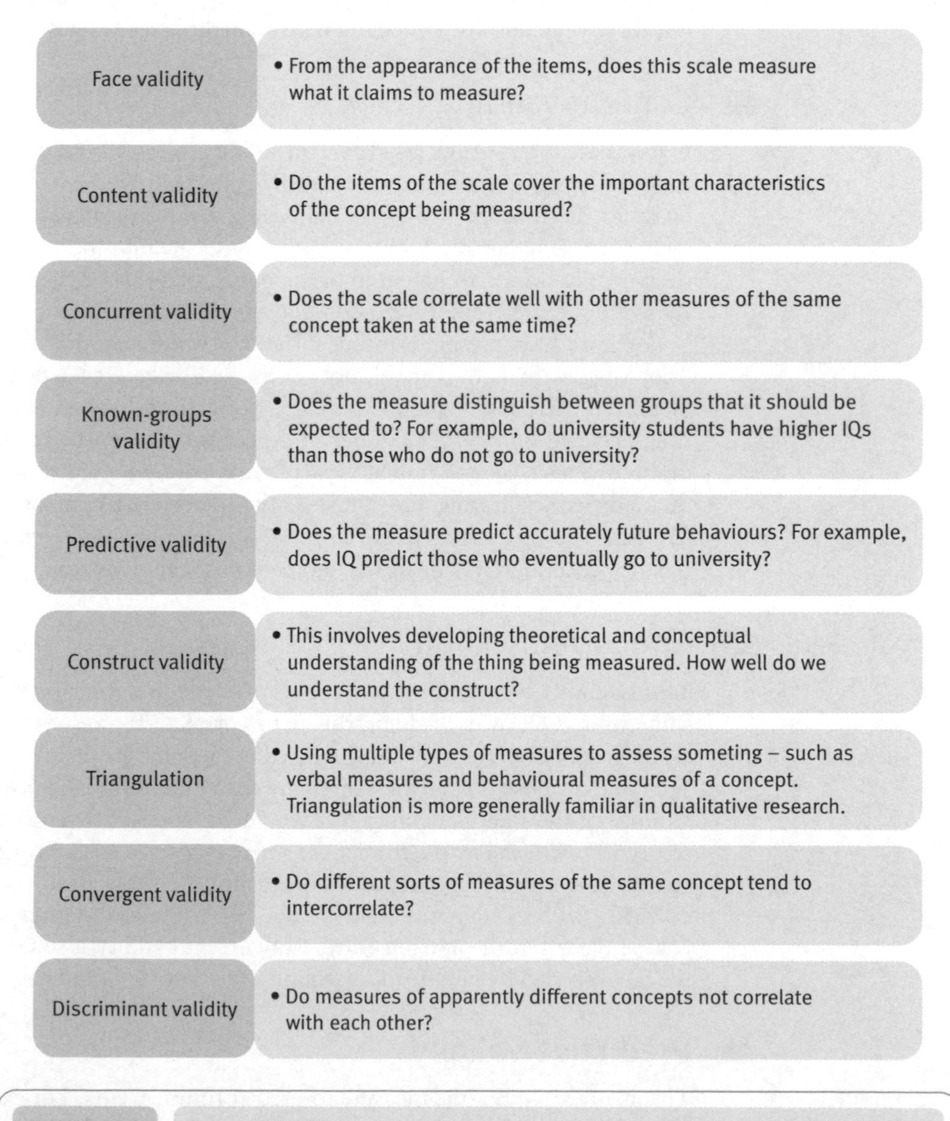

Face validity
• From the appearance of the items, does this scale measure what it claims to measure?
Content validity
• Do the items of the scale cover the important characteristics of the concept being measured?
Concurrent validity
• Does the scale correlate well with other measures of the same concept taken at the same time?
Known-groups validity
• Does the measure distinguish between groups that it should be expected to? For example, do university students have higher IQs than those who do not go to university?
Predictive validity
• Does the measure predict accurately future behaviours? For example, does IQ predict those who eventually go to university?
Construct validity
• This involves developing theoretical and conceptual understanding of the thing being measured. How well do we understand the construct?
Triangulation
• Using multiple types of measures to assess someting – such as verbal measures and behavioural measures of a concept. Triangulation is more generally familiar in qualitative research.
Convergent validity
• Do different sorts of measures of the same concept tend to intercorrelate?
Discriminant validity
• Do measures of apparently different concepts not correlate with each other?

FIGURE 15.2 Types of validity

Face validity

This form of validity can only be assessed informally. One inspects the test items in order to assess whether on the face of things (that is, in terms of the content of the items) the test would appear to be a measure of the psychological concept concerned. Generally speaking, the researcher inevitably applies this form of validity criterion while constructing the test since the measure will include items which the researcher considers to be viable. The problems with face validity are obvious given the need for item-analysis techniques. It would appear that the mere inspection of the items is no guarantee that the retained items form a valid measure. There are lots of reasons for this. For example, items which appear valid to the researcher may be understood very differently by the participants.

Face validity is a very minimum measure of validity which is subjective in that different researchers will come to different conclusions about the face validity of a test. Some tests and measures are constructed to measure what they measure without the researcher being concerned about what the test might correlate with or predict. In these circumstances face validity may be crucial. For example, if a researcher wished to measure opinions about the causes of crime, the content of the items on the measure would be important. Whether or not these opinions are associated with something else might not concern the researchers.

Content validity

In its classic formulation, good content validity follows from the careful creation of a broad range of items. These items are carefully collected together to reflect a wide variety of the facets of the concept being assessed. Using diverse means of eliciting potential items for inclusion is important. Such diversity would include the research literature, interviews with people similar to potential participants, established theory in the field and so forth. By seeking items from a wide domain, the content validity of the measure is enhanced.

Some authors present a rather different version of what constitutes content validity. They suggest that content validity is achieved by reference to experts on the topic being measured. The task of the expert is to offer insight into whether the items cover the range needed or whether significant aspects of the construct being measured have been omitted. This is a very limited view of what content validity is, although it is one aspect of it. By concentrating on such a limited aspect of content validity, authors such as Coolican (2009) misleadingly imply that content validity is merely a version of face validity, although it is a slightly more sophisticated version.

Concurrent validity

This is simply how well the test correlates with an appropriate criterion measured at the same time. One way of doing this is to correlate the test with another (possibly better established) test of the same thing applied to the same group of participants at the same time. So if one proposed replacing examinations with multiple-choice tests then the concurrent validity of the multiple-choice test would be assessed by correlating the multiple-choice test scores with the marks from an examination given at the same time. It stands to reason that a new test which purports to measure the same thing as the examination should correlate with examination marks. The better the correlation, the more confidence that can be placed in the new measure. Concurrent validity is assessed against a criterion, as we have seen. The next form of validity, predictive validity, is equally criterion based.

Predictive validity

This is the ability of the measure to predict future events. For example, does our measure of honesty predict whether a person will be in prison in five years' time? Predictive validity can be measured by the correlation between the test and the future event. Of course, the predictive validity will depend on the nature of the future event that is being predicted. Many psychological tests are not really intended for the prediction of future events so their lack of validity in this respect is of no consequence. It is a bonus if a test does predict future events when it is not intended to. A measure intended to predict future events does not always have to be rich in psychological detail in order to be effective – and there is good reason why it should show internal consistency. For example, if we wished to predict future offending then a measure consisting of variables such as number of previous convictions and gender may be the best way of predicting. Predicting from psychological traits may be relatively ineffective in these circumstances. Again, since this is a criterion-based assessment of validity, the researcher must have expectations that the test or measure will relate highly to future events and also know what these future events might be. In these circumstances, given that prediction is the prime concern, the content of the test as such probably does not matter. The important thing is that it does predict the future event.

Construct validity

For some researchers, life is not quite so simple as the criterion-based validity methods imply. Take a concept such as self-esteem, for example. If we develop what we see as being an effective measure of self-esteem, can we propose criteria against which to assess the concurrent and predictive validity? We might think that self-esteem is inversely related to future suicide attempts though probably very weakly. Alternatively, we might think that the measure of self-esteem should correlate with other measures of self-esteem as in concurrent validity. However, what if we developed a measure of self-esteem which utilised new theoretical conceptualisations of self-esteem? In these circumstances, relating our new measure of self-esteem to existing measures would not be sufficient. Since our new measure is regarded as an improvement then its lack of concurrent validity with older methods of measurement might be a good thing.

Construct validity is generally poorly understood and even more poorly explained in research methods textbooks. The reason is that it is presented as a technical measurement issue whereas, in its original conceptualisation, construct validity should be more about theory development and general progress of knowledge. So it is much more about being able to specify the nature of the psychological construct that underlies our measure than demonstrating that a test measures what it is supposed to measure. In one of the original classic papers on construct validity, Cronbach and Meehl (1955) put the essence of construct validity in the form of a graphic example very apposite for students:

> Suppose measure X correlates .50 with Y, the amount of palmar sweating induced when we tell a student that he has failed a Psychology I exam. Predictive validity of X for Y is adequately described by the coefficient, and a statement of the experimental and sampling conditions. If someone were to ask, 'Isn't there perhaps another way to interpret this correlation?' or 'What other kinds of evidence can you bring to support your interpretation?', we would hardly understand what he [sic] was asking because no interpretation has been made. These questions become relevant when the correlation is advanced as evidence that 'test X measures anxiety proneness.' Alternative interpretations are possible; e.g., perhaps the test measures 'academic aspiration', in which case we will expect different results if we induce palmar sweating by economic threat. It is then reasonable to inquire about other kinds of evidence.
>
> (p. 283)

Cronbach and Meehl then report a variety of 'findings' from other studies which help us understand the nature of test X better:

- Test X has a correlation of .45 with the ratings of the students' 'tenseness' made by other students.
- Test X correlates .55 with the amount of intellectual inefficiency which follows the administration of painful electric shocks.
- Test X correlates .68 with the Taylor anxiety scale.
- The order of means on test X is highest in those diagnosed as having an anxiety state, next highest in those diagnosed with reactive depression, next highest in 'normal' people, and lowest in those with a psychopathic personality.
- There is a correlation of .60 between palmar sweat when threatened with failure in psychology and when threatened with failure in mathematics.
- Test X does not correlate with social class, work aspirations and social values.

The reason Cronbach and Meehl include all of this extra information is that it seems to confirm that academic aspiration is not the explanation of the relationship between test X and palmar sweating. The above pattern of findings better supports the original interpretation that the relationship between test X and palmar sweating is due to anxiety. Cronbach and Meehl go on to suggest that, if the best available theory of anxiety predicts that anxiety should show the pattern of relationships manifested by test X, the idea that test X measures anxiety is even more strongly supported.

So delving into the origins of the concept of construct validity clearly demonstrates it to be a complex process. Test X is assessed in relation to a variety of information about that test, but also other variables associated with it. Put another way, construct validity is a method of developing psychological understanding that seeks to inform the researcher about the underlying psychological construct. In this sense, it is about theory building because we need constructs upon which to build theory. So when authors such as Bryman (2008) suggest that construct validity is about suggesting hypotheses about what a test will correlate with and Coolican (2009) says construct validity is about how a construct explains a network of findings, they are giving only partial accounts and not the totality of construct validity. It should also be clear that construct validity is much more an attitude of mind on the part of the researcher than it is a technical methodological tool. If one likes, construct validity is a methodological approach in the original sense of the term 'methodological', that is, strategies for enhancing and developing knowledge. In modern usage, methodological simply refers to the means of collecting data. There is more to knowledge than just data. So it is possible for a construct to be refined and clarified during the progress of research in a field – the definition of a construct is not fixed for all time.

Construct validity may include a vast range of types of evidence, including some that we have already discussed – the correlations between items, the stability of the measure over time, concurrent and predictive validity findings, and so forth. What constitutes support for the construct depends on what we assume to be the nature of the construct. Finding that a test tends to be stable over time undermines the validity of a test which measures transitory mood, as we have already indicated.

Anyone struggling to understand how construct validity fits in with their work should note the following: 'The investigation of a test's construct validity is not essentially different from the general scientific procedures for developing and confirming theories' (Cronbach and Meehl, 1955, p. 300). This does not particularly help one apply construct validity to one's own research – it merely explains why the task is rather difficult. Construct validity is so important (and difficult) because it basically involves many aspects of the research process.

The following types of validity can be considered to be additional ways of tackling the complex issue of construct validity.

Known-groups validity

If we can establish that scores on a measure differ in predictable ways between two specified groups of individuals then this is evidence of the value of the test – and its known-groups validity. For example, a test of schizoid thought should give higher scores for a group of people with schizophrenia than a group of individuals without schizophrenia. If it does, then the test can be deemed valid by this method. But we need to be careful. All that we have established is that the groups differ on this particular test. It does not, in itself, establish definitively that our test measures schizoid thought. Many factors may differentiate people with schizophrenia from others – our measure might be assessing one of these and not schizoid thought. For example, if people with schizophrenia are more likely to be men than women then the variable gender will differentiate the two groups. Gender is simply not schizoid thought.

A measure which has good known-groups validity is clearly capable of more-or-less accurately differentiating people with schizophrenia from others. Being able to do this may add little to our scientific understanding of the construct of schizoid thought. It is only when it is part of a network of relationships that it is capable of adding to our confidence in the scientific worth of the construct – or not.

Triangulation

Triangulation uses multiple measures of a concept. If a relationship is established using several different types of tests or measures then this is evidence of the validity of the relationship, according to Campbell and Fiske (1959). Ideally, the measures should be quite different, for example, using interviewer ratings of extroversion compared with using a paper and pencil measure of the same concept. In some ways triangulation can be seen as a combination of concurrent and predictive validity. The basic assumption is that if two different measures of ostensibly the same construct both correlate with a variable which the construct might be expected to relate to, then this increases our confidence in the construct.

For triangulation to help establish the construct, all three components of the triangle ought to correlate with each other as in Figure 15.3. All of the arrows represent a relationship between the three aspects of the triangle. If test A and test B correlate then this is the basic requirement of demonstrating concurrent validity. If test A correlates with variable *X*, and test B also correlates with variable *X* then this is evidence of the predictive validity of tests A and B. Notice that in effect this approach is building up the network of associations which Cronbach and Meehl regard as the essence of construct validation. As such, it is much more powerful evidence in favour of the construct than either the evidence of concurrent validity and predictive validity taken separately. Imagine tests A and B both predict variable *X*, but tests A and B do not correlate. The implication of this is that tests A and B are measuring very different things although they clearly predict different aspects of variable *X*.

Triangulation might be regarded as a rudimentary form of construct validation. It is clearly an important improvement over the crude empiricism of criterion validity (predictive and concurrent validity). Nevertheless it is *not* quite the same as construct validity. Its concerns about patterns of relationships are minimal. Similarly, triangulation has only weak allegiances with theory.

Convergent validity

This introduces a further dimension to the concept of validity, that is, measures of, say, honesty should relate irrespective of the nature or mode of the measure. This means that:

- For example, a self-completion honesty scale should correlate with a behavioural measure of honesty (for example, handing in to the police money found in the street)

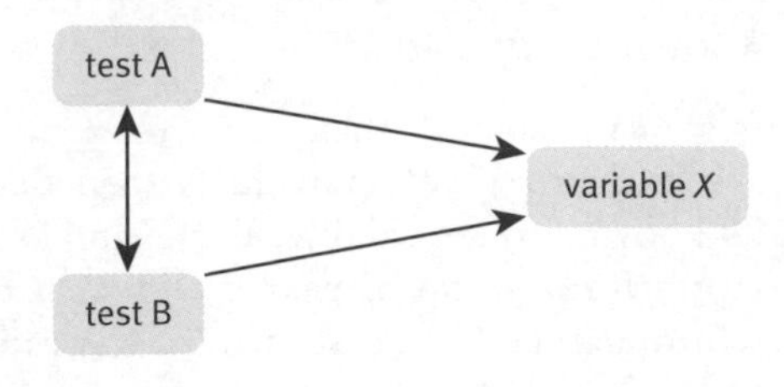

FIGURE 15.3 Triangulation compares several types of test of a concept

and assessments of lying on a polygraph (lie detector) test. All should correlate with each other well and be distinguishable from measures of different but partially related concepts such as religiousness.

- The *type* of measure should not unduly determine the relationships. If we find that our best correlations among measures of honesty and religiosity all use self-completion questionnaires then there may be a validity issue. The domain of measurement (self-completion) seems to be having greater bearing on the relationships than the construct being measured. If self-completion measures of religiousness and honesty have higher correlations with each other than self-completion measures of honesty have with behavioural measures of honesty there is clearly a validity problem. In other words, validity is enhanced by evidence that the concept may be measured by a multiplicity of measures from a wide variety of domain. This is a reasonable criterion of validity, but some of its underlying assumptions might be questioned. One of these assumptions is that there should be a strong relationship between different types of measure of a concept. This is a debatable assumption in some cases. For example, would we expect a strong relationship between racial attitudes as assessed by a self-completion attitude questionnaire and a behavioural measure of racial attitudes such as abusive racist behaviour in the street? Many people with racist attitudes are unlikely to express their attitudes in crude racist chants at a football match simply because 'they do not do that sort of thing' or they fear arrest by the police.

Discriminant validity

Convergent validities indicate that measures which are measuring the same thing ought to correlate with each other substantially. Discriminant validity is just the opposite. If measures are apparently measuring different things then they should not correlate strongly with each other. It should be obvious, then, that convergent and discriminant validity should be considered together when assessing the validity of a measure.

It is fairly obvious that there is a degree of overlap in the different sorts of validity though broadly speaking they are different.

15.5 Conclusion

A crucial feature of the concepts of reliability and validity is that they are tools for thinking and questioning researchers. Employed to their fullest potential they constitute a desire to engage with the subject matter of the research in full. They should not be regarded as kitemark standards which, if exceeded, equate to evidence of the quality of the test or measure. Properly conceived, they invite researchers to explore both the theoretical and the empirical relationships of their subject matter. The value of a measure cannot be assessed simply in terms of possessing both high validity and high reliability. What is reasonable by way of reliability and validity is dependent on the nature of what is being measured as well as what ways are available to measure it.

Key points

- All measures need to be assessed for objectivity, reliability and validity. There is no minimum criterion of acceptable standards of each of these since the nature of what is being measured is also crucial.
- Information about the reliability and validity of some measures is readily available from standard databases, journal articles and the Internet. However, this is not always the case for every measure.
- Reliability is about consistency within the measure or over time.
- Cronbach's alpha, split-half reliability and odd–even reliability are all measures of internal reliability. Cronbach's alpha is readily computed and should probably be the measure of choice.
- Reliability is also assessed at correlating a measure given at different time points (test–retest reliability) and between different versions of the test (alternate-forms reliability).
- Reliability over time should be high for measures of stable psychological characteristics but low for unstable psychological characteristics.
- Validity is often described as assessing whether a test measures what it is supposed to measure. This implies different validities according to the purpose to which a measure is being put.
- Face validity is an examination of the items to see if they appear to be measuring what the test is intended to measure.
- Content validity refers to the processes by which items are developed – sampling from a wide variety of sources, using a wide variety of informants and using a variety of styles of item may all contribute to content validity.
- Construct validity involves a thorough understanding of how the measure operates compared with other measures. Does the pattern of relationships between the measures appear meaningful for a variety of other constructs? Does the measure do what pertinent theory suggests that it should?
- Construct validity is an extensive and complex process which relates more to the process of the development of science than to a single index of validity.
- Known-groups validity, triangulation and convergent validity can all be seen as partial aspects of construct validity. After all, they each examine the complex patterns of relationships between the measure, similar measures, different measures and expectations of what types of person are differentiated by the measure.

ACTIVITY

You ask your partner if they love you. Their reply is 'yes, very much so'. How would you assess the reliability and validity of this measure using the concepts discussed in this chapter?

CHAPTER 16

Coding data

Overview

- Coding is the process of categorising the raw data, usually into descriptive categories.
- Coding is a basic process in both quantitative and qualitative research. Sometimes coding is a hidden process in quantitative research.
- In precoding, the participants are given a limited number of replies to choose from, much as in a multiple-choice test.
- Whenever data are collected qualitatively in an extensive, rich form, coding is essential. However, some researchers will choose to impose their own coding categories whereas others will endeavour to develop categories which match the data as supplied by the participants and which would be meaningful from the participants' perspectives.
- Coding schedules may be subjected to measures of inter-coder reliability and agreement when quantified.

16.1 Introduction

This chapter marks the dividing line between the quantitative and qualitative parts of this book. Perhaps a better way of regarding it is as a major point of intersection between quantitative and qualitative research. More than any other part of this book, coding brings together quantitative and qualitative research.

Coding in some form or another is central to all psychological research. This is true irrespective of the style of research involved – quantitative or qualitative. Inevitably when we research any complex aspect of human activity, we have to simplify this complexity or richness (note the very different ways of describing the same thing) in order to describe and even explain what is happening. No researcher simply videos people's activities and compiles these for publication, no matter how much they abhor quantification. There is always some form of analysis of what is occurring. In this way, all researchers impose structure on the social and psychological world. Coding is quintessentially about how we develop understanding of the nature of the psychological world. Nevertheless, there are radical differences in the way that quantitative and qualitative researchers generally go about the categorisation process. It is important to recognise these differences and to appreciate their relative strengths. Then we may recognise the considerable overlap between the two.

Figure 16.1 indicates the four possible combinations of quantitative and qualitative data collection and analysis methods. Three of the combinations are fairly common approaches. Only the qualitative analysis of quantitative data is rare or non-existent, although some researchers have suggested that it is feasible. The crucial thing about the figure is that it differentiates between the data collection and the data analysis phases of research. The important consequence of this distinction is that it means that quantitative analysis may use exactly the same data collection methods as qualitative analysis. Thus, in-depth interviews, focus groups, biographies and so forth may be amenable to quantitative analysis just as they are amenable to qualitative analysis.

Coding has two different meanings in research:

- The process by which observations, text, recordings and generally any sort of data are categorised according to a classificatory scheme. Indeed, categorisation is a better description of the process than is the word coding.

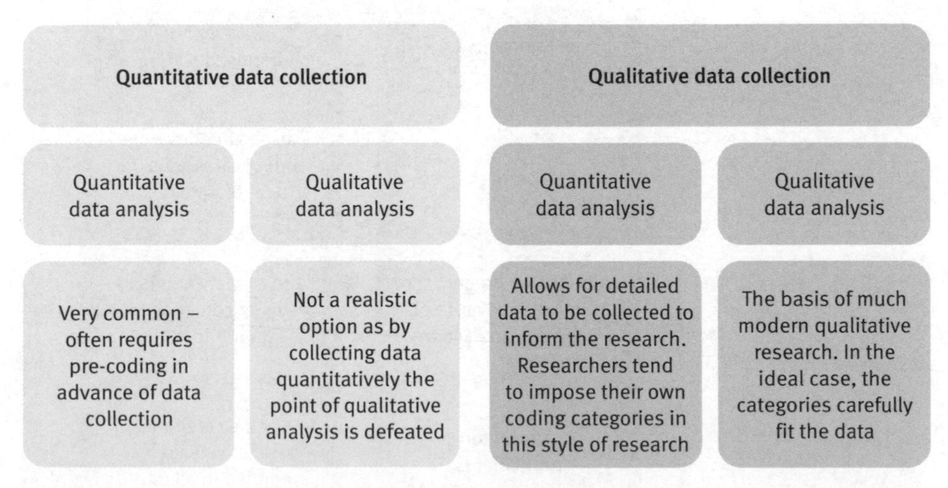

FIGURE 16.1 Relation between quantitative and qualitative data collection, and quantitative and qualitative data analysis

- The process by which items of data are given a number for the purposes of computer analysis. For example, a variable such as gender consists of the two categories: male and female. It simplifies computer applications if these are entered with a numerical code in which, say, 1 represents male and 2 represents female. This seems closer to the dictionary definition of code that suggests that coding is to represent one thing by another.

Often both of these occur in the same research. In highly structured materials such as multiple-choice questionnaires, the categorisation is actually done by the participant, leaving only the computer coding to the researcher. However, coding is a term that is not particularly associated with quantitative data any more than qualitative data. Coding reflects a major step between data collection and the findings of the research. It is really the first stage of the process by which the data are given structure.

Content analysis is a common term which refers to the coding of, especially, mass media content – books, television programmes and so forth. Television programmes provide a source of 'rich' and complex data. It is common for communications researchers to systematically classify or categorise media content. Frequently, they provide rates or frequencies of occurrence of different sorts of content. For example, the researcher might study sexism in television programmes. This would involve, for example, counting the number of times women are portrayed in domestic settings as opposed to work settings, or how often women are used as authoritative voice-overs in advertising. Content analysis typifies the categorisation process needed as part of the analysis of a variety of data.

16.2 Types of coding

There are at least three types of coding (see Figure 16.2):

- pre-coding;
- researcher-imposed coding;
- qualitative coding – coding emerging from the data.

Pre-coding	Researcher-imposed coding	Qualitative coding
Quantitative approach	Researcher develops coding categories out of data	Researcher develops coding system by repeated, back and forth comparisons
Based on theory or piloting in advance of main data collection	The primary aim is to develop categories for quantification	Generally, no intention to carry out any form of quantification
Universal for all participants	Coding categories are universal and applied to all data	May be applied to a single individual (idiographically) or more generally

FIGURE 16.2 Different types of coding

The final form of coding is most characteristic of qualitative data analysis. Aspects of it are covered in Chapter 21 on grounded theory, for example. The other two are rather more characteristic of quantification. Pre-coding is characteristic of highly structured materials such as self-completion questionnaires. Researcher-imposed coding is more typical of circumstances in which the data have been collected in a fairly rich or qualitative form, but the researcher's intention is to carry out a quantitative or statistical analysis. Coding emerging from the data is probably much more typical of research that is not intended to be analysed numerically or statistically.

Pre-coding

This is very familiar since it is the basis of much quantitative research in psychology. So common is it that it has a taken for granted quality. It is also very familiar from magazine and other popular entertainment surveys, so widely has its influence spread. Typical examples of pre-coding are found in attitude and personality questionnaires. In these the participant replies to a series of statements relevant to a particular area of interest. However, typically the respondent only has a very limited predetermined list of alternative replies. In the simplest of cases, for example, participants may be asked to do no more than *agree* or *disagree* with statements like 'Unemployment is the most important issue facing the country today.'

The key feature of pre-coding is, of course, that it occurs prior to data collection. That is, the coding categories are predetermined and are not amenable to change or re-coding (other than the possibility of combining several categories together for the purposes of data analysis). Furthermore, it is participants who code their own 'responses' into one of the available coding categories. As a consequence the researcher is oblivious to how the coding is actually done. Let us examine the reasons for pre-coding by taking an example of a simple research study. A researcher asks a group of young people at a school to answer the following question:

Q 43: What is your favourite television programme? _____

A sample of a hundred young people might write down as many as 100 different television programmes. There may be overlaps, so the actual total may be lower. What can one do with these data which might include football, *Glee*, *The Simpsons* and *X Factor*? Given that a primary purpose of research is to give structure to data, simply listing all of the named programmes is not in itself the solution. The researcher must categorise (or code) the replies of the participants. There is no single and obvious way of doing this. It is important to understand the aims of the research. If the research is to investigate violent television programmes then the researcher could code the favourite programme as violent or not, or even code the level of violence characteristic of the programme. If the researcher is interested in whether children are watching programmes put on late at night for adults then it would be necessary to code the programme in terms of the time at which the programme was transmitted.

If the researcher has a clear conception as to the purpose of the question, coding categories that are both simple and obvious may suggest themselves. So in some circumstances, the researcher may choose to pre-code the range of answers that the respondent may give. Take the following question:

Q 52: Which one of the following is your favourite type of television programme?
(*a*) Sport
(*b*) Films
(*c*) Music
(*d*) Cartoons ☐
(*e*) Other c. 61

In Q 52 the favourite television programme has been pre-coded by asking the respondent to nominate their favourite from a few broad types of programme. Also notice the box c. 61 to the right of the question. This is so that the computer code for the participant's chosen answer may be entered – it also identifies which variable is being referred to on the computer spreadsheet. Apart from putting the respondent's replies in a form suitable for the computer, the researcher's input is minimal (but nevertheless crucial) with pre-coded material. Notice that the pre-coding has limited the information received from the participant quite considerably. That is, in this example nothing at all is known about the specific programmes which the participant likes. In other words, pre-coding of this sort produces very limited information that is of little value to the qualitative researcher. In fact, it misses the point of qualitative data analysis entirely. It is the sort of research that qualitative data analysts often rail against.

Pre-coding is not exclusively a feature of the analysis of self-completion questionnaires. A questionnaire may be used as part of an interview and administered by the researcher or some other interviewer. Furthermore, pre-coding may be employed in studies other than questionnaires. For example, the researcher may be intending to observe behaviour. Those observations may be taken in the form of notes, but there is no reason why the observation categories cannot be pre-coded. Leyens, Camino, Parke and Berkowitz (1975) used a pre-coded observational schedule in order to observe the aggressive behaviour of boys before and after they had seen violent or non-violent films. Such pre-coded observations will obviously greatly facilitate matters such as calculating the degree of agreement between different observers. That is, inter-observer or inter-rater reliability becomes easy to calculate using pre-coded observation schedules.

Pre-coding means that the data collected have largely been limited by the nature of the categories created by the researcher. Unless the participant adds something in the margin or qualifies what they choose, no one would be the wiser. If one is interviewing but only ticking the categories pre-coded on the questionnaire, then much of what the interviewee is saying is likely to simply be disregarded. Pre-coding reduces the richness of the data to a level manageable by the researcher and adequate to meet the purposes of the research. Inevitably, pre-coding means that nuances and subtleties are filtered out. This is neither a good nor a bad thing in any absolute sense. For some purposes, the researcher-imposed broad-brush categories may be perfect; for other purposes, pre-coding is woefully inadequate.

So where do the pre-coded categories come from? There is a range of answers to this question as a perusal of the research literature will demonstrate:

- *Conventional formats* Some pre-coding formats are conventional in the sense that they have been commonly used by researchers over several decades. For example, the yes–no, agree–disagree pre-coded response formats are so established that we tend to take them for granted. However, even using such a simple format raises questions. For example, what if someone does not disagree and does not agree? In consideration of this, some response formats include a 'don't know' or '?' category. Similarly, the five-point Likert scale answer format (strongly agree, agree, ?, disagree and strongly disagree) is conventional and generally works well for its purpose. But there is no reason why seven- or nine-point scales cannot be used instead. Actually, there is no absolutely compelling reason why the middle, neutral or '?' point cannot be left out to leave a four-, six- or eight-point scale. Some argue that the neutral point allows the participants to take the easy option of not making a choice between one end of the scale and the other.

- *Piloting* Many researchers will try out or pilot their materials on a group of individuals similar to the eventual sample. This is done in such a way as to encourage the participants to raise questions and problems which make it difficult to complete the questionnaire. In this way, difficulties with the response format may be raised.

Of course, given the small number of gradations in this sort of answer format, inevitably there may be difficulties in finally making a choice. The danger is that participants may feel forced into making arbitrary and, to them, rather meaningless choices between limited answer categories. As a consequence, they may feel alienated from the research. They may feel that the researcher is not really interested in their opinions, that the researcher is incompetent by not getting at what they really think or feel, or that there is little point in putting any effort into thinking about the issues. On the other hand, they may feel very relieved that completing the questionnaire is so simple and straightforward.

- *Focus groups, etc.* By using focus groups or individual interviews (see Chapter 18), the researcher may develop an understanding of the major types of response to a question. Some themes may be very common and others so rare that they apply to only a few individuals. Some of the themes may appear crucial and others totally mundane. Some may seem irrelevant to the research. Knowing the dominant responses helps the researcher identify useful pre-coded answers.

- Often the pre-coded categories are just the invention of the researchers themselves. What they create will be influenced by the particular set of priorities for the research. The purpose of the research, the consumers of the research and the insightfulness of the researchers are among the factors which will affect what pre-coding categories are selected. There is no reason to expect pre-coding to be inferior or superior to other forms of coding simply because of this reason. The value of research can never be assessed in terms of absolute standards. Fitness for purpose should be the main criterion. Of course, to the extent that the coding categories are created prior to collecting the data, they may match the data poorly for the simple reason that they were created without reference to the data.

Researcher-imposed coding

Sometimes researchers collect data in qualitative form because it is not possible to pre-structure the response categories. For example, the researcher may feel that they cannot anticipate the nature of the data or otherwise identify the likely major issues sufficiently well to allow the pre-structuring of their research instruments. As a consequence the data are collected in a richer, fuller format which can then be used to develop a coding scheme. While this material may be 'ethnographically' coded (see Chapters 17–25 for some of the possibilities), the researcher may prefer to quantify the data. Quantification is achieved through measuring features of the data, that is, imposing coding categories and counting frequencies in the categories.

There are any number of factors which influence the nature of the coding categories developed. The following are some of the possibilities:

- The researcher's interest in a theory which strongly suggests a system of analysis.

- The research may be guided by matters of public policy or may address a politically important topic. Hence the coding categories employed need to reflect those same issues.

- The researcher is interested only in strong trends in the data so simple, very broad categories may be appropriate.

- The research task may be so big that the researchers have to employ workers to do the coding. This tends to demand simpler coding categories.

- Many coding categories do not work (i.e. produce inconsistent or unreliable results across different coders) or are virtually unusable for other reasons.

The coding schedule is a very important document, as is the coding manual. The coding schedule is the list of coding categories applied to the data in question. It is used to categorise the data. The coder would normally have this list of coding categories – usually one schedule per set of data from a participant – and the data would be coded using the schedule. Often it is advantageous to use a coding schedule which is computer-friendly. This is usually achieved by using numbered squares corresponding to the various variables on the computer data spreadsheet. In some respects the coding schedule is rather like a self-completion questionnaire but it consists of a list of implied questions (do the data fit this category?) about which categories fit the data best. Usually the coding schedule is a relatively brief document which lists very simple descriptions of the coding categories.

Sometimes the person drawing-up the coding schedule is not the same person as the coder. The person doing the coding needs to be informed in detail about the meaning of the various categories on the coding schedule. To this end, the coding manual is provided which gives detailed definitions of the coding categories, examples of material which would fit into each category and examples of problematic material. There are no rules about how detailed the manual needs to be, and it may be necessary for coders to get together to resolve difficulties. The basic idea, though, is to develop a coding process which can be delegated to a variety of individuals to enable them to code identical material identically. This is an ideal, of course, and may be only partially met.

Methodologically, another ideal is that each individual's data are coded independently by a minimum of *two* coders. In this way, the patterns of agreement between coders can be assessed. This is generally known as inter-rater or inter-coder reliability. It is probably more common that two coders are used on only a sample of the material rather than all of the data. There are obvious time and expense savings in doing so. There are also disadvantages in that coders may become sloppy if they believe nobody is checking their coding. Another risk is that over the research period the meaning of the coding categories is subtly altered by the coders, but evidence of inter-coder reliability is collected only in the initial stages of the research. Obviously, it might be best if the inter-rater reliability was assessed at various stages in the coding process. However, if the checking is done early on problems can be sorted out; problems identified later on are much more difficult to correct.

Inter-rater reliability is rarely perfect. This raises the question of how disagreements between coders should be handled. The possible solution varies according to circumstances:

- If the second coding is intended just as a reliability check, then nothing needs to be done about the disagreements. The codings supplied by the main coder may be used for the further analysis. Inter-rater reliability assessments should nevertheless be given.
- If coders disagree then they may be required to reach agreement or a compromise whenever this arises. This might involve a revision of the coding manual in serious cases. In minor cases, it may simply be that one of the coders had made an error and the coding schedule or manual needs no amendment.
- Reaching a compromise is a social process in which one coder may be more influential than the other. Consequently, some researchers prefer to resolve disagreements by an equitable procedure. For example, coding disagreements may be resolved by a random process such as the toss of a coin. Randomisation is discussed in Chapters 9 and 13.
- If the coding category is a rating (that is a score variable), then disagreements between raters may be dealt with by averaging. For example, the coders may be required to rate the overall anxiety level of an interviewee. Since the ratings would be scores, then operations such as averaging are appropriate. If the data are simply named categories, then averaging is impossible.

16.3 Reliability and validity

Like any other form of measurement, the issues of reliability and validity apply to coding categories. Validity is a particularly problematic concept in relation to coding categories. Quite what does validity mean in this context? One could apply many of the conceptualisations of validity which were discussed in the previous chapter. Face and content validity are probably the most frequent ways of assessing the validity of coding. It is more difficult to measure things such as concurrent and predictive validity since it is often hard to think of criteria against which to validate the coding categories. Qualitative researchers, however, have a particular problem since the development of categories is often the nub of their research. So this brings about the questions of just what is the value of this sort of research. In response, a wide range of ideas have been put forward concerning how issues such as validity may be tackled in qualitative research. These are dealt with in some detail in Chapter 25 and more extensively in Howitt (2010).

The assessment of the reliability of codings is much more common than the issue of validity. Reliability of codings is actually less straightforward than for scales and measurements. The following goes some way to explaining why this is the case:

- Where ratings (scores) are involved, the reliability of the coding is easily calculated using the correlation coefficient between the two sets of ratings. However, it is harder to establish the degree of agreement between raters on their ratings since the correlation coefficient simply shows that the two sets of ratings correlate – not that they are identical. They may correlate perfectly but both coders' ratings can be entirely different. Correlation only establishes covariance, not overlap or exact agreement.

- The percentage agreement between the two coders in terms of which coding category they use might be considered. However, there is a difficulty. This is because agreement will tend to be high for codings which are very common. The overlap between coders may largely be a consequence of the coders choosing a particular coding category frequently. Rarely used coding categories may appear to have low levels of agreement simply because the categories are so rarely chosen. Suppose two raters, for example, code whether or not people arrested by the police are drunk. If they both always decide that the arrestees were drunk then the agreement between the coders is perfect. But this perfect agreement is not very impressive. Far more impressive are the circumstances in which both raters rate, say, half of the arrestees as being drunk and the other half as sober. If the two raters agree perfectly, then this would be impressive evidence of the inter-rater reliability or agreement between the raters.

- For this reason, indices which are sensitive to the frequency with which a category is checked should be chosen as they are more appropriate. Coefficient kappa (see the companion statistics book, *Introduction to Statistics in Psychology*, Howitt and Cramer, 2011a, Chapter 36) is one such reliability formula. Coefficient kappa is sensitive to situations in which both raters are not varying the coding categories used much or using one coding to the exclusion of the others. Coefficient kappa is available on SPSS Statistics and is discussed in Chapter 31 of the companion computing text *Introduction to SPSS Statistics in Psychology* (Howitt and Cramer, 2011b).

What are the requirements of a good coding schedule and manual? We have seen that a coding schedule can be construed as a questionnaire – albeit one which is addressed to the researcher and not the participant. So all of the requirements of a good set of questions on a questionnaire would be desirable, such as clarity, which might be translated as the appropriateness of the category names and language for the coder, who may not be as familiar with the aims and objectives of the research as the researchers in charge of the research, and so forth. The following are some further considerations:

- For any variable, the number of possible coding categories may be potentially large. All other things being equal it is probably best to err in the direction of using too many categories rather than too few. As computers are almost always used nowadays in research, the computer can be used to amalgamate (collapse) categories if it later proves to be appropriate.

- It is usual, and probably essential, to have an 'other' category for every variable since even the best set of coding categories is unlikely to be able to cope with all of the data provided. Sometimes the 'other' category becomes a dominant category in terms of coding. This is clearly unsatisfactory as it basically means that there is little clarity about many participants' data. Generally, it is best to make a brief note of the essence of data being coded in the 'other' category when that category is used, that is, the sort of material that was being coded as 'other' needs to be recorded. Later on, the 'other' category may be reviewed in order to try to identify any pattern in the other category which might justify re-coding some or all of the data. That is, it becomes apparent that certain 'themes' are very common in the 'other' category. Where the 'other' category just consists of a variety of very different material nothing needs to be done. Unfortunately, if the other category gets large then it may swamp the actual coding categories used.

- Serious consideration should be given to whether or not any variable may be multi-coded. That is, will the coder be confined to using just one of the coding categories for each case or will they be allowed to use more than one category? Generally speaking, we would recommend avoiding multi-coding wherever possible because of the complexity it adds to the statistical analysis. For example, multi-coding may well mean that dummy variables will have to be used. Their use is not too difficult, but is nevertheless probably best avoided by novices.

- Coding categories should be conceptually coherent in themselves. While it is possible to give examples of material which would be coded into a category, this does not, in itself, ensure the coherence of the category. This is because the examples may not be good examples of the category. Hence the coder may be coding into the category on the basis of a bad example rather than that the data actually fits the category.

- The process of developing a coding schedule and coding manual is often best regarded as a drafting and revision process rather than something achieved in a single stage – trial and error might be the best description. The reason is obvious, it is difficult to anticipate the richness of the data in advance of examining the data.

16.4 Qualitative coding

The final type of coding process is that adopted in qualitative data analysis. Characteristically this form of coding seeks to develop coding categories on the basis of an intimate and detailed knowledge of the data. This is an ideal which some qualitative analyses fail to reach. For example, one finds qualitative analyses which do no more than identify a few major themes in the data.

The next part of the book contains a number of chapters on various aspects of qualitative data analysis. Anyone wishing to carry out a qualitative analysis needs to understand its historical roots as well as some of the practicalities of qualitative analysis.

Qualitative analysis is *not* any analysis which does not include statistics. Qualitative analysis has its own theoretical orientation that is highly distinctive from the dominant psychological approaches that value quantification, hypothesis testing, numbers and objectivity.

16.5 Conclusion

The process of turning real life into research involves a number of different stages. Among the most important of these is that of coding or categorisation. All research codes reality in some way or another, but the ways in which the coding is achieved varies radically. For some research, the development of the coding categories is the main purpose of the research endeavour. In other research, coding is much more routine and standardised, as in self-completion questionnaires with a closed answer format. Not only are there differences in the extent and degree of elaboration of the coding process, but there are differences in terms of who does the coding. The participant does the coding in self-completion questionnaires using codes (multiple choices, etc.) developed by the researcher.

Key points

- Through coding, in quantitative research, data are structured into categories which may be quantified in the sense that frequencies of occurrence of categories and the co-occurrence of categories may be studied. In qualitative research, often the end point is the creation of categories which fit the detail of the data.
- Coding is a time-consuming activity which is sometimes economised on by pre-coding the data, that is, the participants in the research are required to choose from a list of response categories. In this way, the data collection and coding phases of research are combined. This is essentially a form of quantitative data collection.
- Coding of data collected qualitatively (that is, through in-depth interview, focus groups, etc.) may be subject to researcher-imposed coding categories which have as their main objective quantification of characteristics judged important by the researcher for pragmatic or theoretical reasons, or because they recognise that certain themes emerge commonly in their data.
- Coding requires a coding schedule which is a sort of questionnaire with which the researcher interrogates the data in order to decide in what categories the data fit. A coding manual details how coding decisions are to be made – what the categories are, how the categories are defined and perhaps examples of the sorts of data that fit into each category.
- Such coding, which is almost always intended for quantitative and statistical analysis, may be subjected to reliability tests by comparing the codings of two or more independent coders. This may involve the use of coefficient kappa if exact correspondence of codings needs to be assessed.
- If the analysis is intended to remain qualitative, then the qualitative analysis procedures discussed in the chapters on Jefferson coding, discourse analysis and conversation analysis, for example, should be consulted (Chapters 19, 22 and 23). The general principles of qualitative research are that the coding or categorisation process is central and that the fit of the categories to the data and additional data is paramount.

ACTIVITIES

1. Create a list of the characteristics that people say make other people sexually attractive to them. You could interview a sample to get a list of ideas. Formulate a smaller number of coding categories which effectively categorise these characteristics. Try half the number of categories as characteristics first, then halve this number, until you have two or three overriding categories. What problems did you have and how did you resolve them? Is there a gender difference in the categories into which the characteristics fall?

2. Obtain a pre-coded questionnaire, work through it yourself (or use a volunteer). What were the major problems in completing the questionnaire using the pre-coded categories? What sorts of things did you feel needed to be communicated but were not because of the format?

PART 4

Qualitative research methods

CHAPTER 17

Why qualitative research?

Overview

- This chapter presents the essential background to doing qualitative research. It is important to understand the underlying theoretical stance common to much qualitative research before looking in detail at methods of data collection, recording and analysis.
- Qualitative research concentrates on describing and categorising the qualities of data. In contrast, quantitative research concentrates on quantifying (giving numbers to) variables.
- Quantitative research is often described as being based on positivism, which is regarded as the basis of the 'hard' sciences such as physics and chemistry.
- The search for general 'laws' of psychology (universalism) is held to stem from positivism and is generally regarded as futile from the qualitative point of view.
- Qualitative researchers attempt to avoid some of the characteristics of positivism by concentrating on data which are much more natural.

17.1 Introduction

Superficially, at least, qualitative research is totally different from quantitative research. *Qualitative* research focuses on the description of the qualities (or characteristics) of data. Probably, historically, case studies of an individual person were the most common qualitative method in psychology though, frequently, they had little in common with modern qualitative methods. By individual case study we mean the detailed description of the psychological characteristics of, usually, one person (or perhaps a single organisation). Sigmund Freud's psychoanalyses of individual patients are early examples of the case study. Other classic examples of case studies would include *The Man Who Mistook His Wife for a Hat* by Oliver W. Sacks (1985). This book includes a detailed account of how a man with neurological problems dealt with his memory and perceptual problems. Such individual case studies often utilising but sometimes helping to develop psychological theory are very different from recent qualitative approaches in psychology. Modern qualitative research generally involves a detailed study of text, speech and conversation (which generically may be termed text) and *not* the specific psychological characteristics of interesting individuals. Text is anything which may be given meaning.

Qualitative research often concentrates on conversational and similar exchanges between people in interviews, the media, counselling and so forth. It is rarely, if ever, concerned with analysis at the level of individual words, phrases or even sentences. It analyses broader units of text, though what the minimum unit of analysis is depends on the theoretical orientation of the qualitative analysis.

One major problem facing anyone wishing to learn to do qualitative research is that it is not fully established as being part of the core of psychological research and theory. That is, one can study certain research fields in psychology for years, and rarely if ever come across qualitative research and theory. This may be changing. In contrast to qualitative approaches, *quantitative* research is undeniably at the centre of psychology. Indeed, quantification characterises most psychology more effectively than the subject matter of the discipline. Take virtually any introductory psychology textbook off the shelf and it is likely to consist entirely of research and theory based on quantitative research methods. References to personality tests of many sorts, intelligence quotients, ability and aptitude measures, attitude scales and similar feature heavily as do physiological measures such as blood pressure, brain rhythms, PET (positron emission tomography) scans and so forth. While all of these measure qualities ascribed to the data (usually called variables in the quantitative context), they are quantified in that they are assigned numerical values or scores. The magnitude of the numerical value indicates the extent to which each individual possesses the characteristic or quality.

Probably the most famous quantitative measure in psychology is the IQ (intelligence quotient) score which assigns a (usually) single number to represent such a complex thing as a person's mental abilities. Comparisons between individuals and groups of individuals are relatively straightforward. Sometimes quantitative data are gathered directly in the form of numbers (such as when age is measured by asking a participant's age in years). Sometimes the quantification is made easy by collecting data in a form which is rapidly and easily transformed into numbers. A good example of this is the Likert attitude scale in which participants are asked to rate their agreement with a statement such as 'University fees should be abolished'. They are asked to indicate their agreement on a scale of strongly agree, agree, neutral, disagree and strongly disagree. These different indicators are assigned the numbers 1 to 5. So common are self-completion questionnaires and scales in psychological research that there is a danger of assuming that such methods are synonymous with quantitative methods and not merely examples of them.

The growth of psychology into a major academic discipline and field of practice was possible because of the growth of quantification. It is also partly responsible for psychology's closeness to scientific disciplines such as biology, physiology and medicine. Historically, many decisive moments in psychology are associated with developments which enabled quantification of the previously unquantifiable:

- Psychophysics was an early such development which found ways of quantifying subjective perceptual experiences.
- The intelligence test developed in France by Alfred Binet at the start of the twentieth century provided a means of integrating a wide variety of nineteenth-century ideas concerning the many qualities of intellectual functioning.
- Louis Thurstone's methods of measuring attitudes in social psychology and the more familiar methods today of Likert were methodological breakthroughs which allowed the development of empirical social psychology.

The development of a good quantitative technique encourages the development of that field of research since it facilitates further research. Examples of research fields spurred on by quantification are topics such as authoritarianism, psychopathy, suggestibility and many other psychological concepts as well as, for example, the MIR (millimetre-wave imaging radiometer) scan. All of this documents a great, collective achievement. Nevertheless, the consequence has been to squeeze out other, less quantifiable, subject matter. Consequently, the history of psychology is dotted with critiques of the focus of psychological knowledge (for example, Hepburn, 2003). Often these critiques are portrayed as 'crises' though probably the term serves the interest of the complainants better than it reflects the view of the majority of researchers and practitioners.

There is a danger of presenting quantitative and qualitative research as almost separate fields of research. This is to neglect the numerous examples of apparently quantitative research which actually include a qualitative aspect. Examples of this are quite common in the history of psychology. Even archetypal, classic laboratory studies sometimes collected significant amounts of qualitative material (for example, Milgram, 1974). It has to be said, though, that recent developments in qualitative methods in psychology would probably eschew this combined approach. Whatever, it is evidence of an underlying view of many psychologists that quantification alone only provides partial answers.

17.2 What is qualitative research?

So is qualitative research that which is concerned with the nature or characteristics of things? One obvious problem with this is that does not *all* research, qualitative or quantitative, seek to understand the nature and characteristics of its subject matter? Perhaps this indicates that the dichotomy between qualitative and quantitative research is more apparent than real. If the distinction is of value then it should be apparent in the relationship between qualitative and quantitative research. Immediately we explore this question, we find that several different claims are made about the interrelationship:

- Qualitative methods are a preliminary stage in the research process which contributes to the eventual development of adequate quantification. Quantification is, in this formulation, the ultimate goal of research. There is a parallel with the physical sciences. In many disciplines (such as physics, biology and chemistry) an early stage involves observation and classification. For example, botanists collected, described and organised plants into 'families' – groups of plants. Members of a 'family' tended to be similar to each other in terms of their characteristics and features. In chemistry, exploration of the characteristics of different elements led to them being organised into the important analysis tool – the periodic table – which allowed their characteristics to be predicted and undiscovered elements to be characterised. This was done on the basis of features such as the chemical reactivity and the electrical conductivity of elements. In many disciplines, qualitative methods (which largely involve categorisation) have led to attempts to quantify the qualities so identified. The model for this is:

Qualitative analysis→Quantitative analysis

This process is not uncommon in psychology. If we return again to the common example of intelligence testing, we can illustrate the process. During the nineteenth century under the influence of an Austrian, Franz Joseph Gall, the idea that different parts of the brain had different mental functions developed. Unfortunately things went wrong in some ways as one of the immediate consequences was the emergence of phrenology as a 'science'. Phrenology holds that the different parts of the brain are different organs of the mind. Furthermore, the degree of development of different parts of the brain (assessed by the size of the 'bumps' on the skull at specific locations) was believed to indicate the degree of development of that mental faculty (or mental ability). Gall believed that the degree of development was innate in individuals. The range of mental faculties included features such as firmness, cautiousness, spirituality and veneration, which are difficult to define, and others such as constructiveness, self-esteem and destructiveness, which have their counterparts in current psychology. The point is that these mental faculties could only be suggested as a result of attempts to describe the characteristics of the mind, that is, a process of categorising what was observed. Phrenology's approach to quantification was immensely crude, that is, the size of different bumps. But the idea that the mind is organised into various faculties was a powerful one, and attempts to identify what they were formed the basis of the conceptualisation of intelligence, which was so influential on Alfred Binet who developed the seminal measure of intelligence. For Binet, intelligence was a variety of abilities, none of which in themselves defined the broader concept of intelligence but all of which were aspects of intelligence. The idea that qualitative research is a first step to quantification is valuable but neglects the fact that the process is not entirely one way. There are many quantitative techniques (for example, factor analysis and cluster analysis) which identify empirical patterns of interrelationships which may help develop theoretical categorisation or classification systems.

- Qualitative methods provide a more complete understanding of the subject matter of the research. Some qualitative researchers argue that quantification fails to come to terms with or misses crucial aspects of what is being studied. Quantification encourages premature abstraction from the subject matter of research and a concentration on numbers and statistics rather than concepts. Because quantification ignores a great deal of the richness of the data, the research instruments often appear to be crude and, possibly, alienating. That is, participants in quantitative research feel that the research is not about them and may even think that the questions being asked of them or tasks being set are simply stupid. Some research is frustrating since, try as the participant may, the questionnaires or other materials cannot be responded to accurately enough. They simply are not convinced that they have provided anything of value to the researcher. Of course, the phrase 'richness of data' might be regarded as a euphemism for unfocused, unstructured, unsystematic, anecdotal twaddle by the critical quantitative researcher. We will return to the issue of richness of data in subsequent chapters.
- A more humanistic view of qualitative data is that human experience and interaction are far too complex to be reduced to a few variables as is typical in quantitative research. This sometimes is clearly the case especially when the research involves the

study of topics involving interactive processes. A good example of this is when one wishes to study the detailed processes involved in conversation; there are simply no available methods for turning many of the complex processes of conversation into numbers or scores. To be sure, one could time the pauses in conversation and similar measures but selecting a variable simply because it is easy to quantify is unsatisfactory. Choosing a measure simply because it is easy and available merely results in the researcher addressing questions other than the one they want to address. What, for example, if the researcher wants to identify the rules which govern turn-taking in conversation? The subtlety of the measurements needed may mean that the researcher has no choice but to choose a qualitative approach. Figure 17.1 gives some of the typical characteristics of the qualitative researcher.

As we have seen, qualitative and quantitative methods are not necessarily stark alternatives. The choice between the two is not simple nor is it always the case that one is to be preferred over the other. Often a similar topic may be tackled qualitatively or quantitatively but with rather different objectives and consequently outcomes. Some

They reject positivism

They adopt relativist position of no fixed 'reality'

They use relatively unstructured data collection methods

They are concerned to capture the individual's perspective

They use highly detailed data analysis methods

They use richly descriptive data

They take the postmodernist perspective in general

They incorporate the constraints of everyday life in studies

They believe that reality is constructed socially/individually

They choose rich and deep data rather than 'hard' data

They tend to be 'closer' to their research participants

They often see themselves as 'insiders' to what is studied

They are concerned with interpretation not causal sequences

They largely reject hypothesis testing

Their theory emerges from close analysis of the data

Their approach, often idiographic, concentrates on the individual

FIGURE 17.1 Some of the characteristics of the typical qualitative researcher

research successfully mixes the two. Many of the classic studies in some fields of psychology took the mixed approach, though often it would not be apparent from modern descriptions of this research. Examples include Stanley Milgram's electric shock/obedience experiments (Milgram, 1974) which we have already mentioned. In these cases, the two approaches are complementary and supplementary. There are certain researchers who invariably choose quantification irrespective of their research question. Equally there are other researchers who avoid quantification when it would be straightforward and appropriate. There are a number of reasons for this diversity:

- Quantification requires a degree of understanding of the subject matter. That is, it is not wise to prematurely quantify that which one cannot describe with any accuracy.
- Quantification may make the collection of naturalistic data difficult or impossible. Quantification (such as the use of questionnaires or the use of laboratory apparatus) by definition implies that the data are 'unnatural'. Quantified data are collected in ways which the researcher has highly structured.
- Some researchers see flaws in either quantification or qualitative research and so are attracted to the other approach.
- Some research areas have had a long tradition of quantification which encourages the further use of quantification. Research in new areas often encourages qualitative methods because measurement techniques have not been developed or because little is known about the topic.

Apart from career quantifiers and career non-quantifiers who will not or can not employ the other method, many researchers tailor their choice of methods to the situation and, in particular, the research question involved. All researchers should have some appreciation of the strengths and weaknesses of each approach. Probably the healthiest situation is where researchers from both perspectives address similar research questions.

17.3 History of the qualitative–quantitative divide in psychology

Laboratory experiments and statistical analyses dominate the contents of most introductory textbooks in psychology. One short explanation of this is that skills in experimentation and quantitative analysis are very marketable commodities in and out of academia. Few disciplines adopted such an approach, though it has the advantage that it suggests that psychologists are detached and objective in their work. These characteristics also tended to position psychology closely to the physical sciences. The setting-up of the psychology laboratory at Leipzig University by Wilhelm Wundt in 1879 was a crucial moment for psychologists according to psychology's historians (Howitt, 1992a). A number of famous American psychologists trained at that laboratory. Wundt, however, did not believe that the laboratory was the place for all psychological research. He regarded the laboratory as a hapless context to study matters related to culture, for example. Nevertheless, the psychological laboratory was regarded as the dominant icon of psychology.

The term *positivism* dominates the quantitative–qualitative debate. Some use it as a pejorative term though it is a word which, seemingly, is often misunderstood. For example, some writers appear to imply that positivism equals statistics. It does not. Positivism is a particular epistemological position. Epistemology is the study of, or theory of, knowledge. It is concerned with the methodology of knowledge (how we go about knowing things) and the validation of knowledge (the value of what we learn). Prior to the emergence of positivism during the nineteenth century, two methods of obtaining knowledge dominated:

- *Theism*, which held that knowledge was grounded in religion which enabled us to know because truth and knowledge were revealed spiritually. Most religious texts contain explanations and descriptions of the nature of the universe, morality and social order. While these are matters studied by psychologists, there is little in modern psychology which could be conceived as being based on theism.

- *Metaphysics*, which held that knowledge was about the nature of our being in the world and was revealed through theoretical philosophising. Relatively little psychology is based on this.

Neither theism nor metaphysics has retained its historical importance. Religious knowledge was central throughout the history of civilisation. Only recently has its pre-eminence faltered in terms of the human timescale. Metaphysics had only a brief period of ascendancy during the period of the Enlightenment (eighteenth century) when reason and individualism were emphasised. Positivism is the third major method of epistemology and directly confronts theism and metaphysics as methods of achieving knowledge. Positivism was first articulated in the philosophy of Auguste Comte in the nineteenth century in France. He stressed the importance of observable (and observed) facts in the valid accumulation of knowledge. It is a small step from this to appreciating how positivism is the basis of the scientific method in general. More importantly in this context, positivism is at the root of so-called scientific psychology.

It should be stressed that positivism applies equally to quantitative methods and to qualitative research methods. It is *not* the province of quantitative psychology alone. There is very little work in either quantitative or qualitative psychology which does not rely on the collection of observed information in some way. Possibly just as a fish probably does not realise that it is swimming in water, qualitative researchers often fail to recognise positivism as the epistemological basis of their work. Silverman (1997) makes a number of points which contradict the orthodox qualitative research view of positivism:

> Unfortunately, 'positivism' is a very slippery and emotive term. Not only is it difficult to define but there are very few quantitative researchers who would accept the label . . . Instead, most quantitative researchers would argue that they do not aim to produce a science of laws (like physics) but simply to produce a set of cumulative, theoretically defined generalizations deriving from the critical sifting of data. So, it became increasingly clear that 'positivists' were made of straw since very few researchers could be found who equated the social and natural worlds or who believed that research was properly theory-free.
>
> (pp. 12–13)

The final sentence is very important. It highlights the difficulty which is that, although positivism stresses the crucial nature of observation, it is the end point or purpose of the observation which is contentious. The real complaint about positivism is that it operates as if there were permanent, unchanging truths to be found. That is, underlying our experiences of the world are consistent, lawful and unchanging principles. The phrase 'the laws of psychology' reflects this universality. The equivalent phrases in the natural sciences are ones like 'the laws of planetary motion', 'the laws of thermodynamics, 'the inverse square law of light', that $e = mc^2$ and so forth. These physical laws are believed to be universally applicable and apply no matter where in the universe. The trouble is that universalism encourages psychologists to seek principles of human nature in, say, New York which they would then apply unchanged in Addis Ababa, Beijing or Cairo and, equally, in 1850 as in 2050.

There were psychologists who were very important in their time who operated more or less according to the positivist maxims and the quest for the laws of human activity in particular. These were members of the Behaviourist School of Psychology which dominated much psychology between the 1920s and the 1960s and beyond. Virtually everything they did reeked of the quest for general laws of psychology. First of all, they argued the basic positivistic position that knowledge comes from observation. So they stressed that psychology should study the links between the incoming stimulus and the outgoing response. There was no point in studying what could not be tested directly through observation. They were primarily interested in the experimental method. If one is seeking universal principles of human behaviour then these should apply in the psychology laboratory just as much as anywhere else. Since the laboratory had other advantages, then why not study human psychology exclusively in such laboratories? They went so far as to wear white coats in the laboratory to emulate scientists from the physical sciences, probably more to enhance their stature by association than because of any direct practical advantage. Famous names in behavioural psychology are B. F. Skinner (1904–1990), Clark Hull (1884–1952) and John Watson (1878–1958), the founder of behaviourism.

Realism would be a term applied to positivism of this sort (that is, there is a reality which research is trying to tap into). *Subjectivism* would take the view that there is no reality to be grasped and in Trochim's (2006) phrase 'we're each making this all up'. Since many psychologists nowadays would not accept the view that universal laws of human psychology are possible or desirable, some would argue that psychology is currently in a *post-positivist* stage. Postmodernism has virtually the same meaning in this context. Psychology's allegiance is still to the importance of observation. However, its aspirations of what knowledge is possible have changed. Probably the failure of the out-and-out positivists to come up with anything which constitutes a worthwhile general law of psychology has led to the present situation. Silverman (1997), in the above quotation, characterises the quest of many modern researchers as being for 'cumulative, theoretically defined generalisations deriving from the critical sifting of data'. Perhaps psychologists, more than some other disciplines, remain inclined towards gross, decontextualised generalisations. They write as if the statements they make concerning their research findings apply beyond the context in which they were studied. Owusu-Bempah and Howitt (2000) are among a number of writers who point out that such a tendency makes psychology practically unworkable beyond limited Western populations and incapable of working with the cultural diversity to be found within modern Western communities.

Qualitative researchers tend to regard the search for the nature of reality as a futile quest. *Critical realism* is the philosophy that can be summed up as accepting that there is a 'reality' out there but we can at best view it through an infinite regress of windows. That is, there is always yet another window that we are looking through and that each window distorts reality in some way. While this implies that there will always be different views of reality depending on which particular window we are looking through, the major problem is the degree of distortion that we are experiencing. Some qualitative analysts will point to the fact that much research in psychology and the social sciences relies on data in the form of language. Language, however, they say, is not reality but just a window on reality. Furthermore, different speakers will give a different view of reality. They conclude that the search for reality is a hopeless task and, to push the metaphor beyond the bounds of endurance, that we should just study the diversity of what is seen through the different windows. Well, that is one approach but not the only one based on critical realism (which only demands that researchers try to get close to reality while realising that they can never achieve that goal). Every method of measuring reality is fallible, but if we use many different measures and they concur, then maybe we are getting towards our goal might be the typical response of a mainstream psychologist. One of the reasons why our data are problematic is that our observations are theory laden. That is, the observer comes to the observation with baggage and expectations. That baggage will include our culture, our vested interests and our general perspective on life, for example.

Psychologists are not born with the ability to see the world without these windows. One strategy to overcome our preconceptions is to throw our observations before others for their critical response as part of the process of analysing our observations.

We have seen that there is no justification for some of the characteristics attributed to positivism. As discussed, some critiques of mainstream psychology labour under the impression that positivism equates to statistical analysis. Yet some of the most important figures in positivistic psychology such as Skinner had little or no time for statistics and did not use them in their work. The use, or not, of statistics does not make for positivism. Similarly, atheoretical empiricism – virtually the collection and analysis of data for their own sake – has nothing to do with positivism which is about knowing the world rather than accumulating data as such.

17.4 The quantification–qualitative methods continuum

The conventional rigid dichotomy of quantitative–qualitative methodologies is inadequate to differentiate different types of research. It implies that research inevitably falls into one or other of these apparently neat boxes. This is not necessarily the case. There is some research which is purely quantitative and other research which is purely qualitative. However, this is to neglect much research that draws on both. Conceptually, research may be differentiated into two major stages:

- data collection;
- data analysis.

Of course, there are other stages but these are the important ones for now.

At the *data collection stage*, there is a range of possibilities. The degree of quantification (assigning of numbers or scores) and qualification (collecting data in terms of rich detail) may vary:

- *Pure quantitative* The data are collected using highly structured materials (such as multiple-choice questionnaires or physiological indexes such as blood pressure levels) in relatively highly structured settings (such as the psychological laboratory). A good example of such a study would be one in which the levels of psychoticism (measured using a self-completion question) were compared in sex offenders versus violent offenders (as assessed by their current conviction).
- *Pure qualitative* The data are collected to be as full and complete a picture as the researcher can possibly make it. This is done, for example, by video or audio-recording extensive amounts of conversation (say between a counsellor and client). There may be no structuring to the data gathered than that, though sometimes the researcher might choose to interview participants in an open-ended manner. Many qualitative researchers try to use as much naturalistic material as possible.
- *Mixed data collection* Between these extremes of quantification and qualitative data gathering are many intermediary possibilities. Some researchers choose to collect data in a quantitative form where there are good means of quantifying variables and concepts but use open-ended and less structured material where the concepts and variables cannot be measured satisfactorily for some reason. Sometimes the researcher will use a mixture of multiple-choice type questions with open-ended questions which may help paint a fuller picture of the data.

However, we ought also to consider the *data analysis stage* of research in terms of the qualitative–quantitative distinction. The same options are available to us:

- *Pure quantification* If data have been collected solely in quantitative form, then there is little option but to analyse the data quantitatively. However, data may have been collected in qualitative form but the researcher wishes to quantify a number of variables or create scores based on the qualitative data. The commonest method of doing this is through a process known as coding (see Chapter 16). In this the researcher develops a categorisation (coding) scheme either based on pre-existing theoretical and conceptual considerations, or develops a categorisation system based on examining the data. This can involve the researcher rating the material on certain characteristics. For example, a global assessment of a participant's hostility to global environmental issues may be obtained by having the researcher rate each participant on a scale. Usually another rater will also independently rate the participant on the same rating scale and the correspondence between the ratings assessed (inter-rater reliability).
- *Pure qualitative* This option is generally available only if the data have been collected in qualitative form (quantitative data are rarely suitable for qualitative analysis, for obvious reasons). Quite what the qualitative analysis should be depends to a degree on the purpose of the research. As conversation (interviews or otherwise) is a common source of qualitative data, then discourse analysis and/or conversation analysis may be helpful. But this is a complex issue, which may best be left until qualitative methods have been studied in a little more depth.
- *Mixed data analysis* This may follow from mixed data collection but equally may be the result of applying qualitative and quantitative methods to qualitative data. This is quite a common approach though it is often fairly informally applied. That is, the researcher often has a primarily quantitative approach which is extended, illustrated or explicated using simple qualitative methods. For example, the researcher may give illustrative quotations from the open-ended material that is collected in addition to the more quantitative main body of the data. Such approaches are unlikely to satisfy the more demanding qualitative researcher.

The main points to emerge out of this are that we should distinguish data collection from data analysis and appreciate that quantitative and qualitative methods may be applied at either stage – this is summarised in Figure 17.2.

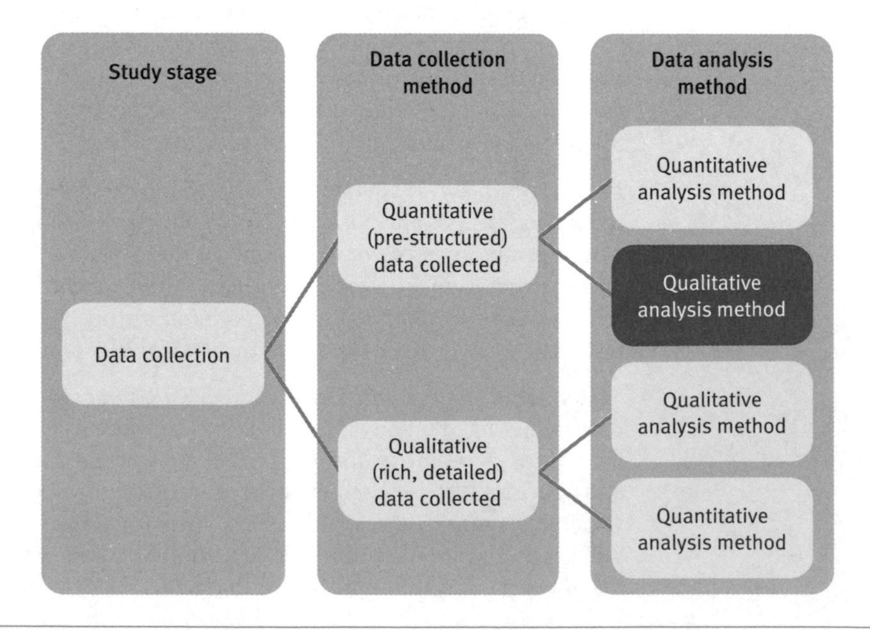

FIGURE 17.2 Varieties of data collection and analysis

17.5 Evaluation of qualitative versus quantitative methods

Denzin and Lincoln (2000) claim that there are five major features distinguishing quantitative from qualitative research *styles*. Some of these have already been touched on in this chapter but they are worth reiterating systematically:

- *Use of positivism and post-positivism* Quantitative and qualitative methods are both based on positivism and many qualitative researchers have applied 'positivist ideals' to messy data. However, qualitative researchers are much more willing to accept the *post-positivist* position that whatever reality there is that might be studied, our knowledge of it can only ever be approximate and never exact. In their actions, quantitative researchers tend to reflect the view that there is a reality that can be captured despite all of the problems. Language data would be regarded by them as reflecting reality whereas the qualitative researcher would take the view that language is incapable of representing reality. Quantitative researchers often treat reality as a system of causes and effects and often appear to regard the quest of research as being generalisable knowledge.
- *Qualitative researchers accept other features of the postmodern sensibility* This really refers to a whole range of matters which the traditional quantitative researcher largely eschewed. Examples of this include verisimilitude, in that the researcher studies things which appear to be real rather than the synthetic product of psychology laboratories for example. The qualitative researcher is represented as having an ethic of caring as well as political action and dialogue with participants in the research. The qualitative researcher has a sense of personal responsibility for their actions and activities.
- *Capturing the individual's point of view* Through the use of in-depth observation and interviewing, the qualitative researcher believes that the remoteness of the research from its subject matter (people) as found in some quantitative research may be overcome.
- *Concern with the richness of description* Quite simply, qualitative researchers value rich description almost for its own sake, whereas quantitative researchers find that such a level of detail actually makes generalisation much more difficult.
- *Examination of the constraints of everyday life* It is argued that quantitative researchers may fail to appreciate the characteristics of the day-to-day social world which then become irrelevant to their findings. On the other hand, being much more wedded in society through their style of research, qualitative researchers tend to have their 'feet on the ground' more.

Probably the majority of these claims would be disputed by most quantitative researchers. For example, the belief that qualitative research is subjective and impressionistic would suggest the lack of grounding of qualitative research in society, not higher levels of it.

The choice between quantitative and qualitative methods when carrying out psychological research is not an easy one to make. The range of considerations is enormous. Sometimes the decision will depend as much on the particular circumstances of the research, such as the resources available, as on profound philosophical debates about the nature of psychological research.

When to use quantification

The circumstances in which quantification is most appropriate include the following:

- When addressing very clearly specified research questions.
- When there is a substantial body of good-quality theory from which hypotheses can be derived and tested.

- In addressing research questions for which there is high-quality research which has typically employed quantification.
- When it can be shown that there are very satisfactory means of collecting information using measures.
- When the researcher has a good understanding of quantitative methods combined with a lack of interest or knowledge concerning qualitative methods.

When to use qualitative research methods

A researcher might consider using qualitative research methods in the following circumstances:

- When the researcher wishes to study the complexity of something in its natural setting.
- When there is a lack of clarity about what research questions should be asked and what the key theoretical issues are.
- When there is generally little or no research into the topic.
- When the research question relates to the complex use of language, such as in extended conversation or other textual material.
- When the researcher has read qualitative research in some depth.
- Where the use of structured materials, such as multiple-choice questionnaires, may discourage individuals from participating in the research.

17.6 Conclusion

The divide between quantitative and qualitative research is not easy to cross. In many ways there are two cultures in psychology and often they are seeking answers to radically different sorts of questions. However, there is more to it than that since if there were simply two camps of psychologists – quantitative and qualitative – who just do totally different things then that would be fine. After all, specialities within psychology are very common. It is virtually unknown to come across psychologists who are well versed in more than a couple of sub-disciplines of psychology. Where it seems an unsatisfactory situation to have quantitative and qualitative camps is in so far as psychologists should be interested in the topic of research and not be straitjacketed within methods. So our preference is for *all* psychologists to have the choice of approaches from which to select when planning their research. This is a convoluted way of saying that the research problem should have primacy. The best possible answer to the question that the researcher is raising cannot lie in any particular method.

This and the next few chapters are our modest answer to uniting the quantitative and qualitative camps in a joint enterprise, not a battle.

Key points

- Qualitative research, especially in the form of case studies, has been a significant but relatively minor aspect in the history of psychological methods. Nowadays, interest in qualitative methods has increased especially in terms of the analysis of language-based data such as conversations, media content and interviews.
- Advances in quantification, nevertheless, have often been significant foci of new psychological research.
- Qualitative research can be regarded as a prior stage to quantitative research. However, there are research questions which are difficult to quantify especially with complex processes such as conversation.
- Positivism is a philosophical position on how knowledge can be obtained which is different from theism (religious basis of knowledge) and metaphysics (knowledge comes from reflecting on issues). Positivism required an empiricist (observational) grounding for knowledge. However, it became equated with relatively crude and quantified methods. Qualitative researchers often overlook their allegiance to positivism.
- Quantification may be applied to data collection or data analysis. Research data collected through the 'rich' methods may be quantified for analysis purposes. Whether or not this is appropriate depends on circumstances.

ACTIVITY

Many psychology students are unfamiliar with examples of qualitative research. Qualitative research needs a positive orientation and a great deal of reading. So now is the time to start. Spend half an hour in the library looking through likely psychology journals for examples to study. Failing that,

MacMartin, C. and Yarmey, A. D. (1998). 'Repression, dissociation, and the recovered memory debate: Constructing scientific evidence and expertise', *Expert Evidence*, *6*, 203–26.

is an excellent example which crosses a range of issues relevant to the work of many psychologists.

CHAPTER 18

Qualitative data collection

Overview

- The commonest qualitative data collection methods are probably the in-depth interview, participant observation and focus groups. These are discussed in this chapter to illustrate the range of concerns of qualitative analysts.
- Virtually all qualitative approaches to data collection have an equivalent structured approach. For example, in-depth or semi-structured interviews may be compared with the structured interview common in market research.
- Qualitative data may, in appropriate circumstances, be analysed quantitatively or qualitatively depending on the objectives of the researcher and the characteristics of the data. Qualitative data collection should not be confused with qualitative data analysis.
- Aspects of observation, focus groups and interviewing as means of collecting qualitative data are presented.

18.1 Introduction

Qualitative data collection is not necessarily followed by qualitative data analysis. Qualitatively collected data may be analysed, sometimes, quantitatively. Qualitative data collection methods essentially provide extensive, detailed and 'rich' data for later analysis. Nevertheless, the primary purpose of the analysis is to turn the complexity of the data into relatively structured numerical analyses. At first sight this may seem a little pointless since we know that researchers often collect data in quantitative form so why bother with qualitative data collection if the analysis is to be quantitative? However, there are circumstances in which it is simply impossible, or undesirable, to collect data quantitatively prior to quantitative analysis:

- It is difficult to design, for example, a self-completion questionnaire which will effectively collect a biographical record of an individual or capture the detail of a complex sequence of events.
- There may be factors that militate against some individuals supplying quantitative data. For example, a researcher wishing to collect accounts of the experience of depression from seriously depressed individuals may find greater success through giving the participants attention by interviewing them than by sending them a questionnaire though the post. Some individuals may not have the intellectual resources or even the writing and reading skills to complete a self-completion question. It would be silly, for example, to carry out research into illiteracy through a questionnaire.
- The researchers may not have sufficient familiarity with the research topic to enable effective structuring of quantitative materials. They may have chosen an entirely novel area of research, for example, so they cannot draw ideas from previous researchers. Some researchers will collect data qualitatively since this allows a degree of exploration of the topic with the participants. Interviews and similar techniques may be part of an exploration process.

The range of methods by which appropriate data for qualitative studies may be obtained is wide. Indeed, any data that are 'rich and detailed' rather than 'abstracted and highly structured' may be candidates for qualitative analysis. Some of the more familiar data collection methods for qualitative analysis include the following:

- Observation: relatively unstructured observation and participation would be typical examples. Observation that involves just a few restricted ratings would probably not be appropriate.
- Biographies (or narratives) which are accounts of people's lives or aspects of their lives.
- Focus groups.
- In-depth interviews.
- Recordings of conversations including research interviews and recordings made for other purposes.
- Mass media output.
- Documentary and historical records.
- Internet sources.

Often qualitative analysis uses material from a range of different types of methods. The material, in general, is overwhelmingly textual. Observations, for example, will be recorded in words. This does not mean that other forms of material (including the visual) cannot be used, but as, in effect, these are transformed into words, then the dominance of words or text is obvious. (Text has a wider meaning in qualitative research – it refers to anything imbued with meaning.) It is not the broad method by which the data are collected which determines whether the data collected are suitable for qualitative analysis. For example, interviews may be used to collect quantitative data only or they may be used to collect qualitative data. It is the detail, expansiveness and richness of the data that determine their suitability for qualitative analysis. Imagine that researchers wish to study violence in television programmes. They might consider two options:

- They could count the number of acts of violence which occur in a sample of television programmes. The relative frequency of such violence in different types of television programme (for example, children's programmes or imported programmes) could be assessed as part of this.
- Episodes of violence on television could be videoed, transcribed and described in prolific detail.

The first version of the research is clearly quantitative as all that has happened is that a total amount of a certain category of content has been obtained. The second version of the research appears to be much more amenable to qualitative analysis strategies. It is the richness of the detail which makes the difference. The researchers may be studying exactly the same television programmes in both cases, but the nature of the data obtained is radically different. The qualitative research approach might allow the researcher to say much more about the context of the violence. However, without counting, the number of violent episodes cannot be assessed.

Generally speaking, although the quality of the research data is of paramount importance, what the best data are depends on a range of factors. These include, for example, the precise nature of the research questions, the nature of participants, the stage of the development of that particular field, the researcher's personal preferences and the resources available, among many other considerations.

18.2 Major qualitative data collection approaches

The key feature of qualitative data is encapsulated in the phrase 'richness of data'. But, as we have seen, there are many associated characteristics, such as unstructured data collection, extensive and interactive textual material, such as that collected in some interviews, the talk of politicians and so forth. Richness does not necessarily relate to interesting or similar ideas. Some qualitative researchers actually like dull, mundane material as this challenges their analytic skills greatly. The range of qualitative data collection methods (and sources of qualitative data) is remarkable. Consequently, it is possible to give only a few examples of the dominant approaches taken to qualitative data collection. We will concentrate on participant observation, focus groups and interviews.

Method 1: Participant observation

Participant observation would seem to offer the opportunity to gather the richly detailed data that qualitative researchers seek. Ethnography is the more modern term in some disciplines such as sociology where participant observation is seen as part of a wider complex of methods for collecting data in the field. Of course, cultural anthropology can be seen as part of the history of participant observation although many early anthropologists did not collect their data by immersion in a culture but from secondary sources such as the accounts of travellers. The origins of ethnography and participant observation in the more modern period are usually attributed to the work of the so-called Chicago School of Sociology, starting in the 1920s. The key aim of participant observation is to describe and explain the social world from the point of view of the actors or participants in that world. By being a participant and not just an observer, access to the point of view of the participant is assured. According to Bryman (2008), the major characteristics of participant observation are as follows:

- The researcher is 'immersed in a social setting' (p. 163) for a considerable period of time. The social setting could be, for example, an informal social group, an organisation or a community.
- The researcher observes the behaviours of members in that social setting.
- The researcher attempts to accurately record activity within that setting.
- The researcher seeks to identify the 'meanings' that members of that setting give to the social environment within which they operate and the behaviour of people within that setting.

In some disciplines, participant observation has been a central research tool. For example, observational research into human social activity is an evident feature of several centuries of cultural or social anthropology. Stereotypically, the cultural anthropologist is a committed researcher who spends years living and working among an aboriginal group isolated from Western culture. The researcher is, therefore, most definitely an alien to the aboriginal culture – a fact which is regarded as part of the strength of the method. After all, it is hard to recognise the distinctive characteristics of routine parts of our lives. This anthropological approach has occasionally found some resonance with psychology. For example, Margaret Mead's *Coming of Age in Samoa* (1944) argued that adolescence is not always a period of upset, rebellion and conflict as it is characterised in Western cultures. It would appear that societies which do not have the West's rigid separation of childhood and adulthood may avoid the typical Western pattern of the adolescent in turmoil; though the adequacy of Mead's study has been questioned.

The term *participant observation* is a blanket term for a variety of related approaches. There are a number of important dimensions which identify the different forms of participant observation (Dereshiwsky, 1999 web pages; also based on Patton, 1986):

- *The observer's role in the setting* Some observers are best described as outsiders with little involvement in the group dynamics whereas others are full members of the group (see Figure 18.1).
- *The group's knowledge of observation process* Overt observation is when the participants know that they are being observed and by whom. Covert observation is when the participants in the study do *not* know that they are being observed and, obviously, cannot know by whom they are being observed (see Figure 18.1).
- *Explication of the study's purpose* This is more than a single dimension and may fall into at least one of the following categories:
  - There is a full explanation given as to the purpose of the research prior to starting the research.
  - Partial explanation means that the participants have some idea of the purpose of the study but this is less than complete for some reason.
  - There is no explanation of the study's purpose because the observation is covert.
  - There is a misleading or false explanation as to the purpose of the study.

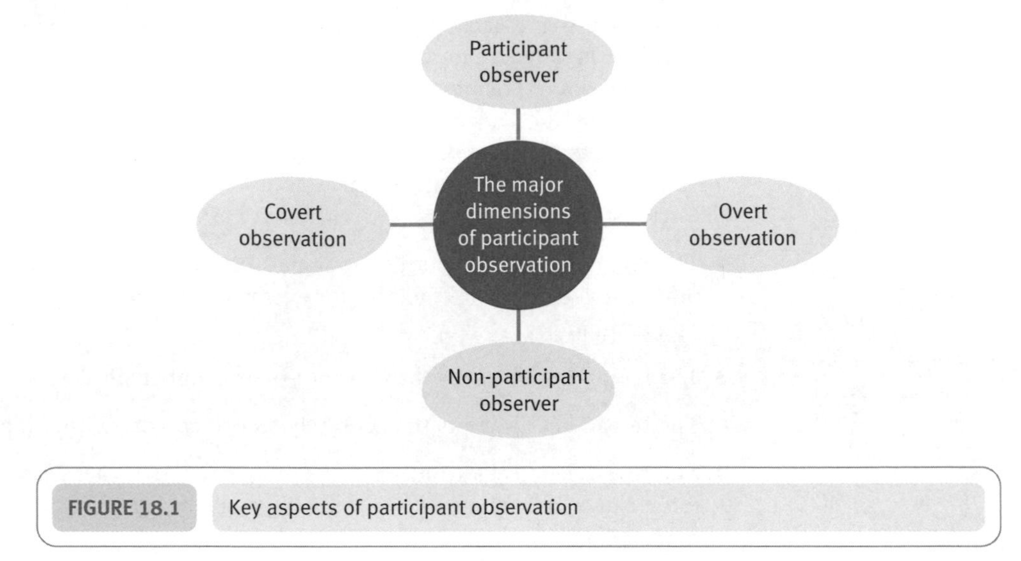

FIGURE 18.1 Key aspects of participant observation

- *Length* The observation may be a single session of a very limited length (for example, a single session of one hour) or there may be multiple observation sessions of considerable length which may continue for weeks or years.
- *Focus* The researcher may focus very narrowly on a single aspect of the situation; there may be an 'expanded' focus on a lengthy but nevertheless predetermined list of variables; there may be a holistic or 'rich data' approach which involves the observation of a wide variety of aspects in depth.

It is very difficult to set out the minimum requirements for a participant observation study. For example, what is required to justify the observation being described as a *participant* observation? Participant observation is uncommon in psychological research though it is frequently a topic for research methods modules – and textbooks. One of its major difficulties as a means of psychological research lies in its frequent dependency on the observations of a single individual. That is, participant observation may be accused of subjectivity because it is dependent on uncorroborated observations. It would be regarded as more objective if the strength of the agreement between different participant observers could be established, which is rarely the case.

Method 2: Focus groups

In some respects, focus groups are like the daytime television discussion shows in which the presenter throws in a few issues and questions, and the audience debates them among themselves. It is the dynamic quality of the focus group situation which differentiates it from interviews and is the main advantage of the method. Focus groups generate data which are patently the product of a group situation and so may, to some extent, generate different findings from individual interviews. Focus groups originated in the work of the famous sociologist Robert Merton when he researched the effectiveness of propaganda using a method he termed *focused interviewing* (Merton and Kendall, 1946). In subsequent decades it was taken up by advertising and market researchers until eventually becoming more accepted in academic research. Focus groups allowed the researcher to concentrate on matters which market research interviews fail to assess adequately. In recent years, researchers have increasingly regarded focus groups as a means of generating ideas and understanding, especially for new research topics, perhaps prior to another more quantitative approach. In effect, the members of the focus group are given the task of making sense of the issue. This is achieved through the group dynamics, that is, through the relatively normal processes of discussion and debate among ordinary people. This is very difficult to achieve through conventional interviewing techniques involving a single interviewee.

Focus groups may be used in at least three different ways:

- As an early stage of research in order to explore and identify what the significant issues are.
- To generate broadly conversational data on a topic to be analysed in its own right. This is a controversial area and lately qualitative researchers have preferred more naturalistic conversation sources.
- To evaluate the findings of research in the eyes of the people that the research is about. That is, discussion of research conclusions.

For the researcher, the focus group has other advantages; that is, most of the resources come from the participants. The researcher generally 'facilitates' the group processes in order to ensure that a pre-planned range of issues is covered but at the same time allowing unexpected material to enter the discussion. So, ideally, the researcher does not dominate the proceedings. If necessary, the researcher steers the discussion along more productive lines if the group seems to be 'running out of steam'. The researcher running the focus group is known as the moderator or the facilitator.

In order to organise focus group research effectively, the following need some consideration:

- Allow up to about two hours running time for a focus group. Short running times may indicate an unsatisfactory methodology.
- A single focus group is rarely if ever sufficient even if the group seems very productive in terms of ideas and discussion. The researcher will need to run several groups in order to ensure that a good range of viewpoints has been covered. It is difficult to say just how many groups are needed without some knowledge of the purpose of the research. Indeed, the researcher may consider running groups until it appears that nothing new is emerging. In a sense this is subjective, but it is also practical within the ethos of qualitative methodology.
- The size of a focus group is important. If there are too many participants some will be inhibited from talking or unable to find the opportunity to participate; too few and the stimulation of a limited range of viewpoints will risk stultifying the proceedings. Generally it appears that the ideal is six to ten individuals, though this is not a rule.
- Participants in focus groups are not intended to be representative of anything other than variety. They should be chosen in order to maximise the productivity of the discussion. This is, of course, a matter of judgement which will get better with experience in focus group methodology. However, Gibbs (1997) offers the following practical advice, which is worthwhile considering:
    - Don't tell focus group members too much in advance of the meeting. If you do there is a risk that they will figure out their own particular thoughts and attitudes on the topic of the focus group and, consequently, they may be unresponsive to the input of others in the group.
    - Unless there is a very good reason for doing otherwise, ensure that the focus group members are strangers to each other prior to the meeting.
    - Focus group members should generally be varied (heterogeneous) in terms of obvious factors. That is, they should vary in educational level, race and ethnicity,

gender and social economic status. However, it should be appreciated that some of these factors in some circumstances may be inhibitory. For example, a discussion of race may be affected by having different races present.

The tasks of the focus group moderator include (Gibbs, 1997):

- explaining the purpose and objectives of the focus group session;
- creating a positive experience for the group members and making them feel comfortable in the situation;
- prompting discussion by posing questions that may open up the debate or by focusing on an issue;
- enabling participation by all members of the group;
- highlighting differences in perspective between people so that they are encouraged to engage in the nature of this difference in the discussion;
- stopping conversational drifts from the point of the topic of the focus group.

Among the characteristics required of the focus group moderator are:

- the ability *not* to appear judgemental;
- the ability to keep their personal opinions to themselves.

It is nonsensical to evaluate a focus group in the same terms as, say, an individual interview. A focus group is not intended to be a convenient substitute for the individual interview and cannot compete with it in all respects. In particular, focus groups cannot be used to estimate population characteristics should these be a focus of the study. Any attempt to use focus group data as indicative of the typical attitudes, beliefs or opinions of people in general is mistaken. Focus groups do not involve, say, random sampling from the population so they are not indicative of population characteristics. Focus groups have a number of disadvantages, which mean that they should not be undertaken without clear reasons:

- They take a great deal of time and effort to organise, run and transcribe. For example, bringing a group of strangers together is not always straightforward logistically.
- The focus group takes away power from the researcher to direct the research process and the sorts of data collected. Consequently, it is difficult to imagine a profitable use of the focus group as a method of collecting data for the typical laboratory experiment.

Among the advantages of the focus group is the motivation aroused in the participants simply through being in a group situation. The participant is not a somewhat alienated individual filling in a rather tedious questionnaire in isolation. Instead the participant is a member of a group being stimulated by other members of the group. So the experience is social, interesting and to a degree fun. Furthermore, membership of a focus group can be, in itself, empowering. Members of a focus group are given a voice to, perhaps, communicate to the management of their organisation via the focus group and the researcher.

The analysis of focus group data may follow a number of routes. The route chosen will largely be dependent on why the focus group approach was selected for data collection. If the focus group is largely to generate ideas for further research or as a preliminary to more structured research, the researcher may be satisfied simply by listing the major and significant themes emerging in the focus group discussions. On the other hand, the focus group may have served as a means of generating verbal data for detailed textual analysis of some sort. Detailed data analysis of this sort requires transcriptions to be made of the group discussion from the audio or video-recording. The Jefferson transcription system (Chapter 19), for example, may be appropriate in many cases. However, the level of detail recorded in Jefferson transcription may be too much for some research purposes. Appropriately transcribed data may be analysed using the broad principles of grounded theory, discourse analysis or conversation analysis in particular (see Chapters 21, 22 and 23). Of these, grounded theory analysis may suit more researchers than the other more specific approaches. In other words, the analysis, as ever, needs to be tailored to the purpose of the research.

Method 3: Interviews

The interview is a very diverse situation with very little evidence of a common strategy being used by the majority of researchers. Our short coverage can only give some indication of the range of activities that constitute the interview. Interviews can be highly structured (little different in many ways from a self-completion questionnaire). These would be known as structured interviews. Alternatively, interviews may be unstructured such that emerging issues can be explored rather than questions asked and answers recorded. These are qualitative interviews.

Structured interviews

Market research interviewers are everywhere – in the streets, on our phones, etc. Few of us have not been subjected to their questions. Characteristically the questions are highly structured and a range of response alternatives provided from which we choose. The interviewer mostly tries to stick to the 'script' of the questionnaire. Such interviews have a number of advantages so far as the researcher is concerned:

- Since the interviewers have quotas of persons to interview, the approach ensures satisfactory numbers of completed questionnaires are obtained. There is usually little or nothing in the interview that could not have been achieved by the questionnaire being completed by the interviewee alone.
- Probably the main reason why interviews are used is that the participants are recruited on the spot. Mailing questionnaires to a sample of people is likely to result in derisory return rates and derisory sample sizes as a consequence.
- The pre-coded, multiple-choice format allows quick computer analysis of the data.
- The process is quick and it is perfectly feasible to plan research and have some sort of report ready for clients in a very short period of time – even just a few days.

Variants on structured interviewing are employed by academic researchers, and the strengths and weaknesses remain much the same. Nevertheless, if the structured approach is adequate for the purposes of one's research, then it should be considered if only for reasons of economy.

In-depth interviews

Sometimes also referred to as semi-structured interviews, these reverse the principles of the structured interview. Consequently, the qualitative ethos pervades research using such interviews. Some researchers are attracted to in-depth interviews because of their conversational characteristics. However, it is wrong to view them as normal conversation. They are a highly specialised form of conversation which occur in a very different context from normal conversation. For one thing, they are intended to be much more one-sided in terms of input. That is, the rule is that the interviewee is talking about themselves whereas the interviewer will spend little or no time doing this. Most conversation

Table 18.1 Structured and qualitative interviewing contrasted

Structured interview	Qualitative interview
Researcher has highly specific and well-formulated questions that require answers.	The researcher has a less clear agenda in terms of content and the agenda is less clearly researcher-led.
The format allows ready assessment of reliability and validity.	Reliability and validity are rather problematic or complex concepts in this context.
The research addresses concerns that emerge from the status of the researcher – which has research-based knowledge and theory as part of the components.	The research normally is led in part by the agenda of concerns as felt by the participant. The researcher has a broader agenda which accommodates this.
Participants are 'forced' to stick to the point and there is little or no scope for them to express idiosyncratic points of view. Sometimes token questions such as 'Is there anything that you think should be mentioned but has not been?' are appended.	According to some, rambling accounts are to be encouraged in qualitative interviewing as this pushes the data far wider than the researcher may have anticipated.
Structured interviews allow little or no departure of the interviewer from the questionnaire in the interests of standardisation.	Qualitative interviewers expect to rephrase questions appropriately, formulate new questions and probes in response to what occurs in the interview, and generally to engage in a relatively relaxed approach to standardisation.
Inflexible.	Flexible.
Answers generated are supposed to be readily and quickly coded with the minimum of labour.	The researcher is looking for rich and detailed answers which result in extensive and labour-intensive coding processes (for example, see Chapter 21 on grounded theory).
Repeat interviewing is rare except in longitudinal studies in which participants may be interviewed on a number of separate occasions.	Repeat interviewing is not uncommon since it allows the researcher to 'regroup' – to reformulate their ideas during the course of the research. Checking and gathering data that had previously been omitted from the first interview, etc. are among these characteristics.

taxes neither of the participants. In-depth interviews are likely to be difficult for interviewee and interviewer. The interviewee will be pressed on detail about matters beyond what is normal in everyday conversation. The interviewer will have prepared extensively for the interview; in addition the interviewer of necessity must absorb a lot of information during the course of the interview in order to question and probe effectively. Having a recorder does not do away with this demand since the recording cannot be referred to during the course of the interview. In other words, one should expect in-depth interviews to be taxing. Table 18.1 extends the comparison of structured interviewing and qualitative interviewing (drawing on Bryman and Bell, 2007).

Almost invariably, the interviewer in qualitative interviewing will have at the minimum the skeleton of the interview in the form of a list of topics or questions to be covered. This is known as the *interview guide*. The guide may be added to as the researcher interviews more participants and becomes aware of issues which could not or had not been anticipated at the start of the research. The guide is often little more than a memory aid which gives the basics of what the researcher intends to cover and probably is a poor reflection of the contents of the interviews themselves. This is only to be expected if the ideals of qualitative research are met by the researcher. That is, the topics are partially formulated by the participant, the enterprise is very exploratory, and rich detail (which by definition is not routine) is the aim. Experienced researchers will probably refer very little to the interview guide – perhaps only using it as a check at the end of the interview in order to ensure that the major issues have been covered.

Certain considerations need to be addressed in preparing the interview guide:

- The researcher may wish to record some routine, basic information in a simple structured form. Matters such as the participant's age, gender, qualifications, job and so forth may be dealt with by using a simple standardised list of answer categories, for example, the highest level of academic qualification obtained.
- The formulation of questions and topics should not simply be a list of obvious questions or questions included because the replies *just might* be interesting. The questions need to be developed in terms of the requirements of the research. Just what sorts of information would help the researcher address what they regard as the important things about the research topic? The interview guide may need modifying part-way through the research to take account of things learnt during the earlier interviews.
- The questions or topics should be structured in a sensible and helpful order. This makes them easier for the interviewer and interviewee to deal with. There is a lot of memory work and other thinking for both participants so a logical structure is important.
- Frame the interview schedule using the appropriate language for the participant group. Children will require different language from adults, for example. However, this is also true for adult groups. What is appropriate may not be known to the researcher without talking to members of that group or without piloting the methodology.

If in-depth interviewing sounds easy then the point has been missed. This is probably best illustrated by asking what the researcher is actually doing when conducting the interview. We can begin by suggesting what they do *not* do:

- The researcher is *not* taking detailed notes. A high-quality audio or video-recording of the interview is the main record. Some researchers may make simple notes but this is not a requisite of the sort that the recording is. These notes are more useful as a memory aid during the course of the interview rather than as data for future analysis. It is very easy to be overawed by the interview situation and forget one's place in the schedule or forget what has been said.

So what is the interviewer doing? The following are the ideal from the perspective of qualitative methods though difficult to achieve:

- The interviewer is actively building as best they can an understanding of what they are being told. In contrast, it hardly matters in a structured interview whether the interviewer gets an overview of this sort since they merely write down the answers to individual questions. However, without concentrating intensely on the content of the interview, the qualitative researcher simply cannot function effectively.
- The interviewer formulates questions and probes in a way which clarifies and extends the detail of the account being provided by the participant. Why did they say this? Who is the person being described by the participant? Does what is being said make sense in terms of what has been said before? Is what is being communicated unclear? The list of questions is, of course, virtually endless. But this aspect of the task is very demanding on the cognitive and memory resources of the interviewer as it may also be for the participant.
- The interviewer is cognisant of other interviews which they (and possibly co-workers) have conducted with other participants. Issues may have emerged in those which

appear missing in the current interview. Why is that? How does the participant respond when specifically asked about these issues?

- The objective of the interviewer's activity is to expand the detail and to interrogate the information as it is being collected. This is very much in keeping with the view that qualitative data analysis *starts* at the stage of data collection. It also reflects the qualitative ideal that progress in research depends on the early and repeated processing of the data.

In addition to all of this, there are practical issues which are too easily overlooked by the researcher but may have a significant impact on the quality of the research generated by the in-depth interview:

- Just how many different researchers will be conducting the interviews? Using two or more different interviewers produces problems in terms of ensuring similarity and evenness of coverage across interviews.

- How are developments communicated between the interviewers? It is probably worth considering the use of semi-structured interviews if the logistics of using several interviewers become too complex.

- The data are usually no more than whatever is on the recording. As a consequence, it is important to obtain the best possible recording as this greatly facilitates the speed and quality of the final transcription (for example, see Chapter 19). Beginners tend to assume that a recorder that functions well enough when spoken into by the researcher will be adequate to pick-up an interview between two people, physically set apart, in perhaps a noisy environment. Consider the best equipment available as an investment in terms of the quality of recording commensurate with the saving in transcription time. A recorder which allows the recording to be monitored through an earphone as it is being made will help ensure that the recording quality is optimised.

- The physical setting of the interview needs to be considered. Sometimes privacy will be regarded as essential for the material in question. In other circumstances privacy may not be so important (that is, if the topic is in no way sensitive). Taking the research to the home or workplace of the participants may be the preferred option over inviting the participants along to the researcher's office, for example. Interviews at home may unexpectedly turn into family interviews if one does not take care to ensure that it is understood that this is an individual interview. Many homes will have just a couple of places in which the interview may take place so be prepared to improvise.

Much of the available advice to assist planning an interview is somewhat over general. What is appropriate in one sort of interview may be inappropriate in another. What may be appropriate with adults may not work with youngsters with learning difficulty. If one gets the impression that good interviewing requires social skills, quickness of thought or a great deal of concentration, and resourcefulness, then that is just about right. For example, *Child Abuse Errors* (Howitt, 1992b) contains psychological research based on in-depth qualitative interview methods. The research essentially addresses the question of the processes by which parents become falsely accused of child abuse. This was partly stimulated by the cases in Cleveland in England where a number of parents were accused of child sexual abuse against their own children. The children were given a simple medical test which was erroneously believed by some doctors to be indicative of anal abuse. But these are not the only circumstances in which parents are accused, apparently falsely, of child abuse. The problems of this research in many ways are the ones which stimulate in-depth interviewing in general. That is, at the time there was virtually no research on the topic, indeed there was virtually nothing known about such cases. So inevitably the task was to collect a wide variety of accounts of the parents' experiences from a wide variety of circumstances. The initial interviews were, consequently, 'stabs in the dark'. The parents taking part in the study were participants with much, in general, to say about their experiences. Consequently, there was a complex account to absorb very quickly as the participants spoke. Furthermore, these were, of course, emotional matters for the parents who essentially had their identity as good parents and their role of parent removed. The complex and demanding nature of the in-depth interviewer's task in such circumstances is obvious.

18.3 Conclusion

The main criterion for an effective qualitative data collection method is the richness of the data it provides. Richness is difficult to define but it refers to the lack of constraint on the data which would come from a highly structured data collection method. Part of the richness of data is a consequence of qualitative data collection methods being suitable for exploring unknown or previously unresearched research topics. In these circumstances the researcher needs to explore a wide variety of aspects of the topic, not selected features. Some of the qualitative data collected by researchers using methods like those described in this chapter will be analysed in a traditional positivist way with the participants' contributions being used as something akin to representing reality. Other data collected using these self-same methods might be subjected to discourse analysis, for example, which would eschew the representational nature of the material in favour of the language acts that are to be seen in the text.

Key points

- Qualitative data analysis is *not* the same thing as qualitative data collection. Qualitative data collection may in some cases become a quantitative analysis if quantifiable coding techniques are developed for the data.
- All qualitative data collection methods vary in terms of their degrees of structuring across different research studies. That is, there is no agreed standard of structuring which is applied in every case.
- Participant observation essentially has the researcher immersed as a member of a social environment. It has its origins in anthropology as a means of studying cultures. There is no strong research tradition of its use in psychology.
- Focus groups are increasingly popular in psychology and other disciplines as a means of collecting rich, textual material. It is a rather social research method in which the participants actively interact with others, under the gentle steering of the researcher. Because it highlights similarities and differences between group members, it is very useful for generating ideas about the topic under research as part of the pilot work, though it is equally suitable for addressing more developed research questions.
- Interviewing may be structured or unstructured. Generally, somewhat unstructured interviews are most likely to be the foundation of qualitative research simply because the lack of structure provides 'richer' unconstrained textual data. In-depth interviewing places a lot of responsibility on the interviewer in terms of the questioning process, coping with the emotions of the interviewee, and ensuring that the issues have been covered exhaustively.

ACTIVITIES

1. Write a schedule for a structured interview on text messaging. Interview a volunteer using the schedule. Re-interview them using the schedule as a guide for a qualitative interview. What additional useful information did the structured interview uncover?

2. Get a group of volunteers together for a focus group on text messaging. What did you learn from the focus group compared with the interviews above?

CHAPTER 19

Transcribing language data

The Jefferson system

Overview

- Transcription is the process by which recordings are transformed into written text.
- The transcription of auditory and visual recordings is a vital stage in analysing much qualitative data.
- Transcription techniques are much better developed for auditory than for visual recordings.
- Transcription inevitably loses information from the original recording. Methods and transcribers differ in the extent that they can deal with the nuances of the material on the original recording.
- The detail required of the transcription is dependent on the purposes of the research and the resources available.
- The Jefferson transcription method places some emphasis on pauses, errors of speech and people talking over each other or at the same time.
- It is evident from research that 'errors' are not uncommon in transcriptions.

19.1 Introduction

Research imposes structure on its subject matter. The structuring occurs at all stages of the research process. For example, the way the researcher decides to collect research data affects the nature of the research's outcome. If the researcher takes notes during an interview, what is recorded depends on a complex process of questioning, listening, interpreting and summarising. It could not be otherwise. Research is the activity of humans, *not* super-humans. If a researcher audio records conversation then all that is available on permanent record is the recording. Visual information such as body posture and facial expression are not recorded. Once the recording is transcribed as text, further aspects of the original events are lost. The intonation of the speaker, errors of speech and other features cannot be retained if the recording is merely transcribed from the spoken word to the written word. This does not make the transcription bad, it just means that it may be useless for certain purposes. If the researcher wishes to obtain 'factual' accounts of a typical day in the life of a police officer, the literal transcription may be adequate. (That is, the researcher is using language as a representation of reality and would have no problems with such a transcription. Qualitative researchers who argue that this view is wrong would regard such a transcription as useless.)

Research may have a vast range of valid purposes. Take, for example, the needs of a researcher who is interested in the process of conversation. The literal words used are inadequate to understand the nuances of conversation. On the other hand, a speech therapist might well be interested in transcribing particular features of speech which are most pertinent to a speech therapist's professional activities. Thus pronunciation of words may be critical as may be recording speech impediments such as stuttering. In other words, the speech therapist may be disinclined to record information which helps to understand the structuring of conversation as opposed to the speech of a single individual (Potter, 1997). So there is a test of 'fitness for purpose' which should be applied when planning transcription.

An example may be helpful. Take the following sentence as an example of 'literal' text:

Dave has gone on his holidays.

Strictly grammatically and literally, this sentence may mean something quite different in the context of a real-life conversation. Perhaps the researcher has actually transcribed the sentence as:

Dave has gone on errrrr [pause] his holidays.

This second version could be understood to mean that Dave is in prison. The 'errrrr' is not a word and the pause is not a word. They are paralinguistic features which help us to revise what the meaning of the sentence is. Given this, researchers studying language in its social context need to incorporate paralinguistic elements since they provide evidence of how the words are interpreted by participants in conversation. The paralinguistic features of language often have subtle implications. For example, 'errrr', which is a longer version of 'er', may often imply different things. 'Errrr' implies a deliberate search for an appropriate meaning, whereas 'Er' may often simply signal that one has forgotten the word. The experts on the subtle use of language are ordinary, native speakers of the language. One may describe this as an ethnographic approach to social interaction since we need to understand the conversation much as the participants in the conversation would. Of course, there is no *fixed* link between paralinguistic features of language and the meaning they add. So the presence of a particular feature should be regarded as informative rather than indicative.

19.2 Jefferson transcription

One popular system for transcribing speech is the system developed by Gail Jefferson. This has its origins in her work with Harvey Sacks, the 'founding-parent' of conversation analysis (see Chapter 23). The Jefferson system can appear a little confusing to novices – and not easy for those familiar with it – but using it is a skill which will improve with practice.

Jefferson's system has *no* special characters so it can be used by anyone using a standard computer or typewriter keyboard. Consequently, some familiar keystrokes have a distinctive meaning in Jefferson transcription. These keystrokes are used as symbols to indicate the way in which the words are delivered in the recording. This means, for example, that conventional punctuation may have its conventional meaning or may have a distinctive Jefferson meaning. Thus *capital* letters may indicate the start of a sentence or a proper noun, but they may indicate that the speaker has said a word with considerable emphasis using greater emphasis than the surrounding words. The main Jefferson conventions are given in Table 19.1. There are symbols which are used to indicate pauses, elongations of word sounds, where two speakers are overlapping and so forth. Refer back to this table whenever necessary to understand what is happening in a transcript. You may also spot that there are slight differences between transcribers on certain matters of detail.

Jefferson transcription is not unproblematic in every instance. To illustrate this, take the following:

For:::get it

The ::: indicates that the For should be extended in length. However, just what is the standard length of 'for' in 'forget'? In some dialects, the For will be longer than in others. And just what is the difference between for:::get and for::get?

Example of Jefferson transcription

The excerpt on p. 323 is from a study of police interviews with paedophile suspects (Benneworth, 2004). The researcher had access to police recordings of such interviews. As a consequence, the data consist solely of audio-recorded text without any visual information. Of course, the researcher might have wanted to video-record the interviews in order to get evidence of facial expression, etc., but this was not an available option. Sound recordings are routine for British police interviews so the transcription may be regarded as being of a naturally produced conversation – a recorded police interview – not an artefact of the research process. The people involved in the transcribed material below are a detective constable (DC) and the suspect being interviewed (Susp). The issue is about the suspect's use of pornography in his dealings with a young girl. Transcripts can vary markedly in terms of how closely they adopt the Jefferson system and just what features are regarded as of significance in the recording. Furthermore, the Jefferson system has evolved over the years as Jefferson developed it. So transcriptions from different periods may show varying conventions and characteristics. Benneworth's transcription seems to us to be well balanced in that it is clear to read even by relative novices to transcription:

Table 19.1 Main features of the Jefferson transcription system

Jefferson symbol	Meaning
CAPITALS	Indicate that the word(s) is louder than the surrounding words.
Underlining	Indicates emphasis such as on a particular syllable.
Aster*isk	The speaker's voice becomes squeaky.
Numbers in brackets (1.2)	Placed in text to indicate the length of a pause between words.
A dot (.) in brackets	This is a micropause – a noticeable but very short pause in the speech.
[]	Square brackets are used when two (or more) speakers are talking together. The speakers are given different lines and the brackets should be in line where the speech overlaps.
//	Another way of indicating the start of the second overlapping speaker's utterance.
; or :	Used to separate the speaker's name from their utterances.
?;	Indicates that the speaker is not recognisable to the analyst of the transcript.
?Janet;	Indicates a strong likelihood that Janet is the speaker.
. . .	Three dots are used to indicate a pause of untimed length.
??;	Two or more ? marks indicate that this is a new unidentified speaker from the last unidentified speaker.
[. . .]	Indicates material has been omitted at that point.
°I agree°	Words between signs ° are spoken more quietly by the speaker.
→	This is not part of the transcription. It is placed next to lines which the analyst wishes to bring to the reader's attention.
↑↓	Used to indicate substantial movements in pitch. They indicate out of the ordinary changes, not those characteristic of a particular dialect, for instance.
Heh heh	Indicates laughter which is voiced rather almost as if it were a spoken word rather than the uncontrolled noises that may constitute laughter in some circumstances.
I've wai::ted	The preceded sound is extended proportionate to the number of colons.
Hhh	Expiration – breathing out sounds such as when signalling annoyance.
(what about)	Words in brackets are the analyst's best guess as to somewhat inaudible passages.
((smiles))	Material in double brackets refers to non-linguistic aspects of the exchange.
(? ?)	Inaudible passage approximately the length between the brackets.
I don't accept your argument = and another thing I don't think you are talking sense	Placed between two utterances to indicate that there is no identifiable pause between the two. Also known as latching.
= signs placed vertically on successive lines by different speakers	In this context, the = sign is an indication that two (or more) speakers are overlapping on the text between the = signs.
[ ] placed vertically on successive lines by different speakers	As above, but instead the [ ] brackets are used to indicate that two (or more) speakers are overlapping on the text between the brackets.
>that's all I'm saying<	Talk between > and < signs is speeded up.
< that's it>	Talk between < and > signs is slowed down.

For more details of Jefferson coding see Hutchby and Wooffitt (1998).

363 DC: What made you feel okay about showing them to a
364 [eleven year old girl]
365 Susp: [accidentally] she first saw them when
366 I opened my boot one day I forgot they were
367 there and then she (1.8) °expressed an interest
368 in them and like looking at them and that's how
369 it developed.°
370 DC: So you felt confident about showing them to (.)
371 Lucy whereas you wouldn't have shown them to
372 [your wife].
373 Susp: [yeah I was] I was (3.8) s::::exually (0.8) umm
374 (4.0) unconfident anymore about sex and Lucy
375 showing an interest in me and that was
376 flattering in itself and .hhh cos there was no
377 sexual relationships with my wife.
378 DC: Was it easier to feel confident with Lucy
379 because she was so young? <And you were an
380 adult and [more in control.>]
381 Susp: [no it's just that] >it was the first
382 .hhh first young lady that's ever expressed an
383 interest in me during my troubled (.) marriage
384 over the past three years< (.) °I said°.

Source: Benneworth (2004)

You will probably have noted a number of features of the transcript:

- Each line is numbered 363, 364, etc., so it is clear that this is just an excerpt from a much longer transcript. The numbering is fairly arbitrary in the sense that another researcher may have produced lines of a different length and hence different numbers would be applied in their transcriptions. Notice that the lines do not correspond to sentences or any other linguistic unit.
- Look at lines 364 and 365. The words enclosed in square brackets [ ] are parts of the conversation where the two participants overlap. It would not be possible to transcribe this if the system did not utilise arbitrary line lengths.
- The use of Jefferson notation is not only time-consuming for the researcher, but it makes it difficult for readers unskilled in the Jefferson system. Simply attempting to read the literal text ignoring the transcription conventions is not easy.
- Jefferson transcription cannot be done by untrained personnel such as secretaries.
- The researcher would almost certainly have used a transcription machine which allows rapid replays of short sections of the recording. In other words, transcription is a slow, detailed process that should only be undertaken if the aims and objectives of the research study require it.

It is also worthwhile noting that Jefferson's transcriptions of the identical material will vary from researcher to researcher. It is probably fair to say that this transcription is at an intermediate level of transcription detail. In other words, by this stage the researcher has made a contribution to the nature of the data available for further analysis. Anyone carrying out Jefferson transcription will experience a degree of uncertainty as to whether they have achieved an appropriate level of transcription detail. It should be remembered that qualitative researchers tend to be very familiar with their texts before the transcription is complete. This familiarity will help them set the level of detail that is appropriate for their purposes.

An important question is what the Jefferson transcription enables which a secretary's word-by-word transcription might miss out. The following gives the above transcription with notation omitted. Often a secretary would fail to give the overlapping talk so the dominant voice at the time of the overlap would be transcribed and the other voice perhaps noted as inaudible. This is possibly because a secretary would regard text as linear much as when taking dictation from the secretary's boss. The following is a guess as to what a typical secretary's transcription of the same type might be. Two people talking together would probably be regarded as inaudible:

DC: What made you feel okay about showing them to a eleven-year-old girl?

Susp: (inaudible) she first saw them when I opened my boot one day I forgot they were there and then she expressed an interest in them and like looking at them and that's how it developed.

DC: So you felt confident about showing them to Lucy whereas you wouldn't have shown them to your wife.

Susp: (inaudible) I was sexually unconfident anymore about sex and Lucy showing an interest in me and that was flattering in itself and cos there was no sexual relationships with my wife.

DC: Was it easier to feel confident with Lucy because she was so young? And you were an adult and more in control?

Susp: (inaudible) it was the first first young lady that's ever expressed an interest in me during my troubled marriage over the past three years I said.

Source: Benneworth (2004)

It has to be said that even in this version of the text there is a great deal that strikes one as important. For example, the following:

- The way in which the suspect presents the pornography as something that the girl happened on by chance and as a result of her actions. There is no indication that the suspect actively created a situation in which she was exposed to the pornography.
- The way in which the suspect excuses his offending by blaming his troubled marriage and his wife.
- The way in which the offender represents his non-normative relationship (an adult man with an 11-year-old girl) as if two adults were involved. So the 11-year-old girl is represented as a 'young lady' indicating maturity rather than a 'young girl' which represents an immature person.

Researchers and practitioners with knowledge and experience of paedophiles and other sex offenders have themselves noted such 'denial' and 'cognitive distortion' strategies (Howitt, 1995). Indeed, much of the therapy for sex offenders involves group therapy methods of modifying such cognitions. The above excerpt is perhaps not altogether typical of the sort of text used by qualitative researchers. For one thing, it is the sort of text which is unfamiliar to many and so is quite different from the everyday, routine conversation studied by many qualitative researchers. The implication is that unfamiliar subject matter is likely to reveal a lot because it contrasts markedly with more familiar sorts of conversation from everyday life. It is likely that at least some researchers would find in this simple transcription all that they require for their research purposes. For example, if a researcher was interested in the types of denial and cognitive distortions demonstrated by offenders, the transcription process may not need the Jefferson-style of elaboration.

So what does the Jefferson transcription system add which a secretary's transcription omits? There are a few obvious things:

- The Jefferson transcription gives a lot more information about what happened in the conversation. The secretary's version gives the impression of a smooth, unproblematic conversational flow. The Jefferson transcription demonstrates a variety of turn-taking errors, quite lengthy pauses in utterances, and dynamic qualities of the way that the conversation is structured, for example, the very quiet passages.
- The Jefferson transcription allows parts of the conversation to be rapidly referred to.
- Even by carefully reading the transcription, let alone doing the transcription, a reader has a more intimate knowledge of the text. Consequently, the extra detailed work done in order to produce a Jefferson transcript means that the researcher becomes very familiar with the material. They may begin to conceptualise what is happening in the text sooner. This early and detailed familiarity with the data is claimed to be one of the analytic virtues of qualitative research, though this is greatly undermined if researchers do not do their own transcription.

What else does the researcher gain by using the Jefferson transcription system? After all, some may regard the Jefferson system as merely providing irrelevant and obscuring detail. Suggestions include:

- If we look carefully for what the Jefferson transcription adds to the information available to the researcher, we find in line 373/4 the following: Susp: [yeah I was] I was (3.8) s::::exually (0.8) umm (4.0) unconfident anymore about sex. Not only is the word sexually highlighted in speech by the elongation into s::::exually but it is also isolated by lengthy gaps of four or so seconds on each side. Benneworth (2004) refers to this as a 'conversational difficulty' which takes the form of 'hesitant speech' and 'prolonged pauses'. This may have led her to pointing out that the term 'sexually' is part of a particular language repertoire which the suspect only applies to relationships with an adult. When speaking of the child, he employs what Benneworth (2004) describes as 'relationship discourse and euphemism'. '*Lucy showing an interest in me . . . that was flattering in itself*'. The offender does not use the term 'victim' of the child, though it is a term that most of us would use. That is, the offender is not using language repertoire that would indicate the child has been abused sexually as opposed to the language repertoire used to indicate a mutual relationship.
- The use of Jefferson transcription clearly encourages the researcher to concentrate closely on the text as a matter of a social exchange rather than information requested and supplied. For example, Benneworth notes how the detective constable constantly brings the youth of the girl into the conversation which contrasts with the offender's representation of the girl as if she were a mature female rather than a child. Furthermore, by taking this excerpt and contrasting it with other excerpts from other interviews, the researcher was able to explore different interview styles – one is more confrontational and challenging of the suspect, whereas the other almost colludes with the offender's 'distorted' cognitions.

- Similarly, the use of the Jefferson transcription facilitates the linking of the text under consideration with established theory in the field. So Benneworth argues that the words 'I opened my boot one day' in line 366 grounds the offender's account in common day-to-day experience rather than the more extraordinary abuse of a child. Such a device may be seen as a discursive device for creating a sense of ordinariness (Jefferson, 1984) and essentially creates a distance from the suspect's actions and the criminal consequences ensuing from them.

19.3 Advice for transcribers

It should be emphasised that the Jefferson system is only one of a number of systems of transcription that can be employed. Indeed, there is no reason why a researcher should not contemplate developing their own system if circumstances require it. O'Connell and Kowal (1995) evaluated a number of text transcription systems employed by researchers, including that of Jefferson. They suggest that transcription is not and cannot be 'a genuine photograph of the spoken word' (p. 105). Generally all transcription systems attempt to record the exact verbal words said. Nevertheless, transcription systems vary considerably in terms of transcribing other features of speech, indeed in some cases other features are not included. So some systems include prosodic features such as how a word was spoken (loudly, softly, part emphasised, etc.), paralinguistic features (such as words said with a laugh or sigh) and extralinguistic features (facial expressions, gestures, etc.) – some systems exclude some or all of them.

The following is some of the generic advice offered to transcribers by O'Connell and Kowal:

- The principle of parsimony: only those features of speech which are to be analysed should be transcribed. That is, there is little point in including extralinguistic features such as gestures in the transcription if they will not be part of the analysis.
- Similarly, the transcriptions provided in reports should only include whatever is necessary to make the analysis intelligible to the reader.
- Subjectively assessed aspects of conversation should not be included in the transcription as if they are objective measurements. For example, transcribers may subjectively estimate the lengths of short pauses (0.2) but enter them as if they are precise measures. O'Connell and Kowal report that transcribers omitted almost four out of five of such pauses in radio interviews. This begs the question why the other pauses were included.
- Transcribers make frequent, uncorrected errors. For example, verbal additions, deletions, relocations and substitutions are commonly found when a transcript is compared with the original recording. Qualitative researchers often stress the importance of checking the transcription against the original source to minimise this problem.

There is other, perhaps more routine, advice available to transcribers. For example, Potter (2004) suggests that technological advances have made transcription easier. Transcription is labour-intensive and 20 hours of transcription may be necessary for 1 hour of recording. It is obvious that high-quality digital recordings using, say, a mini-disc player will be enormously beneficial – and result in fewer errors. Furthermore, there are digital editing programs (for example, Cool Edit) which allow the transcription of recordings on screen. As the recording is held as a file on the computer, the system allows frequent checking against the original. Since the recording may be displayed as a visual waveform, it becomes easier to measure precisely gaps and pauses in conversation and speech.

19.4 Conclusion

Although normally described as a *transcription* system, Jefferson's approach is also a low-level coding or categorisation system. If researchers want a perfect transcription of the recording then what better than their original recording? Of course, what they want is a simplified or more manageable version of the recording. Inevitably this means coding or categorising the material and one can only capture what the system of coding can capture. Conversational difficulties, for example, are highlighted by the Jefferson system so these are likely to receive analytic consideration. Facial expression is not usually included so facial expressions (which may totally negate the impression created in a conversation) are overlooked.

Transcription is a very time-consuming process. The Jefferson system is more detailed than most and takes up even more time. So there is little point in using any system of transcription unless it adds something to achieving the sort of analysis the researcher requires. Transcription is generally regarded by qualitative researchers as a task for the researcher themselves. Ideally it is not something farmed out to junior assistants or clerical workers. Qualitative researchers need intimate familiarity with their material – this is facilitated by doing one's own transcriptions.

Key points

- Transcription is the stage between the collection of data in verbal form and analysis. Usually it is producing a written version of audio-recordings, but video material may also be transcribed.
- In qualitative data analysis, transcription may take into account more than the words spoken by indicating how the words are spoken. That is, errors of speech are included, pauses are indexed, and times when people are speaking at the same time are noted.
- Transcription is not regarded as a necessary chore but one of the ways in which the researcher becomes increasingly familiar with their data. Transcription is not usually passed over to others.
- Inevitably transcription omits aspects of the original and there is the risk that the transcription is inadequate. It is normally recommended that the researcher refers back to the original recording when the transcription appears complete or considers asking another researcher to assess the veracity of it.
- The commonest form of transcription is the Jefferson method which has its roots in conversation analysis. It is very commonly used by qualitative researchers but can be unnecessarily time-consuming if the analysis is only of the words.

ACTIVITIES

1. In pairs, act out the conversation which was subject to Jefferson transcription in the main body of the chapter between the police and the suspect. Record the conversation if you can and compare the product of the attempts of different pairs of actors.

2. Record a conversation, select an interesting part, transcribe it and annotate it with Jefferson transcription symbols. List the difficulties you experience for discussion.

CHAPTER 20

Thematic analysis

Overview

- Thematic analysis is one of the most commonly used methods of qualitative analysis. However, as a method it has received little detailed attention and accounts of how to carry out a thematic analysis are scarce. Furthermore, many researchers gloss over what they actually did when carrying out a thematic analysis. This means that the method is not so easily accessed by novices as some other approaches.
- Thematic analysis is not as dependent on specialised theory as some other qualitative techniques such as discourse analysis (Chapter 22) and conversation analysis (Chapter 23). As a consequence, thematic analysis is more accessible to novices unfamiliar with the relevant theory in depth.
- In thematic analysis the task of the researcher is to identify a limited number of themes which adequately reflect their textual data. This is not so easy to do well though the identification of a few superficial themes is generally quite simple but does not reflect the required level of analysis adequately.
- As with all qualitative analysis, it is vitally important that the researcher is extremely familiar with their data if the analysis is to be expedited and insightful. Thus data familiarisation is a key to thematic analysis as it is for other qualitative methods. For this reason, it is generally recommended that researchers carry out their data collection themselves (for example, conduct their own in-depth interviews) and also transcribe the data themselves. Otherwise, the researcher is at quite a disadvantage.
- Following data familiarisation, the researcher will normally code their data. That is, they apply brief verbal descriptions to small chunks of data. The detail of this process will vary according to circumstances including the researcher's expectations about the direction in which the analysis will proceed. Probably the analyst will be making

codings every two or three lines of text but there are no rules about this and some analyses may be more densely coded than others.

- At every stage of the analysis, the researcher will alter and modify the analysis in the light of experience and as ideas develop. Thus the researcher may adjust earlier codings in the light of the full picture of the data. The idea is really to get as close a fit of the codings to the data as possible without having a plethora of idiosyncratic codings.
- On the basis of the codings, the researcher then tries to identify themes which integrate substantial sets of these codings. Again this is something of a trial-and-error process in which change and adjustment will be a regular feature. The researcher needs to be able to define each theme sufficiently so that it is clear to others exactly what the theme is.
- The researcher needs to identify examples of each theme to illustrate what the analysis has achieved.
- As in all report writing, the process of writing up the analysis and the results of the analysis is part of the analysis process and a good researcher may re-think and re-do parts of their analysis in the course of the write-up.
- There is no reason why researchers cannot give numerical indications of the incidence and prevalence of each theme in their data. For example, what percentage of participants mention things which refer to a particular theme?

20.1 Introduction

Almost certainly, thematic analysis is the approach to qualitative analysis most likely to be adopted by newcomers to qualitative analysis. There are good reasons for this since thematic analysis needs less knowledge of the intricacies of the theoretical foundations of qualitative research than most other qualitative techniques. Compared with, say, discourse analysis or conversation analysis, thematic analysis does not require the subtle and sophisticated appreciation of a great deal of the theory underlying the method. Hence, it is amenable to novices. No particular theoretical orientation is associated with thematic analysis and it is flexible in terms of how and why it is carried out. So one will see thematic analyses carried out by researchers who would not seem to have any particularly strong affinity to qualitative research. In a sense, it is at entry level a somewhat undemanding approach to the analysis of qualitative data – interviews in particular. Thematic analysis does not demand the intensely closely detailed analysis which typifies conversation analysis, for example. All of this adds up to strong praise for thematic analysis or damning criticism, depending on one's point of view. Like anything else in research, thematic analysis can be well done or poorly done. It is important for you to know the difference – until you do then you cannot expect to do good work. All-in-all, with a little care, it can be recommended as a useful initiation for students into qualitative research.

There is a downside to all of this. Thematic analysis is not a single, identifiable approach to the analysis of qualitative data. There is no accepted, standardised approach to carrying out a thematic analysis, so different researchers do things differently. While this is typical of qualitative methods in general, it clearly is an obstacle to carrying out thematic analysis. So it is impossible to provide a universally acceptable set of guidelines which, effortlessly, will lead to a good thematic analysis. Actually this is true for many different aspects of research, including the analysis of data using statistical methods. As understanding of quantitative techniques develops and the amount of data the researcher collects becomes extensive, it becomes clear that there is no simple set of 'rules' which can be followed to carry out a standard analysis. There are many ways of carrying out a statistical analysis of complex data. Similarly, there are many ways of doing thematic analysis and one simply has to make choices. Nevertheless, the key aspects of thematic analysis can be identified.

Sometimes very basic and unsystematic approaches form the basis of thematic analysis. The researcher simply reads through their data in transcribed form and tries to identify, say, half a dozen themes which appear fairly commonly in the transcripts. Then the researcher writes a report of their data analysis in which they lace together the themes that they have identified with illustrative excerpts from the transcripts. So what is wrong with this? The problem with such an approach is that the researcher is not actually doing a great deal of analytic work. The task is too easy in the sense that so long as the researcher can suggest some themes and provide illustrative support for them from the transcripts then there is little intellectual demand on the researcher. So long as the excerpt matches the theme then this is evidence in support of the theme. Who is to say that the themes are 'wrong' since there is no criterion to establish that they are wrong? But think about it. The process involved in this analysis lacks a great deal in terms of transparency. It is unclear how the researcher processed their data to come up with the themes; it is unclear the extent to which the themes encompass the data – do the themes exhaust the data or merely cover a small amount of the transcribed material? Generally, such reports do not establish the amount of the data dealt with by the themes. Furthermore, the task need not be very onerous for the researcher who once he or she has thought of a handful of themes has little more work to do apart from writing up the report. They have not had the tougher task of developing themes to cover the entirety of the data which would require them to do more and more intensive analytic work. The likelihood is that by increasing the analytic demands on the researcher, there would be an increased likelihood that new, different and more subtle ways of looking at the data would work. The more work that goes into the analysis, the better the analytic outcome would be one way of putting this. Figure 20.1 gives some indications of the roots of thematic analysis.

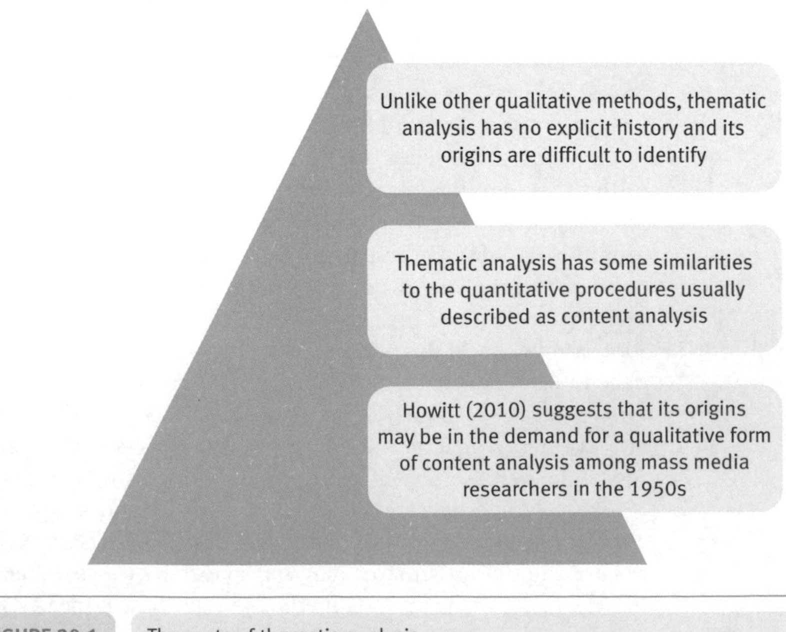

Unlike other qualitative methods, thematic analysis has no explicit history and its origins are difficult to identify
Thematic analysis has some similarities to the quantitative procedures usually described as content analysis
Howitt (2010) suggests that its origins may be in the demand for a qualitative form of content analysis among mass media researchers in the 1950s

FIGURE 20.1 The roots of thematic analysis

20.2 What is thematic analysis?

The phrase thematic analysis first appeared in the psychological journals in 1943 but is much more common now. Nevertheless, thematic analysis is something of the poor relative in the family of qualitative methods. It has few high-profile advocates and, possibly as a consequence, has not been formalised as a method. Users of thematic analysis pay scant attention to the method in their reports and provide very few details about what it is they do. As a result, there is very little available by way of systematic instruction into how to carry out a thematic analysis. Since the method tends to be glossed over in reports, it is difficult to use published papers as a guide to how to do thematic analysis. Typically, instead of describing in detail how the analysis was done, thematic analysts simply write something like 'a thematic analysis was carried out on the data'. In other cases, reports which describe themes identified in qualitative data may make no reference at all to thematic analysis; for example, Gee, Ward and Eccleston (2003) report 'A data-driven approach to model development (grounded theory) was undertaken to analyse the interview transcripts' (p. 44). Thematic analysis is also a poor relative of other qualitative methods since it often appears to be sloppily carried out and very subjective in terms of the findings which emerge. Such claims are easy to make since in thematic analysis the detail of the analysis process is usually omitted so the reader of the report may be forgiven for thinking that the researcher merely perused a few transcripts and then identified a number of themes suggested by the data. The only support provided for the analysis is that each of the themes is illustrated by quotes taken from the data which one assumes are among the most convincing examples that can be found. Put this way, thematic analysis does not amount to much and, to be frank, there do seem to be some published thematic analyses to which these comments would apply. However, carried out properly, thematic analysis is quite an exacting process requiring a considerable investment of time and effort by the researchers.

Just as the label says, thematic analysis is the analysis of textual material (newspapers, interviews and so forth) in order to indicate the major themes to be found in it. A theme, according to *The Concise Oxford Dictionary* is 'a subject or topic on which a person speaks, writes, or thinks'. This is not quite the sense of the word 'theme' used in thematic analysis. When a lecturer stands up and talks about, say, eyewitness testimony for an hour, the theme of the lecture would be eyewitness testimony according to the dictionary definition. However, in thematic analysis the researcher does not identify the overall topic of text. Instead the researcher would dig deeper into the text of the lecture to identify a variety of themes which describe significant aspects of the text. For example, the following themes may be present in the lecture: the unreliability of eyewitness testimony, the ways of improving the accuracy of testimony, and methodological problems with the research into eyewitness testimony. This may not be the most scintillating thematic analysis ever carried out, but nevertheless it does give us some understanding of this particular lecture as an example of text. Of course, a lecture is normally a highly organised piece of textual material which has been split up by the lecturer into several different components and given a structure so that everything is clear to the student. This is not the case with many texts such as in-depth interviews or transcripts of focus groups. People talking in these circumstances simply do not produce highly systematic and organised speech. Thus the analytic work is there for the researcher to organise the textual material by defining the main themes which seem to represent the text effectively. While it is possible to carry out thematic analysis on a single piece of text, more generally researchers use material from a wider range of individuals or focus groups, for example.

There are other methods of qualitative research which seem to compete with thematic analysis in the sense that they take text and, often, identify themes. Grounded theory

(Chapter 21) is a case in point. Indeed, if the basic processes involved in carrying out a grounded theory analysis are compared with those of thematic analysis then differentiating between the two is difficult. But there is a crucial difference: grounded theory is intended as a way of generating theory which is closely tied to the data. Theory development is not the intention of thematic analysis. Of course, any process which leads to a better understanding of data may lead subsequently to the development of theories.

Thematic analysis is not aligned with any particular theory or method though overwhelmingly it is presented from a qualitative perspective which is data-led. However, sometimes the approach taken is to develop themes based on theory and then test the themes against the actual data – though this violates basic assumptions from most qualitative perspectives. One also sees from time to time thematic analyses quantified in the sense that the researcher counts the number of interviews, for example, in which each theme is to be found. Thematic analysis, used in this way, is difficult to distinguish from some forms of content analysis described in Chapter 16. The lack of a clear theoretical basis to thematic analysis does not mean that theory is not appropriate to your research – it merely means that the researcher needs to identify the theoretical allegiance of his or her research. For example, is the research informed by feminist thinking, is it phenomenological in nature, or does it relate to some other theory? Purely empirical thematic analyses may be appropriate in some cases but they may not be academically very satisfying as a consequence.

Given all of these comments, it should be obvious that the term 'thematic analysis' refers to a wide range of different sorts of analysis ranging from the atheoretical to the theoretically sophisticated, the relatively casual to the procedurally exacting, and the superficial to the sophisticated in terms of the themes suggested. At the most basic level, thematic analysis can be described as merely empirical as the researcher creates the themes simply from what is in the text before him or her; this may be described as an inductive approach. On the other hand, the researcher may be informed by theory in terms of what aspects of the text to examine and in terms of the sorts of themes that should be identified and how they should be described and labelled. If there is a theoretical position which informs the analysis, then this should be discussed by the researcher in the report of their analysis; in this sense, the analysis may be theory driven.

20.3 A basic approach to thematic analysis

The basic essential components of a thematic analysis are shown in Figure 20.2. They are transcription, analytic effort and theme identification. It is important to note that the three stages are only conceptually distinct: in practice they overlap considerably. Briefly, the components can be described as follows:

- *Transcribing textual material* This can be based on any qualitative data collection method including in-depth interviews and focus groups. The level of transcription may vary from a straightforward literal transcript much as a secretary would produce to, for example, a Jeffersoned-version of the text which contains a great deal more information than the literal transcription (see Chapter 19). Generally speaking, there would appear to be no reason for using Jefferson transcription with thematic analysis but, by the same token, if a researcher sees a place for it then there is nothing to prevent that. No qualitative researcher should regard transcription as an unfortunate but necessary chore since the work of transcribing increases the familiarity of the

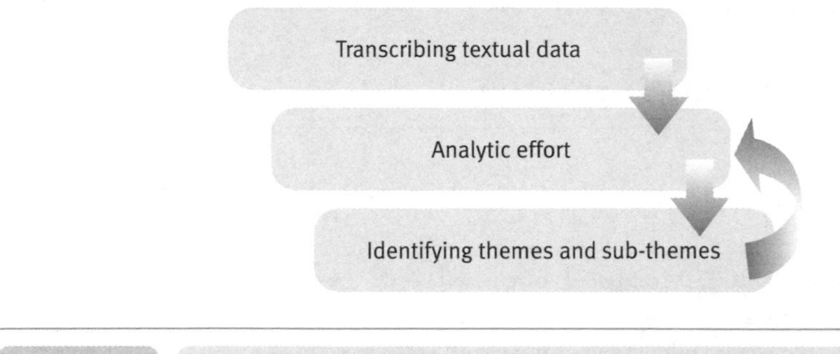

FIGURE 20.2 Basic thematic analysis

researcher with his or her material. In other words, the transcription process is part of the process of analysis. In the best case circumstances, the researcher would have conducted the interviews or focus groups themselves and then transcribed the data themselves. Thus the process of becoming familiar with the text starts early and probably continues throughout the analysis.

- ***Analytic effort*** This refers to the amount of work or processing that the researcher applies to the text in order to generate the final themes which are the end point of thematic analysis. There are several components to analytic effort: (a) the process of becoming increasingly familiar with the text so that understanding can be achieved and is not based on partial knowledge of the data; (b) the detail with which the researcher studies his or her data which may range from a line-by-line analysis to a much broader brush approach which merely seeks to summarise the overall themes; (c) the extent to which the researcher is prepared to process and reprocess the data in order to achieve as close a fit of the analysis to the data as possible; (d) the extent to which the researcher is presented with difficulties during the course of the analysis which have to be resolved; and (e) the willingness of the researcher to check and recheck the fit of his or her analysis to the original data.
- ***Identifying themes and sub-themes*** While this appears to be the end point of a thematic analysis, researchers will differ considerably in terms of how carefully or fully they choose to refine the themes which they suggest on the basis of their analysis. The researcher may be rapidly satisfied with the set of themes since they seem to do a 'good enough' job of describing what they see as key features of the data. Another researcher may be dissatisfied at this stage with the same themes because they realise that the themes, for example, describe only a part of the data and there is a lot of material which could not be coded under these themes. Hence the latter researcher may seek to refine the list of themes in some way, for example, by adding themes and removing those which seem to do a particularly poor job of describing the data. Of course, by being demanding in terms of the analysis, the researcher may find that they need to refine all of the themes and may find that for some of the themes substantial sub-themes emerge. Also, again as a consequence of being demanding, the researcher may find it harder to name and describe the new or refined themes accurately. All of this continues the analytic work through to the end of the total thematic analysis.

On the basis of this, the flow diagram of the process perhaps is as shown in Figure 20.2. In the next section, we go on to provide a more sophisticated version of thematic analysis. An example of thematic analysis is described in Box 20.1.

Box 20.1 Research Example

Thematic analysis

Sheldon and Howitt (2007) compared offenders convicted of using the Internet for sexual offending purposes (for example, downloading child pornography) with child molesters (the traditional paedophile). They were interested in (a) the 'function(s)' of Internet child pornography for Internet sex offenders and (b) the concept of desistance from child molestation. Internet offenders have a strong sexual proclivity towards children (for example, they are sexually aroused by children) but mainly do not go on to sexually molest children. Despite their close similarities to traditional paedophiles, Internet offenders were desisting from offending against children. How do Internet offenders explain why they do not express their paedophilic orientation towards children by directly assaulting children sexually? The researchers carried out a thematic analysis of what the offenders told them about the functions of Internet child pornography in their lives and their desistance from offending directly against children. The offenders provided detailed data on a topic which has not been extensively researched.

So during the course of lengthy interviews, Internet offenders were asked why they did not contact offend (i.e. physically offend) against children and contact paedophiles were asked why they used child pornography on the Internet as a substitute for contact offending. All of the fieldwork for this study was conducted by one researcher who therefore had (a) interviewed all of the participants in the study herself and (b) transcribed in full all of the interviews using direct literal (secretarial) methods. The transcriptions were not 'Jeffersoned' (see Chapter 19) since the researchers simply wanted to study broadly how offenders accounted for these aspects of their offending.

Of course, the interviews and transcripts contained much data irrelevant to the question of desistance (for example, matters such as childhood experiences, details of the offending behaviour and their cognitive distortions). Hence, the researchers needed to identify relevant material for this aspect of the study which was confined to answers to specific questions (for example, their reasons for not engaging in a particular sort of offending behaviour). This was done by copying and pasting the material from the computer files of the transcripts into a new file but it could have been done by highlighting the relevant text with a highlighter pen or highlighting the material on the computer with a different font or font colour. Because of the sheer volume of data in this study coming from over 50 offenders it was best to put the pertinent material into a relatively compact computer file. In this way, the material can easily be perused for the coding process.

The phases of thematic analysis are very similar to those of other forms of qualitative analysis. The process began with a descriptive level of coding with minimal interpretation. The researchers applied codes to 'chunks' of data, that is, a word, phrase, sentence or even a paragraph. For example, one of the functions of child pornography according to offenders was to avoid negative feelings/moods encountered in their everyday lives and so was coded as 'negavoidance' each time this occurred in the transcripts. Coding was not a static process so initial codes were revised as the researcher proceeded through the transcript. Some codes became subdivided or revised if the initial codes were not adequate or some codes were combined as there was too much overlap in meaning. Jotting down of ideas and codes was an integral part of this early stage. As the researcher had conducted the interviews, she was also very familiar with the material.

The next formal level of coding involved a greater degree of interpretation. More superordinate constructs were identified which captured the overall meaning of some of the initial descriptive codes used at the earlier stage. Throughout the entire process of analysis the researcher moved constantly backwards and forwards between the interview extracts and the codes. This stage also involved an early search for themes. This process of moving towards identifying themes involved writing the codings onto different postcards (together with a brief description of them) and organising them into 'theme piles'. This allowed the researcher to check whether the themes worked in relation to the coded extracts.

In the final stage of this particular thematic analysis, psychological theories were drawn upon to aid interpretation of the codings and to identify the overarching themes. At the same time, it was essential that the analysis remained grounded in the actual data. Engaging with previous research and theory was very important in this particular study as it helped in understanding the meaning and implications of the patterns in the codings or the themes identified. At the same time, the researcher was engaged in the process of generating clear definitions and names for each theme. Overall, this thematic analysis generated only a few themes but these themes represented more general concepts within the analysis and subsumed the lower level codes.

If themes are clearly defined then it is possible within a qualitative analysis to add a quantitative component. Just how common are the themes in the data? There are different ways of doing this. It can be asked just how prevalent a theme is, meaning just how many of the participants mention a particular theme in their individual accounts. Alternatively, one might ask how many incidents of a particular theme occur in a particular account. In this study, following the thematic analysis, each interview was studied again and the percentage of each type of sex offender mentioning a particular theme at least once was assessed. Ideally, there should be several instances of a theme across the data but more instances of a theme does not necessarily mean the theme is any more crucial. Key themes capture something important in terms of the research question and this is not entirely dependent on their frequency of occurrence in the data.

There were three strong themes identified in what the offenders told the researchers about desistance: (a) focus on fantasy contact, (b) moral/ethical reasoning and (c) fear of consequences. These are very different themes and probably not entirely predictable. Certainly the idea of moral/ethical reasoning in terms of child pornography and child molestation is not a common-sense notion. The themes identified by the study were illustrated by excerpts such as the following:

- *Focus on fantasy contact* 'I never got to the point where I would want to touch . . . looking at the images is enough, though a lot of people will disagree . . . I mean I've met people in prisons . . . who are in for the same thing and . . . their talk was never of actual sexual contact. Definitely. No. No. I would never.'
- *Moral/ethical reasoning* 'No . . . because as an adult you've got to be thinking for the child . . . they've got to live with it for the rest of their life.'
- *Fear of consequences* 'Partly because I wouldn't want the guy to go "Ahh! This man's trying to grope me!" . . . and I'd have his big brothers' mates coming with baseball bats.'

Notice that if one checks the excerpts against the name of the theme then only the one theme seems to deal with the data in each case. Try to switch around the names of the themes with the different excerpts and they simply do not fit. This is an illustration of back-checking the themes against the data, though in the study proper the researchers were far more comprehensive in this checking process.

We are grateful to Kerry Sheldon for her help with this box.

20.4 A more sophisticated version of thematic analysis

Braun and Clarke (2006) provide what is probably the most systematic introduction to doing thematic analysis to date. This is a fully fledged account of thematic analysis which seeks to impose high standards on the analyst such that more exacting and sophisticated thematic analyses are developed. They write of the 'process' of doing a thematic analysis which they divide into six separate aspects that very roughly describe the sequence of the analysis, though there may be a lot of backtracking to the earlier aspects of the process in order to achieve the best possible analysis. The simple approach as described previously includes some elements similar to the Braun–Clarke approach but they are aiming for a somewhat more comprehensive and demanding kind of thematic analysis which, to date, has only been rarely approached. Their six aspects or steps are:

- familiarisation with the data;
- initial coding generation;
- searching for themes based on the initial coding;
- review of the themes;
- theme definition and labelling;
- report writing.

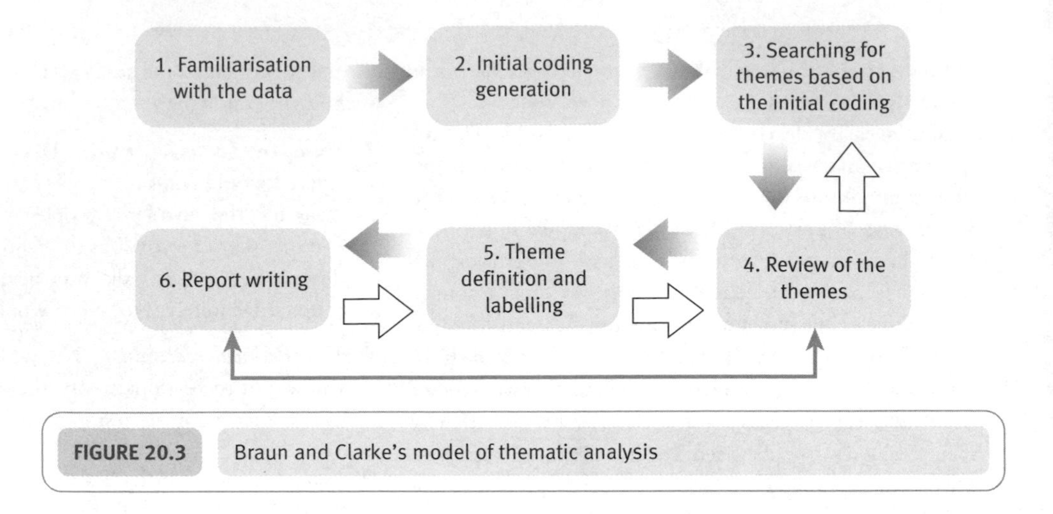

FIGURE 20.3 Braun and Clarke's model of thematic analysis

The entire process is summarised in Figure 20.3. Notice that the figure indicates a sort of flow from one aspect to the next but there are many loops back to the earlier aspects of the analysis should circumstances demand it. In truth, at practically any stage of the process the analyst may go back to any of the earlier stages for purposes of refinement and clarification. The six steps in the analysis not only loop back to earlier stages but the stages are best regarded as conceptually distinct since in practice there may be considerable overlap.

Step 1 Familiarisation with the data

This is the early stage in which the researcher becomes involved actively with the data. The familiarisation process depends partly on the nature of the text to be analysed. If the text is interview data, for example, the researcher has probably been actively involved in interviewing the participants in the research. Inevitably, while interviewing the participants the interviewer will gain familiarity with what is being said. Unless the interviewer is so overwhelmed by the interview situation they fail to pay proper attention to what the participant is saying, features of what each interviewee is saying will become familiar to the researcher. Equally, over a series of interviews, most interviewers will begin to formulate ideas about what is being said in the interview just as we get ideas about this in ordinary conversation. In the research context, the researcher will be well aware that they will eventually have to produce some sort of analysis of the interviews. Thus, more than in ordinary conversation, there is an imperative to absorb as much of what is being said as possible and to develop ideas for the analysis. Of course, the more interviews that have been carried out the easier it is to begin to recognise some patterns. These may stimulate very preliminary ideas about how the data will be coded and, perhaps, ideas of the themes apparent in the data.

Furthermore, interview data has to be transcribed from the recording, partly because this facilitates more intense processing of the text by the researcher at the later stage when the text needs to be read and re-read but also because excerpts of text are usually included in the final report to illustrate the themes. Usually in thematic analysis the transcription is a literal transcription of the text much as a secretary would do. It is far less common to use Jefferson transcription with thematic analysis (Chapter 19). Jefferson transcription is more laborious than literal transcription. The choice of how the transcribing is done depends partly on whether the thematic analysis can effectively utilise the additional information incorporated in the Jefferson transcription. In thematic analysis, the researcher tends to have a realist perspective on the text, that is, the belief that the text represents a basic reality and so can largely be understood literally – hence there is little need for the Jefferson system. The process of transcription in qualitative analysis should be regarded as a positive thing despite the tedium that may be involved. Ideally, doing the transcription will make the researcher even more familiar with the research data. There are limitations to this because transcription proceeds slowly and usually involves just a few words at a time which makes getting the full picture more difficult. Finally, the transcriptions will be read and re-read a number of times to further familiarise the researcher with the material and as an aide memoire. Researchers who do not themselves actively collect and transcribe the text they intend to analyse will be at a disadvantage; they would need to spend much more time on reading the transcripts. There are no shortcuts in this familiarisation process if a worthwhile analysis is to be performed. All other things being equal, a researcher who is well immersed in their data will have better ideas about later stages of the process and may, early on, have ideas about the direction in which the analysis will go. Writing notes to oneself about what one is reading is part of the process of increasing familiarity with the data but also constitutes an earlier stage in the coding process which technically comes next.

Useful tip

As a novice researcher it is likely that you will have to do all of the data collection and transcriptions yourself. This is an asset, not a hindrance.

Step 2

Initial coding generation

Initial coding is a step in the process by which themes are generated. The research suggests codings for the aspects of the data which seem interesting or important. The initial coding process involves the analyst working through the entirety of the data in a systematic way, making suggestions as to what is happening in the data. Probably this is best done on a line-by-line basis but sometimes this will be too small a unit of analysis. The decision about how frequently to make a coding depends partly on the particular data in question but also on the broader purpose of the analysis. As a rule of thumb, a coding should be made at fairly regular intervals – every line may be too frequent, every two or three lines would probably be acceptable. The chunk of the text being coded does not have to be exactly the same number of lines each time a coding is made. We are analysing people's talk, which does not have precise regularity. The initial codings are intended to capture the essence of a segment of the text and, at this stage, the objective is *not* to develop broader themes. At first, the initial codings may seem like jottings or notes rather than a sophisticated analysis of the data. If so, all well and good, because this is precisely what you should be aiming for. Of course, the analyst will be pretty familiar with the text already so the initial codings will be on the interviews that the researcher conducted and the transcripts which the researcher made of the interviews in the first stage of data familiarisation. As a consequence, they will already have an overview of the material and so they are not simply responding to a short piece of the text in isolation.

There may be two different approaches depending on whether the data are data-led or theory-led, according to Braun and Clarke (2006):

- ***The data-led approach*** This is dominated by the characteristics of the data and the codings are primarily guided by a careful analysis of what is in the data.
- ***The theory-led approach*** The structure for the initial codings is suggested by the key elements of the theory being applied by the researcher. Feminist theory, for example, stresses that relationships between men and women are dominated by the power and dominance of the male gender over the female gender in a wide variety of aspects of the social world including employment, domestic life and the law. Thus a thematic analysis based on feminist theory would be oriented to the expression of power relationships in any textual material.

Of course, there is something of a dilemma here since it is unclear how a researcher can avoid applying elements of a theoretical perspective during the analysis process. Just how would it be possible to differentiate between a theory-led coding and a data-led coding unless the researcher makes this explicit in their writings?

Usually in reports of thematic analyses, such initial codings are not included by the researcher for the obvious reason that there is rarely sufficient space to include all of the data let alone the codings in addition. Consequently, those new to thematic analysis may assume that the initial codings are more sophisticated than they actually are. The following is a brief piece of transcript provided by Clarke, Burns and Burgoyne (2006) which includes some initial codings for a few lines of text:

	Initial coding
it's too much like hard work I mean how much paper have you got to sign to change a flippin' name no I I mean no I no we we have thought about it ((inaudible)) half heartedly and thought no no I jus' – I can't be bothered, it's too much like hard work. (Kate F07a)	1. Talked about with partner 2. Too much hassle to change name

The initial codings can be seen to be little more than a fairly mundane summary of a few lines of text. Thus, the coding 'too much hassle to change name' is not very different from 'it's too much like hard work' and 'I can't be bothered, it's too much like hard work' which occur in the text at this point. So the initial coding stage is not really about generating substantial insights into the data but merely a process of identifying and summarising the key things about what is going on in the text. Of course, this same piece of text could be coded in any number of different ways. For example, 'half heartedly' in the text might have been coded 'lack of commitment'. Of course, the researcher will normally have some ideas about the direction in which the analysis is going by this stage in the analysis. Consequently, the codings do not have to be exhaustive of all possibilities and, indeed, over-coding at this stage may make it difficult for the analyst to move on to the later phases of the analysis because too much coding obscures what is going on; it is important to remember that the initial codings are brief summaries of a chunk of text and not the minutiae of the text expressed in a different way. The researcher is trying to simplify the text not complicate it.

Also notice that in the example, the same segment of the text is coded in more than one way. This is more likely where one is coding bigger chunks of text than if the coding is line by line.

At this stage, the analyst will typically wish to collate the data which have so far been given a particular initial code. In this way, the researcher is essentially linking a particular code with the parts of the text to which the code has been applied. So, for example, the researcher would bring together the different parts of the text which have been coded as 'Talked about with partner' in the same place. A simple way of doing this is to copy and paste the relevant text material under the title of that particular initial coding. It is possible at this stage that the analyst will feel it appropriate to change the initial codings name to fit better with the pattern of textual material which has received that particular code. Furthermore, it is likely that the researcher will notice that two or more initial codings mean much the same thing despite being expressed in different words. For example, 'discussed matter with husband' may be the same as 'talked about with partner' so should not be regarded as a distinct coding.

Initial coding development (and the later development of themes) is an active process on the part of the researcher. Braun and Clarke (2006) are extremely dismissive of the idea that codings (and themes) 'emerge', that is, suddenly appear to the researcher as part of the analysis process. Codings and themes are synthesised actively from the data by the researcher; they are not located in the data as such but created by the minds and imaginations of researchers.

Useful tip

As a novice, it would be very daunting to write down codings without some practice so why not select a page of transcript which you found particularly interesting and try to code that material first?

Step 3

Searching for themes based on the initial coding

The relationship between text, codings and themes is illustrated in Figure 20.4. The initial codings, of course, are likely to be used quite frequently in the coding of the text though we illustrate them as occurring only once. Then the themes are essentially obtained by joining together (or collapsing together) several of the codings in a meaningful way. Thus the process of initial coding has involved the researcher in formulating descriptive suggestions for the interesting aspects of their data. As we have seen, these codings are fairly close to the text itself. So, if you like, a theme can be seen as a coding of codings. Thus themes identify major patterns in the initial codings and so are a sort of second level of interpretation of the text where the analyst focuses on the relationships between the codings. In some instances, it is possible that a theme is based simply on one of the initial codings. Of course, it is difficult for the analyst to separate the coding phase from the theme-generation phase so one might expect the occasional close correspondence between a single coding and a theme.

This begs the question of how an analyst suggests the themes which bring together the initial codings in a meaningful way. Of course, this may be instantly obvious but not always. One way of identifying themes would be to write each of the different initial codings onto a separate piece of paper or card. Then the initial codings may be sorted into separate piles of codings which seem to be similar. The remaining task would be to put words to the ways in which the similar codings are, indeed, similar. Since the sorting process has an element of trial and error, this procedure will allow the analyst to change the groupings (piles) as their analytic ideas develop. Alternatively, one could place the slips

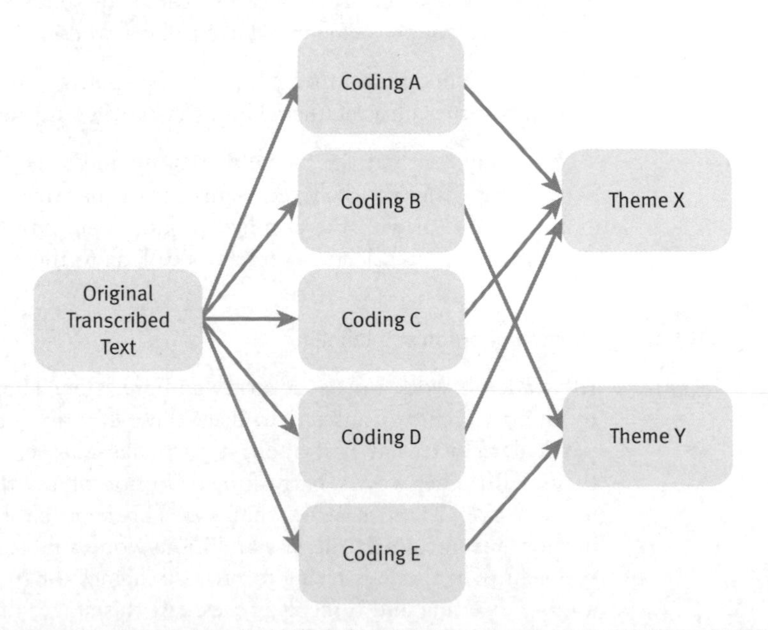

FIGURE 20.4 Relationship between text, codings and themes

of paper on a table top and physically move them around so that initial codings which are similar are next to each other and those which are dissimilar are physically apart. In this way, the relationships between the codings may be made more apparent. It may be that it becomes clear that some apparently very different codings are merely the opposites of each other. So maybe they actually should be part of the same theme.

The entire process is one of trying to understand just what are the overarching themes which bring together the individual codings in a meaningful way. Of course, the themes have to be related back to the original data so the data associated with each theme need to be compiled/collated; in this way, the themes can be related back easily to the original textual data. Moreover, the more systematic the analysis is the greater the data management tasks involved in collating the themes with the original material. The use of computers – if only word-processing programs – should greatly facilitate the process of linking together the data for the themes that the researcher is developing. There are specialist computer programs which also do much the same job.

Useful tip

Unfortunately, developing themes will be easier the harder that you work. Just sitting and staring at a computer screen or the coded transcript will waste time. So any of the active procedures we have suggested in this section are recommended. Spreading the themes on a table top and actively moving them together or apart depending on how similar they are is likely to lead to dividends.

Step 4

Review of the themes

By this stage you will have a set of tentative themes which help one understand what is in the transcriptions. However, these themes probably are not very refined at this stage and need to be tested against the original data once again. There are a number of possibilities:

- You may find that there is very little in the data to support a theme that you have identified so the theme may have to be abandoned or modified in the light of this.
- You may find that a theme needs to be split up since the data which are supposed to link together into the theme imply two different themes or sub-themes.
- You may feel that the theme works, by and large, but does not fit some of the data which initially you believed were part of that theme, so you may have to find a new theme to deal with the non-fitting data. You may need to check the applicability of your themes to selected extracts as well as to the entire dataset.

Step 5

Theme definition and labelling

All academic work aims at accuracy and precision. The definition and labelling of themes by the researcher is unlikely to meet these criteria without considerable refinement. In particular, just what is it about a particular theme which differentiates it from other themes? In other words, part of the definition of any theme includes the issue of what it is not as well as the issue of what it is. This is probably not so complicated as it sounds in most instances and the less ambitious your analysis then the less likely it is to be a problem; where one is trying to provide themes for the entire data then it is likely to be a more exacting and difficult process. At this stage, the analyst may find it appropriate to identify sub-themes within a theme which adds to the task of defining and labelling these accurately. Of course, defining themes and sub-themes precisely cannot take place in a vacuum but needs to be done in relation to the data too. So the researcher would have to go through the data once again to ensure that the themes (and sub-themes) which have been defined in this stage actually still effectively account for the data since the definition imposes a structure and clarity that may not have been present in the initial coding process and the identification of themes. As you do this, you may well find that there are data which have not previously been coded which can be coded now using your refined themes and better level of understanding of the material.

Useful tip

At this stage, you might wish to go 'public' with your ideas. By this we mean that discussing your analysis with others may pay dividends as you have to explain your themes clearly to what may be a sceptical friend or colleague. You have, in this way, a challenge to your theme definition and labelling which may stimulate further thought or revision.

Step 6

Report writing

All research reports tell a story that you want to tell about your data and this applies equally to reports of thematic analysis. Of course, the story being told relates back to the research question which initiated your research – and the stronger the research question, then, all other things being equal, the more coherent a story you can tell. One should not regard report writing as merely telling a story about the steps in your research; the report-writing stage is a further opportunity for reflecting on one's data, one's analysis, and the adequacy of both with respect to each other. So what emerges at the end of the report-writing process may be a somewhat different and probably more refined story than was possible before starting the report. In other words, report writing is another stage in the analysis and not just a chore to be completed to get one's work published or the grades one wants for a psychology degree.

The final report requires that you illustrate your analysis using extracts from your data. Of course, it is more than appropriate to choose the most apposite extracts to illustrate the outcome of your analysis. But, in addition, the selected extracts may be the most vivid of the instances that you have. The final report also provides the opportunity to discuss your analysis in the light of the previous research literature. This may be either (a) the literature that you choose to discuss in order to justify why you have chosen to research a particular research question in a particular way or (b) relating your analyses to the findings and conceptualisations of other analysts. In what way does your analysis progress things beyond theirs? What distinguishes your analysis from theirs? Is it possible to resolve substantial differences?

Useful tip

Most reports of thematic analysis avoid describing in any detail just how the analysis was carried out. Do not emulate this but instead try to be as systematic as you possibly can be about just how the analysis was done. If there are problems defining a theme then identify these and do not simply sweep difficulties under the carpet.

Thematic analysis involves three crucial elements – the data, the coding of data and the identification of themes. The procedure described above essentially stresses the way in which the researcher constantly loops back to the earlier stages in the process to check and to refine the analysis. In other words, the researcher constantly juxtaposes the data and the analysis of the data to establish the adequacy of the analysis and to help refine the analysis. A good analysis requires a considerable investment of time and effort.

20.5 Conclusion

Probably thematic analysis can best be seen as a preferred introduction to qualitative data analysis. The lack of bewildering amounts of theoretical baggage makes it relatively user-friendly to novices to qualitative analysis. Nevertheless, it is an approach which can fail to be convincing if not performed in sufficient depth. The simplicity of thematic analysis is superficial and disguises the considerable efforts that the analyst needs to make in order to produce something that goes beyond the mundane (or, perhaps, what merely states what the researcher 'knew' already). While the temptation may be to pick out a few themes which then become 'the analysis', the researcher must push further than this. Simple notions such as ensuring that as much of the material in the data is covered by the themes help ensure that the analysis challenges the researcher. So thematic analysis is as demanding as any other form of analysis in psychology. The important thing is that the researcher does not stint on the analytic effort required to produce an outcome which is stimulating and moves our understanding of the topic on from the common-sensical notions which sometimes pass as thematic analysis. But this is no different from the challenge facing most researchers irrespective of the method they employ.

Key points

- The secret of a good thematic analysis lies in the amount of analytic work that the researcher contributes to the process. Researchers unwilling to spend the time and effort required in terms of familiarisation with the data, coding, recoding, theme development and so forth will tend to produce weaker and less convincing analyses. They have only superficially analysed their data so produce less insightful and comprehensive themes.
- It improves a report of a thematic analysis if detail of the method used by the researcher is included. It is insufficient (and perhaps misleading) to merely say that a thematic analysis was carried out and that certain themes 'emerged' during the course of the analysis. This gives no real indication of how the analysis was carried out or the degree to which the researcher is active in constructing the themes which their report describes.
- A good thematic analysis can be quantified in terms of the rates of the prevalence and incidents of each of the themes. Prevalence is the number of participants who say things relevant to a particular theme and incidence is the frequency of occurrence of the theme throughout the dataset or the average number of times it occurs in each participant's data.

ACTIVITY

Thematic analysis can be carried out on any text. For example, it could be tried out on two or three pages of a novel you are reading (or a magazine article for that matter). Try to develop initial codes of each line of a few pages of a novel or some other text. What themes can these be sorted into? How good is the fit of the set of themes to the actual text? Do some lines of text fail to appear in at least one theme?

CHAPTER 21

Grounded theory

Overview

- Grounded theory basically involves a number of techniques which enable researchers to effectively analyse 'rich' (detailed) qualitative data effectively.
- It reverses the classic hypothesis-testing approach to theory development (favoured by some quantitative researchers) by defining data collection as the primary stage and requiring that theory is closely linked to the entirety of the data.
- The researcher keeps close to the data when developing theoretical analyses – in this way the analysis is 'grounded' in the data rather than being based on speculative theory which is then tested using hypotheses derived from the theory.
- It employs a constant process of comparison back and forwards between the different aspects of the analysis and also the data.
- Grounded theory does not mean that there are theoretical concepts just waiting in the data to be discovered. It means that the theory is anchored in the data.
- In grounded theory, categories are developed and refined by the researcher in order to explain whatever the researcher regards as the significant features of the data.

21.1 Introduction

Sometimes qualitative data analysis is regarded as being an easy route to doing research. After all, it does not involve writing questionnaire items, planning experimental designs or even doing statistics. All of these tasks are difficult and, if they can be avoided, are best avoided. Or so the argument goes. Superficially, qualitative data analysis does seem to avoid most of the problems of quantification and statistical analysis. Carry out an unstructured interview or conduct a focus group or get a politician's speech off the Internet or something of the sort. Record it using an audio-recorder or video-recorder, or just use the written text grabbed from the World Wide Web. Sounds like a piece of cake. You are probably familiar with the caricature of quantitative researchers as boffins in white coats in laboratories. The qualitative researcher may similarly be caricatured. The qualitative researcher is more like a manic newspaper reporter or television reporter who asks a few questions or takes a bit of video and then writes an article about it.

What is the difference between the qualitative researcher and the TV reporter with the audio-recorder or camera crew? The answer to this question will take most of this chapter.

We can begin with one of the most important and seminal publications in qualitative research. The book, *Discovery of Grounded Theory* (Glaser and Strauss, 1967), is regarded as a classic and remains a major source on the topic of grounded theory despite numerous developments since then. Historically, Glaser and Strauss's approach was as much a reaction to the dominant sociology of the time as it was radically innovative. Basically, the book takes objection to the largely abstract sociological theory of the time which seemed divorced from any social or empirical reality. Indeed, empirical research was as atheoretical as the theoretical research was unempirical in sociology at the time. In its place was offered a new, data-based method of theory development. Grounded theory reversed many of the axioms of conventional research in an attempt to systematise many aspects of qualitative research. As such, it should be of interest to quantitative researchers since it highlights the characteristics of their methods.

However, many readers of this chapter will not yet have read any research that involves the use of grounded theory. So what are the characteristics of a grounded theory analysis? Ultimately the aim is to produce a set of categories into which the data fit closely and which amounts to a theoretical description of the data. Since the data are almost certain to be textual or spoken language the major features of most grounded theory analyses are fairly similar. A word of warning: to carry out a grounded theory analysis is a somewhat exacting task. Sometimes authors claim to have used grounded theory though perusal of their work reveals no signs of the rigours of the method. Sometimes the categories developed fit the data because they are so broad that anything in the data is bound to fit into one or other of the coding categories. Like all forms of research, there are excellent grounded theory analyses, but also inadequate or mundane ones.

Like properly done qualitative data analyses in general, grounded theory approaches are held to be time-consuming, arguably because of the need for great familiarity with the data but also because the process of analysis can be quite exacting. Grounded theory employs a variety of techniques designed to ensure that researchers enter into the required intimate contact with their data as well as bringing into juxtaposition different aspects of the data. The approach has a lot of aficionados across the wide cross-section of qualitative research – though its use is less than universal.

Just to stress, grounded theory methods result in categories which encompass the data (text or speech almost invariably) as completely and unproblematically as the researcher can manage. In this context, theory and effective categorisation are virtually synonymous. This causes some confusion among those better versed in quantitative methods who tend to assume that theory means an elaborate conjectural system from which specific hypotheses are derived for testing. That is not what grounded theory provides – the categorisation system is basically the theory though the method does involve attempts to generalise the theory beyond the immediate data. Furthermore, researchers seeking a theory that yields precise predictions will be disappointed. While grounded theory may generalise to new sets of data, it is normally *in*capable of making predictions of a more precise sort. Charmaz (2000) explains:

> . . . grounded theory methods consist of systematic inductive guidelines for collecting and analyzing data to build middle-range theoretical frameworks that explain the collected data. Throughout the research process, grounded theorists develop analytic interpretations of their data to focus further data collection, which they use in turn to inform and refine their developing theoretical analyses.
>
> (p. 509)

Several elements of this description of grounded theory warrant highlighting:

- Grounded theory consists of *guidelines* for conducting data collection, data analysis and theory building, which may lead to research which is closely integrated to social reality as represented in the data.
- Grounded theory is *systematic*. In other words, the analysis of data to generate theory is not dependent on a stroke of genius or divine inspiration, but on perspiration and application of general principles or methods.
- Grounded theory involves *inductive* guidelines rather than deductive processes. This is very different from what is often regarded as conventional theory building (sometimes described as the 'hypothetico-deductive method'). In the hypothetico-deductive method, theory is developed from which hypotheses are derived. In turn, these hypotheses may be put to an empirical test. Research is important because it allows researchers to test these hypotheses and, consequently, the theory. The hypothetico-deductive method characterised psychology for much of its modern history. Without the link between theory building and hypothesis testing, quantitative research in psychology probably deserves the epithet of 'empiricism gone mad'. Particularly good illustrative examples of the hypothetico-deductive approach are to be found in the writings of psychologists such as Hans Eysenck (for example, Eysenck, 1980). However, grounded theory, itself, was not really a reaction against the hypothetico-deductive method but one against overly abstracted and untestable social theory.
- Grounded theory requires that theory should develop out of an understanding of the complexity of the subject matter. Theories (that is, coding schemes) knit the complexity of the data into a coherent whole. Primarily, such theories may be tested effectively only in terms of the fit between the categories and the data, and by applying the categories to new data. In many ways this contrasts markedly with mainstream quantitative psychology where there is no requirement that the analysis fits all of the data closely – merely that there are statistically significant trends, irrespective of magnitude, which confirm the hypothesis derived from the theory. The unfitting data are regarded as measurement error rather than a reason to explore the data further in order to produce a better analysis, as it may be in qualitative research.
- The theory-building process is a continuous one rather than a sequence of critical tests of the theory through testing hypotheses. In many ways, it is impossible to separate the different phases of the research into discrete components such as theory development, hypothesis testing, followed by refining the theory. The data collection phase, the transcription phase and the analysis phase all share the common intent of building theory by matching the analysis closely to the complexity of the topic of interest.

21.2 Development of grounded theory

Grounded theory is usually described as being a reaction against the dominant sociology of the twentieth century, specifically the Chicago School of Sociology. Some of the founders of this school specifically argued that human communities were made up of sub-populations, each of which operated almost on a natural science model – they were like ecological populations. For example, sub-populations showed a pattern whereby they began to invade a territory, eventually reaching dominance, and finally receding as another sub-population became dominant. This was used to explain population changes in major and developing cities such as Chicago. Large-scale social processes and not the experiences of individuals came to be the subject of study. The characteristics which are attributed to the Chicago School are redolent of a lot of psychology from the same period. In particular, the Chicago School sought to develop exact and standard measuring instruments to measure a small number of key variables that were readily quantified. In sociology, research in natural contexts began to be unimportant in the first half of twentieth century – the corresponding change in psychology was the increased importance of the psychological laboratory as a research base. In sociology, researchers undertook field research mainly in order to develop their measuring instruments. Once developed, they became the focus of interest themselves. So social processes are ignored in favour of broad measures such as social class and alienation, which are abstractions. The theorist and the researcher were often different people, so much so that much research became alienated from theory, that is, atheoretical (Charmaz, 1995).

Grounded theory methodology basically mirror-imaged or reversed features of the dominant sociology of the 1960s in a number of ways:

- Qualitative research came to be seen as a legitimate domain in its own right. It was not a preliminary or preparatory stage for refining one's research instruments prior to quantitative research.
- The division between research and theory was undermined by requiring that theory comes after or as part of the data collection and is tied to the data collected. Furthermore, data collection and their analysis were reconstrued as being virtually inseparable. That is, analysis of the data was encouraged early in the collection of data and this early analysis could be used to guide the later collection of data.

In order to achieve these ends, grounded theory had to demonstrate that quantitative research could be made rigorous, systematic and structured. The idea that quantitative data analysis is no more than a few superficial impressions of the researcher was no part of grounded theory. Equally, case studies are considered in themselves *not* to achieve the full potential of qualitative research.

Despite being the mirror image of mainstream research, grounded theory analysis does not share all of the features of other qualitative methods such as discourse analysis and conversation analysis. In particular, some users of grounded theory reject *realism* (the idea that out there somewhere is a social reality which researchers will eventually uncover) whereas others accept it. Similarly, some grounded theorists aim for objective measures and theory development that does not depend on the researcher's subjectivity. Others regard this as a futile and inappropriate aim. See Figure 21.1 for some of the key aspects of the development of grounded theory.

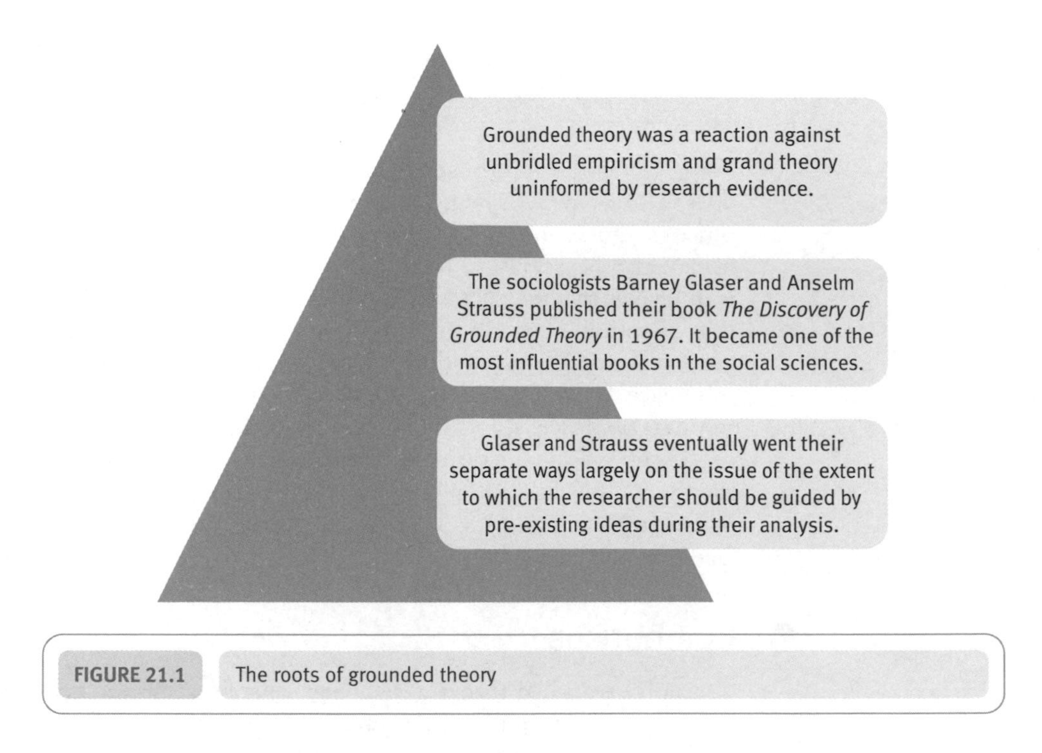

Grounded theory was a reaction against unbridled empiricism and grand theory uninformed by research evidence.
The sociologists Barney Glaser and Anselm Strauss published their book *The Discovery of Grounded Theory* in 1967. It became one of the most influential books in the social sciences.
Glaser and Strauss eventually went their separate ways largely on the issue of the extent to which the researcher should be guided by pre-existing ideas during their analysis.

FIGURE 21.1 The roots of grounded theory

21.3 Data in grounded theory

Grounded theory is not primarily a means of collecting data but the means of data analysis. However, grounded theory does have things to say about the way in which data should be collected in a manner guided by the needs of the developing grounded theory. Grounded theory does not require any particular type of data although some types of data are better for it than others. There is no requirement that the data are qualitative, especially in the early formulations of grounded theory. So, for example, grounded theory can be applied to interviews, biographical data, media content, observations, conversations and so forth or anything else which can usefully inform the developing theory. All of these sources potentially may be introduced into any study. The key thing is, of course, that the primary data should be as richly detailed as possible, that is, not simple or simplified. Charmaz (1995, p. 33) suggests that richly detailed data involve 'full' or 'thick' written descriptions. So, by this criterion, much of the data collected by quantitative researchers in the quantitative approach would be unsuitable as the primary data for analysis. There is little that a grounded theory researcher could do with answers to a multiple-choice questionnaire or personality scale. Yes–no and similar response formats do not provide detailed data – though the findings of such studies may contribute more generally to theory building in grounded theory. The data for grounded theory analysis mostly consist of words, but this is typical of much data in psychology and related disciplines. As such, usually data are initially transcribed using a transcription system though normally Jefferson's elaborate method (Chapter 19) would be unnecessary. Some lessons from grounded theory could be useful to all sorts of researchers. In particular, the need for richness of data, knowing one's data intimately and developing theory closely in line with the data would benefit a great deal of research.

21.4 How to do grounded theory analysis

Potter (1998) likens grounded theory to a sophisticated filing system. This filing system does not merely put things under headings, there is also cross-referencing to a range of other categories. It is a bit like a library book that may be classified as a biography, but it may also be a political book. Keep this analogy in mind as otherwise the temptation is to believe that the data are filed under only one category in grounded theory analysis. It is notorious that Glaser and Strauss did not see eye-to-eye academically speaking later in their careers so rather different versions of grounded theory evolved. The main difference between them was in the extent to which the researcher should come to the data with ideas and thoughts already developed or, as far as possible, with no preconceptions about the data. There seems to be a general acceptance that grounded theory analysis has a number of key components and the following summarises some of the important analytic principles that broadly can be described as grounded theory. These are outlined below.

Comparison

Crucially, grounded theory development involves constant comparisons at all stages of the data collection and analysis process – without comparing categories with each other and with the data, categories cannot evolve and become more refined:

- People may be compared in terms of what they have said or done or how they have accounted for their actions or events, for example.
- Comparisons are made of what a person does or says in one context with what they do and say in another context.
- Comparisons are made of what someone has said or done at a particular time with a similar situation at a different time.
- Comparisons of the data with the category which the researcher suggests may account for the data.
- Comparisons are made of categories used in the analysis with other categories used in the analysis.

So, for example, it is a common criticism of quantitative research that the researcher forces observations into ill-fitting categories for the purpose of analysis; in grounded theory the categories are changed and adjusted to fit the data better. This is often referred to as the method of *constant comparisons*. Much of the following is based on Charmaz's (1995, 2000) recommendations about how to proceed.

Coding/naming

Grounded theory principles require that the researcher repeatedly examines the data closely. The lines of data will be numbered at some stage to aid comparison and reference. In the initial stage of the analysis, the day-to-day work involves coding or describing the data line-by-line. It is as straightforward as that – and as difficult. (Actually, there is no requirement that a line be the unit of analysis and a researcher may choose to operate at the level of the sentence or the paragraph, for example.) The line is examined and a description (it could be more than one) is provided by the researcher to describe what is happening in that line or what is 'represented' by that line. In other words, a name is being given to each line of data. These names or codings should be generated out of what

Table 21.1 A modified extract of grounded theory coding based on Charmaz (1995, p. 39)

Interview transcript	Coding by researcher
If you have lupus, I mean one day it's my liver	Shifting symptoms
One day it's in my joints; one day it's in my head, and	Inconsistent days
It's like people really think you're a hypochondriac if you keep	Interpreting images of self
. . . It's like you don't want to say anything because people are going to start thinking	Avoiding disclosure

Source: Charmaz (1995)

is in that particular line of data. In many ways, describing this as coding is a little misleading, because it implies a pre-existing system, which is not the case. Others describe the process in slightly different terms. For example, Potter (1997) describes the process as being one of giving labels to the key concepts that appear in the line or paragraph.

The point of the coding is that it keeps the researcher's feet firmly in the grounds of the data. Without coding, the researcher may be tempted to over-interpret the data by inappropriately attributing 'motives, fears or unresolved personal issues' (Charmaz, 1995, p. 37) to the participants. At the end of this stage, we are left with numerous codings or descriptions of the contents of many lines of text.

It is difficult to give a brief representative extract of grounded theory style codings. Table 21.1 reproduces a part of such codings from Charmaz (1995) which illustrates aspects of the process reasonably well. Take care though since Table 21.1 contains a very short extract from just one out of nearly two hundred interviews conducted by her. It can be seen that the codings/categories are fairly close to the data in this example. It should be noted that hers are not the only codings which would work with the data.

Categorisation

Quite clearly, the analyst has to try to organise these codings. Remember that codings are part of the analysis process and the first tentative steps in developing theory. These are the smallest formal units in the grounded theory analysis. While they may describe the data more-or-less well, by organising them we may increase the likelihood that we will be able to effectively revise them. This is a sort of reverse filtering process: we are starting with the smallest units of analysis and working back to the larger theoretical descriptions. So the next stage is to build the codings or namings of lines of data into categories. This is a basic strategy in many sorts of research. In quantitative research, there are statistical methods which are commonly used in categorising variables into groupings of variables (for example, factor analysis and cluster analysis). These statistical methods are not generally available to the grounded theorist, so the categorisation process relies on other methods. Once again, the process of constant comparison is crucial, of course. The analyst essentially has to compare as many of the codings with the other codings as possible. That is, is the coding for line 62 really the same as that for line 30 since both lines are described in very similar words? Is it possible to justify coding lines 88 and 109 in identical fashion since when these data lines are examined they appear to be very different?

The constant comparing goes beyond this. For example, does there seem to be a different pattern of codings for Mr X than for Mrs Y? That is, does the way that they talk about things seem to be different? We might not be surprised to find different patterns for Mr X and Mrs Y when we know that this is a couple attending relationship counselling or that one is the boss of a company and the other an employee. The data from a person at a particular point in time or in a particular context may be compared with data from the same person at a later point in time or in different contexts.

It need not stop there. Since the process is one of generating categories for the codings of the data which fit the data well and are coherent, one must also compare the categories with each other as they emerge or are developed. After all, it may become evident, for example, that two of the categories cannot be differentiated – or you may have given identical titles to categories which actually are radically different. The process of categorisation may be facilitated by putting the data or codings or both onto index cards which can be physically moved around on a desk or table in order to place similar items close together and dissimilar items further apart. In this way, relationships can begin to be identified in a more active visual way.

Memo writing

The stages in grounded theory analysis are not as distinct as they first appear. The process of analysis is not sequential, although explaining grounded theory analysis makes it appear so. It is a back-and-forward process. Memo writing describes the aspect of the research in which the data are explored rather than described and categorised. The memo may be just as one imagines – a notebook – in which the researcher notes suggestions as to how the categories may be linked together in the sense that they have relationships and interdependencies. But the memo does not have to be a purely textual thing. A diagram – perhaps a flow diagram – could be used in which the key concepts are placed in boxes and the links between them identified by annotated arrows. What do we mean by relationships and interdependencies? Imagine the case of male and female. They are conceptually distinct categories but they have interdependencies and relationships. One cannot understand the concept of male without the concept of female.

The memo should not be totally separated from the data. Within the memo one should include the most crucial and significant examples from the data which are indicative and typical of the more general examples. So the memo should be replete with illustrative instances as well as potentially ill-fitting or problematic instances of ideas, conceptualisations and relationships that are under development as part of the eventual grounded theory:

> If you are at a loss about what to write about, look for the codes that you have used repeatedly in your data collection. Then start elaborating on these codes. Keep collecting data, keep coding and keep refining your ideas through writing more and further developed memos.
>
> (Charmaz, 1995, p. 43)

In a sense, this advice should not be necessary with grounded theory since the processes of data collection, coding and categorisation of the codes are designed to make the researcher so familiar with their data that it is very obvious what the frequently occurring codings are. However, it is inevitable that those unaccustomed to qualitative analysis will have writing and thinking blocks much the same as a quantitative researcher may have problems writing questionnaire items or formulating hypotheses.

Sometimes the memo is regarded as an intermediary step between the data and the final written report. As ever in grounded theory, though, in practice the distinction between the different stages is not rigid. Often the advice is to start memo writing just as soon as anything strikes one as interesting in the data, the coding or categorisation. The sooner the better would seem to be the general consensus. This is very different from the approach taken by quantitative researchers. Also bear in mind that the process of theory development in grounded theory is not conventional in that the use of a small number of parsimonious concepts is not a major aim. (This is essentially Occam's razor which is the logical principle that no more than the minimum number of concepts or assumptions is necessary. This is also referred to as the principle of parsimony.) Strauss and Corbin (1999) write of conceptual density which they describe as a richness of concept development and relationship identification. This is clearly intended to be very different from reducing the analysis to the very minimum number of concepts as is characteristic of much quantitative research.

Theoretical sampling

Theoretical sampling is about how to validate the ideas developed within the memo. If the ideas in the memo have validity then they should apply to some samples of data but not to others. The task of the researcher is partly to suggest which samples the categories apply to and which they should not apply to. This will help the researcher identify new sources of data which may be used to validate the analysis to that point. As a consequence of the analysis of such additional data, subsequent memo writing may be more closely grounded in the data which it is intended to explain.

Literature review

In conventional methodological terms, the literature review is largely carried out in advance of planning the detailed research. That is, the new research builds on the accumulated previous knowledge. In grounded theory, the literature review should be carried out after the memo-writing process is over – signed, sealed and delivered. In this way, the grounded theory has its origins in the data collected *not* the previous research and theoretical studies. So why bother with the literature review? The best answer is that the literature review should be seen as part of the process of assessing the adequacy of the grounded theory analysis. If the new analysis fails to deal adequately with the older research then a reformulation may be necessary. On the other hand, it is feasible that the new analysis helps integrate past grounded theory analyses. In some respects this can be regarded as an extension of the grounded theory to other domains of applicability.

That is what some grounded theorists claim. Strauss and Corbin (1999) add that the grounded theory methodology may begin in existing grounded theory so long as they 'seem appropriate to the area of investigation' and then these grounded theories 'may be elaborated and modified as incoming data are meticulously played against them' (pp. 72–3). An overall picture of the stages of grounded theory are shown in Figure 21.2. This includes an additional stages of theory development which do not characterise all grounded theory studies in practice.

21.5 Computer grounded theory analysis

A number of commercially available grounded theory analysis programs are available. Generically they are known as CAQDAS (Computer-Assisted Qualitative Data Analysis Software). NUD*IST was the market leader but it has been replaced by NVivo, which is very similar, and there are others. These programs may help with the following aspects of a grounded theory analysis:

- There is a lot of paperwork with grounded theory analysis. Line-numbered transcripts are produced, coding categories are developed, and there is much copying and pasting of parts of the analysis in order to finely tune the categories to the data. There is

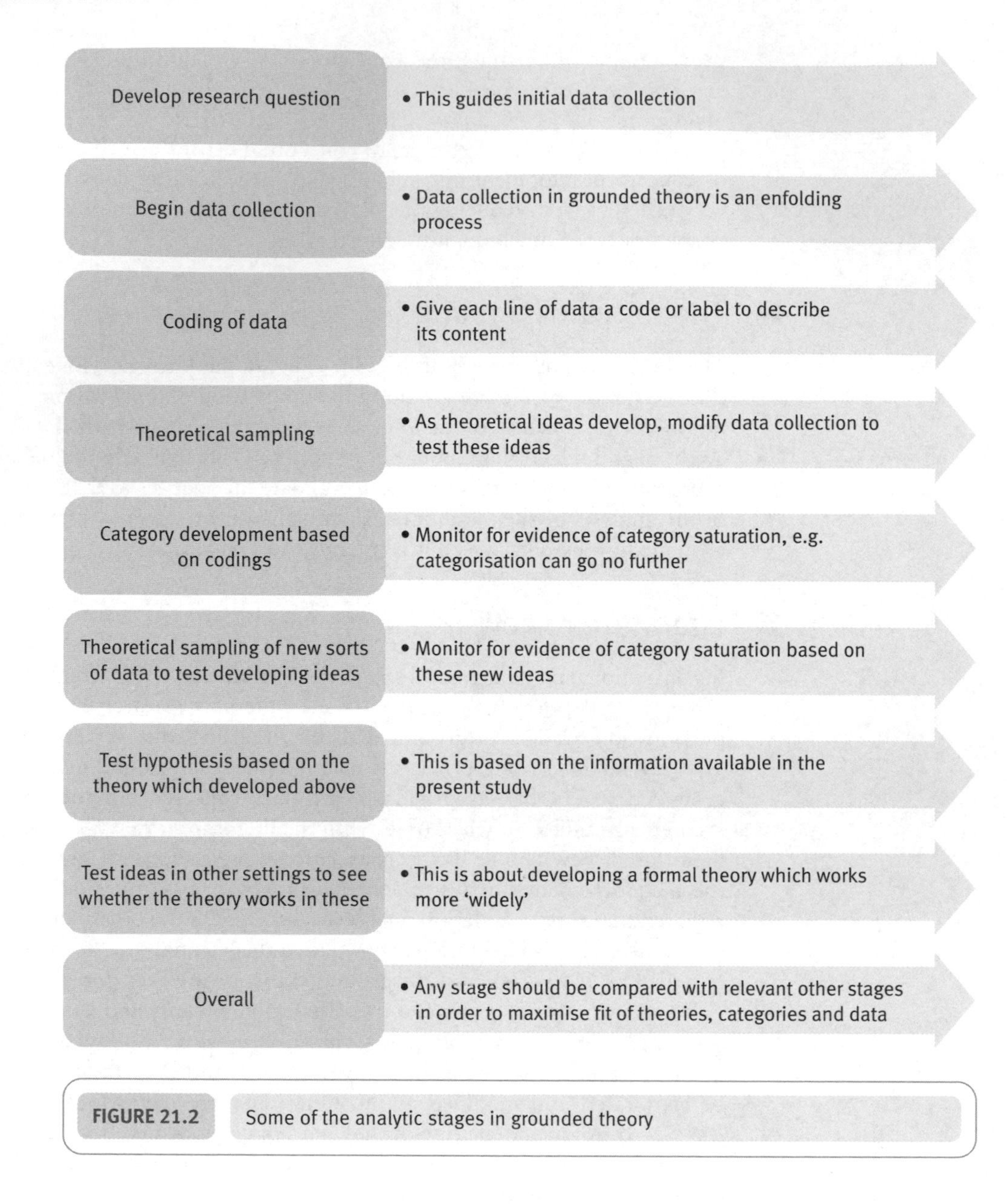

Develop research question
• This guides initial data collection
Begin data collection
• Data collection in grounded theory is an enfolding process
Coding of data
• Give each line of data a code or label to describe its content
Theoretical sampling
• As theoretical ideas develop, modify data collection to test these ideas
Category development based on codings
• Monitor for evidence of category saturation, e.g. categorisation can go no further
Theoretical sampling of new sorts of data to test developing ideas
• Monitor for evidence of category saturation based on these new ideas
Test hypothesis based on the theory which developed above
• This is based on the information available in the present study
Test ideas in other settings to see whether the theory works in these
• This is about developing a formal theory which works more 'widely'
Overall
• Any stage should be compared with relevant other stages in order to maximise fit of theories, categories and data

FIGURE 21.2 Some of the analytic stages in grounded theory

almost inevitably a large amount of textual material to deal with – a single focus group, for example, might generate 10 or 20 transcribed pages. Computers, as everyone knows, are excellent for cutting down on paper when drafting and shifting text around. That is, the computer may act as a sort of electronic office for grounded analyses.

- One key method in grounded theory is searching for linkages between different aspects of the data. A computer program is eminently suitable for making, maintaining and changing linkages between parts of a document and between different documents.
- Coding categories are developed but frequently need regular change, refinement and redefinition in order for them to fit the data better and further data that may be introduced perhaps to test the categories. Using computer programs, it is possible to recode the data more quickly, combine categories and the like.

Box 21.1 discusses computer-based support for grounded theory analysis.

Box 21.1 Practical Advice

Computers and qualitative data analysis: Computer-Assisted Qualitative Data Analysis Software (CAQDAS)

Using computer programs for the analysis of qualitative data is something of a mixed blessing for students new to this form of analysis. The major drawback is the investment of time needed to learn the software. This is made more of a problem because no qualitative analysis program does all of the tasks that a qualitative analyst might require. Thus it is not like doing a quantitative analysis on a computer program such as SPSS Statistics where you can do something useful with just a few minutes of training or just by following a text. Furthermore, qualitative analysis software is much more of a tool to help the researcher whereas SPSS Statistics, certainly for simple analyses, does virtually all of the analysis. So think carefully before seeking computer programs to help with your qualitative analysis, especially if time is short, as it usually is for student projects. There is little or nothing that can as yet be done by computers which cannot be done by a researcher using more mundane resources such as scissors, glue, index cards and the like. The only major drawback to such basic methods is that they become unwieldy with very substantial amounts of data. In these circumstances, a computer may be a major boon in that it keeps everything neat and tidy and much more readily accessible on future occasions.

There are two main stages for which computer programs may prove helpful: data entry and data analysis.

Data entry

All students will have some word-processing skills which may prove helpful for a qualitative analysis. The first major task after data have been collected is to transcribe it. This is probably best done by a word processing program such as Microsoft's Word which is by far the most commonly used of all such programs. Not only is such a program the best way of getting a legible transcript of the data but it is also useful as a resource of text to be used in illustrating aspects of the analysis in the final research report. Word-processing programs can be used for different sorts of transcription including Jefferson transcription (see Chapter 19) which utilises keyboard symbols universally available. Of course, the other big advantage of word-processing programs is that they allow for easy text manipulation. For example, one can usually search for (find) strings of text quickly. More importantly, perhaps, one can copy-and-paste any amount of text into new locations or folders. In other words, key bits of text can be brought together to aid the analysis process simply by having the key aspects of the text next to each other.

Computers can aid data entry in another way. If the data to be entered are already in text form (e.g. magazine articles or newspaper reports) then it may be possible to scan the text directly into the computer using programs such as TextBridge. Alternatively, some may find it helpful to use voice recognition software to dictate such text into a computer as an alternative to typing. Of course, such programs are still error-prone but then so is transcribing words from the recording by hand. All transcripts require checking for accuracy no matter how they are produced.

In the case of discourse analysis or conversation analysis (see Chapters 22 and 23) features of the data such as pauses, voice inflections and so forth need to be recorded. So editing software such as CoolEdit (now known as Adobe Audition) are useful for these specialised transcription purposes. This program, for example, has the big advantage that it shows features of the sound in a sort of continuous graphical form which allows for the careful measurement of times of silences and so forth. There is a free-to-download computer program which helps one transcribe sound files. It is known as SoundScriber and can be obtained at http://www-personal.umich.edu/~ebreck/sscriber.html.

The downside is that by saving the researcher time, the computer program reduces their familiarity with their data. This undermines one of the main strategies of qualitative analysis which is to encourage the researcher to repeatedly work through the analysis in ways which encourage greater familiarity.

Data analysis

There are many different forms of analysis of qualitative data so no single program is available to cope with all of these. For example, some analyses simply involve counts of how frequently particular words or phrases (or types

of words or phrases) occur in the data or how commonly they occur in close physical proximity. Such an analysis is not typical of what is done in psychology and, of course, it is really a type of quantitative analysis of qualitative data rather than a qualitative data analysis as such. The most common forms of qualitative analysis tend to involve the researcher labelling the textual data in some way (coding) and then linking the codings together to form broader categories which constitute the bedrock of the analysis. The most famous of the computer programs helping the researcher handle grounded theory analyses is NUD*IST, which was developed in the 1980s, and NVivo, which was developed a decade or so later but is closely related to NUD*IST. The researcher transcribes their data (usually these will be interviews) and enters the transcription into one of these programs usually using RTF (rich text format) files which Word can produce. The programs then allow you to take the text and code or label small pieces of it. Once this is complete the codes or labels can be grouped into categories (analogous to themes in thematic analysis – Chapter 20). The software company which owns NVivo suggests that it is useful for researchers to deal with rich-text data at a deep level of analysis. They identify some of the qualitative methods discussed in this book as being aided by NVivo such as grounded theory, conversation analysis, discourse analysis and phenomenology. The software package is available in a student version at a moderate price but you may find that a trial download is sufficient for your purposes. This was available at the following address at the time of writing: http://www.qsrinternational.com/products_free-trial-software.aspx.

The system allows the user to go through the data coding the material a small amount at a time (the unit of analysis can be flexible) or one can develop some coding categories in a structured form before beginning to apply them to the data. In NVivo there is the concept of *nodes* which it defines as 'places where you store ideas and categories'. There are two important types of nodes in the program which are worth mentioning here:

- *Free nodes* These can most simply be seen as codings or brief verbal descriptions or distillations of a chunk of text. These are probably best used at the start of the analysis before the researcher has developed clear ideas about the data and the way the analysis is going.

- *Tree nodes* These are much more organised than the free nodes (and may be the consequence of joining together free nodes). They are in the form of a hierarchy with the parent node leading to the children nodes which may lead to the grandchildren nodes. The hierarchy is given a numerical sequence such as 4 2 3 where the parent node is here given the address 4, one of the child nodes is given the address 2 and where the grandchild is given the address 3. Thus 4 2 3 uniquely identifies a particular location within the tree node. So, for example, a researcher may have as the parent node *problems at work*, one of the child nodes may be *interpersonal relationships* (another child node might be *redundancy*, for example), and one of the grandchild nodes might be *sexual harassment*.

These nodes are not fixed until the researcher is finally satisfied but can be changed, merged together or even removed should the researcher see fit. This is the typical process of checking and reviewing that makes qualitative research both flexible and time-consuming.

NVivo has other useful features such as a 'modeller' which allows the researcher to link the ideas/concepts developed in the analysis together using connectors which the researcher labels. There is also a search tool which allows the researcher to isolate text with particular contents or which has been coded in a particular way.

An alternative to NUD*IST/NVivo is to use CDC EZ-Text which is free-to-download at http://www.cdc.gov/hiv/SOFTWARE/ez-text.htm if you want to access a qualitative analysis program without the expense of the commercial alternatives. EZ-Text is available for researchers to create and manage databases for semi-structured qualitative interviews and then analyse the data. The user acts interactively with the computer during the process of developing a list of codes to be applied to the data (i.e. creating a codebook) which can then be used to give specific codes to passages in the material. The researcher can also search the data for passages which meet the researcher's own preset requirements. In many respects, this is similar to NVivo.

There is no quick fix for learning any of these systems. There are training courses for NVivo, for example, lasting several days, which suggests that the systems cannot be mastered quickly.

Example of a NVivo/NUD*IST analysis

Pitcher and her colleagues (2006) studied sex work by using focus group methodology (see Chapter 18) in which residents in a particular area talked together under the supervision of a facilitator. NUD*IST was used to analyse the data originally. In order to demonstrate NVivo, we have taken a small section of their data and re-analysed it. This is shown in the screenshot (Figure 21.3). Of course, different researchers with different purposes may analyse the same qualitative data very differently. We have done the most basic coding by entering free nodes for the interview

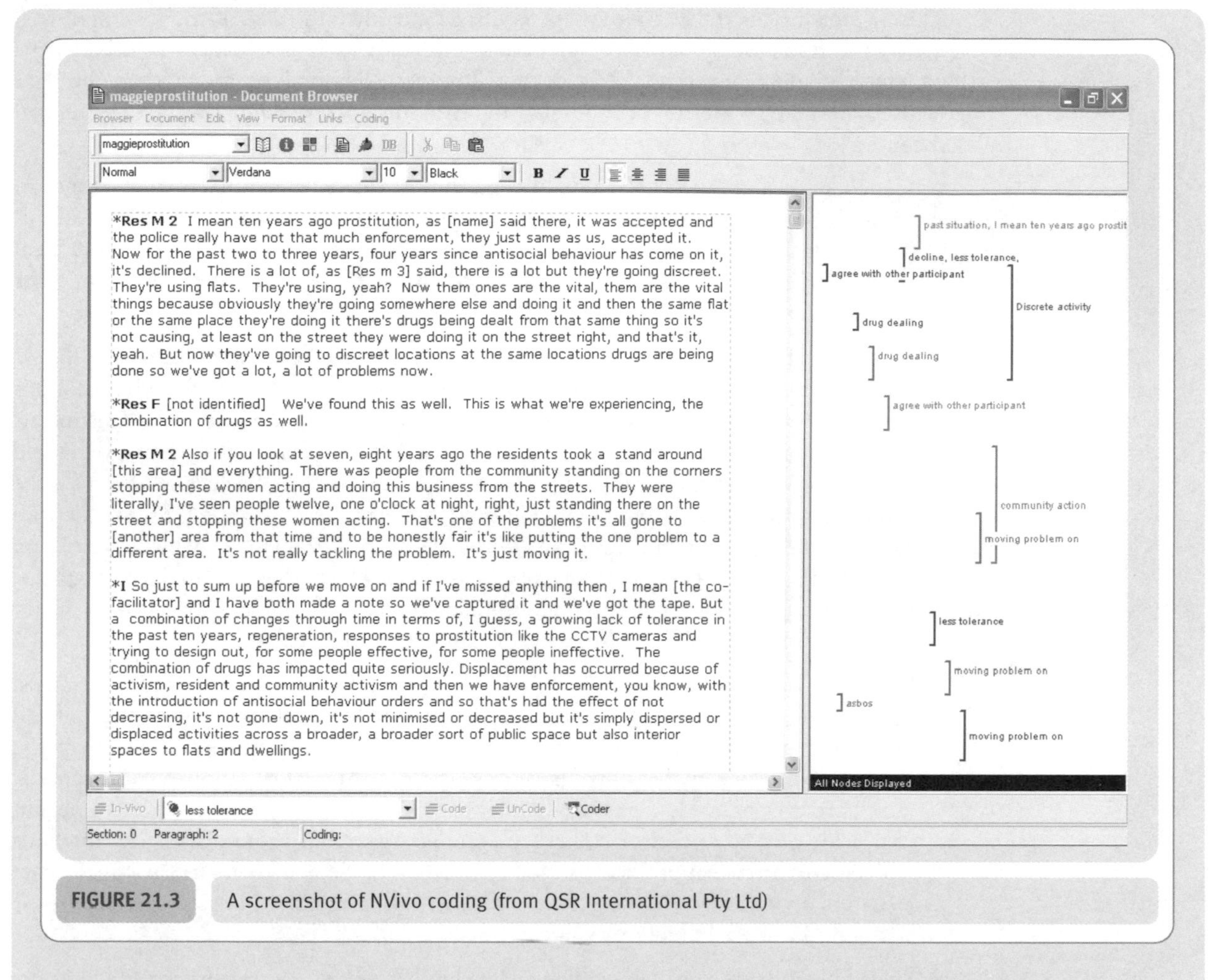

FIGURE 21.3 A screenshot of NVivo coding (from QSR International Pty Ltd)

passage. This will give you some idea of how complex even this initial coding can be with NVivo – notice the pane at the side of the screenshot where the sections coded are identified between horizontal square brackets. Also, notice how the sections coded can overlap. It is possible to give several distinct codings or free nodes to the same selection of text. Basically, the researcher highlights a section of text, chooses an existing coding for that section or adds a new coding by typing in the lower box, and then selects the code. Of course, this is just the start since the researcher may wish to revise the codings, put the codings (free nodes) into a tree node structure, identify all of the text with a particular coding and so forth.

We are grateful to Maggie O'Neil and Jane Pitcher for help with this box.

21.6 Evaluation of grounded theory

Potter (1998) points out that central to its virtues is that grounded theory:

> . . . encourages a slow-motion reading of texts and transcripts that should avoid the common qualitative research trap of trawling a set of transcripts for quotes to illustrate preconceived ideas.
>
> (p. 127)

This is probably as much a weakness as a strength since the size of the task may well defeat the resources of novices and others. Certainly it is not always possible to be convinced that preconceived ideas do not dominate the analysis rather than the data leading the analysis. There are a number of criticisms which seem to apply to grounded theory:

- It encourages a pointless collection of data, that is, virtually anything textual or spoken could be subject to a grounded theory analysis. There are no clear criteria for deciding, in advance, what topics to research on the basis of their theoretical or practical relevance. Indeed, the procedures tend to encourage the delay of theoretical and other considerations until after the research has been initiated.
- Potter (1998) suggests that 'The method is at its best where there is an issue that is tractable from a relatively common sense actor's perspective . . . the theoretical notions developed are close to the everyday notions of the participants' (p. 127). This means that commonsensical explanations are at a premium – explanations which go beyond common sense may be squeezed out. Potter puts it another way elsewhere 'how far is the grounding derived not from theorizing but from reproducing common sense theories as if they were analytic conclusions?' (Potter, 1998, p. 127). This may be fair criticism. The difficulty is that it applies to any form of research which gives voice to its participants. Ultimately, this tendency means that grounded theory may simply codify how ordinary people ordinarily understand the activities in which they engage.
- There is a risk that grounded theory, which is generally founded on admirable ideals, is used to excuse inadequate qualitative analyses. It is a matter of faith that grounded theory will generate anything of significant value, yet at the same time, done properly, a grounded theory analysis may have involved a great deal of labour. Consequently, it is hard to put aside a research endeavour which may have generated little but cost a lot of time and effort. There are similar risks that grounded theory methods will be employed simply because the researcher has failed to focus on appropriate research questions, so leaving themself with few available analysis options. These risks are particularly high for student work.
- Since talk and text are analysed line by line (and these are arbitrarily divided – they are not sentences, for example) the researcher may be encouraged to focus on small units rather than the larger units of conversation as, for example, favoured by discourse analysts (Potter, 1998). Nevertheless, grounded theory is often mentioned by such analysts as part of their strategy or orientation.

So it is likely that grounded theory works best when dealing with issues that are amenable to common-sense insights from participants. Medical illness and interpersonal relationships are such topics where the theoretical ideas that grounded theory may develop are close to the ways in which the participants think about these issues. This may enhance the practicality of grounded theory in terms of policy implementation. The categories used and the theoretical contribution are likely to be in terms which are relatively easy for the practitioner or policymaker to access.

21.7 Conclusion

Especially pertinent to psychologists is the question of whether grounded theory is really a sort of Trojan horse which has been cunningly brought into psychology, but is really the enemy of advancement in psychology. Particularly troubling is the following from Strauss and Corbin (1999):

> . . . grounded theory researchers are interested in patterns of action and interaction between and among various types of social units (i.e., 'actors'). So they are not especially interested in creating theory about individual actors as such (unless perhaps they are psychologists or psychiatrists).
>
> (p. 81)

Researchers such as Strauss and Corbin are willing to allow a place for quantitative data in grounded theory. So the question may be one of how closely psychological concepts could ever fit with grounded theory analysis which is much more about the social (interactive) than the psychological.

Key points

- Grounded theory is an approach to analysing (usually textual) data designed to maximise the fit of emerging theory (categories) to the data and additional data of relevance.
- The aim is to produce 'middle range' theories which are closely fitting qualitative descriptions (categories) rather than, say, cause-and-effect or predictive theories.
- Grounded theory is 'inductive' (that is, does not deduce outcomes from theoretical postulates). It is systematic in that an analysis of some sort will almost always result from adopting the system. It is a continuous process of development of ideas – it does not depend on a critical test as in the case of classic psychological theory.
- Comparison is the key to the approach – all elements of the research and the analysis are constantly compared and contrasted.
- Coding (or naming or describing) is the process by which lines of the data are given a short description (or descriptions) to identify the nature of their content.
- Categorisation is the process by which the codings are amalgamated into categories. The process helps find categories which fit the codings in their entirety, not simply a few pragmatic ideas which only partially represent the codings.
- Memo writing is the process by which the researcher records their ideas about the analysis throughout the research process. The memo may include ideas about categorisation but it may extend to embrace the main themes of the final report.
- Computer programs are available which help the researcher organise the materials for the analysis and effectively alter the codings and categories.
- A grounded theory analysis may be extended to further critical samples of data which should be pertinent to the categories developed in the analysis. This is known as theoretical sampling.
- The theoretical product of grounded theory analysis is not intended to be the same as conventional psychological theorisation and so should not be judged on those terms.

ACTIVITY

Grounded theory involves the bringing of elements together to try to forge categories which unite them. So choose a favourite poem, song or any textual material, and write each sentence on a separate sheet of paper. Choose two at random. What unites these two sentences? Then choose another sentence. Can this be united with the previous two sentences? Continue the exercise until you cease coming up with new ideas. Then start again.

CHAPTER 22

Discourse analysis

Overview

- There are two main forms of discourse analysis in psychology. One is the social constructionist approach of Potter and Wetherell (1995) which this chapter concentrates upon as it is the more student friendly. The other approach is that of Foucauldian discourse analysis which is rather more demanding and is not accessible so readily by newcomers. Discourse analysis refers to a variety of ways of studying and understanding talk (or text) as social action.
- The intellectual roots of discourse analysis are largely in the linguistic philosophy of the 1960s. Most significant in this regard is the conceptualisation that discourse is designed to do things linguistically and that the role of the discourse analysts is to understand what is being done and how it is done through speech and text.
- At one level, discourse analysis may be regarded as a body of ideas about the nature of talk and text which can be applied to the great mass of data collected by psychologists in the form of language.
- A number of themes are common in discourse analysis – these include rhetoric, voice, footing, discursive repertoires and the dialogical nature of talk.
- The practice of discourse analysis involves a variety of procedures designed to encourage the researcher to process and reprocess their material. These include transcription, coding and re-coding.

22.1 Introduction

There are two major types of discourse analysis in psychology. They are equally important but one is less amenable to the needs of students new to qualitative research whereas the other has numerous applications to student work throughout psychology. We shall, of course, devote more of this chapter to the latter than to the former. The first sort of discourse analysis originates in the work of Michel Foucault (1926–1984), the French academic. Foucault was trained in psychology at one stage though he is better described as a philosopher and historian. His work focused on critical studies of social institutions such as psychiatry, prisons, human sciences and medicine. Foucault saw in language the way in which institutions enforce their power. So for example, he argued that the nineteenth century medical treatment of mentally ill people was not a real advance on the crudity and brutality which characterised the treatment of 'mad' people before this time. Mental illness had become a method of controlling those who challenged the morality of bourgeois society. Madness is not a permanently fixed social category but something that is created through discourse to achieve social objectives. Foucauldian discourse analysis largely made its first appearance in psychology in the book *Changing the Subject: Psychology, social regulation, and subjectivity* (1984) by Henriques, Hollway, Urwin, Venn and Walkerdine. Other examples of the influence of Foucauldian ideas on psychology can be found in the work of Ian Parker, particularly in the field of mental health, in the books *Deconstructing Psychopathology* (Parker, Georgaca, Harper, McLaughlin and Stowell-Smith, 1995) and *Deconstructing Psychotherapy* (Parker, 1999). This branch of discourse analysis is closely linked to critical discourse analysis.

The other approach to discourse analysis also has its roots outside psychology. It can be described as a social constructionist approach to discourse analysis and came into psychology through the work of Jonathan Potter and Margaret Wetherell (1987). It is this that we will concentrate upon. According to *The Oxford Companion to the English Language* (McArthur, 1992) its early origins lie in the work which extended linguistics into the study of language beyond the individual sentence. Perhaps the earliest of these was Zellig Harris who worked in the USA in the 1950s on issues such as how text is related to the social situation within which it was created. A little later, a linguistic anthropologist, Dell Hymes, investigated forms of address between people – that is speech as it relates to the social setting. In the 1960s, a group of British linguistic philosophers (J. L. Austin, J. R. Searle and H. P. Grice) began to regard language as social *action* – a key idea in discourse analysis. Potter (2001) takes the roots back earlier to the first half of the twentieth century by adding the philosopher Ludwig Wittgenstein into the melting pot of influences on discourse analysis theory. Particularly important is Wittgenstein's idea that language is a toolkit for *doing* things rather than a means of merely representing things (see Figure 22.1).

It is a largely wasted effort to search for publications on discourse analysis by psychologists before the 1980s. The term first appears in a psychology journal in 1954 but applied to ideas very different from its current meaning. However, despite its late appearance in psychology, the study of discourse began to grow rapidly in the 1960s in other disciplines. Consequently, discourse analysis draws from a variety of disciplines, each of which has a different take or perspective on what discourse analysis is. Because of the pan-disciplinary base for discourse analysis, a psychologist, no matter how well versed in other aspects of psychology, may feel overawed when first exploring the approach. Not all psychologists have, say, the detailed knowledge of linguistics, sociology and other disciplines that contribute to the field. Furthermore the distinctive contribution made by psychologists to discourse analysis is not always clear.

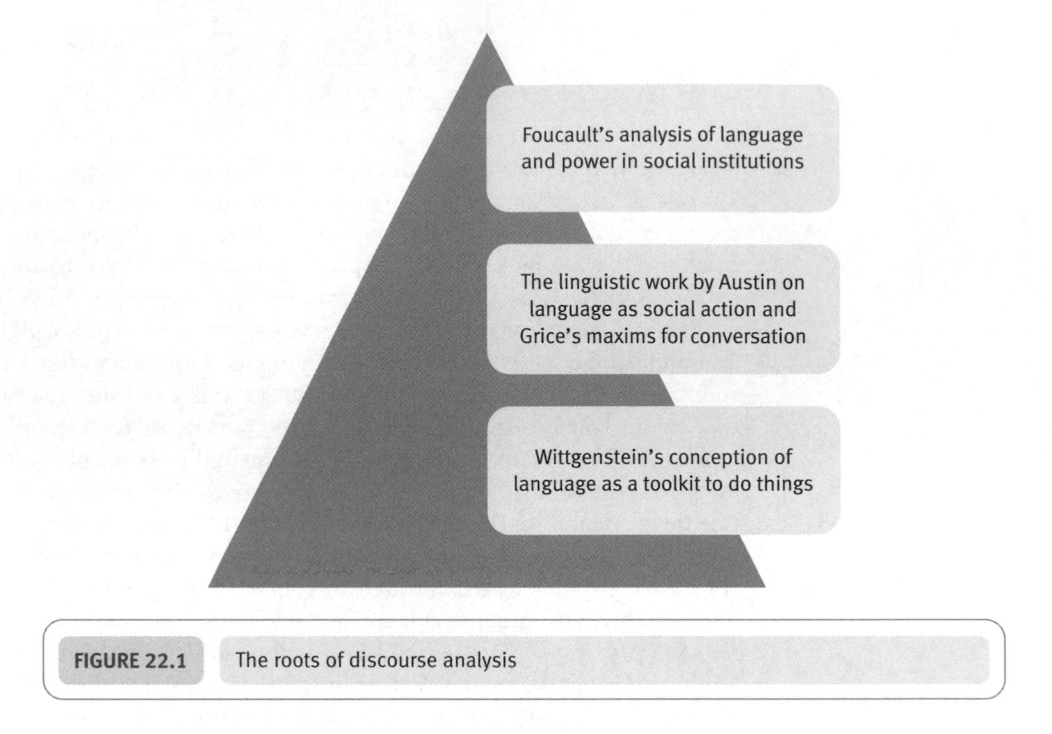

FIGURE 22.1 The roots of discourse analysis

So discourse analysis studies language in ways very different from traditional linguistics. The latter is largely concerned with:

- word sounds (phonetics and phonology);
- units which make up words (morphology);
- meaning (semantics);
- word order within sentences (syntax).

Beaugrande (1996) refers to this traditional style of linguistics, in a somewhat derogatory fashion, as 'language science'. For him, traditional linguistics was at fault for profoundly disconnecting language from real-life communications – to just study sounds, words and sentences. In discourse analysis, language is regarded as being much more complex. Discourse is how language operates in real-life communicative events. Discourse analysis involves the analysis of speech, text and conversation so its concerns are with analyses beyond the level of the sentence. Hence, Stubbs (1983, p. 1) defines discourse analysis as being about the way in which language is used at the broader level than the sentence and other immediate utterances. A good illustration of this (Tannen, 2007) are the two signs at a swimming pool:

Please use the toilet, not the pool.
Pool for members only.

Considered separately, just as sentences, the signs convey clear messages. However, if they are read as a unit of two sentences, then they either constitute a request for non-members to swim in the toilet or an indication that members have exclusive rights to urinate in the swimming pool! Analysing just two sentences at a time as in this example causes us to re-write the meaning of both sentences.

Discourse analysis emphasises the ways in which language interacts with society, especially in the nature of dialogue in ordinary conversation. Above all, then, discourse analysis is a perspective on the nature of language. Discourse analysis does *not* treat language as if it were essentially representational – language is not simply the means of articulating internal mental reality. Quite the reverse, discourse analysis is built on the idea that truth and reality are *not* identifiable or reachable through language.

Language for the discourse analyst is socially situated and it matters little whether the text under consideration is natural conversation or written text in the form, say, of newspaper headlines. Each of these provides suitable material for analysis. Since discourse analysis is a shift in the way language is conceptualised, what language does becomes more important than what is represented by language. Researchers in this field have different emphases, but it is typically the case that language is regarded as doing things and especially it is regarded as doing things in relation to other people in a conversation. As such, discourse analysis employs 'speech act theory' (Austin, 1975) which regards language as part of a social performance. In the course of this performance, social identities are formed and maintained, power relations are created, exercised and maintained, and generally a lot of social work is done. None of these can be effectively construed as language simply and directly communicating internal thought. Language in this context becomes a practice, not a matter of informal or formal structures.

The phrase discourse analysis may arouse pictures of a well-established empirical method of analysing people talking, interviews and so forth. In a sense it is. However, discourse analysis is *not* merely a number of relatively simple analytical skills that can quickly be learnt and easily applied. Some aspects are relatively simple. For example, transcription methods such as Jefferson's (Chapter 19) are relatively easily assimilated. Discourse analysts themselves have frequently presented the definition and practice of discourse analysis as problematic. For example, Edley (2001) wrote:

> . . . there is no simple way of defining discourse analysis. It has become an ever broadening church, an umbrella term for a wide variety of different analytic principles and practices.
>
> (p. 189)

and Potter (2004) adds to the sense of incoherence:

> . . . in the mid 80s it was possible to find different books called Discourse Analysis with almost no overlap in subject matter; the situation at the start of the 00s is, if anything, even more fragmented.
>
> (p. 607)

and Taylor (2001) reinforces the view that discourse analysis should not be regarded as merely another variant or sub-discipline of methods:

> . . . to understand what kind of research discourse analysis is, it is not enough to study what the researcher does (like following a recipe!). We also need to refer back to these epistemological debates and their wider implications.
>
> (p. 12)

Even discourse analysts with backgrounds in psychology do not offer a united front on its nature. Different viewpoints exist partly because they draw on different intellectual roots. Consequently, one needs to be aware that there is no consensus position which can be identified as the core of discourse analysis. This would help heighten and facilitate the theoretical debate, it is not a criticism.

22.2 Important characteristics of discourse

Within discourse analysis, there is a variety of ideas, observations and concepts that broadly suggest ways in which language and text should be analysed. These are discussed in this section.

Speech acts

Of particular importance to psychologists working in discourse analysis is the theory of *speech acts*. This originated in the work of Austin (1975). Performatives was his term for utterances which have a particular social effect. For Austin, all words perform social acts:

- *Locution* This is simply the act of speaking.
- *Illocution* This is what is done by saying these words.
- *Perlocution* This is the effect or consequence on the hearer who hears the words.

To speak the words is essentially to do the act (speech act). For the words to have an effect certain conditions have to be met. These were illustrated in Searle (1969) using the following utterances:

> Sam smokes habitually
> Does Sam smoke habitually?
> Sam. Smoke habitually!
> Would that Sam smoked habitually.

Saying the words constitutes an *utterance act* or *locutory act*. They constitute the act of uttering something. At the same time, they are also propositional acts because they refer to something and predicate something. Try introducing any of them into a conversation with a friend. It is unlikely that the sentence could be said without some sort of response from the other person on the topic of Sam, smoking or both. Each of the sentences also constitutes an *illocutory act* because they do things like state, question, command, promise, warn, etc.

For example, in some contexts each of these sentences may contain an unstated indication that the speaker wishes to do something about Sam's smoking. 'Would that Sam smoked habitually' may be a slightly sardonic way of suggesting that it would be great if Sam could be persuaded to smoke as infrequently as habitually! Equally, if they were uttered by a shopkeeper about his customer Sam who spends large amounts on chocolates for his wife every time he calls in for a packet of cigarettes, the same words would constitute a different *illocutory act*. The sentences are also *perlocutionary* acts in that they have an effect or consequence on the hearer though what this effect is depends on circumstances. For example, the locution 'Sam smokes habitually' might be taken as a warning that the landlord is unhappy about Sam smoking in violation of his tenancy contract if said by Sam's landlord to Sam's wife. In speech act theory, the indirect nature of speech is inevitably emphasised since the interaction between language and the social context is partially responsible for meaning.

Grice's maxims of cooperative speech

Another contribution which comes from the philosophy of language are Grice's (1975) maxims. These indicate something of the rule-based nature of exchanges between people. The overriding principle is conversational cooperativeness over what is being communicated at the time. Cooperativeness is achieved by obeying four maxims:

- Quality, which involves making truthful and sincere contributions.
- Quantity, which involves the provision of sufficient information.
- Manner, which involves making one's contributions brief, clear and orderly.
- Relation, which involves making relevant contributions.

These maxims contribute to effective communications. They are not the only principles underlying social exchanges. For example, unrestrained truthfulness may offend when it breaches politeness standards in conversation.

Face

Similarly, language often includes strategies for protecting the status of the various participants in the exchange. The notion of 'face' is taken from the work of Goffman (1959) to indicate the valued persona that individuals have. We speak of saving face in everyday conversation. Saving face is a collective phenomenon in which all members of the conversation may contribute, not simply the person at risk of losing face.

Register

The concept of register highlights the fact that when we speak the language style that we use varies according to the activity being carried out. The language used in a lecture is not the same as that used when a sermon is being given. This may be considered a matter of style but it is described as register. Register has a number of components including:

- field of activity (for example, police interview, radio interview);
- medium used (for example, spoken language, written language, dictated language);
- tenor of the role relationship in a particular situation (for example, parent–child, police officer–witness).

22.3 The agenda of discourse analysis

It should be clear by now that discourse analysis is not a simple to learn, readily applied technique. Discourse analysis is a body of theory and knowledge accumulated over a period of 50 years or more. It is not even a single, integrated theory. Instead it provides an analytical focus for a range of theories contributed by a variety of disciplines such as philosophy, linguistics, sociology and anthropology. Psychology as somewhat a latecomer to the field is in the process of setting its own distinctive stamp on discourse analysis. What is the agenda for a discourse analysis-based approach to psychology? To re-stress the point, there are no short cuts to successful discourse analysis. Intellectual and theoretical roots are more apparent in the writings of discourse analysts than practically any other field of psychology. In other words, the most important *practical* step in using discourse analysis is to immerse oneself in its theoretical and research literature. One cannot just get on with doing discourse analysis. Without understanding in some depth the constituent parts of the discourse analytic tradition, the objectives of discourse analysis cannot be appreciated fully. This would be equally true of designing an experiment – we need to understand what it does, how it does it, why it does it, and when it is inappropriate.

The agenda of psychological discourse analysis according to Potter and Wetherell (1995) includes the following:

- *Practices and resources* Discourse analysis is not simply an approach to the social use of language. It focuses on discourse practices – the things that people do in talk and writings. But it also focuses on the resources that people employ when achieving these ends. For example, the strategies employed in their discourse, the systems of categorisation they use, and the interpretative repertoires that are used. Interpretative repertoires are the 'broadly discernible clusters of terms, descriptions, and figures of speech often assembled around metaphors or vivid images' (Potter and Wetherell, 1995, p. 89). So, for example, when newspapers and politicians write and talk about drugs they often use the language repertoire of war and battle. Hence the 'war on drugs', 'the enemy drugs' and so forth. Discourse analysis seeks also to provide greater understanding of traditional (socio-)psychological topics such as the nature of individual and collective identity, how to conceive social action and interaction, the nature of the human mind, and constructions of the self, others and the world.
- *Construction and description* During conversation and other forms of text, people create and construct 'versions' of the world. Discourse analysis attempts to understand and describe this constructive process.
- *Content* Talk and other forms of discourse are regarded as the important site of psychological phenomena. No attempt is made to postulate 'underlying' psychological

Box 22.1 Key Ideas

Critical discourse analysis

The term 'critical discourse analysis' has a rather narrower focus than the words might imply. Critical does *not* mean crucial in this context and neither does it imply a generally radical stance within the field of discourse analysis. Critical discourse analysis is simply a school of thought which emphasises *power and social inequality* in the interpretation of discourse. Critical discourse analysis studies the way in which power is attained and maintained through language. According to van Dijk (2001), *dominance* is the exercise of social power by elites, institutions and social groups. Consequences of the exercise of power include varieties of social inequality – ethnic, racial and gender inequality, for example. Equally, language has the potential to serve the interests of disadvantaged groups in order to redress and change the situation. The phrase 'black is beautiful', for example, was used to counter destructive effects on the self-image of black children of racist views of themselves.

Dominance is achieved using language in a variety of ways. For example, the way a group is represented in language, how language is used to reinforce dominance, and the means by which language is used to deny and conceal dominance:

> . . . critical discourse analysts want to know what structures, strategies or other properties of text, talk, verbal interaction or communicative events play a role in these modes of reproduction.
>
> (van Dijk, 2001, p. 300)

Brute force is an unacceptable means of achieving power in modern society. Persuasion and manipulation through language provide a substitute for violence, and so power is achieved and maintained through language.

Others extend the notion of 'critical' to include a wider set of concerns than van Dijk. Hepburn (2003) suggests that it also includes issues of politics, morality and social change. These are all components or facets of van Dijk's notion of power. Concerns such as these link to a long tradition of concerned psychology in which power, politics, morality and social change are staple ingredients (for example, Howitt, 1992a).

mechanism to explain that talk. So racist talk is regarded as the means by which discrimination is put into practice. There is no interest in 'psychological mechanisms' such as authoritarian personalities or racist attitudes.

- *Rhetoric* Discourse analysis is concerned with how talk can be organised so as to be argumentatively successful or persuasive.
- *Stake and accountability* People regard others as having a vested interest (stake) in what they do. Hence they impute motives to the actions of others which may justify dismissing what these others say. Discourse analysis studies these processes.
- *Cognition in action* Discourse analysis actively rejects the use of cognitive concepts such as traits, motives, attitudes and memory stores. Instead it concentrates on the text by emphasising, for example, how memory is socially constructed by people, such as when reminiscing over old photographs.

This agenda might be considered a broad framework for psychological discourse analysis. It points to aspects of text which the analyst may take into account. At the same time, the hostility of discourse analysis to traditional forms of psychology is apparent.

22.4 Doing discourse analysis

It cannot be stressed too much that the objectives of discourse analysis are limited in a number of ways – especially the focus on the socially interactive use of language. In a nutshell, there is little point in doing a discourse analysis to achieve ends not shared by discourse analysis. Potter put the issue as follows:

> To attempt to ask a question formulated in more traditional terms ('what are the factors that lead to condom use among HIV+ gay males') and then use discourse analytic methods to answer it is a recipe for incoherence.
>
> (Potter, 2004, p. 607)

Only when the researcher is interested in language as social action is discourse analysis appropriate. Some discourse analysts have contributed to the confusion by offering it as a radical and new way of understanding psychological phenomena; for example, when they suggest that discourse analysis is anti-cognitive psychology. This, taken superficially, may imply that discourse analysis supersedes other forms of psychology. It is more accurate to suggest that discourse analysis performs a different task. A discourse analysis would be useless for biological research on genetic engineering but an excellent choice to look at the ways in which the moral and ethical issues associated with genetic engineering are dealt with in speech and conversation.

Although a total novice can easily learn some of the skills of discourse analysis (for example, Jefferson transcription), doing a good discourse analysis is far harder. Reading is crucial but experience and practice also play their parts. The best way to get a feel of what a discourse analysis is would be to read the work of the best discourse analysts. That would apply to all forms of research. A report of a discourse analysis makes frequent reference to research and theory in the field. We have outlined some of this but ideas are developing continually.

There is a degree of smog surrounding the steps by which a discourse analysis is done. It is not like calculating a *t*-test or chi-square in a step-by-step fashion. No set of procedures exists which, if applied, guarantee a successful discourse analysis. Potter (2004) puts it as follows:

> There is no single recipe for doing discourse analysis. Different kinds of studies involve different procedures, sometimes working intensively with a single transcript, other times drawing on a large corpus. Analysis is a craft that can be developed with different degrees of skill. It can be thought of as the development of sensitivity to the occasioned and action-oriented, situated, and constructed nature of discourse. Nevertheless, there are a number of ingredients which, when combined together are likely to produce something satisfying.
>
> (p. 611)

The use of the word 'craft' suggests the carpenter's workshop in which nothing of worth can be produced until the novice has learnt to sharpen tools, select wood, mark out joints, saw straight and so forth. Likewise in discourse analysis, the tools are slowly mastered. Actually, Potter (1997) put it even more graphically when he wrote that for the most part:

> . . . doing discourse analysis is a craft skill, more like bike riding or chicken sexing than following the recipe for a mild chicken rogan josh.
>
> (p. 95)

Publications which seek to explain the process of discourse analysis often resort to a simple strategy: students are encouraged to apply the concepts developed by major contributors to discourse theory. In other words, tacitly the approach encourages the novice discourse analyst to find instances in their data of key discourse analytic concepts. For example, they are encouraged to identify what register is being employed, what awkwardness is shown in the conversation, what rhetorical devices are being used, what discourse repertoires are being employed and so forth. Some of this is redolent of the work of expert discourse analysts. If one initially attempts to understand the text under consideration using standard discourse analytic concepts, the later refinement and extension of these concepts will be facilitated. Areas which are problematic in terms of the application of standard concepts will encourage the revision of those concepts. At this stage, the analysis may begin to take a novel turn.

The features of language which need to be examined in a discourse analysis can be listed. This may be described as an itinerary for discourse analysis (Wetherell and Taylor, 2001). Where to go and what to look for, as well as the dead-ends that should be ignored, are part of this. So one may interrogate one's data to find out which of the following are recognisable or applicable to the particular text in question:

- Language is an inappropriate way of accessing reality. Instead language should be regarded as constructive or constitutive of social life. Through discourse, individuals and groups of individuals build social relations, objects and worlds.

- Since discourse constructs versions of social reality, an important question of any text is why a particular version of reality is being constructed through language and what does this particular version accomplish?

- Meaning is produced in the context of speech and does not solely reside in a cultural storehouse of agreed definitions. Discourse analysts refer to the co-production of meaning. The analysis partly seeks to understand the processes by which meaning is created. Meaning, for example, is a 'joint production' of two or more individuals in conversation.

- Discursive practice refers to the things which happen in language to achieve particular outcomes.

- Discursive genres are the type of language extract under consideration. So the discourse analyst may study the particular features of news and how it differs from other types of language. There are cues in news speech which provide indicators that it is news speech rather than, say, a sermon (contextualisation cues).
- Footing (a concept taken from the sociologist Goffman) refers to whether the speaker talks as if they are the author of what is being said, the subject of the words that are being said, or whether they are presenting or animating the words of someone else. These different footings are not mutually exclusive and all may be present in any text.
- Speech is dialogical. That is, when we talk we combine or incorporate things from other conversations. Sometimes this is in the form of reporting what 'he said' or what 'she replied'. More often, however, the dialogical elements are indirect and not highlighted directly as such. For example, children who say something like 'I mustn't go out on my own' reflect previous conversations with their parents and teachers.

Taylor (2001) characterises discourse analysis as an 'open-ended' and circular (or iterative) process. The task of the researcher is to find patterns without a clear idea of what the patterns will be like. She writes of the 'blind faith' that there must be something there worthy of the considerable effort of analysis. The researcher will need to go through the data repeatedly 'working up' the analysis on the basis of what fits and does not fit tentative patterns. 'Data analysis is not accomplished in one or two sessions' (pp. 38–9). Taylor indicates that the process of examination and re-examination may not fit comfortably with conventional research timescales. The direction or end point of the analysis is also difficult to anticipate. She feels that qualitative data are so 'rich' (that is detailed) that there may be more worth studying in the data even when the possibilities seem to be exhausted.

The process of carrying out a discourse analysis can be summarised in the six steps illustrated in Figure 22.2. The steps are superficially straightforward but this belies the need for a level of intensity in examining and re-examining the data. What the analyst is looking for are features which stand out on reading and re-reading the transcript. These are then marked (coded) systematically throughout the transcript to be collected together later – simply by copying and pasting the excerpts from the transcript into one file, perhaps. It is this collection that is subject to the analytic scrutiny of the researcher concentrating on things like deviant cases which do not seem to fit the general pattern. When the analysis is complete, then the issue of validity may be addressed by, for example, getting the participants in the research to comment on the analysis from their perspective.

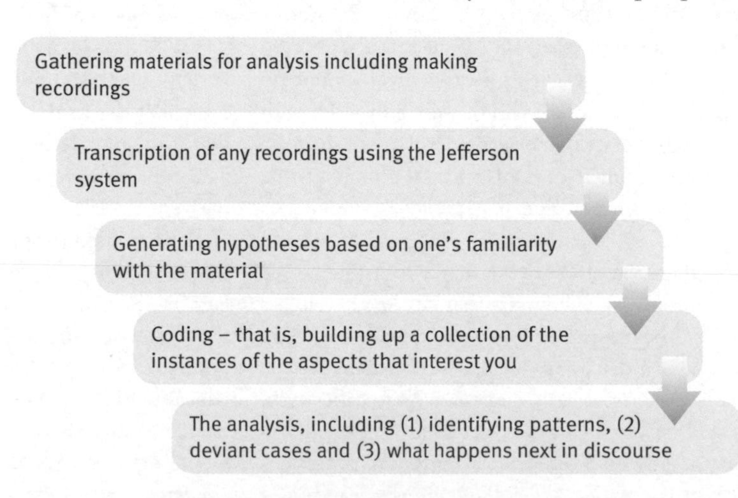
Gathering materials for analysis including making recordings
Transcription of any recordings using the Jefferson system
Generating hypotheses based on one's familiarity with the material
Coding – that is, building up a collection of the instances of the aspects that interest you
The analysis, including (1) identifying patterns, (2) deviant cases and (3) what happens next in discourse

FIGURE 22.2 The steps involved in a typical discourse analysis

Box 22.2 Research Example

Discourse analysis

In research on menstruation, Lovering (1995) talked with 11- and 12-year-old boys and girls in discussion groups. Among a range of topics included on her guide for conducting the discussions were issues to do with menstruation. These included questions such as: 'Have you heard of menstruation?'; 'What have you been told about it?'; 'What do you think happens when a woman menstruates?'; 'Why does it happen?'; and 'Who has told you?' (Lovering, 1995, p. 17). In this way relatively systematic material could be gathered in ways closer to ordinary conversation than would be generated by one-on-one interviews. She took detailed notes of her experiences as soon as possible after the discussion groups using a number of headings (p. 17):

- How she (Lovering) felt
- General emotional tone and reactions
- Non-verbal behaviour
- Content recalled
- Implications and thoughts.

This is a form of diary writing of the sort discussed already in relation to grounded theory. The difference perhaps is that she applied it to the data collection phase rather than the transcription phase. Lovering transcribed the tape-recording herself – eventually using the Jefferson system described in Chapter 19. She also employed a computer-based analysis program (of the sort that NVivo is the modern equivalent). Such a program does not do the analysis for you; it allows you to store and work with a lot of text, highlight or mark particular parts of the text, sort the text and print it out. All of these things can be achieved just using pencil and paper, but a computer is more convenient.

The next stage was to sort the text into a number of categories – initially, she had more than 50. She developed an analysis of the transcribed material partly based on her awareness of a debate about the ways in which male and female bodies are socially construed quite differently. Boys' physical development is regarded as a gradual and unproblematic process, whereas in girls the process is much more problematic. The following excerpts from a transcript illustrate this:

A: They [school teachers] don't talk about the boys very much only the girls = yes = yes.

A: It doesn't seem fair. They are laughing at us. Not much seems to happen to boys.

A: Girl all go funny shapes = yes = like that = yes.

A: Because the boys, they don't really . . . change very much. They just get a little bit bigger.

A: It feels like the girls go through all the changes because we are not taught anything about the boys REALLY.

(Lovering, 1995, pp. 23–4)

Menstruation was learnt about from other people – predominantly female teachers or mothers. Embarrassment dominated, and the impression created was that menstruation was not to be discussed or even mentioned as a consequence. Talk of female bodies and bodily functions by the youngsters features a great deal of sniggering. In contrast, when discussing male bodies things become more ordinary and more matter of fact. Furthermore, boys are also likely to use menstruation as a psychological weapon against girls. That is, menstruation is used to make jokes about and ridicule girls. In Lovering's analysis, this is part of male oppression of females: even in sex education lessons learning about menstruation is associated in girls' minds as being 'laughing at girls'.

Of course, many more findings emerged in this study. Perhaps what is important is the complexity of the process by which the analysis proceeds. It is not possible to say that if the researcher does this and then does that, a good analysis will follow. Nevertheless, it is easy to see how the researcher's ideas relate to key aspects of discourse analytic thinking. For example, the idea that menstruation is used as a weapon of oppression of females clearly has its roots in feminist sexual politics which suggests that males attempt to control females in many ways from domestic violence through rape to, in this example, sex education lessons. One could equally construe this as part of Edwards and Potter's (1993) discursive action model. This suggests, among other things, that in talk, conversation or text, one can see social action unfolding before one's eyes. One does not have to regard talk, text or conversation as the external manifestation or symptom of an underlying mental state such as an attitude. A topic such as menstruation may be seen to generate not merely hostility in the form of laughter towards the female body, but also as a means of accessing concepts of femininity in which the female body is construed as mysterious, to be embarrassed about, and sniggered over by both sexes. Horton-Salway (2001) uses the same sort of model to analyse how the medical profession went about presenting the medical condition ME in a variety of ways through language.

Discourse analysis, like other forms of qualitative analysis, is not amenable to short summaries. This might be expected given that a discourse analysis seeks to provide appropriate analytic categories for a wide range of texts. The consequence is that anyone wishing to understand discourse analytic approaches will need to read original analyses in detail.

22.5 Conclusion

One of the significant achievements of discourse analysis is that of bringing to psychology a theoretically fairly coherent set of procedures for the analysis of the very significant amounts of textual material which forms the basis of much psychological data. It has to be understood that discourse analysis has a limited perspective on the nature of this data. In particular, discourse analysts reject the idea that language is representational, that is, in language there is a representation of, say, the individual's internal psychological state. Instead they replace it with the idea that language is action and is designed to do something, not represent something. As a consequence, discourse analysis is primarily of use to researchers who wish to study language as an active thing. In this way, discourse analysis may contribute a different perspective on many psychological processes, but it does not replace or supersede the more traditional viewpoints within psychology.

Key points

- Discourse analysis is based on early work carried out by linguists, especially during the 1950s and 1960s, which reconstrued language much as a working set of resources to allow things to be done rather than regarding language as merely being a representation of something else.
- The analysis uses larger units of speech than words or sentences such as a sequence of conversational exchanges.
- Precisely how discourse analysis is defined is somewhat uncertain as it encompasses a wide range of practices as well as theoretical orientations.
- Central to most discourse analysis is the idea of speech as doing things – such as constructing and construing meaning.
- Discourse analysis has its own roadmap which should be considered when planning an analysis. Discourse practices and resources, for example, are the things that people do in constructing conversations, texts and writings. Rhetoric is about the wider organisation of language in ways that facilitate its effectiveness. Content is regarded for what it is rather than what underlying psychological states it represents.
- Critical discourse analysis has as its central focus the concept of power. It primarily concerns how social power is created, reaffirmed and challenged through language.

ACTIVITIES

1. When having a coffee with friends, note occasions when the conversation might be better seen as speech acts rather than taken literally. Better still, if you can record such a conversation, transcribe a few minutes' worth and highlight in red where literal interpretations would be appropriate and in yellow where the concept of speech act might be more appropriate.

2. Study day-to-day conversations of which you are part. Is there any evidence that participants spare other participants in the conversation embarrassment over errors? That is, to what extent does face-saving occur in your day-to-day experience?

3. Choose a chapter of your favourite novel: does discourse analysis have more implications for its contents than more traditional psychology?

CHAPTER 23

Conversation analysis

Overview

- Conversation analysis studies the structure of conversation by the detailed examination of successive turns or contributions to a conversation.
- It is based on ethnomethodological approaches derived from sociology that stress the importance of participants' understandings of the nature of the world.
- Many of the conventions of psychological research are turned on their head in conversation analysis. The primacy of theory in developing research questions and hypotheses is replaced by an emphasis on the importance of the data in generating explanations.
- Because of its reversal of conventional psychological research methodology, conversation analysis warrants the careful attention of all psychologists since it helps define the nature of psychological research.

23.1 Introduction

Conversation analysis has its intellectual roots in ethnomethodology championed in the 1960s by the American sociologist Harold Garfinkel (1967). He wanted to understand the way in which interactions in everyday life are conducted. In particular, ethnomethodologists were concerned with ordinary everyday conversation. The term 'ethnomethodology' signifies Garfinkel's method of studying the common-sense 'methodology' used by ordinary conversationalists to conduct social interactions. Just how is interaction constructed and managed into largely unproblematic sequences?

One of Garfinkel's major contributions was to show that everyday interaction between people involves a *search for meaning*. Care is needed because this is *not* saying that everyday interaction is meaningful as such – only that participants in that interaction regard it as meaningful. To demonstrate this he relied on a form of experimental research. In one example of this, students attended a 'counselling' session in a university's psychiatry department (McHugh, 1968). The situation was such that participants communicated only indirectly with the 'counsellor' who, totally at random, replied to the client with either 'yes' or 'no'. In this instance, essentially the real world was chaotic and meaningless (unless, of course, one is aware of the random process and the purpose of the study). Nevertheless the participants in the research dealt with this chaotic situation of random responses by imposing a meaningful and organised view of the situation. The concern of ethnomethodologists such as Garfinkel and Aaron Cicourel with the fine detail of this sense-making process influenced others – the most important of whom was Harvey Sacks, who is regarded as the founder of conversation analysis. Also influential on Sacks was the work of Erving Goffman who stressed the nature of social interaction as a social institution which imposed norms and obligations on members (see Figure 23.1).

During the 1960s, Sacks became interested in the telephone calls made to hospital emergency departments (Sacks, 1992). While some researchers might have sought to classify the types of call made, for example, Sacks had a much more profound approach to his chosen subject matter. Substantial numbers of callers to the emergency department

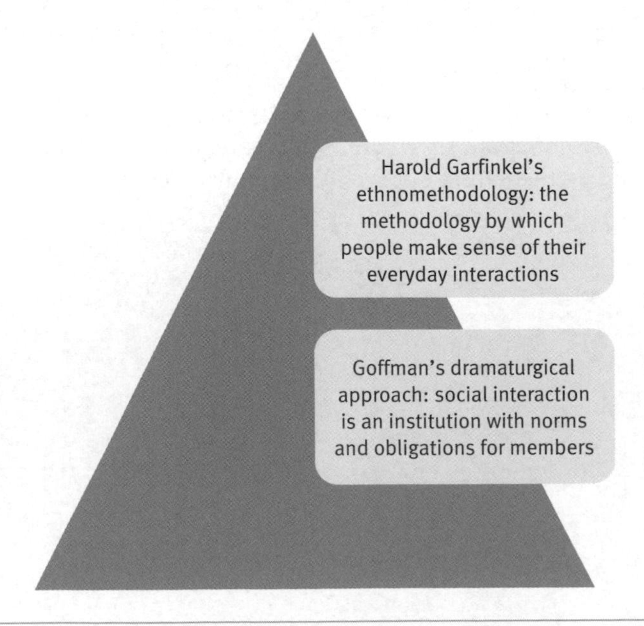

Harold Garfinkel's ethnomethodology: the methodology by which people make sense of their everyday interactions
Goffman's dramaturgical approach: social interaction is an institution with norms and obligations for members

FIGURE 23.1 The roots of conversation analysis

would wind up not providing their name. Obviously, this limited the possible response of the hospital once the conversation ceased – in those days, without a name it would be virtually impossible to track down the caller. Sacks wanted to know by what point in a telephone conversation one could know that the caller would be unlikely to give their name. That is, what features of the telephone conversation were associated with withheld names?

His research strategy involved an intense and meticulous examination of the detail of such conversations. In conversation analysis, emphasis is placed on the analysis of turn-taking – members of conversations take turns to speak and these turns provide the basic unit for examining a conversation. Look at the following opening turns in a telephone conversation:

Member of staff: Hello, this is Mr Smith. May I help you?

Caller: Yes this is Mr Brown.

The first turn is the member of staff's 'Hello, this is Mr Smith. May I help you?' In this case, by the second turn in the conversation (the contribution of the caller), the caller's name is known. However, if the *second* turn in the conversation was something like:

Caller: Speak up please – I can't hear you.

or

Caller: Spell your name please.

there would be the greatest difficulty in getting the caller's name. Often the name would not be obtained at all. One reason for this is that the phrase 'May I help you?' may be interpreted by the caller as indicative of a member of staff who is just following the procedures laid down by the training scheme for staff at the hospital. In other words, if the caller believes that they have special needs or that their circumstances are special, 'May I help you?' merely serves to indicate that they are being treated as another routine case.

Two conversation turns (such as in the above examples) in sequence are known as 'adjacency pairs' and essentially constitute one of the major analytic features in conversation analysis. Without the emphasis on adjacency pairs, the structured, turn-taking nature of conversation would be obscured. A prime objective of conversation analysis is to understand how an utterance is 'designed' to fit in with the previous utterances and the likely nature of subsequent turns in the conversation. In other words, conversation analysis explores the coherence of patterns in turns. By concentrating on adjacent pairs of turns, the researcher works with much the same building blocks as the participants in the conversation use themselves in their task of giving coherence to the conversation.

Telephone conversations illustrate conversation analysis principles well in that they indicate something of the organised nature of even the simplest of conversational acts:

a: Hello

b: Hello it's me

a: Hi Jenny

b: You took a long time to answer the phone

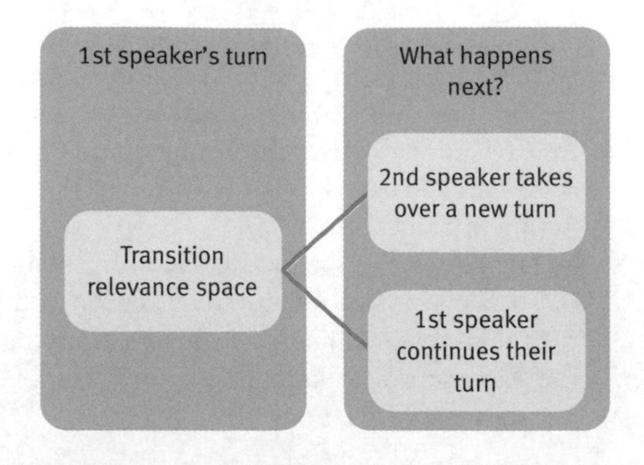

FIGURE 23.2 The turn relevance space in a conversation

It is quite easy to see in this the use of greetings such as *Hello* to allow for the voice identification of the speaker. There is also an assumption that the hearer should be able to recognise the voice. Finally there is an expectation that the ring of a telephone should initiate the response fairly rapidly in normal circumstances. It does not take a great deal to figure out that this is likely to be a call between close friends rather than, say, a business call. This interpretation may not be sophisticated in terms of the principles of conversation analysis, but it does say a good deal about the nature of turn-taking in general and especially in telephone calls.

It is one of the assumptions of conversation that the speaker will either indicate to another speaker that it is their turn to speak or provide the opportunity for another to take their turn. So it is incumbent on the speaker to include a 'transition relevance space' which provides the opportunity for another speaker to take over speaking. The approach in conversation analysis is to study the various things which can occur next and not the most likely thing to occur next as probably would be the objective of a mainstream quantitative psychologist. Figure 23.2 illustrates the basic situation and the possible broad outcomes. Although there is a conversationally presented opportunity for another person to take over the conversation, there is no requirement that they do. Hence there are two possible things which can happen next – one is that the conversation shifts to another speaker and the other is that the original speaker carries on speaking. The question is just how this happens and the consequences for the later conversation of its happening.

For anyone grounded in the mainstream methods of psychology, there is a major 'culture shock' when confronted with a research paper on conversation analysis. It is almost as if one is faced with a research paper stripped bare. The typical features of a psychology report may well be missing. For example, the literature review may be absent or minimal, details of sampling, or participant or interaction selection are often sparse, and generally detail of the social context in which the conversation took place is largely missing. Some of the reasons for this are the traditions and precepts of conversation analysis. More important though to understanding conversation analysis is the realisation that what is crucial in terms of understanding everyday conversation is the way in which the conversation is understood structurally by the participants. As such, theoretical discussions would obstruct ethnomethodological understanding since the theory is not being used by the participants in the conversation. Similarly, details of the social context of the conversation, while important from some perspectives, for a conversation analyst miss the point. The key idea is that in conversation analysis the principles of conversation are regarded as directly governing the conversation. The consideration of factors extraneous to the conversation merely diverts attention away from this.

In conversation analysis the study of how ordinary life conversation is conducted involves the following stages:

- The *recording* stage, in which conversation is put on video or audio-only recording equipment.
- *Transcription*, in which the recording or parts of the recording are transcribed using the minutely detailed methods of Gail Jefferson's transcription system (see Chapter 19). This system does not merely include what has been said but, crucially, a great deal of information about the *way* in which it has been said.
- *Analysis* consists of the researcher identifying aspects of the transcription of note for some reason, then offering suggestions as to the nature of the conversational devices, etc. which may be responsible for the significant features. But things do not come quite as easily as this implies.

Conversation analysis essentially eschews the postulate that there are key psychological mechanisms which underlie conversation. Interaction through talk is not regarded as the external manifestation of inner cognitive processes. In this sense, whether or not participants in conversation have intentions, motives or interests, personality and character is irrelevant. The domain of interest in conversation analysis is the structure of conversation (Wooffitt, 2001). This, of course, refers to the psychological theorising of the researcher – the participants in the conversation may well incorporate psychological factors into their understanding of the conversation and, more importantly, refer to them in the conversation.

23.2 Precepts of conversation analysis

So the major objective of conversation analysis is the identification of repeated patterns in conversation that arise from the joint endeavour of the speakers in the production of conversation. One example of such a pattern is the preference of members of a conversation to allow conversational errors to be self-corrected (as opposed to being corrected by other participants in the conversation). The next person to speak after the error may offer some device to prompt or initiate the 'repair' without actually making the repair. For example, a brief silence may provide the opportunity for the person who has said the wrong thing to correct themselves.

Drew (1995) provided a list of methodological precepts or principles for doing a conversation analysis:

- A participant's contribution (turn) is regarded as the product of the sequence of turns preceding it in the conversation. Turns are basically subject to a requirement that they fit appropriately and coherently with the prior turn. That is, adjacency pairs fit together effectively and meaningfully. Of course, there will be deviant cases when this does not happen.
- Participants develop analyses of each other's verbal conduct. The nature of these analyses is to be found in the detail of each participant's utterances. Contributors to a conversation interpret one another's intentions and attribute intention and meaning to each other's turns as talk. (Notice that intention and meaning are *not* being provided by the researcher but by the participants in the conversation.)
- Conversation analysts study the design of each turn in the conversation. That is, they seek to understand the activity that a turn is designed to perform in terms of the details of its verbal construction.

- The principal objective of conversational analysis is to identify the sequential organisation or patterns in conversation.
- The recurrence and systematic nature of patterns in conversation are demonstrated and tested by the researcher. This is done by reference to collections of cases of the conversational feature under examination.
- Data extracts are presented in such a way as to enable others to assess or challenge the researcher's analysis. That is, detailed transcriptions are made available in a conventional transcription format (Jefferson transcription).

To make things more concrete, conversation analysts have a special interest in:

- how turn-taking is achieved in conversation;
- how the utterances within a person's turn in the conversation are constructed;
- how difficulties in the flow of a conversation are identified and sorted out.

Having addressed these basic questions or issues, conversation analysts then attempt to apply them to other domains of conversation. So they might study what happens in hearings in courts of law, telephone conversations, when playing games or during interviews. In this way, the structure of conversation can be explored more deeply and more comparatively.

23.3 Stages in conversation analysis

Unlike most other approaches to research, conversation analysis rejects prior theoretical speculation about the significant aspects of conversation. A conversation analysis does *not* start with theory which is explored or tested against conversation. The conversation analyst's way of studying conversation is to understand the rules that ordinary people are using in conversation. So the ethnomethodological orientation of conversation analysis strategy stresses the importance of the participant's interpretations of the interaction as demonstrated in the conversation – the priorities are not to be laid down by the researcher in this sense. The participants' interpretations as revealed in the conversation are assumed to be much more salient than any arbitrary, theory-led speculative set of priorities established by researchers (Wooffitt, 2001). That is, conversation analysis does *not* involve hypothesis testing based on cumulative and all-embracing theory.

The conversation analyst's fundamental strategy is to work through the fragments of conversation, making notes of anything that seems interesting or significant. There is no limit to the number of observations that may be written down as part of the analysis process. However, it is crucial that the conversation analyst confines themselves solely to the data in question. They must not move beyond the data to speculate whether, for example, the participant has made a revealing Freudian slip or that in reality they meant to say something quite different. If you like, this is the conversation analysis mindset – a single-minded focus on seemingly irrelevant, trivial detail. Nothing in the interaction being studied can be discarded or disregarded as inconsequential. Consequently, the transcripts used in conversation analysis are messy in that they contain many non-linguistic features. In order to ensure the transcript's fidelity to the original recording, transcripts contain such things as false starts to words and the gaps between words and participants' turns. In conversation analysis, tidying up transcripts is something of a cardinal sin.

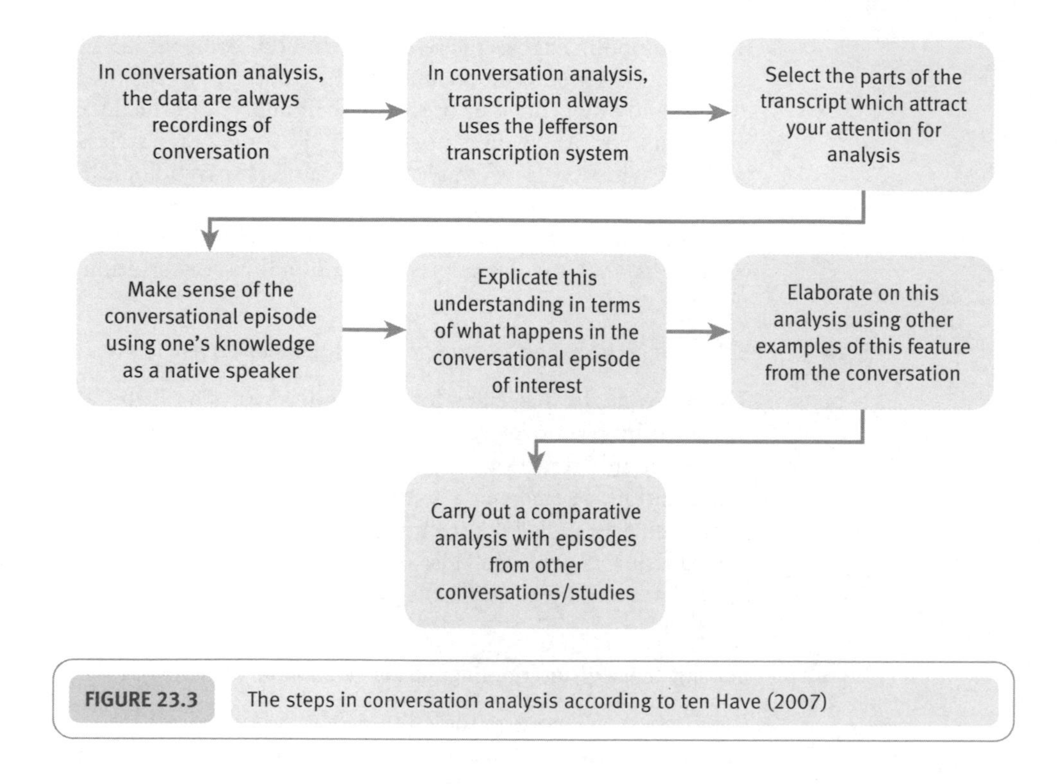

FIGURE 23.3 The steps in conversation analysis according to ten Have (2007)

Paul ten Have (2007) has provided a seven-step model of conversation analysis research practices. Practice implies the way things are done rather than some idealised concept of a research method. Conversation analysts frequently claim that there are no set ways of proceeding. Nevertheless, analysis is confined by fairly strict parameters, knowledge of which should help the novice to avoid straying too far beyond the purview of conversation analysis. Ten Have's steps are best regarded as an ideal for researchers in the field. They are not necessarily characteristic of any particular analyst's work in their entirety (see also Figure 23.3).

Step 1

Mechanical production of the primary database (materials to be analysed)

Data recording is done by a machine. A human decides what and when to record, but the recording is basically unselective, so that what is recorded is not filtered through human thinking systems, tidied up and the like. This original recording may be returned to at any point. For that reason, the recording remains available continually throughout the analysis for checking purposes by the researcher or even by others (Drew, 1995).

Step 2

Transcription

The production of the transcript is ideally free from considerations of the expectations of the researcher. That is, the transcription should be as unsullied by the transcriber as possible. It is all too easy for a transcriber to make systematic errors, some of which fall into line with their expectations (Chapter 19). To this end, the transcript may be checked against the mechanical recording by the transcriber or other researchers. There are, of course, several possible 'hearings' for any mechanical recording. Each transcriber needs to be alert to this possibility. Many regard it as essential that the researcher themselves transcribes the recording. This ensures the close familiarity that is needed for effective analysis. However, no matter how good the transcription is, it cannot be complete and something must be lost in the process compared with the original recording. Nevertheless, the transcription enables the researcher to both confront but also cope with the rich detail of the conversation. The presence of the transcript in the research report means that the analyst and others must be precise in their analysis of the detail of the conversation. This direct link with the data is not possible in some other styles of research. The summary tables of statistics, for example, found in much psychological research cannot be related directly back to the data that were collected by the reader. The links are too deeply buried in the analysis process for this to happen. Few researchers present their data in such a relatively raw form as is conventional in conversational analysis.

Step 3

Selection of the aspects of the transcript to be analysed

There are no formal rules for this and it can simply be a case of the analyst being intrigued by certain aspects of the conversation. Others may be interested in particular aspects of an interaction – as Sacks was when he investigated name-giving in telephone conversations to emergency units. Others may wish to concentrate on the contributions of particularly skilled contributors to a conversation such as where a major shift in the conversation is achieved. Just one adjacency pair (the minimum unit that makes up a conversation) may be sufficient to proceed.

Step 4

Making sense of/interpreting the conversational episode

Researchers are part of the culture that produced the conversational episode. Hence, they can use their own common-sense knowledge of language to make sense of the episode. This is appropriate since it reflects exactly what participants in the interaction do when producing and responding in the conversation (constructing adjacency pairs). Typically the analyst may ask what aspects of the conversation do or achieve during specific conversational exchanges. The relation between different aspects of conversation can then be assessed.

Step 5

Explication of the interpretation

Because the conversation analyst is a member of the broad community whose conversations he or she studies, the researcher's native or common-sense understanding of what happens in an episode is clearly a useful resource as we have seen. Nevertheless, this is insufficient as an explication without bringing the links between this resource and the detail of the conversational episode together. That is, the analyst may feel they know what is happening in the conversation, but they need to demonstrate how their understanding links to the detail of the conversation. Just what is it which happens in each turn of the conversation which leads to what follows and why?

Step 6

Elaboration of the analysis

Once a particular episode has been analysed, the rest of the transcription can be used to elaborate on it. Later sequences in the conversation may in some way, directly or indirectly, relate back to the analyst's central conversational episode. The conversationalists may hark back to the earlier episode and reveal ways in which they understood it. As a consequence, the analyst may need to reformulate the analysis in some way or even substitute a completely different analysis.

Step 7

Comparison with episodes from other conversations

The analysis process does not end with a particular transcription and its analysis. It continues to other instances of conversation which are apparently similar. This is vital because a particular conversational episode is *not* considered to be unique since the devices or means by which a conversational episode is both recognised and produced by the conversationalists are the same for other conversationalists and conversations. Some studies specifically aim to collect together different 'samples' of conversation in order that these different samples may be compared one with the other. In this way, similarities and dissimilarities may encourage or demand refinement of the analysis.

Steps 4 to 7 may seem less distinct in the analysis process itself than is warranted by describing them as separate steps. Ten Have's steps are schematic. They do not always constitute a precise and invariant sequence of steps which analysts must follow invariably and rigidly. Ultimately the aim of most conversational analysts is not the interpretation of any particular episode of conversation. Conversation analysis delves into its subject matter thoroughly and deeply, which means that a psychologist more familiar with mainstream psychological research may find the attention to detail somewhat daunting. Instead of establishing broad trends, conversation analysis seeks to provide a full account of the phenomena in conversation which are studied. So the ill-fitting case may be given as much emphasis as common occurrences.

Box 23.1 Research Example

Conversation analysis

There are a number of good examples of conversation analysis in the work of psychologists – even though it is difficult to specify how the work of a sociologist, for example, analysing the same conversation would be different. The following are particularly useful in that they clearly adopt some of the sensibilities of mainstream psychology.

Cold reading

Psychic phenomena are controversial. Wooffitt (2001) studied 'cold reading', which is the situation when, on the basis of very little conversation with a client, the medium seems to have gathered a great deal of information from beyond the grave. Some commentators suggest that the medium simply uses the limited interaction with the client to gather information about the client. This information is then fed back to the client as evidence of the medium's psychic powers.

Conversation analysts might be expected to have something to say about these 'conversations' between a medium and a client. The following excerpt from such a conversation is typical of what goes on between a psychic (P) and a client (S) (Wooffitt, 2001):

Extract 31

P: h Ty'ever though(t) o(f) .h did you want to go into a caring profession early *on*, when you were choosing which way you were gonna go.

(.)

S: yeah I wanted to: go into child care actually when I

P: MMMmmm. . . .

S: =when I left school

P: That's right yeah >well< h (.) 'm being shown that>but (t)-< h it's (0.2) it's not your way ye(t) actually but i(t) y'y may be caring for (t-)ch- children or whatever later on okay?

Although this excerpt is written using Jefferson transcription methods (Chapter 19 will help you decode this), in this case the importance of the excerpt can be understood simply on the basis of the sequence of words.

What happens? Well first of all the psychic asks a question about caring professions. Then the client replies but the reply is fairly extended as the client explains that they wanted to go into child care. The psychic then interrupts with 'MMMmmm'. The client carries on talking but quickly at the first appropriate stage the psychic takes over the conversation again with 'That's right yeah'. The information is then characterised by the psychic as emanating from the spiritual world. In terms of the conventional explanation of cold reading phenomena, this is a little surprising. One would expect that the psychic would allow the client to reveal more about themselves if the purpose of the conversation is to extract information which is then fed back to the client as if coming from the spiritual world.

In contrast, what appears to be happening is that the psychic rapidly moves to close down the turn by the client. Wooffitt (2001) argues that the first turn in the sequence (when the medium asks the question) is constructed so as to elicit a relatively short turn by the client (otherwise the psychic might just as well have asked the client to tell the story of their life). So ideally the client will simply agree with what the psychic says in the second turn of the sequence. This does not happen in the above example. If it had, then the floor (turn to speak) would have returned quickly to the psychic. As it happens, the psychic has to interrupt the client's turn as quickly as possible.

By analysing many such excerpts from psychic–client conversations, it was possible to show that if the client gives more than a minimal acceptance of what the psychic says, then the psychic will begin to overlap the client's turn, eventually forcing the client's turn to come to an end. Once this is done, the psychic can attribute their first statement (question) to paranormal sources. Wooffitt describes this as a three-turn series of utterances: a proposal is made about the sitter, this is accepted by the sitter, and then it is portrayed in terms of being supernatural in origin. Without the emphasis on the structure of turns which comes from conversation analysis, Wooffitt may well have overlooked this significant pattern.

Date rape

One notable feature of conversation analysis is that it dwells on the everyday and mundane. Much of the data for conversation analysis lack any lustre or intrinsic interest. This is another way of saying that conversation analysis has explored ordinary conversation as a legitimate research goal. However, it is intriguing to find that the principles originating out of everyday conversation can find resonance in the context of more extraordinary situations – date rape, for example. Training courses to help young women prevent date rape often include sessions teaching participants how to say 'no', that is, refusal skills in this case for unwanted sexual intercourse (Kitzinger and Frith, 2001).

Research by conversation analysts about refusals following invitations has revealed the problematic nature of saying 'no' irrespective of the type of invitation. That is, saying 'yes' to an invitation tends to be an extremely smooth sequence without gaps or any other indication of conversational hiccups. Refusing an invitation, on the other hand, produces problems in the conversation. For example, there is likely to be a measurable delay of half a second between invitation and refusal. Similarly, 'umm' or 'well' is likely to come before the refusal. Palliative phrases such as 'That's really kind of you but . . .' may be used. Finally, refusal is likely to be followed by justifications or excuses for the refusal whereas acceptance requires no justification. Kitzinger and Frith (2001) argue that date rape prevention programmes fail by not recognising the everyday difficulties inherent in refusing an invitation. That it is problematic (even when refusing sexual intercourse) can be seen in the following exchange between two young women:

Liz: It just doesn't seem right to say no when you're up there in the situation.

Sara: It's not rude, it's not rude – just sounds awful to say this, doesn't it.

Liz: I know.

Sara: It's not rude, but it's the same sort of feeling. It's like, 'oh my god, I can't say no now, can I?'

(Kitzinger and Frith, 2001, p. 175)

Using a focus group methodology, Kitzinger and Frith (2001) found that young women who participated in these groups deal with the problem of refusing sexual intercourse on dates by 'softening the blow' with a 'more acceptable excuse':

> . . . young women talk about good excuses as being those which assert their inability (rather than their unwillingness) to comply with the demand that they engage in sexual intercourse: from the vague (and perhaps, for that reason, irrefutable) statement that they are 'not ready', through to sickness and menstruation.
>
> (p. 176)

23.4 Conclusion

Conversation analysis provides psychology with an array of analytical tools and methods that may benefit a range of fields of application. Nevertheless, essentially conversation analysis springs from rather different intellectual roots from the bulk of mainstream psychology and specifically excludes from consideration many quintessential psychological approaches. Furthermore, in terms of detailed methodological considerations, conversation analysis reverses many of the conventional principles of mainstream research methods. For example, the context of the conversation studied is not a particular concern of conversation analysts so detail about sampling and so forth may appear inadequate. This reversal of many of the assumptions of conventional psychological research methods warrants the attention of all researchers as it helps define the assumptions of conventional research. That is, understanding something about conversation analysis is to understand more about the characteristics of mainstream psychology.

Like other areas of qualitative research, some practitioners are gradually beginning to advocate the use of quantification in the analysis of data. This is obviously not popular with all qualitative researchers. However, if a conversation analyst makes claims that imply that there are certain relationships between one feature of conversation and another, it might seem perverse to a mainstream psychologist not to examine the likelihood that one feature of conversation will follow another.

Key points

- Conversation analysis emerged in the 1960s in the context of developments in sociological theory.
- Ethnomethodology was developed by Garfinkel almost as the reversal of the grand-scale sociological theories of the time. Ethnomethodology concerned itself with everyday understandings of ordinary events constructed by ordinary people.
- Harvey Sacks is considered to be the founder of conversation analysis. His interest was in the way conversation is structured around turns and how one turn melds with the earlier and later turns.
- Conversation analysis requires a detailed analysis and comparison of the minutiae of conversation as conversation. It draws little on resources outside the conversation (such as the social context, psychological characteristics of the individuals and so forth).
- Superficially, some of the features characteristic of conversation analysis may seem extremely sloppy. For example, the downplaying of the established research literature in the field prior to the study of data, the lack of contextual material on the conversation, and the apparent lack of concern over such matters as sampling are reversals of the usual standards of psychological research.
- Carrying out conversation analysis involves the researcher in close analysis of the data in a number of ways. In particular, the Jefferson conversation transcription system encourages the researcher to examine the detail rather than the broad thrust of conversation. The transcription is interpreted, reinterpreted, checked and compared with other transcriptions of similar material in the belief that there is something 'there' for the analyst.
- Conversation analysis is commonly applied to the most mundane of material. Its primary concern, after all, is understanding the structure of ordinary or routine conversation. However, the insights concerning ordinary conversation highlight issues for researchers attempting to understand less ordinary situations.

ACTIVITIES

1. Collect samples of conversation by recording 'natural' conversations. Does turn-taking differ between genders? Does turn-taking differ cross-gender? Is there any evidence that 'repairs' to 'errors' in the conversation tend to be left to the 'error-maker'?

2. Make a list of points which explain what is going on in an episode of conversation. Which of your points would be acceptable to a conversation analyst? Which of your points involve considerations beyond the episode such as psychological motives or intentions, or sociological factors such as social class?

3. Charles Antaki has a conversation analysis tutorial at the following web address: http://www-staff.lboro.ac.uk/~ssca1/sitemenu.htm. Go to this site and work through the exercises there to get an interesting, practical, hands-on insight into this form of analysis, which includes the source material in the form of a video.

CHAPTER 24

Interpretative phenomenological analysis

Overview

- Interpretative phenomenological analysis (IPA) is a recent psychology-based qualitative method which is gaining popularity among researchers.
- It is primarily concerned with describing people's personal experiences of a particular phenomenon. Additionally, it seeks to interpret the psychological processes that may underlie these experiences. In other words, it aims to explain people's accounts of their experiences in psychological terms.
- IPA assumes that people try to make sense of their experiences and the method describes how they do this and what it may mean.
- The method has its roots in phenomenology which was a major branch of philosophy during the twentieth century and it also has close links with hermeneutics and symbolic interactionism.
- The data for IPA generally come from semi-structured interviews in which people freely recall their experiences, although other sources of accounts can be used. The questioning style used ideally encourages participants to talk about their experiences at length.
- The whole interview is usually sound recorded and then transcribed in a literal, secretarial style though other information may be included if appropriate.
- One account is usually analysed first before other accounts are looked at in detail. Subsequent accounts may be examined in relation to this first account which is a way of exploring the adequacy of the initial analysis. Subsequent accounts may be analysed in terms of the themes of the initial account or each account may be examined afresh. Similarities and differences between accounts may be noted.

- Each account is read several times to enable the researcher to become familiar with the material. Any impressions may be noted in the left-hand margin of the account as it is being read. There is no set way of doing this and no rules that must be followed.
- After familiarising themselves with the account, the researcher looks for themes in the material. Although themes are clearly related to what was said, they are usually expressed at a slightly more abstract or theoretical level than the original words used by the participant in the research. Themes are usually described in terms of short phrases of only a few words and these are written in the right-hand margin of the account.
- Once the main themes have been identified, the researcher tries to group them together in broader and more encompassing superordinate themes. These superordinate themes and their subordinate components may be listed in a table in order of their assumed importance starting with the most important. Next to each theme may be a short verbatim example which illustrates it together with a note of its location in the account.
- The themes that have been identified are discussed in terms of the existing literature on that topic in the report.
- Interpretative phenomenological analysis, unlike some other forms of qualitative analysis, deals with internal psychological processes and does not eschew the use of psychology in general as part of the understanding of people's experiences.

24.1 Introduction

Interpretative phenomenological analysis (IPA) is a recent qualitative approach which has rapidly grown in popularity since Jonathan Smith first outlined it (Smith, 1996). It has been used in numerous research studies and articles in peer-reviewed journals and books. Applications of IPA have been largely in the fields of health and clinical psychology though it can be applied more generally than that. As its name implies, its primary concern is with providing a detailed description and interpretation of the accounts of particular experiences or phenomena as told by an individual or a small number of individuals. A good example is research on the experience of chronic back pain (Smith and Osborn, 2007). A basic assumption of the approach is that people try to make sense of their experiences and understanding this is part of the aims of an IPA study. So the researcher needs to (a) describe people's experiences effectively and (b) try to make sense of these experiences. In other words, the researcher attempts to interpret the interpretations of the individual. Interpretative phenomenological analysis acknowledges, however, that the researcher's own conceptions form the basis of the understanding of the phenomenological world of the person that is being studied. This means that the researcher can never, entirely, know this personal world but can only approach somewhere towards accessing it.

The approach has been used to address the following questions among others:

- What does paranoia feel like (Campbell and Morrison, 2007)?
- How do feelings affect the use of private and public transport (Mann and Abraham, 2006)?

- What does it feel like to opt to have surgery to control obesity (Ogden, Clementi and Aylwin, 2006)?
- How does alcohol drinking in adolescents result in having unprotected sex (Coleman and Cater, 2005)?

Notice that these are fairly open and general research questions and specific hypotheses are not involved.

24.2 Philosophical foundations of interpretative phenomenological analysis

It is important when trying to understand the variety of research methods to appreciate precisely what set of ideas one is 'buying into' if a particular method is adopted. This can be difficult since the underlying assumptions of the major psychological methods are rarely directly spelt out by their advocates. This is particularly the case for much of the psychology which dominates introductory psychology textbooks and lectures. In Chapter 17 we explained the assumptions of logical positivism which is at the root of much of what is written and taught as psychology. These assumptions are not shared by all psychological methods, as was explained. Qualitative methods, in particular, generally reject most if not all of the assumptions of logical positivism and positivism more generally. So a mature understanding of research methods such as interpretative phenomenological analysis requires that their philosophical basis is clear to us.

Just like all other forms of qualitative analysis, interpretative phenomenological analysis has its own philosophical and theoretical roots. It may share some of the assumptions of other forms of qualitative analysis but does not necessarily give the same weight to each as other qualitative approaches. Not surprisingly, phenomenology contributes, as do symbolic interactionism and hermeneutics (see Figure 24.1):

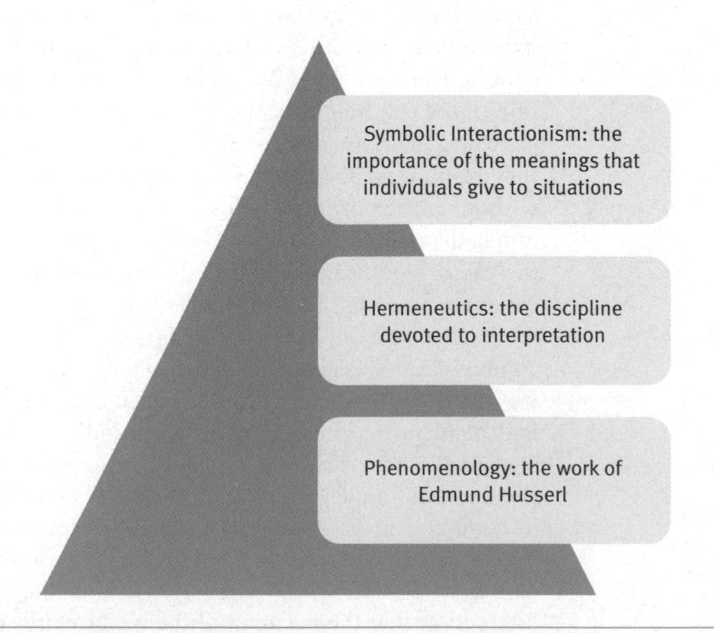
Symbolic Interactionism: the importance of the meanings that individuals give to situations
Hermeneutics: the discipline devoted to interpretation
Phenomenology: the work of Edmund Husserl

FIGURE 24.1 The roots of interpretative phenomenological analysis

- *Phenomenology* is the study of conscious experiences. The origins of phenomenology lie in the work of the Austrian-born German philosopher Edmund Husserl in *Logische Untersuchungen* (1900/1970) though he used the term first in a later book published in 1913 (*Ideen zu einer reinen Phänomenolgie und phänomenologischen Philosophie*). While Husserl is usually described as a philosopher, the distinction between philosophy and psychology was not so strong when he was writing as now. So important early psychologists such as Franz Brentano were particularly influential on Husserl. The basic assumption of phenomenology is that reality is not something which is independent of human experience but is made up of things and events as perceived by conscious experience. In other words, it eschews the idea of an objective reality. It is thus a way of understanding consciousness from the point of view of the person who has the experience. The phenomena studied from a phenomenological perspective include some familiar and some less familiar psychological concepts such as thought, memory, social action, desire and volition. The structure of experience involves conscious intentionality, that is to say, experience involves something in terms of particular ideas and images which together constitute the meaning of a particular experience. Phenomenology in different guises had major influences on twentieth century academic thinking including the existentialism of such people as Jean-Paul Sartre and, in American sociology, major developments such as ethnomethodology.

- *Symbolic interactionism* is based on the idea that the mind and the self emerge out of social interactions involving significant communications. It is a *sociological approach* to small-scale social phenomena rather than the major structures of society. It has in the past been influential on social psychological thinking, especially what is sometimes termed sociological social psychology. Probably the best example of this is the work of Erving Goffman which has been influential on social psychology. Goffman's highly influential book *Asylums* published in 1961 examined institutionalisation which is the patient's reaction to the structures of total institutions. In order to understand interactions in these contexts Goffman adopted the basic phenomenological perspective. George Herbert Mead was the major influence on symbolic interactionism though the term is actually that of Herbert Blumer. The approach is to regard mind and self as developing through social interaction which constitutes the dominant aspect of the individual's experiences of the world. Of course, these social processes and social communication exist prior to any individual so, in this sense, the social is the explanation of the psychology of the individual. The conversation of gestures is an early stage in which the individual (such as a young child) is communicating with others since they respond to the gesture but the child is unaware of this. This is communication without conscious intent. However, out of this the individual progresses towards the more advanced forms of social communication whereby they can communicate through the use of significant symbols. Significant symbols are those where the sender of the communication has the same understanding of the communication as those who receive the communication. Language consists of communication using such significant symbols. Communication is not an individual act but one involving two individuals at a minimum. It provides the basic unit through which meaning is learnt and established and meaning is dependent on interactions between individuals. The process is one in which there is a sender, a receiver *and a* consequence of the communication. It is in this way that the mind and understanding of self arise. Of course, there develops an intentionality in communication because the individual learns to anticipate the responses of other individuals to the communication and can use these to achieve the desired response of others. So the self is purposive. It is in the context of communication or social interaction that the meaning of the social world comes about.

- *Hermeneutics* is, according to its Greek roots, the analysis of messages. It is about how we study and understand texts in particular. As a consequence of the influential Algerian/French philosopher Jacques Derrida, a text in this context is not merely something written down but can include anything which people interpret in their day-to-day lives which includes their experiences. It is relevant to interpretative phenomenological analysis because of its emphasis on understanding things from the point of view of others. Meaning is a social and a cultural product and hermeneutics applies this basic conceptualisation to anything which has meaning. So it can be applied to many aspects of human activity which seem far removed from the biblical texts to which the term 'hermeneutics' originally applied. But the wider usage of the term 'hermeneutics' gives a primacy to matters to which tradition makes an important contribution. So hermeneutics studies the meaning and importance of a wide range of human activity primarily from the first-person perspective. Looking at parts of the text in relation to the entirety of the text in a sort of looping process leads to understanding the meaning of the text. Hermeneutics is also responsible for originating the term 'deconstruction'. This was introduced by the German philosopher Martin Heidegger but with a different emphasis from its modern usage. Basically he realised that the interpretation of texts tended to be influenced by the person interpreting the text. In other words, the interpreter was constructing a meaning of the text which may be different in some respects from the original meaning of the text. So in order to understand the influence of these interpretations one needs to deconstruct the interpretations to reveal the contributing constructions of the interpreters. Religious texts are clearly examples where constructions by interpreters essentially alter the meanings of texts. Thus there are various constructions of Islam though the original texts on which each is based are the same. However, deconstruction under the influence of Derrida has come to mean a form of criticism of the interpreter's influence on the meaning of the text, whereas it was originally merely the identification of traditions of the understanding of text.

It is easy to see aspects of interpretative phenomenological analysis in phenomenology, symbolic interactionism and hermeneutics. However, analysts using the method do not simply take what the studied individual has to say as their interpretation. The analyst adds to the interpretation and does not act with straightforward empathy to the individual being studied. The approach includes what Smith and Osborn (2003) describe as a questioning hermeneutics (see later). To illustrate this they offer the following comment:

> . . . IPA is concerned with trying to understand what it is like, from the point of view of the participants, to take their side. At the same time, a detailed IPA analysis can also involve asking critical questions of the texts from participants, such as the following: What is the person trying to achieve here? Is something leaking out here that wasn't intended? Do I have a sense of something going on here that maybe the participants themselves are less aware of?
>
> (p. 51)

In this, the need for the use of the word 'interpretative' in interpretative phenomenological analysis becomes apparent since the researcher is being encouraged not simply to take the interpretative side of the participant in the research but to question that interpretation in various ways. These are tantamount to critical deconstructions.

24.3 Stages in interpretative phenomenological analysis

The key thing when planning an IPA analysis is to remember that it is primarily concerned with describing and understanding people's experiences in a specified area of interest. So whatever the textual material used, it needs to involve detailed accounts of such experiences.

This rules out a lot of textual material simply because they are not or are only tangentially concerned with people's perceptions of things which happen to them. Of course, the easiest way to get suitable rich textual material is to ask people to discuss in an interview things which happen in their lives. So long as the researcher takes care to maximise the richness of the description obtained by using carefully thought out and relevant questions, then a semi-structured interview will generally be the appropriate form of data collection though not exclusively so. In other words, the primary thing with IPA data collection is to remember what sort of information one is collecting. This is quite different from other forms of qualitative analysis where a particular domain of content may not be so important.

Smith and his colleagues have described how an IPA study may be carried out (Smith and Eatough, 2006; Smith and Osborn, 2003). They acknowledge that other researchers may adapt the method to suit their own particular interests, that is, the method is not highly prescriptive in terms of how a study should be carried out. A crucial part of interpretative phenomenological analysis is getting the semi-structured interviews right as this is the main source of data in this form of analysis. The interview consists of a series of open questions designed to enable participants to provide lengthy and detailed answers in their own words to the questions asked by the researcher. As with any other study, piloting of the research instruments is advisable, so the IPA researcher should try out their questions on a few participants. In this way, the researcher can check to make sure that the questions are suitable for their purpose, that is, the participants answer freely in terms of their experiences and views. Other forms of personal account such as diaries or autobiographic material could be used if their content is appropriate.

There are two major aspects of interpretative phenomenological analysis:

- data collection;
- data analysis.

We will deal with each of these in turn (see Figure 24.2).

Data collection

Smith and Osborn (2003) go into detail about the formulation of research questions in interpretative phenomenological analysis. There is no specific hypothesis as such since

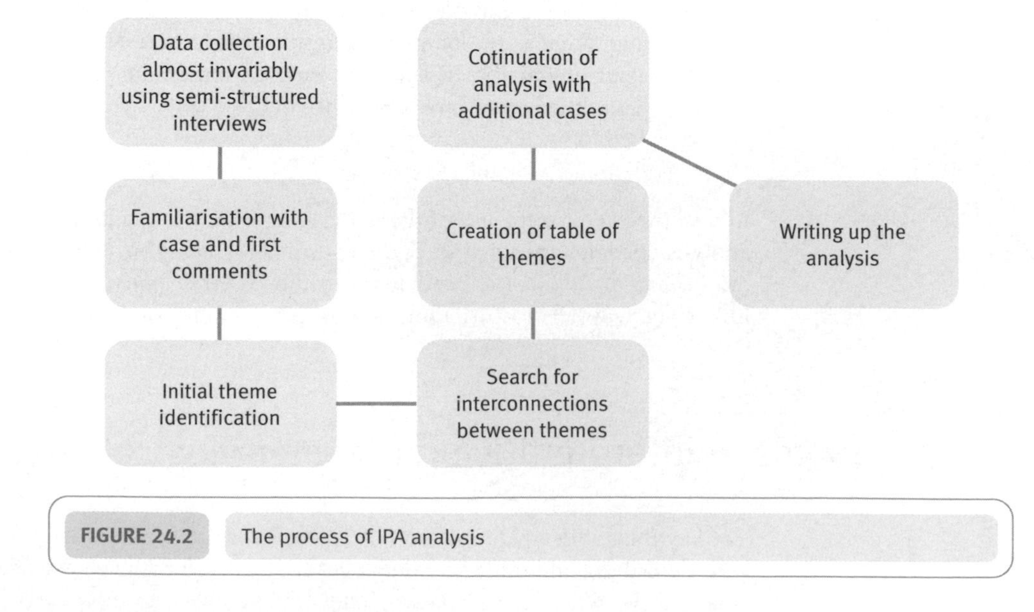

FIGURE 24.2 The process of IPA analysis

the approach is exploratory of the area of experience that the researcher is concerned with. However, generally the IPA research question is to find out the perceptions that the individual has concerning a given situation they experience and how they make sense of these experiences.

The IPA procedures involve almost exclusively the use of semi-structured interviews to provide data. Interviews are intended to be flexible in their application and the questions are not read to the participant in a fixed order since the intention is that the interviewer should be free to probe matters of interest which arise during the course of the interview. In particular, the interview can be led by the participant's particular issues rather than simply being imposed by the researcher. To some extent, the researcher can pre-plan the sorts of additional probes which are asked of participants in order to get them to supply more information on a particular topic. So these probes can be included in the interview schedule. There is also advice provided by Smith and Osborn about how to construct the interview questions (pp. 61–2):

- Questions should be neutral rather than value-laden or leading.
- Avoid jargon or assumptions of technical proficiency.
- Use open, not closed, questions.

Generally this is the sort of advice appropriate for framing questions for any in-depth interviewing strategy aimed at eliciting rich data (see Chapter 18).

The semi-structured interview usually opens with a general question which is normally followed by more specific questions. For example, in a study on back pain the researcher may begin by asking a participant to describe their pain before asking how it started and whether or not anything affects it (Smith and Osborn, 2007). The researcher should memorise the interview schedule so that the interview can flow more smoothly and naturally. The order in which the questions are asked and the nature of the questions asked may vary according to what the participant says. So if the participant has already provided information to a question that has yet to be asked, there is no need to obtain that information again by asking that question. For example, if the participant in answer to the first question on describing their pain also said how it started, it would not be appropriate to ask the question on how it had started as this question has already been answered. The participant may raise issues which the researcher had not anticipated and which seem of interest and relevance to the topic. Where this happens, the researcher may wish to question the participant about these matters even though questions on these issues were not part of the original interview schedule. The researcher may wish to include questions on this issue when interviewing subsequent cases. In other words, researchers should be sensitive to the material that participants provide and should not necessarily be bound by their original set of questions.

However, Smith and Osborn (2003) suggest that good interviewing technique in interpretative phenomenological analysis would comply with the following (p. 63):

- Avoid rushing to the main area of interest too quickly as this may be quite personal and sensitive. It takes time for the appropriate trust and rapport to build up.
- While the effective use of probes ensures good-quality data, the overuse of probes can be distracting to the participant and disrupt the quality of the narrative.
- Ask just one question at a time and allow the participant time to answer it properly.
- Be aware of the effect that the interview is having on the participant. Adjustments may need to be made to the style of questioning, etc. should there appear to be problems or difficulties.

The interview is usually sound-recorded so that the researcher has a full record of what has been said. Recording the interview also has the advantage of allowing the researcher to pay closer attention to what is being said as the participant is speaking since the interviewer is not preoccupied with the task of taking detailed notes as the interview progresses. Generally, the advice is to transcribe the interviews prior to analysis since the resulting transcript is quicker to read and check than it is to locate and replay parts of the interview. Furthermore, a transcript makes it easier for the researcher to see the relation between the material and the analysis which is to be carried out. With interpretative phenomenological analysis, the transcription may be the literal secretarial-style transcription which simply consists of a record of what was said. There is no need for the Jefferson-style transcription (Chapter 19), which includes other features of the interview such as voice inflections and pauses, though it is not debarred. However, in some circumstances it may be worthwhile to note some of these additional features such as expressions of emotions if these help convey better just what the participant in the interview has said. It would be usual to have wide margins on either side of the pages of the transcript where one can put one's comments as the transcribed material is being analysed. While making the transcription, the researcher should make a note of any thoughts or impressions they have about what the interviewee is saying since otherwise these may be forgotten or overlooked subsequently. These comments may be put in the left-hand margin of the transcription next to the text to which it refers (the right-hand margin is used for identifying themes). This sort of transcription of the recording can take up to about eight times the time taken to play the recorded material and it is something which cannot be rushed if the material is to be transcribed accurately.

Smith and his co-workers suggest, as do many other qualitative researchers, that because the process of data collection, transcription and analysis is time-consuming, it is possible to interview only a small number of participants. Nevertheless, the number of cases in published studies has varied from 1 (Eatough and Smith, 2006) to as many as 64 (Coleman and Cater, 2005) though the latter is exceptional. The size of sample thought suitable for a study will vary according to its aims and the resources the researcher has. So, for student projects, there may be time and resources available to deal with only three to six cases. It is recommended by Smith and Osborn (2003) that the sample should consist of relatively similar (homogeneous) cases rather than extremely different ones. It should be recognised that interpretative phenomenological analysis is at its roots idiographic and primarily focused on the individual (as in any case study) as someone to be understood. That is one reason why single case studies are common and acceptable in this sort of analysis. Of course, research may move from what has been learnt of the one individual to others but primarily focuses on individuals to be understood in their own right. The distinction between idiographic and nomothetic approaches to knowledge was introduced into psychology in the 1930s by Gordon Allport, though the concepts were originally those of the German philosopher Wilhelm Windelband. Idiographic understanding concerns the individual as an individual in his or her own right and emphasises the ways in which that individual is different from other individuals. Nomothetic understanding is based on the study of groups of individuals who are seen as representing all individuals in that class. Hence it is possible in nomothetic approaches to formulate abstract laws or generalisations about people in general.

■ Data analysis

The analysis of the data is seen as consisting of four to six main stages depending on the number and duration of interviews carried out. Many of these steps are very similar to those of other forms of qualitative analysis:

Step 1

Initial familiarisation with a case and initial comments

The researcher should become as familiar as possible with what a particular participant has said by reading and re-reading the account or transcript a number of times. The researcher may use the left-hand margin of the document containing the account to write down anything of interest about the account that occurs to them. There are no rules about how this should be done. For example, the account does not have to be broken down into units of a specified size and there is no need to comment on all parts of the account. Some of the comments may be attempts at summarising or interpreting what was said. Later on the comments may refer to confirmation, changes or inconsistencies in what was said.

Step 2

Initial identification of themes

The researcher needs to re-read the transcript to make a note of the major themes that are identified in the words of the participant. Each theme is summarised in as few words as necessary and this brief phrase may be written down in the right-hand margin of the transcription. The theme should be clearly related to what the participant has said but should express this material at a somewhat more abstract or theoretical level.

Step 3

Looking for connections between themes

The researcher needs to consider how the themes that have been identified can be grouped together in clusters to form broader or superordinate themes by looking at the connections between the original themes. So themes which seem to be similar may be listed together and given a more inclusive title. This process may be carried out electronically by 'copying and pasting' the names of the themes into a separate document. Alternatively the names of the themes may be printed or written down on cards or slips of paper, placed on a large flat surface such as a table or floor and moved around to illustrate spatially the connections between them. It is important to make sure that the themes relate to what participants have said. This may be done by selecting a short phrase the participant used which exemplifies the original theme and noting the page and line number in the document where this is recorded. Other themes may be omitted because they do not readily fit into these larger clusters or there is little evidence for them.

Step 4

Producing a table of themes

This involves listing the groups of themes together with their subordinate component themes in a table. They are ordered in terms of their overall importance to what the participant was seen to have said, starting off with the most important superordinate theme. This listing may include a short phrase from a participant's account to illustrate the theme and noting the information about where this phrase is to be found as was done in the previous stage. This is illustrated in Table 24.1.

Step 5

Continuing with further cases

Where there is more than one case, the analysis proceeds with the other cases in a similar way. Themes from the first case may be used to look for similar themes in the ensuing cases or each case can be looked at anew. It is important for the analyst to be aware of themes that are similar between participants as well as those that differ between participants as these may give an indication of the variation in the analysis. Once all the accounts have been analysed a final table(s) containing all the themes needs to be produced.

Step 6 **Writing up the analysis**

The final stage is to write up the results of the analysis. The themes seen as being important to the analysis need to be described and illustrated with verbatim extracts which provide a clear and sufficient example of these themes. The researcher tries in the analysis write-up to interpret or make sense of what a participant has said. It should be made clear where interpretation is being provided and the basis on which it has been done. There are two ways of presenting the results in a report. One way is to divide the report into a separate 'Results' and 'Discussion' section. The 'Results' section should describe and illustrate the themes while the 'Discussion' section should relate the themes to the existing literature on the topic. The other way is to have a single 'Results and Discussion' section where the presentation of each theme is followed by a discussion of the literature that is relevant to that theme.

Before attempting to carry out an interpretative phenomenological analysis, it is important to familiarise yourself with the method by reading reports of other studies that have used this approach. There are an increasing number of such reports to draw upon and you should choose those that seem most relevant to what you want to do. As you are most probably unlikely to be able to anticipate the themes that will emerge in your study, you will need to spend some time after the analysis has been completed seeing what the relevant literature is on the themes that you have found.

IPA researchers have provided relatively detailed and clear explanations of their methods including examples of the questions used to collect data and the stages in the analysis of the data giving examples of codings and theme developments (Smith and Osborn, 2003; Smith and Eatough, 2006). These can be consulted in order to develop a finer-tuned understanding of the method.

Box 24.1 Research Example

Interpretative phenomenological analysis

Campbell and Morrison (2007) studied how people experience paranoia. They point out that it has recently been established that the sort of persecutory ideas that characterise paranoia are exaggerations of normal psychological processes. For example, individuals who show non-clinical levels of paranoia also tend to demonstrate self-consciousness in both public and private situations. One possible consequence of this sort of self-examination process is that self-recognised shortcomings may be projected onto other people who are then seen as threatening in situations which are in some way threatening. Negative beliefs that sufferers have about the condition of paranoia, the world in general, and the self are responsible for the distress caused by psychoses such as paranoia. Campbell and Morrison point out that there have been no previous studies that have investigated the subjective experience of paranoia.

Based on these considerations, Campbell and Morrison designed a study to explore subjective experiences of paranoia by comparing patients and non-patients. They had a group of six clinical patients and a group of six other individuals who had no clinical history although they had endorsed two questions on the Peters Delusions Inventory. One of these asked whether they ever feel like they are being persecuted in some way and the other asked whether they ever feel that there is a conspiracy against them.

The participants were interviewed using a semi-structured method. Questions were asked about a number of issues including the following (p. 77):

- *Content of paranoia* For example, 'Can you tell me what sort of things you have been paranoid about?'
- *Beliefs about paranoia* For example, 'What are your thoughts about your paranoid ideas?'
- *Functions of paranoia* For example, 'Do you think that your paranoid ideas have any purpose?'

- *Traumatic life experiences* For example, 'Have you ever experienced anything very upsetting or distressing?'
- *Trauma and paranoia* For example, 'Do you think that your paranoid ideas relate to any of your past experiences?'

Through a process of reading and re-reading each of the transcripts, initial thoughts and ideas about the data were noted in the left-hand margin of the transcripts and these led to the identification of themes which were noted in the right-hand margin of the transcripts. Following this, the researchers compiled a list of the themes which had been identified. Superordinate themes were then identified which brought together a number of themes. These superordinate themes may have been themes already identified but sometimes they were new concepts. Interestingly, the researchers checked their analysis with the participants in the research as a form of validity assessment, which led to some updating and revision of the themes where the analysis was not accepted by the participants.

The researchers suggest that there were four superordinate themes of note which emerged in the analysis and which they define as:

- the phenomenon of paranoia;
- beliefs about paranoia;
- factors that influence paranoia;
- the consequences of paranoia.

They produce tables which illustrate these superordinate themes, the 'master' themes which are grouped together under this heading, and the subcategories of each 'master' theme. So, by way of illustration, we can take the superordinate theme described as the phenomenon of paranoia. This includes three master themes: (A) the content of paranoia, (B) the nature of paranoia and (C) insight into paranoia. Again by way of illustration, we can take the first of these 'master' themes, the content of paranoia, which breaks down into the following subcategories: (a) perception of harm, (b) type of harm, (c) intention of harm and (d) acceptability of belief. In their discussion, Campbell and Morrison illustrate each of the subcategories by a representative quotation taken from the transcripts of the data. This is done in the form of tables – there is one for each of the superordinate themes. Each master theme is presented and each subcategory listed under that heading. It is the subcategories which are illustrated by a quotation. Although this is a simple procedure, it is highly effective because the reader has each subcategory illustrated but by looking at the material for all of the subcategories the 'master' theme is also illustrated. The general format of this is illustrated in Table 24.1. We have only partially given detail in the table to keep it as simple as possible in appearance and we have used fictitious quotes. Of course, this tabular presentation limits the lengths of quotations used and, inevitably, results in numerous tables in some cases. Thus it is only suitable when the number of superordinate themes is relatively small since this determines the number of tables. Nevertheless, the systematic nature of the tabular presentations adds clarity to the presentation of the analysis.

A further feature of the analysis, not entirely typical of qualitative analyses in general, was the comparison between the clinical group and the normal group in terms of paranoia. For example, in terms of 'intention of harm' it was found that there was a difference between the two groups. For the normal group the harm tended to be social harm whereas for the patient group it tended to be physical or psychological harm.

Table 24.1 The structure of the illustrative quotations table for theme (1) The phenomenon of paranoia

(A) The content of paranoia	(B) Another master theme	(C) Another master theme
(a) Perception of harm *'People would sit around talking about me, I thought.'* **Victor 1**	(a) Subcategory *Illustrative quote*	(a) Subcategory *Illustrative quote*
(b) Type of harm *'It felt like people I hardly knew were backstabbing me.'* **Janet 1**	(b) Subcategory *Illustrative quote*	(b) Subcategory *Illustrative quote*
(c) Intention of harm *'It felt like I was being deliberately persecuted.'* **Norman 2**	(c) Subcategory *Illustrative quote*	
(d) Acceptability of the belief *'It was MI5 that was behind all of the plotting and telephone tapping.'* **Mary 1**		

24.4 Conclusion

Interpretative phenomenological analysis can be seen as being a much more specific approach to qualitative research than, say, thematic analysis or grounded theory and discourse analysis or conversation analysis. This is because its focus is quite different from these. Discourse analysis is really a theory of language-as-action and so can be seen as part of a theory of language use and its application. It focuses on how we talk about things. Conversation analysis is a very fine-grained approach to the study of how conversation proceeds and is organised. In contrast, interpretative phenomenological analysis is not about how we talk about our experiences but, instead, it concentrates on what our experiences are. It is not particularly interested in how language is used but it is interested in what people can tell us about their experiences through language. Indeed, in great contrast to discourse analysis, in particular, interpretative phenomenological analysis is about internal psychological states since it is about the conscious experience of events. There seems to be a clarity and transparency about data presentation in interpretative phenomenological analysis which is not always emulated in other forms of qualitative research. The use of tables of data in interpretative phenomenological analysis is in many ways redolent of the use of tables in quantitative analysis. The tables are systematic in interpretative phenomenological analysis but they are different from quantitative tables in that they only provide illustrations of the themes by illustrative quotations from the transcriptions.

Key points

- Interpretative phenomenological analysis was first introduced as an analytic technique in the 1990s. It draws heavily on some of the more important developments in philosophy, psychology and sociology, *inter alia*, in the twentieth century. These have been identified as phenomenology, hermeneutics and symbolic interactionism.
- Interpretative phenomenological analysis is a variant of phenomenological analysis, though, to date, it has been second-party research in which a researcher guides the data collection and analysis, though phenomenological analysis can be solely first-party research in its original form. The aims of IPA research are to describe people's experiences in a particular aspect of life and draw together explanations of these experiences. Much of the research to date has been in the field of health psychology.
- Interpretative phenomenological analysis shares many of the techniques of other qualitative methods. In particular, the primary aim of the analysis is to identify themes in what participants have to say about their experiences. The main processes of analysis involve the literal transcription of pertinent interview data which is then processed by suggesting themes which draw together aspects of the data. Further to this, the researcher may seek to identify superordinate themes which embrace a number of the major themes emerging in the analysis.
- The precise ways in which interpretative phenomenological analysis differs from other forms of qualitative analysis are complex. It departs from, say, discourse analysis, for example, in having little interest in language as such other than the medium through which the researcher can learn about how individuals experience particular phenomena. In other words, language helps to reveal the subjective realities of consciousness. Thus it refers to internal psychological states of the sort which are often eschewed by the qualitative researcher. Of course, it shares with these other approaches the rejection of physical reality as the focus of research but, at the same time, it assumes that in the data provided by participants lies their reality of their experiences.

ACTIVITIES

1. Although interpretative phenomenological analysis has not been used in this way, phenomenological research can involve the researcher investigating his or her own experiences. Write a narrative account in the first person about your experiences of exams. Then explore your narrative using interpretative phenomenological analysis. What are the major themes you can identify? What are the superordinate themes and what are the subordinate themes?

2. Plan a semi-structured interview on a topic such as childbirth, going to a doctor for a consultation or a turn on a fairground ride. Carry out and record an interview on this topic. Draw a table of superordinate themes, subordinate themes and illustrative quotes.

CHAPTER 25

Evaluating and writing up qualitative research

Overview

- The evaluation of qualitative research emphasises the value of the analysis – that is the coding and theory-building process typically rather than the data collection instruments.
- Evaluating qualitative research requires a clear understanding of the intellectual roots and origins of qualitative research in psychology. Hence you need to study Chapters 17 to 24 for this chapter to be most helpful.
- Many criteria are similar in some ways to those applied to quantitative analysis. However, great emphasis is placed on ensuring that the analysis corresponds closely to qualitative ideals.
- Suggestions are made as to how a newcomer to qualitative research should tackle self-evaluation of their work.

25.1 Introduction

Surely qualitative research is evaluated in much the same ways as quantitative research? This is not so. Qualitative research may be evaluated in a number of ways, none of which can be regarded as a final seal of approval. Some of the criteria are quite close to the positivistic position (reliability and validity) but, as we saw in Chapter 17, this is eschewed by at least some qualitative researchers. Some of these prefer to emphasise the radically different philosophies which straddle the quantitative–qualitative divide if not continuum. For example, quantitative researchers take it for granted that observations should be reliable in the sense that different researchers' observations of the same events are expected to be similar. That is, different observers should observe the same thing if the data are of value. In contrast, some qualitative researchers argue that this is an inappropriate criterion for evaluating qualitative data. They point out that different readers of any text will have a different interpretation (reading) of the text. The diversity of interpretations, they argue, is the nature of textual material and should be welcomed by qualitative analysts. As a consequence, different 'readings' of the data should not be regarded as methodological flaws. The underlying difference between quantitative and qualitative researchers is not a matter of numbers and statistics. It is much more fundamental. At its most extreme, quantitative and qualitative research are alternative ways of seeing the world, not just different ways of carrying out research. It is, after all, the difference between the modern (scientific) approach with its emphasis on cause and the postmodern approach with its emphasis on interpretation.

It may be useful to consider that, according to Denscombe (2002), there are a number of features that distinguish good research *of all types* from not so good research (see Figure 25.1). Among the features that he lists are the following:

- The contribution of new knowledge.
- The use of precise and valid data.
- The data are collected and used in a justifiable way.
- The production of findings from which generalisations can be made.

These are tantalisingly simple criteria which are hard to question. Perhaps the difficulty is that they are so readily accepted. Some might suggest that we are all so imbued with positivist ideas that we no longer recognise them in our thinking. Phrases such as 'new knowledge', 'precise/valid data', 'justifiable' and 'generalisation' may be more problematic than at first appears. What is new knowledge for example? By what criteria do we decide that research has contributed new knowledge? What is precise and valid data? How precise need data be to make them acceptable in research? For what purposes do the data need to be valid to make them worthwhile? Why should worthwhile knowledge be generalisable? Should knowledge that works for New York City be generalisable to a village in Mali?

This boils down to the problematic nature of evaluation criteria. If it is difficult to suggest workable criteria for quantitative research, just what criteria should be applied to qualitative research? One approach is to recognise that much qualitative research has its intellectual roots in postmodernist ideas which, in themselves, are a reaction against the modernist ideas of traditional science and positivism. That is, it would seem that the criteria should be different for qualitative and quantitative research given this. Neverthless, in some ways, it would seem better to seek criteria which are equally applicable to both qualitative and quantitative research. One such set of criteria is that which determines high standards of scholarship in any field. What are these criteria? Careful analysis, detachment, accuracy, questioning and insight are among the suggestions.

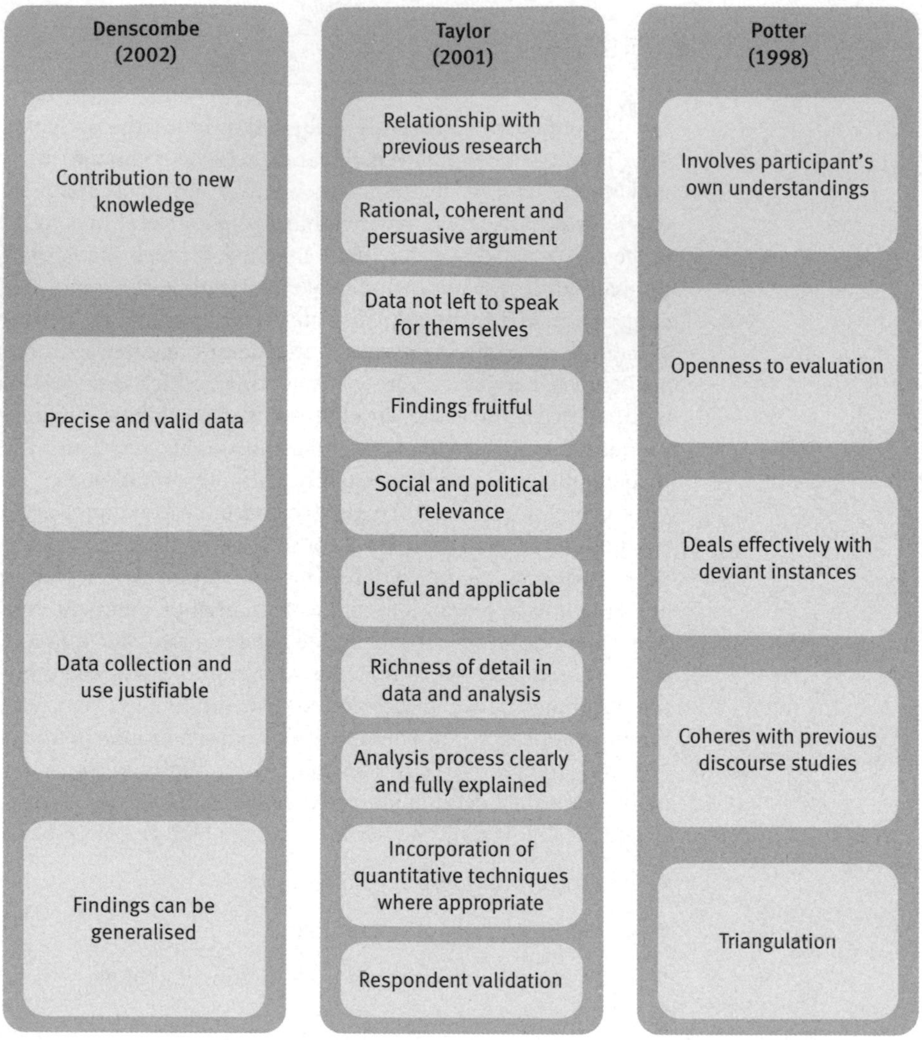

Denscombe (2002)
- Contribution to new knowledge
- Precise and valid data
- Data collection and use justifiable
- Findings can be generalised

Taylor (2001)
- Relationship with previous research
- Rational, coherent and persuasive argument
- Data not left to speak for themselves
- Findings fruitful
- Social and political relevance
- Useful and applicable
- Richness of detail in data and analysis
- Analysis process clearly and fully explained
- Incorporation of quantitative techniques where appropriate
- Respondent validation

Potter (1998)
- Involves participant's own understandings
- Openness to evaluation
- Deals effectively with deviant instances
- Coheres with previous discourse studies
- Triangulation

FIGURE 25.1 Validity criteria in qualitative research

But there is nothing in such criteria which clarifies what is good psychology and what is bad. As we saw in Chapter 17, the intellectual roots of qualitative analysis are outside psychology, where different priorities exist. And why is detachment a useful criterion, for example? It hints that the researcher ideally is an almost alienated figure. Indeed, criteria such as detachment have been criticised for encouraging research to be anodyne (for example, Howitt, 1992a).

Universal criteria for evaluating what is good psychology may be a futile quest and possibly an undesirable one. Such an endeavour would seem to miss the point since epistemological bases of qualitative and quantitative research are in many ways incompatible. Many of the precepts of quantitative research are systematically reversed by coteries of qualitative researchers. For example, when qualitative researchers reject psychological states as explanatory principles, they reject much psychology. The alternative to finding universal evaluation criteria is to evaluate qualitative and quantitative methods by their own criteria, that is, in many respects differently.

25.2 Evaluating qualitative research

The distinction between qualitative data collection and qualitative data analysis is paramount (Chapter 18). If the researcher seeks to quantify 'rich' data collected through in-depth methods such as open-ended interviewing then criteria appropriate to qualitative analyses may not always apply. It is fair to say that qualitative researchers do not speak with one voice about what the evaluative criteria should be. Qualitative research is an umbrella term covering a multitude of viewpoints, just as quantitative research is.

Taylor (2001) discusses a number of evaluative criteria for qualitative research. Some of them apply to research in general but often they take on a special significance in qualitative research. Others are criteria which best make sense only when considering qualitative research. The following are some of Taylor's more general criteria for evaluating qualitative research. We will discuss her more specific criteria and those of others later:

- ***How the research is located in the light of previously published research*** Traditionally in psychological research, knowledge grows cumulatively through a process which begins with a literature search, through development of an idea based on this search and data collection, to finally reporting one's findings and conclusions. This is not the case in all forms of qualitative research. Some qualitative researchers begin with textual material that they wish to analyse, and delay referring back to previous research until after their analysis is completed. The idea is that the previous literature is an additional resource, more text if one likes, with which to explore the adequacy of the current analysis and its fit with other circumstances. The delay also means that the researcher is not so tempted to take categories off the peg and apply them to their data. In some forms of qualitative analysis – especially conversation analysis – reference to the published research can be notably sparse. So the research literature in qualitative research is used very differently from its role in quantitative research. In quantitative research, knowledge is regarded as building out of previous knowledge, so one reviews the state of one's chosen research and uses it as a base from which to build further research. Traditionally, quantitative research demands the support of previous research in order to demonstrate the robustness and replicability of findings across samples and circumstances. Often, in the quantitative tradition, researchers resort to methodological considerations as an explanation of the variability in past and current findings. In the qualitative tradition, any disparity between studies is regarded much more positively and as less of a problem. Disparity is seen as a stimulus to refining the analytic categories used, which is the central activity of qualitative research anyway.
- ***How coherent and persuasive the argument is rather than emotional*** Argumentation and conclusion-drawing in psychology are typically regarded as dependent on precise logical sequences. It is generally not considered appropriate to express oneself emotionally or to engage in rhetorical devices in psychological report writing. This is quite a different matter from being dispassionate or uninvolved in one's subject matter. A great deal of fine psychological writing has been built on the commitment of the researcher to the outcomes of the research. Nevertheless, the expectation is that the researcher is restrained by the data and logic. In this way, the researcher is less likely to be dismissed as merely expressing personal opinions. This is the case no matter what research tradition is being considered.
- ***Data should not be 'left to speak for themselves' and the analysis should involve systematic investigation*** The meaning of data does not reside entirely in the data themselves. Data needs to be interpreted in the light of a variety of considerations of both a methodological and a theoretical nature. Few if any data have intrinsic,

indisputable and unambiguous meanings. Hence the role of the researcher as interpreter of the data has to be part of the process. This interpretation has to be done with subtlety. There is a temptation among newcomers to qualitative analysis to feel that the set of data should speak 'for itself'. So large amounts of text are reproduced and little by way of analysis or interpretation offered. To do so, however, is to ignore a central requirement of research which is to draw together the data to tell a story in detail. In qualitative research this is through the development of closely fitting coding categories in which the data fit precisely but in a way which synthesises aspects of the data. Qualitative research may cause problems for novice researchers because they substitute the data for analysis of the data. Of course, ethnographically meaningful data should carry its meaning for all members of that community. Unfortunately, to push this version of the ethnographic viewpoint too far leaves no scope for the input of the psychologist. If total fidelity to the data is more important than the analysis of the data, then researchers may just as well publish recordings or videos of their interviews, for example. Indeed, there would be no role for the researcher as anyone could generate a psychological analysis. But, of course, this is not true. It takes training to be capable of quality analyses.

- ***Fruitfulness of findings*** Assessing the fruitfulness of any research is not easy. There are so many ways in which research may be fruitful and little research is fruitful in every respect. Most research, however, can be judged only in the short term, and longer-term matters such as impact on the public or other researchers may simply be inappropriate. Fruitfulness is probably best judged in terms of the immediate pay-off from the research in terms of the number of new ideas and insights it generates. Now it is very difficult to catalogue just what are new ideas and insights but rather easier to recognise work which lacks these qualities.

- ***Relevance to social issues/political events*** Qualitative research in psychology often claims an interest in social issues and politics. There are a number of well-known studies which deal with social and political issues. The question needs to be asked, however, just how social and political issues need to be addressed in psychology, and from what perspective? Mainstream psychology has a long tradition of interest in much the same issue. Institutionally, the Society for the Study of Social Issues in the USA has actively related psychology to social problems (Howitt, 1992a), for example, for most of psychology's modern history. There is a distinct tradition of socially relevant quantitative research. Given the insistence of many qualitative researchers that their data are grounded in the mundane texts of the social world, one might expect that qualitative research is firmly socially grounded. One criticism of qualitative research though is that it has a tendency to regard the political and social as simply being more text that can be subjected to qualitative analysis. As such the social and political text has no special status other than as an interesting topic for textual analysis. Some qualitative researchers have been very critical of the failure of much qualitative research to deal effectively with the social and political – concepts such as power, for instance (Parker, 1989). Since power is exercised through social institutions then one can question the extent to which analysis of text in isolation is sufficient analysis.

- ***Usefulness and applicability*** The relevance of psychological research of any sort is a vexed question. Part of the difficulty is that many researchers regard their work as one aspect of an attempt to understand its subject matter for its own sake without the constraints of application. Indeed, applied research in psychology has often been seen as a separate entity from academic research and often somewhat derided as ordinary or pedestrian. Nevertheless, this point of view seems to have reduced in recent years and it is increasingly acceptable to consider the application of research findings as an indication of the value of, at least, some research. It is fairly easy to point to examples from mainstream psychology of the direct application of psychological research – clinical, forensic and educational psychology all demonstrate this in abundance. Part of the success of psychology in these areas is in finding ways of dealing with the practical problems of institutions such as prisons, schools and the mental health system. Success in the application of psychology stems partly from the power of research findings to support practical activities. Qualitative researchers have begun to overlap some of these traditional fields of the application of psychology. Unfortunately the claim that qualitative research is subjective tends to undermine its impact from the point of view of mainstream psychology. Nevertheless, topics such as counselling/psychotherapy sessions and medical interview are to be found in the qualitative psychology literature. As yet, it is difficult to give examples of the direct application of the findings of such psychological research.

The above criteria are in some ways similar to those which we might apply to quantitative research. They are important in the present context since it clarifies their importance in qualitative research too. They sometimes take a slightly different form in the two types of research.

25.3 Validity

The concept of validity is difficult to apply to qualitative research. Traditionally validity in psychology refers to an assessment of whether a measure actually measures what it is intended to measure. This implies there is something fixed which can be measured. The emphasis is really on the validity of the measures employed as indicators of corresponding variables in the actual world. So the validity of a measure of schizophrenia is the extent to which it corresponds with schizophrenia in the actual world beyond that measure. This is not usually an assumption of qualitative research. In qualitative research, the emphasis of validity assessment is in terms of the question of how well the analysis fits the data. A good analysis fits the data very well. In quantitative research, often a very modest fit of the hypothesis to the data is acceptable – so long as the minimum criterion of statistical significance is met.

As we saw in Chapter 15, there are a number of ways of assessing validity in quantitative research. They all imply that there is something in actuality that can be measured by our techniques. This is unlikely to be the case with qualitative research for a number of reasons. One is the insistence by some qualitative researchers that text has a multiplicity of readings and that extends to the readings by researchers. In other words, given the postmodernist emphasis on the impossibility of observing 'reality' other than through a looking glass of subjectivity, validity cannot truly be assessed as a general aspect of measurement.

Discussions of validity by qualitative researchers take two forms:

- It is very common to question the validity of quantitative research. That is, to encourage the view that qualitative research is the better means to obtaining understanding of the social and psychological world.
- The tendency among qualitative researchers to treat any text (written or spoken) as worthwhile data means that the validity of the data is not questioned. The validity of the transcription is sometimes considered, but emphasis is placed on ways in which the fidelity of the transcription, say to the original audio-recording, may be maximised. The greatest emphasis is placed on ways in which the validity of the qualitative analysis

as a qualitative analysis may be maximised. This really is the primary meaning of validity in qualitative research. So many of the criteria listed by qualitative researchers are ones which are only meaningful if we understand the epistemological origins of qualitative research. That is, there are some considerations about the worth of qualitative data which do not normally apply to quantitative research.

Potter (1998) uses the phrase 'justification of analytic claims' alongside using the word validity. The phrase 'justification of analytic claims' emphasises the value of the analysis rather than the nature of the data. He suggests four considerations which form the 'repertoire' with which to judge qualitative research. Different researchers may emphasise different combinations of these:

- ***Participant's own understandings*** When the qualitative material is conversation or similar text, we need to remember that speakers actually interpret the previous contributions by previous speakers. So the new speaker's understanding of what went before is often built into what they say in their turn. For example, a long pause and a change of subject may indicate that the speaker disagrees with what went before but does not wish to express that disagreement directly. Potter argues that by very carefully paying attention to such details in the analysis, the analyst can more precisely analyse the conversation in ways which are relevant to the participant's understandings. It is a way of checking the researcher's analysis.
- ***Openness to evaluation*** Sometimes it is argued that the readers of a qualitative analysis are more in contact with the data than typically is the case in quantitative research in which tables and descriptive statistics are presented but none of the original data directly. Qualitative analyses often incorporate substantial amounts of textual material in support of the analytic interpretation. Because of this, the qualitative analysis may be more open to challenge and questioning by the reader than other forms of research. Relatively little qualitative research is open in this way, however. Potter suggests that for much reported grounded theory and ethnographic research, very little is presented in a challengeable form and a great deal has to be taken on trust, just as with quantitative research. Even where detailed transcripts are provided, however, Potter's ideal may not be met. For example, what checking can be done if the researcher does not report the full transcript but rather selected highlights? Furthermore, what avenues are open to the reader who disagrees with an analysis to challenge the analysis?
- ***Deviant instances*** In quantitative research, deviant cases are largely treated as irrelevant. The participant who bucks the trend of the data is largely ignored – as 'noise' or randomness. Sometimes this is known as 'experimental error', but it is really an indicator of how much of the data is actually being ignored in terms of explanation. Often no attempt is made to explain why some participants are not representative of the trend. In qualitative research, partly because of the insistence on detailed analysis of sequences, the deviant case may be much more evident. Consequently the analysis needs to be modified to include what is truly deviant about it. It may be discovered that the seemingly deviant case is not really deviant – or it may become apparent why it 'breaks the rules'. It may also prove a decisive reason for abandoning a cherished analytic interpretation.
- ***Coherence with previous discourse studies*** Basically the idea here is that qualitative studies which cohere with previous studies are more convincing than ones which are in some way at odds with previous research. There is a sense in which this is a replicability issue since not only does coherence add conviction to the new study but it also adds conviction to the older studies. This is also the case with quantitative studies. But there are difficulties with this form of validity indicator. Qualitative research

varies in terms of its fidelity to previous research when a replication is carried out. Some research will be close to the original and some may be substantially different. In this context, if a qualitative study is merely designed to apply the theoretical concepts derived from an earlier study then the findings are more likely to cohere with the earlier studies. Studies not conceived in this way will be a more effective challenge to what has gone before – and provide greater support if they confirm what went before.

Additional criteria for the evaluation of qualitative research are available (Taylor, 2001). These are not matters of validity, but do offer means of evaluating the relative worth of different qualitative studies:

- *Richness of detail in the data and analysis* The whole point of qualitative analysis is to develop descriptive categories which fit the data well. So one criterion of the quality of a study is the amount of detail in the treatment of the data and its analysis. Qualitative research requires endless processing of the material to meet its aims. Consequently, if the researcher just presents a few broad categories and a few broad indications of what sorts of material fit that category, then one will be less convinced of the quality of the analysis. Of course, richness of detail is not a concept which is readily tallied so it begs the question of how much detail is richness. Should it be assessed in terms of numbers of words, the range of different sources of text, the verbal complexity of the data, or how? Similar questions may be applied to the issue of the richness of detail in the analysis. Just what does this mean? Is this a matter of the complexity of the analysis and why should a complex analysis be regarded as a virtue in its own right? In quantitative research, in contrast, the simplicity of the analysis is regarded as a virtue if it accounts for the detail of the data well. It is the easiest thing in the world to produce coding categories which fit the data well – if one has a lot of coding categories then all data are easily fitted. The fact that each of these categories fits only a very small part of the data means that the categories may not be very useful.
- *Explication of the process of analysis* If judged by the claims of qualitative analysts alone, the process of producing an adequate qualitative analysis is time-consuming, meticulous and demanding. As a consequence of all of this effort, the product is both subtle and true to the data. The only way that the reader can fully appreciate the quality of the effort is if the researcher gives details of the stages of the analysis process. This does not amount to evidence of validity in the traditional sense, but is a quality assurance indicator of the processes that went into developing the analysis.
- *Using selected quantitative techniques* Some qualitative researchers are not against using some of the techniques of quantitative analysis. There is no reason in their view why qualitative research should not use systematic sampling to ensure that the data are representative. Others would stress the role of the deviant or inconsistent case in that it presents the greatest challenge to the categorisation process. The failure of more traditional quantitative methods to deal with deviant cases other than as 'noise', error or simply irrelevant should be stressed again in this context.
- *Respondent validation* Given the origins of much qualitative research in ethnomethodology, the congruence of the interpretations of the researcher with those of the members of the group being studied may be seen as a form of validity check. This is almost a matter of definition – the meanings arrived at through research are intended to be close to those of the people being studied in ethnomethodology. Sometimes it is suggested that there is a premium in the researcher having 'insider status'. That is, a researcher who is actually a member of the group being studied is at an advantage. This is another reflection of the tenets of ethnomethodology. There is always the counter-argument to this that such closeness actually stands in the way of insightful research. However, there is no way of deciding which is correct. Owusu-Bempah and

Howitt (2000) give examples from cross-cultural research of such insider perspectives. Of course, the importance of these criteria is largely the consequence of allegiance to a particular theoretical stance. It is difficult to argue for universality of this criterion.

- *Triangulation* This concerns the validity of findings. When researchers use very different methods of collecting data yet reach the same findings on a group of participants, this is evidence of the validity of the findings. Or, in other words, their robustness across different methods of data collection or analysis. The replication of the findings occurs within settings and not across settings. This is then very different from triangulation when it is applied to quantitative data. In that case, the replication is carried out in widely different studies from those of the original study. The underlying assumption is that of positivist universality, an anathema to qualitative researchers.

Box 25.1 Practical Advice

Writing up a qualitative report

In some ways, writing up a report of qualitative research is potentially beset with problems. There are many reasons for this especially because no set format has yet emerged which deals effectively with the structure of qualitative practical reports. The conventional structure explained in Chapter 5 is clearly aimed at quantitative research and, at first sight, there may be questions about its relevance to qualitative research. However, they both have as their overriding consideration the need for the utmost academic rigour and, in part, that is what the standard report structure in psychology helps to achieve. However, we have already explained in Chapter 5 that the conventional report structure often needs some modification when quantitative research departs from the basic laboratory experiment model. By modifying the basic structure, many of its advantages are retained in terms of clarity of structure and reader-friendliness resulting from its basic familiarity. Our recommendation is that you write up qualitative research studies using the traditional laboratory report structure which you modify by adding additional headings or leaving out some as necessary. Of course, you would probably wish to consult journal articles which employ similar methods to your own for ideas about how to structure your report. These can, if chosen wisely and used intelligently, provide an excellent model for your report and are an easy way of accessing ideas about how to modify the conventional laboratory report structure for your purposes. Occasionally, you will come across a qualitative journal article which is somewhat 'off the wall' in terms of what you are used to but we would not recommend that you adopt such extreme styles. You are writing a qualitative report in the context of the psychological tradition of academic work and you will do best by respecting the academic pedigree of this.

By adopting but adapting the conventional laboratory report structure you are doing yourself a favour. Quantitative report writing is likely to be familiar to you and you will have had some opportunity to develop your skills in this regard in all probability. Everyone has difficulties writing quantitative reports but this partly reflects the academic rigour that writing such reports demands. The reader of your report will benefit from the fact that they are reading something which has a more-or-less familiar structure where most of the material is where it is expected to be in the report. There will be differences, of course. In particular, it is unlikely (but possible) that you would include hypotheses in a qualitative report just as it is fairly unlikely that you would include any statistical analysis (but again possibly especially with techniques such as thematic analysis). Many forms of qualitative research are methodologically demanding and the analysis equally so. It would not be helpful to you to produce sloppy reports given this. Bear the following in mind when writing your qualitative report:

- The *introduction* is likely to discuss in some length conceptual issues concerning the type of analysis that you are performing. This is most likely to be the case when conducting a discourse analysis which is highly interdependent with certain theories of language. You probably will spend little time discussing conceptual issues like these when conducting a thematic analysis which is not based particularly on any theory.

- The *literature review* is generally as important in qualitative write-ups as quantitative ones. Indeed, especially when using qualitative methods in relation to applied topics, you may find that you need to refer to research and theory based on quantitative methods as well as qualitative research. While it is not common for quantitative methods to be looking at exactly the same issues as qualitative studies, there are circumstances in which each can inform the other. Although professional publications using conversation analysis often have very few references (as conversation analysis sees itself as data-driven and not theory-driven in terms of analysis), we would not recommend that students emulate this. As in any other writing you do as a student, we would recommend that you demonstrate the depth and extent of your reading of the relevant literature in your writings. You cannot expect credit for something that you have not done.
- Although preliminary *hypotheses* are inappropriate for most qualitative analyses (since hypotheses come from a different tradition in psychological research), you should be very clear about the aims of your research in your report. This helps to focus the reader in terms of your data collection and analysis as well as demonstrating the purposive nature of your research. In other words, clearly stated *aims* are a helpful part of telling the 'story' of your research.
- The *method* section for a qualitative report should be comparable to one for a quantitative report in scope and level of detail. There are numerous methods of data collection in qualitative research so it is impossible to give detailed suggestions which apply to each of these. Nevertheless, there is a temptation to give too little detail when reporting qualitative methods since often the methods are quite simple compared with the procedures adopted in some laboratory studies, for example. So it is best to be precise about the procedures used even though these may at times appear to be relatively simple and straightforward compared with other forms of research.
- Too frequently qualitative analysts fail to give sufficient detail about how they carried out their analysis. Writing things like 'a grounded theory analysis was performed' or 'thematic analysis was employed' is to say too little. There is more to qualitative analysis than this and great variation in how analyses are carried out. To the reader, such brief statements may read more like an attempt to mystify the analysis process than to elucidate important detail. It is especially important for students to explain in some detail about how they went about their analysis since, by doing so, not only does the reader get a clearer idea about the analytic procedures employed but the student demonstrates their understanding and mastery of the method. As ever in report writing, it is very difficult to stipulate just how much detail should be given – judgement is involved in this rather than rules – but we would suggest that it is best to err on the side of too much detail.
- There is a difficulty in deciding just how much data should be presented in a report. A few in-depth interviews can add up to quite a bulky number of pages of transcripts. However, in terms of self-presentation, these transcripts (especially if they involved Jefferson transcription methods) are a testament to how carefully and thoroughly the researcher carried out the analysis. Not to include them in your report as an appendix means that the reader has no idea of the amount of effort that went into your analysis but, also, the reader is denied the opportunity to check the analysis or to get a full picture of what happened in the interviews. Normally transcriptions do not count towards word limits though you might wish to check this locally with your lecturers.
- You should make sure that you include analysis in your report – sometimes researchers simply reproduce numerous quotations from their data which are weakly linked together by a simple narrative. This may not constitute an analysis at all in a meaningful sense of the term. A commentary on a few quotations is not what is meant by qualitative analysis.
- Your analytic claims should actually be supported by the data. So you need to check that your interpretation of your data actually is reflected in the excerpts that you use.
- It is possible to be systematic in the presentation of your analysis of qualitative data. A good example of this is the way in which IPA analysts (see Chapter 24) produce tables to illustrate the themes which they identify. In this way, themes can be linked hierarchically and illustrative excerpts from the data included for each theme in a systematic manner.
- Furthermore, with thematic analysis especially, it can be very helpful to give some basic statistical information about the number of interviews, for example, in which the theme was to be found or some other indication of their rates of occurrence.
- When it comes to discussing the findings from your qualitative research, you will find numerous criteria by which the adequacy of a qualitative study can be assessed in this chapter. Why not incorporate some of these criteria when evaluating your research findings?

25.4 Criteria for novices

There is probably no qualitative study that effectively embraces all of the criteria of quality that we have discussed. The criteria are not normally discussed within a qualitative report and are more often referred to in theoretical discussions of qualitative methodology. Hence, it is difficult to provide researchers new to qualitative research with a well-established set of procedures which serve as routine quality assurance checks. In this way, quantitative research is very different. Significance testing, reliability estimates, validity coefficients and so forth are minimum quality indicators. Similarly, the literature review is part of the process of assessing the worth of the new findings. Of course, many other indicators of quality are neglected in quantitative reports, just as they often are in qualitative ones.

While these criteria of the worth of a qualitative study can be seen to be intrinsically of value (once the intellectual roots of qualitative research are understood), it is likely that the complexity of the criteria will defeat some novice researchers in the field. They certainly do not gel as a set of principles to help launch good-quality qualitative research by newcomers. So in this section we will suggest some of the criteria which beginners might wish to adopt as a more pragmatic pathway to successful qualitative research (see also Figure 25.2):

- Have you immersed yourself in the qualitative research literature or undergone training in qualitative research? Analytic success is a long journey and you need to understand where you are heading.
- Why are you not doing a quantitative analysis? Have you really done a quantitative analysis badly and called it qualitative research?
- Can you identify the specific qualitative method that you are using and why? Qualitative research is not an amorphous mass but a set of sometimes interlinking approaches.

Quality criteria for novice researcher

- Have you studied qualitative research methods in some depth?
- Can you justify not doing quantitative research?
- Do you know what specific qualitative method to use and why it is appropriate?
- Have you got the personal resources and skills to do qualitative research?
- Have you coded all of your data? If not, can you explain why not?
- Have you spent a great deal of effort refining your codings and categories?
- Precisely what parts of your data are accounted for by your analytic categories?
- How deeply were you engaged in your

FIGURE 25.2 Some quality indicators for novice researchers

- What resources are you devoting to your data collection and analysis? Qualitative data analysis probably requires more personal research skills than much quantitative data analysis. It requires a good interviewing technique, for example, to obtain the richness of data required. Qualitative data require transcription (or quantitative coding), which is time-consuming and exacting. If you do not understand the point of this then your research is almost certainly of dubious quality.
- Have you coded or categorised *all* your data? If not, why not? How do you know that your categories work unless you have tested them thoroughly against the entirety of what you want to understand? If you can only point to instances of categories you wish to use then how do you know that you have a satisfactory fit of your categories with the data?
- Has there been a process of refining your categories? Or have you merely used categories from other research or thought of a few categories without these being worked up through revisions of the data?
- Can you say precisely what parts of your data fit your categories? Phrases such as 'Many participants . . .', 'Frequently . . .' and 'Some . . .' should not be used to cover up woolliness about how your data are coded.
- How deeply engaged were you in the analysis? Did it come easily? If so, have you taken advantage of the special gains which may result from qualitative research?

25.5 Conclusion

Very few of the traditional criteria which we apply to quantitative research apply to qualitative research directly. They simply do not have the same intellectual roots and, to some extent, they are in conflict. There are a number of criteria for evaluating qualitative research, but these largely concentrate on evaluating the quality of the coding or categorisation process (the qualitative analysis). These criteria can be applied but, as yet, there is no way of deciding whether the study is of sufficient quality. They are merely indicators. This contrasts markedly with quantitative and statistical research where there are rules of thumb which may be applied to decide on the worth of the research. Significance testing is one obvious example of this when we apply a test of whether the data are likely to have been obtained simply by sampling fluctuations. Internal consistency measures of reliability such as alpha also have such cut-off rules. This leaves it a little uncertain how inexperienced qualitative researchers can evaluate their research. It is clearly the case that qualitative researchers need to reflect on the value of their analysis as much as any other researcher.

Key points

- Since qualitative research is a reaction to positivism and its influence on research, qualitative research needs to be evaluated in part in its own terms.
- Some criteria apply to both quantitative and qualitative research. The criteria include how the research is located in relation to previously published research, the coherence and persuasiveness of the argument, the strength of the analysis to impose structure on the data, the potential of the research to stimulate further research or the originality and quantity of new insights arising from the research, and the usefulness of applicability of the research.
- Yet other criteria which may be applied are much more specific to qualitative research. These include the correspondence of the analysis with the participant's own understandings, the openness of the report to evaluation, the ability of the analysis to deal with otherwise deviant instances in the data, the richness of detail in the analysis, which is dependent on the richness of the data in part, and how clearly the process of developing the analysis is presented.
- The criteria that novice researchers use to evaluate their own research may be a little more routine. Considerations include factors related to the amount of effort devoted to developing the analysis, the degree to which the analysis embraces the totality of the data, and even questioning whether a quantitative study would have been more appropriate anyway.

ACTIVITIES

1. Could a qualitative researcher simply make up their analysis and get away with it? List the factors that stop this happening.

2. Develop a set of principles by which all research could be evaluated.

PART 5

Research for projects, dissertations and theses

CHAPTER 26

Developing ideas for research

Overview

- If you are planning a research project, ideally this will be firmly based on your knowledge of psychology in general and the pertinent research literature in particular. Rarely is it personally satisfactory to simply reproduce (replicate) what other researchers have done. It is better to use their work creatively and intelligently to produce a valuable variation or extension of what has already been achieved. In this way, one's work is more appreciated by lecturers. Sometimes, student researchers hit upon ideas which have not been effectively researched previously. Occasionally, their research may be publishable.
- Typically, one only has a rudimentary research idea for a project. This idea will be 'knocked into shape' by a process of reading, discussion with a supervisor or peers, and exploring the possibilities in a systematic, disciplined fashion.
- Initially, try drawing up a list of ideas which may then be honed down into a list of manageable and feasible research ideas. There are practical limits of time and other resources which mean that sometimes very good ideas have to be set aside. Although psychology is a fascinating subject, be careful to concentrate on just a few ideas since the time involved in reading pertinent material can be considerable. Hence, do not spread your resources too thinly over too many possibilities.
- The research idea you ultimately focus upon ought to be satisfying to you in different ways. It should be of interest to you, it should be capable of making a contribution to the research field in question, and it should be feasible within your limits of time and available resource. Student research is confined by a fixed time schedule and the most brilliant student research project ever is a waste if essential deadlines are missed.

- It is not uncommon for students to find it difficult to come up with ideas for a research project or dissertation. Don't worry too much if you are one of these. Having good ideas comes with practice and learning to have ideas is part of the process of learning to become a researcher. There are a number of ways of helping yourself through this difficult stage.
- After a number of years at university and having lived for a while, there will be some topics that you have studied or some life experiences that you have had which have interested you. It is likely that you have even clearer ideas of the sorts of thing which you find boring and which you would find it hard to generate interest in. Our best advice is to avoid these even if they could otherwise make good studies.
- At some stage, and the sooner the better, every researcher has to start reading the recently published literature on the topics which interest them the most. This may be simply to 'keep up with the field' but it is more likely to be to survey an area that one is or is becoming interested in. This is not easy and takes time – one has to try to understand what others have done, how they have done it, and why they have done it. It is not unknown for other researchers to have made life difficult in this respect. Once one has understood what others have done, it may remain necessary to appreciate why they have done it.
- You should also bear in mind just what you are trying to emulate. Think of what you believe a professor of psychology should be. Are they not expected to adopt a curious, questioning and critical attitude to whatever they read? Furthermore, they are very cautious people unwilling to take things for granted and demand evidence for everything – even things which seem self-evident to regular people. Reading like a professor should help you come up with a number of ideas about what needs to be done further.
- The more one reads, the more ideas come to one. It is a bit like writing a tune. Most of us would struggle to write a tune, whereas a skilled musician who has listened to and studied innumerable melodies would do so easily. Having heard and played thousands of tunes curiously makes it easier not harder to write a tune.
- Substantial student research projects are largely modelled on the style of academic publications – final year dissertations, for example. Consequently, some of the better student projects may be worthy of publication although this is not their prime purpose. While there is no guarantee that the results of your study will be publishable, it is a goal worth aiming at.

26.1 Introduction

The research conducted by students has as its primary purpose demonstrating what the student has achieved. Does their work demonstrate the necessary skills involved in designing, planning, analysing and reporting their psychological research? Demonstrating such a level of achievement is the first requirement. To this is then added an assessment of the layer of extra finesse that relates to the quality of the research ideas involved and the execution of the research. It is almost universal that psychology students have to carry out a research project as part of their training – at the undergraduate level, the postgraduate level or both.

It is only natural that students vary considerably in the extent that they can use their own ideas as the basis of their work. Departments of psychology will vary in the extent to which they expect this of student research as will members of staff within the department in question. Some departments require students to carry out research on topics outlined already by members of staff in a sort of apprenticeship system. At the other end of the range, other departments will positively encourage students to come up with their own research ideas. Both of these are reasonable options and have their own advantages and disadvantages. This situation is very much like academic research in general. For example, many junior research staff are employed simply to carry out the research plans of more senior staff – such as when they are employed as research assistants. In other situations you may be offered a rough idea of what to do, which you need to develop into a project yourself. Whatever the case where you study, remember that the quality of the outcome may have an important bearing on your future and so you should satisfy yourself that what you choose to do is worthwhile.

There are three main broad considerations that student researchers need to reflect upon when they plan to carry out a research project (see Figure 26.1):

- *Motivation* We all differ to some extent in terms of what motivates us best. Some students work best in fields which are especially pertinent to their experiences. For example, many students draw on their personal experiences as a basis for planning research – they want psychology to be relevant to their everyday life. Research into the experience of cancer, alcoholism, dyslexia and relationships may be firmly wedded to things which have happened in their lives. While it is often argued that academic researchers should be dispassionate, it does not follow from this that this excludes topics for research which are of personal relevance. Other students may be attracted to topics which are solely of intellectual interest to them – they may not expect or require the grounding of their research in real life. Given that a research project is a long-term investment of time and energy, it is a mistake to adopt a topic for research that cannot sustain one over a period of months. Many students find research projects and dissertations a major trial of stamina and character.

- *Practicality* There is little point in risking failure with a student research project. So be very cautious about planning research which is dependent on unknown or unpredictable contingencies for completion. For example, it might be a wonderful idea to study cognitive processes in a group of serial killers and, if properly done, the research might make a major contribution and even save a few lives. But what are the

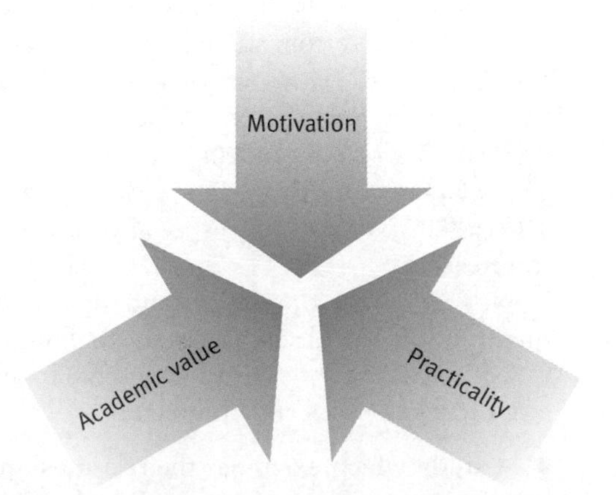

FIGURE 26.1 Some primary considerations when planning research

practicalities of something like this for a student research project? Fortunately, most of us do not know any serial killers and probably would need to resort to getting the cooperation of the prison service in order to obtain a sample. The likelihood of the prison service cooperating ought to be assessed seriously alongside the seriousness of the consequences should the prison service says no – as it is likely to in this case. Be very cautious of vague or even seemingly firm and well-intentioned promises of cooperation – we have seen cases where cooperation has been withdrawn at a late stage, leaving the student having to revamp their plans completely.

- *Academic value* Student research is most likely to be judged using conventional academic criteria. Research can be valuable for many other reasons but this does not necessarily mean that its weight in academic content is strong. For example, it may be very important for practitioners to know young people's attitudes to safe sex and AIDS (auto-immunodeficiency syndrome). The information gathered in a survey of young people may be highly valued by such practitioners. On the other hand, in terms of academic weight such a survey may meet few of the requirements of academics. The research might be seen by them as being atheoretical and merely a simple data-gathering task. Many academics would prefer research which helps develop new theories, validates theories, or is simply very smart or clever. So a student should try to ensure that their research is operating in the right playing field. Usually, the issue is one of ensuring that the theoretical concerns of the dissertation are made sufficiently strong – that is, there should be evidence that the research has an orientation towards theory.

26.2 Why not a replication study?

Although research projects such as final year projects and dissertations are in part judged in terms of their technical competence, they are also judged in the same terms as any other research work; for example, on the extent to which the research potentially makes a useful or interesting contribution. One must be realistic about what any single research study can contribute, of course. Many published papers make only a small contribution though, it should be stressed, there will probably be some disagreement as to the actual worth of any particular study. Excellence is often in the eye of the beholder. We have already seen that even experts can disagree widely as to whether a particular paper is of publishable quality (Cicchetti, 1991). Major theoretical or conceptual breakthroughs are *not* expected from student research or the run of the mill professional research paper for that matter. However, it is not unknown for student projects, if they are of top quality, to be published in academic journals. For example, the undergraduate projects of a number of our students have been published (for example, Cramer and Buckland, 1995; Cramer and Fong, 1991; Medway and Howitt, 2003; Murphy, Cramer and Lillie, 1984). This is excellent, especially where the student has ambitions towards a career in research.

Not all research projects stand an even chance of being regarded as of good or high quality. Some projects are likely to find less favour than others simply because they reflect a low level of aspiration and fail to appreciate the qualities of good research. Here are a few examples and comments:

- A study which examines the relationship between a commercial (ready-made) test of creativity and another commercially available test measuring intelligence. This study, even if it could be related to the theoretical literature on creativity and intelligence, does not allow the student to demonstrate any special skill in terms of method or analysis. It has also probably been carried out many times before. The value of the study might be improved if the variables measured were assessed using newly developed measuring instruments created by the researcher.

- A study which looks at, say, gender differences on a variable or age differences. Some research questions are mundane. Gender differences or age differences may well be important but it is difficult to establish their importance without an elaborate context which demonstrates why they are important. Sometimes the technicalities of demonstrating the gender difference are challenging and would compensate for the lack of complexity of the research question. Simply showing a gender difference for an easily measured variable has probably little going for it in terms of demonstrating a student's ability.

Replication studies are an interesting case in point. It is important to understand why some replication studies would be highly valuable whereas others would be regarded as rather mundane. A replication study that simply repeats what has already been done will probably be regarded as demonstrating technical and organisational proficiency at best. What it does not show is evidence of conceptual ability, creativity and originality – that extra little spark. Replications do have an important part to play in research – they are crucial to the question of the replicability of the findings. For example, if it were found that eating lettuce was associated with reductions in the risk of cancer then one priority would be to replicate this finding. Regrettably, replications are not accorded the high status that they warrant even in professional psychological research. No matter how important replication is in research work, it is not particularly effective at demonstrating the full range of a researcher's skills. This does not mean that a straightforward replication is easy – the information in a journal article, for example, may well be insufficient and the researcher doing the replication may have to contribute a great many ideas of their own. Even simple things such as the sorts of participant are difficult to replicate in replication research.

Relatively few straight or direct replication studies are to be found in the psychology research literature despite the great emphasis placed on replicability in the physical sciences. One reasonable rule of thumb suggests that direct replication is only valued to the extent that the original study was especially important or controversial – and that in some way additional value has been added by the inclusion of additional manipulations, checks or measures. For example, you might consider the circumstances in which the original findings are likely to apply and those where they do not. Extra data could be collected to assess this possibility.

However, as soon as one begins to think of a replication study in this way then the replication becomes something very different from a simple, direct or straight replication. One is including conditions which were not part of the original study and one is also thinking psychologically and conceptually. So we are talking about a part or partial replication here. A replication study can only confirm the original findings wholly or to some extent disconfirm them. Built into a partial replication is the likelihood that something new will be learnt over and above this. They are worthwhile because of this extra value which is added: they provide information about the conditions under which a finding holds in addition to showing the extent to which the original finding is replicable.

Examples of straight replication and partial replication may help:

- ***Straight replication*** Suppose a study found that women were more accurate at recognising emotions than men. We need not be concerned with the exact details of how this study was done. One approach would be to video people acting various emotions, play these videos to women and men, and ask them to report what emotions were expressed. The outcome of the replication would simply establish the extent to which the original findings were reliable.

- *Partial replication* What if we noticed that the people asked to act out the emotions were all, or predominantly, women? We may then be inclined to think that the results simply showed that women were more accurate than men at judging the emotions of women; we would not assume that women were generally more accurate at judging emotion than men. In order to extend our understanding we may want to know whether women are also more accurate than men at recognising the emotions of men. This could be achieved simply by ensuring that the video material included both women acting out emotions (as in the original study) as well as men acting out the emotions (unlike the original study). Obviously it would be important to ensure that the emotions acted by the men and women were the same ones, for example. This new research design is a partial replication since it actually accurately reproduces the original study when using videos of women acting emotionally but extends it to cover men acting emotionally. Why is this more worthwhile? Simply because it allows us to answer more questions such as:

    - Are women better at recognising emotions in general?
    - Are women only better at recognising emotions exhibited by members of their own gender?

  Now knowing this may not seem to be a huge amount of progress but it begins to open up theoretical and other issues about emotion recognition between and within the genders. Just what would account for the research findings? Has something important been established which warrants careful future research? (It is one of the curiosities about research in psychology that the study that answers a research question definitively seems to have a lower status than one that stimulates a plethora of further studies to sort out a satisfactory answer to the original research question.)

Although a straight replication increases our confidence in the original findings, it does nothing to further our understanding of the topic. If the new findings do not reflect the original findings, then this is of interest but does nothing in itself to explain the discrepancy between the findings of the original study and the replication. We could speculate as to the reasons why this is the case but this is sound evidence of nothing. Always there is more to be gained from investigating the new questions generated by the original study than merely replicating it. So, with care, a creative replication has a lot to commend it as a basis for student research.

26.3 Choosing a research topic

Most researchers are, at times, stuck for ideas for their future research if their expectations are high. They may be extremely well known and expert in their fields, yet research ideas do not flow simply because of this. It can be hard work to generate a good idea for research no matter one's level of expertise. It takes even more effort to convince others that you have a good idea! Once one has a good idea, there is a great deal of intellectual sweat and labour to turn it into a feasible, detailed plan for research. Consequently, do not expect to wake up one morning with a fully formed research question and plan in your mind. At first there is a vague idea that one would like to do research on a particular topic or research question and, perhaps, a glimmering recognition that the idea is researchable. The process is then one of discussing one's tentative ideas with anyone willing to listen and chip in thoughts, reading a lot around the topic and discarding what ideas do not seem to be working well. Usually some ideas with potential will establish themselves worthy of development. Sometimes one's ideas are too productive and not practicable as a consequence and so it is necessary to limit them in some way.

It is a good idea to think about the styles of research which appeal to you. These can have an impact on what is possible in terms of research. For example, if you have been particularly interested in in-depth interviewing as a means of data collection, you might ask yourself what can be done on the topic using this method. On the other hand, if you think that a well-designed laboratory experiment is your preferred mode of data collection then you can ask yourself what limitations this puts on the sorts of research questions you can ask.

Towards the end of their degree course, most students have found some topics from their lecture courses which are of interest to them. The research project is an opportunity to tackle something that interests you but in depth. Perhaps you will be spoilt for choice since there seem to be too many different things which intrigue you. There are several ways in which you may try to narrow down this choice:

- Try focusing on the topic that first aroused your interest in psychology. Does it still interest you? Have you been unable to satisfy your interest in the topic, perhaps because it was not covered in any of the courses you took?

- Try choosing a topic that may be relevant to the kind of work you intend to go into after graduating from university. For example, if you intend to go into teaching it may be useful to look at some aspect of teaching, such as what makes for effective teaching. This is a really good idea as not only is it relevant to your future career but it is a way of establishing that you have an interest in matters to do with that profession. It can work wonders at interviews, for instance.

- Choose a topic that interests you, which is part of a lecture course that you will be taking at the same time as doing the research project. In this way, the research and your studies will complement each other and you are likely to have a greater in-depth knowledge to bring to the lecture course as a consequence.

- If your attempts to focus down to a topic are not helping, try brainstorming a range of topics which you have some interest in. Try reading in depth into these, possibly starting with what you see as your best bet. Does one of them emerge as a front-runner for your interests? Does your reading on one topic have anything that might be transferred to another topic?

- If all else fails, try spending a couple of hours on a computer terminal simply skimming through the latest research abstracts irrespective of topic. Out of what you read does anything stand out as especially interesting? This is a quick way of getting an idea of the range of topics psychologists have studied and how they study them.

Regardless of how you approach selecting a topic, it is best to start thinking about the research possibilities a topic offers as soon as possible. Make a note of any ideas that come to you as you listen to lectures or read the literature. Although ideas may spring from your own experience and your observation of what happens around you, the greatest source of ideas is likely to come from reading and thinking about the ideas of others. Without studying the work of others, it is very difficult to develop your own ideas. This reading explains how researchers have conceived the topic of interest. It would be undesirable to ignore all of this past work since it amounts to a repository of hard thinking, good analysis and ways of conceptualising the important issues. Authors may propose in the *discussion* section of their paper one or more specific suggestions about further work that may be worth carrying out on the topic that they have been investigating. The basic formula is that reading makes ideas that work (see Figure 26.2).

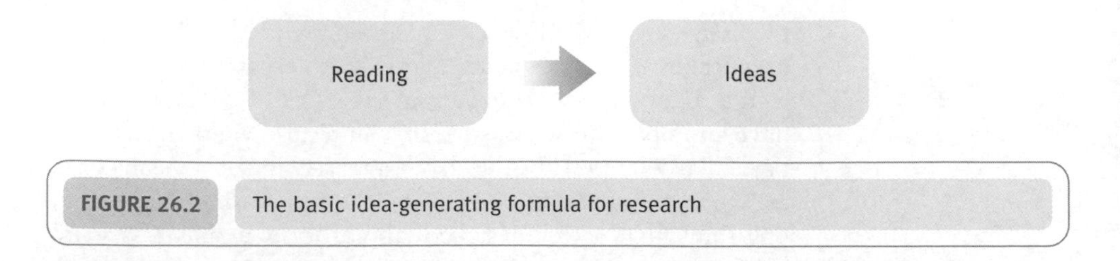

FIGURE 26.2 The basic idea-generating formula for research

It is surprising that there are big gaps in psychological knowledge and many areas of research simply have received little or no previous coverage. Nevertheless, sometimes students get disconcerted when they come across research which is similar to that which they are planning or doing. Of course, there is always the chance that someone else publishes work similar to yours before your project is completed. This seems to occur very infrequently, however. Perhaps this is because the way we think about a topic is usually very different from the way that other people think about it. Whatever the reason, it is unlikely that someone will be about to publish the study that you are currently thinking of doing. However, if this does occur, simply acknowledge it in your report and remember to evaluate the two studies and describe their similarities and differences.

26.4 Sources of research ideas

Research into how psychologists get their research ideas seems conspicuously absent – a good research for a dissertation?! So there is little to be written based on this. McGuire (1997) suggested 49 different ways or heuristics of generating hypotheses which is one less than the number of ways to leave your lover! Our list of suggestions about sources of research ideas is more modest than this. Really our suggestions are of things to think about and they are not mutually exclusive. Several different aspects of our list might be adopted in order to come up with ideas. Ours is not an exhaustive list either. Others will have other ideas and if they work for you then they have done their job. We will illustrate our potential sources of ideas with a brief example or two of the kind of ideas they might generate wherever possible (see Figure 26.3).

- ***Detailed description of a phenomenon*** It is often important to try to obtain a thorough and accurate description of what occurs before trying to understand why it occurs. It is possible that previous studies may have done this to some extent but their descriptions may omit what appears to you to be certain critical aspects of the phenomenon. For example, we do not know why psychotherapy works. One way of trying to understand what makes it effective is to ask patients in detail how it has helped them to cope with or to overcome their problem.
- ***Theory*** No matter what your chosen field of research, attention to the relevant pertinent theories is invaluable. Remember that the purpose of research is *not* primarily to produce more data but to extend our conceptual understanding of our chosen subject matter. Researchers are well advised to emphasise the relevant theory in their chosen field as a consequence. An absence of theory means that the conceptualisation of the relevant issues is much harder for the researcher. After all, the purpose of theory is to present a conceptual scheme to describe and understand a phenomenon. If there is an absence of theory in the published writings in the field then are there theories in other, perhaps similar, fields which can be used? These may help illuminate

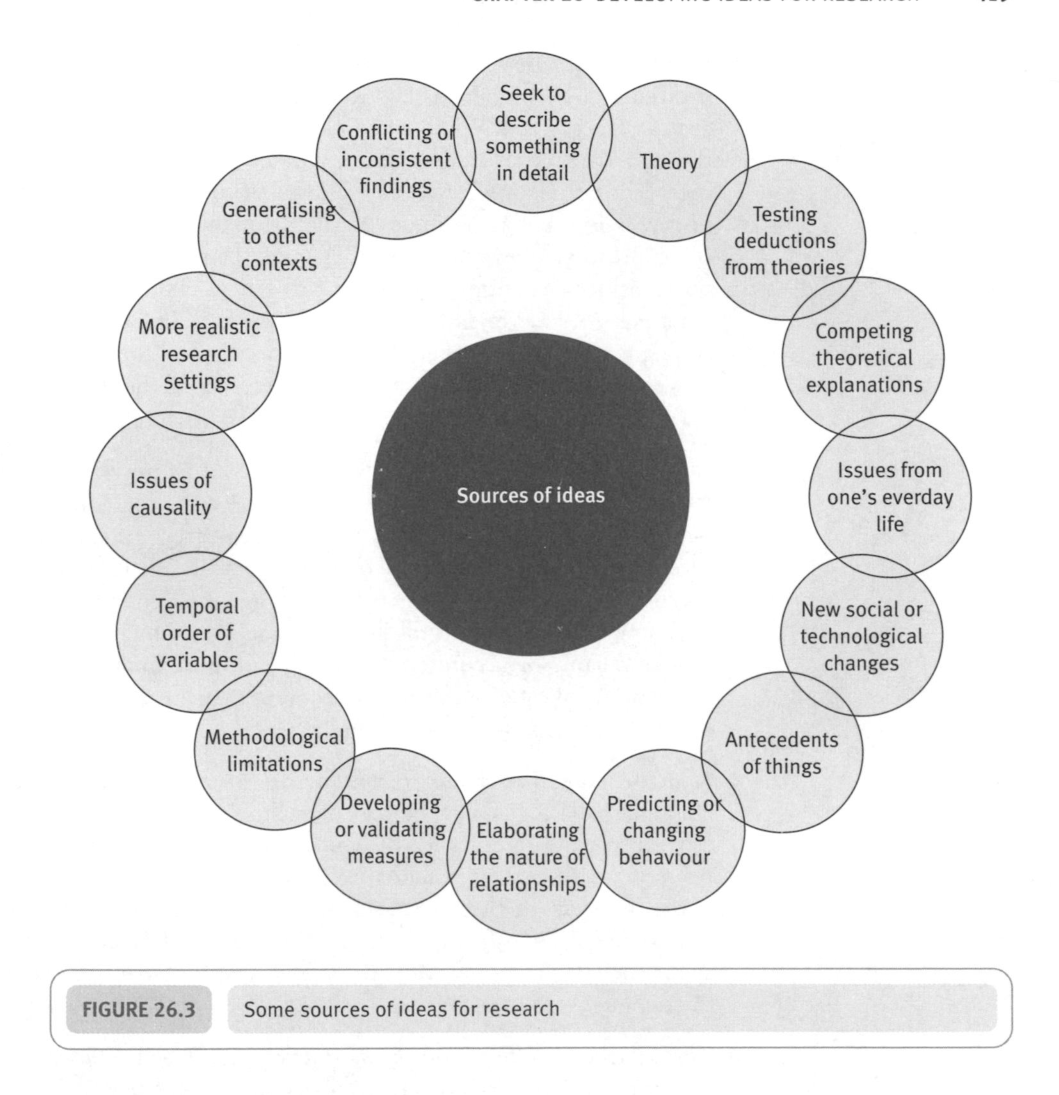

FIGURE 26.3 Some sources of ideas for research

the field better than most purely empirical studies would. The integration of theory with empirical work is the best combination. It is more than a matter of testing theory in your study since many theories are too imprecise for such a test. On the other hand, the theory may have potential for integrating various aspects of your analysis and report in general. It is a useful minimum requirement that you seek to introduce relevant theory into your writings. If you can achieve some integration of theory beyond this minimum then you are doing very well in your report. If your research can explore the application of theory to a particular topic then this generally has a powerful effect on your report's quality. The big limitation is that theory in psychology tends to deal with a modest level of generalisation which can make it difficult to apply in new contexts. Deductions from theories are discussed later. Box 26.1 shows that psychologists who make the biggest impact on the discipline are overwhelmingly those with a lot to say theoretically.

- *Deductions from theories* Much of psychology is concerned with developing theories to explain behaviour. Theories which attempt to explain a wider variety of behaviours are generally more useful than those which have a narrower focus for that very reason. While some aspects of these theories may have been extensively tested, other aspects of them may have received little or no attention but may be worth investigating.

- *Competing theoretical explanations* In the classic view of scientific progress, there is an idea that competing theories used to describe a phenomenon may be put to a critical decisive test. While many psychologists would believe that few psychological theories are so precise that this is possible, nevertheless attempts to do so are well regarded. For example, there are a number of theories for explaining the occurrence of depression. It may be an intellectually satisfying challenge to take the main theories of depression and examine how they might be put to some sort of crucial test. So why not consider evaluating competing theoretical explanations of your chosen topic as the basis of your research project? While it is unlikely that a death blow will be struck against one of the competing theories, your contribution would be part of the longer-term process of evaluating the tenability of the theory.

- *Everyday issues* Frequently there are a number of different ways in which something can be done in our everyday lives. There may be no research to help choose which is the most effective procedure to adopt or what the consequences might generally be of using a particular approach. We could carry out some research to help answer these questions. For example, if we concentrate on research itself for the moment, are potential participants less likely to agree to complete a longer than a shorter questionnaire? If they do fill in the longer questionnaire, are their answers likely to be less reliable or accurate than if they complete a shorter questionnaire? Does the order of the questions affect how accurate they are when disclosing more sensitive or personal information about themselves?

- *New or potential social, technological or biological developments* We live in changing times when new biological, technological and social developments are being introduced, the significance of which we are not sure. Their effects or what people perceive as being their potential effects may be a topic of great public concern. What is the influence of the Internet or text messaging on our social behaviour? Has the development and spread of the human immunodeficiency virus (HIV) affected our sexual behaviour in any way? Do student fees influence the occupational aspirations of students?

- *The antecedent–behaviour–consequences (A–B–C) model* It may be useful to remember that any behaviour (B) that we are interested in often has both antecedents (A) and consequences (C). For example, the antecedents of depressive behaviour may include unsatisfactory childhood relationships and negative experiences. The consequences may be unsatisfactory personal relationships and poor school or work attendance. Deciding whether we are more interested in the antecedents than the consequences of a particular behaviour may also help us focus our minds when developing a research question. Once this decision has been made, we might try to investigate neglected antecedents of depression.

- *Predicting or changing behaviour* How could you go about changing a particular behaviour? Do you think that you could predict when that behaviour is likely to occur? Addressing these questions would require you to think about the variables which are most likely to be, or have been found to be, most strongly associated with the behaviour in question. These variables would form part of an explanation for this behaviour. You could investigate whether these variables are in fact related to the behaviour in question. This approach may encourage you to think of how your knowledge of psychology could be applied to practical problems.

- *Elaborating relationships* There may be considerable research showing that a relationship exists between two variables. For example, there is probably substantial amounts of research which demonstrates a gender difference such as females being less aggressive than males. The next step once this has been established would be to

Box 26.1 Talking Point

Top of the citations

It is intriguing to find that many of the most cited psychologists in psychology journals are very familiar names to most psychology students. Many of the most cited psychologists include those who have made highly influential theoretical contributions. Freud, for example, was a major theorist but a minor contributor of research. Remember that the theories referred to in journals are those which are influential on research. Haggbloom and his colleagues (2002) generated a list which ranked the 100 psychologists most frequently cited in journals according to how often they have been cited. The first 25 of these are shown in Table 26.1. (Beside the name of each psychologist we have given one major contribution that this person is known for although we do not know whether this contribution is the reason why they have been cited.) Not all the people on this list have put forward a major theory. For example, Ben Winer who is ranked fourth is most probably cited for his writings on statistical analysis. There are, of course, many other theories which are not listed which may be of greater interest to you.

Table 26.1 The 25 psychologists cited most often in journals

Rank	Psychologist	Citations	Contribution
1	Sigmund Freud	13 890	Psychoanalytic theory
2	Jean Piaget	8821	Developmental theory
3	Hans J. Eysenck	6212	Personality theory; behaviour therapy
4	Ben J. Winer	6206	Statistics
5	Albert Bandura	5831	Social learning theory
6	Sidney Siegel	4861	Statistics
7	Raymond B. Cattell	4828	Personality theory
8	Burrhus F. Skinner	4339	Operant conditioning theory
9	Charles E. Osgood	4061	Semantic differential scale
10	Joy P. Guilford	4006	Intelligence and personality models
11	Donald T. Campbell	3969	Methodology
12	Leon Festinger	3536	Cognitive dissonance theory
13	George A. Miller	3394	Memory
14	Jerome Bruner	3279	Cognitive theory
15	Lee J. Cronbach	3253	Reliability and validity
16	Erik H. Erikson	3060	Psychosocial developmental theory
17	Allen L. Edwards	3007	Social desirability
18	Julian B. Rotter	3001	Social learning theory
19	Don Byrne	2904	Reinforcement-affect theory
20	Jerome Kagan	2901	Children's temperaments
21	Joseph Wolpe	2879	Behaviour therapy
22	Robert Rosenthal	2739	Experimenter expectancy effect
23	Benton J. Underwood	2686	Verbal learning
24	Allan Paivio	2678	Verbal learning
25	Milton Rokeach	2676	Values

Adapted from The 100 most eminent psychologists of the 20th century, *Review of General Psychology*, 6, 139–52 (Haggbloom, S. J., Warnick, R., Warnick, J. E., Jones, V. K., Yarbrough, G. L., Russell, T. M. *et al.* 2002), American Psychological Association.

try to understand why this gender difference exists – what the factors are which are responsible for this difference. If you believe that the factors are likely to be biological, you look for biological differences which may explain the differences in aggression. If you think that the factors are probably psychological or social, you investigate these kinds of factors. If you are interested in simply finding out whether there are gender differences in some behaviour, then it is more useful and interesting to include one or more factors which you think may explain this difference. In other words, it is important to test for explanations of differences or relationships rather than merely establish empirically that a relationship exists.

- *Developing and validating measures* Where there is no measure for assessing a variable that you are interested in studying, then it is not too difficult to develop your own. This will almost certainly involve collecting evidence of the measure's reliability and, probably, its validity. For some variables there may be a number of different measures that already exist for assessing them, but it may be unclear which is the most appropriate in particular circumstances or for a particular purpose. Alternatively, is it possible to develop a more satisfactory measure than the ones which are currently available? Your new measure may include more relevant aspects of the variable that you want to assess. Would a shorter or more convenient measure be as good a measure as a longer or less convenient one? Do measures which seem to assess different variables actually assess the same variable? For example, is a measure of loneliness distinguishable from a measure of depression in practice?
- *Alternative explanations of findings* A researcher may favour a particular explanation for their findings but there may be others which have not been considered or tested. Have you come across a research publication which has intrigued you but you are not absolutely convinced that the researcher has come up with the best explanation? Do you have alternative ideas which would account for the findings? If so, why not try to plan research which might be decisive in helping you choose between the original researcher's explanation and yours? This is not quite as easy as it sounds. One of the reasons is that you need to read journal articles in a questioning way rather than as being information which should be accepted and digested. For much of your education, you have probably read uncritically merely to gain information. To be a researcher, you need a rather different mindset which says 'convince me' to the author of a research paper rather than 'you're the expert so I accept what you say'.
- *Methodological limitations* Any study, including important ones, may suffer from a variety of methodological limitations. An obvious one for much research is the issue of high internal validity but low external validity. Basically this is a consequence of using contrived laboratory experiments to investigate psychological processes.These experiments may be extremely well designed in their own terms, but have little relevance to what happens in the real world. For example, there are numerous laboratory studies of jury decision-making but would one be willing to accept their findings as relevant in the world beyond the psychology laboratory? A study which explores in naturalistic settings phenomena which have been extensively studied in the psychology laboratory would be a welcome addition in many fields of research. Any variable may be operationalised in a number of different ways. How sound are the measures used in the research? For example there are numerous studies of aggression which rely on the (apparent) delivery of a noxious electrical shock as a measure of aggressiveness. As you probably know from the famous Stanley Milgram studies of obedience, this same measure would be used by some researchers as an indicator of obedience rather than a measure of aggressiveness. Yet these are identical measures but are claimed to measure different things. Clearly there is the opportunity to question many measures including those used in classic research studies.

- *Temporal precedence or order of variables* Studies that involve a dynamic component of change over a period of time are relatively uncommon in psychology despite many researchers advocating their use. They are also uncommon in student research. Obviously time constraints apply, which may account partially for their rarity. Longitudinal or panel designs, such as those outlined in Chapter 12 that measure the same variables on the same individuals at two or more points in time, enable the temporal relationships between variables to be examined and compared. Often this sort of research takes place in a less contrived setting than is possible using an experimental design. Consider the possibility of using a longitudinal design since not only is it more challenging than other types of research but it may also generate new possibilities for research in fields which are perhaps otherwise heavily researched.

- *Causality* Student researchers can often benefit from concentrating on the possibility of carrying out an experimental study in their chosen field. Despite there being limitations to the experimental method for some purposes, it remains the quintessential research method in psychology. Consequently, an experimental design will garner favour. Remember that the main purpose of student research is for the student to demonstrate that they have mastered the crucial skills of research. Experimental designs are a good way of doing this.

- *More realistic settings* One of the common criticisms of psychology, especially that which is taught at university, is that it dwells on the laboratory experiment too much and neglects research carried out in more naturalistic settings. Is it possible to take one of these somewhat contrived laboratory experiments and recast it in a more naturalistic and less contrived fashion? Often the way in which we study a phenomenon may be contrived in order to control for variables which may affect our findings. But this is not the only reason. For example, take the example once again of Stanley Milgram's famous study of obedience in which participants ostensibly gave massive electric shocks to another person in the context of a study of learning. One might ask about obedience in real-life settings, for example. Just what are the determinants of obedience to authority in real-life settings such as a sports team? What determines whether the captain's instructions are adhered to? Sometimes it can be useful to see if similar findings can be obtained in less contrived circumstances than the original study in order to assess just how robust the original findings are.

- *Generalising to other contexts* Theories or findings in one area may be applicable to those in other areas. For example, theories which have been developed to explain personal relationships may also apply to work relationships and may be tested in these contexts.

- *Conflicting or inconsistent findings* It is a common comment that psychological research on a given topic has some studies finding one outcome and other studies finding the reverse outcome. That is, the findings of studies on a particular topic are less than consistent with each other. For example, some studies may find no difference between females and males for a particular characteristic, others may find females show more of this characteristic, while others may find females show less of this characteristic. Why might this be the case? If the majority of studies obtain a similar kind of finding, the findings of the inconsistent remaining studies may be due to sampling error and many psychologists will ignore the inconsistent studies as a consequence. However, it is often better to regard the inconsistency as something to be explained and taken into account. Is it possible to speculate about the differences between the studies that might account for the difference in outcomes? In this situation it is not very fruitful to repeat the study to see what the results will be because we already know what the possible outcomes are. While it may not be easy to do this, it is better to think of variables which may explain these inconsistent findings and to see whether or not this is the case. One could perhaps consider carrying out a meta-analysis of the studies in order to explore a number of differences which might account for the variation in the outcomes of the studies. Meta-analysis can be carried out using quite simple procedures. The technique is described in detail in the companion statistics text, *Introduction to Statistics in Psychology* (Howitt and Cramer, 2011a) at a level which is within the capabilities of most students.

26.5 Conclusion

Research projects are intended to be major means to develop a student's intellectual development and, at the same time, to assess this. While most psychology students do some research at each stage of their education and training, early on it is likely that they carry out a study according to a plan more or less given to them in finished form by academic staff. Individual research projects come at the end of a degree programme simply because a student needs to have mastered many skills before they can do a good job of planning and carrying out research of their own. The better that everything that has gone before is mastered, the more likely a student is to make a good job of independent research. For example, unless you have read examples of how researchers formulate and justify their research ideas, you will not know how to formulate psychological research questions. At its root, this process is one of reading and study. Even then, students can find it difficult to come up with a good idea for a research project. It is something that cannot be done in a hurry and adequate time needs to be laid aside in order to develop ideas. Usually students who have had a positive approach to reading and studying will have fewer problems in generating research ideas. They have started the ground work after all.

It is never too early to start thinking about research projects. While it may be exceptional to find a student who thinks a whole year ahead, the sooner that you can find time to think about research ideas the better. If you have given yourself enough time, you may find it helpful to keep a range of possible research topics on the table for consideration. This will minimise the damage if it should happen that the topic that you set your heart on does not turn into a workable idea. One learns to think psychologically about things through a fairly lengthy process of reading and actively studying and the same applies to thinking like a researcher.

You can get to understand how psychologists actually do research by reading a variety of research papers in any field that interests you. When you are planning research, however, you need to focus on and familiarise yourself with the established research literature on the topic – especially recent research. By reading in this way you will learn something about what are sensible research questions to pose. You will gain insight into how people interested in the same issue as yourself have construed the topic and planned the research. You will know about what sorts of measures are typically taken and what procedures for doing research seem to work well in this field.

However, do not prevent yourself from asking what might seem to be obvious questions about the topic that do not seem to have been addressed. These obvious questions may have escaped the attention of researchers and may form the basis of your own research. Think also of situations in which the findings may not apply. We may have a tendency to seek instances which confirm what we know rather than instances which disconfirm our preconceptions. Thinking of situations in which the findings may not apply will make us aware of the extent to which we cannot generalise these findings to all contexts.

Practicalities may prevent you doing what you really want to do. Always try to anticipate these and consider more modest possibilities. For example, it is unlikely that you would be able to evaluate the effectiveness of, say, a substantial therapeutic intervention, but you may be able to investigate people's preference for or attitude to that intervention. Doing one's own research provides one of the few in-depth opportunities to learn about a topic and to make a contribution, however modest, to understanding that topic. It allows you to show and to see just what you are capable of intellectually. For many students, it will be the most fulfilling and possibly the most frustrating part of their studies. Hopefully what you have learnt from this will provide you with a sound understanding of research methods and with resources to help you develop further. Nevertheless you cannot learn to be a researcher just from a book. Conducting research is a skill which requires practice. We would be delighted to know that this book has stimulated your appetite for research.

Alternatively, there is always Plan B!

Key points

- Developing good research ideas is not easy. Even once you have the necessary academic skills, it takes time to choose a topic and to familiarise yourself with the research on that topic. The sooner you start thinking about your research ideas the better.
- Be realistic about what you can achieve with the limited resources that are available to you. Much of the research that you read about has probably taken a great deal more time to conceive and to carry out than you have available. Nonetheless, students at all levels can and have carried out research of value – some of which has been published.
- Simply replicating research that has already been done is unlikely to advance our understanding of that topic. It is also unlikely to impress those assessing your work. While replication is important to determine how reliable a finding is, it is sensible to do more than just replicate the original study. At the same time as doing the replication, it is often possible to address new questions which emerge from the original research report. It is better to distinguish this kind of replication by calling it part or partial replication. Much research is partial replication.
- Choose a topic that interests you or that may be relevant to what you want to do in your future career or further studies. Think of a few topics that interest you and start reading around your favourite. If necessary go on to the next topic once you understand why your first choice has not been productive.
- There is no single, foolproof way of generating a good idea for research. However, ideas are most likely to be generated as one reads and ponders over the research that other people have reported. Often what you do will involve a variation of what has been previously done and may clear up an unresolved issue. It is also likely to raise a number of other issues that in turn need to be answered.
- Carrying out a piece of original research is your opportunity to make a contribution to the topic that interests you. Make the most of this opportunity.

ACTIVITY

Think of three topics that you would like to do research on. Decide which of these three is the most promising for you and start thinking what you would like to know about it or what you think should be known about it. Do a literature search preferably using an electronic database such as Web of Science or PsycINFO. Try to locate the most recent research on the topic and read it. When reading the research adopt a critical and questioning attitude to what has been written. Try to read some of the references cited to see whether they have been accurately cited and are relevant to the point being made in the paper you are currently reading. Are the views in these references generally consistent with those in the paper? If they are not consistent, what seem to be the reasons for this? Which view do you generally support? In terms of the study, how were the main variables operationalised? Are there better ways of operationalising them? What suggestions were made for further research and do these seem worth pursuing? What questions remain to be answered? How would you design a study to test these?

GLOSSARY

***a priori* comparison:** An analysis between two group means from several means in which the direction of the difference has been predicted on the basis of strong grounds and before the data have been collected.

abstract: The summary of a publication.

action: Behaviour which is meaningful rather than reflexive.

adjacency pairs: A unit consisting of two turns in a conversation which follow a standard pattern.

alpha reliability (Cronbach's alpha): A measure of internal reliability (consistency) of items. It is effectively the mean of all possible split-half reliabilities adjusted for the smaller number of variables making up the two halves.

alternate (alternative) hypothesis: A statement or expression of a proposed relationship between two or more variables.

alternate forms reliability: The correlation between different versions (forms) of a measure designed to assess the same variable. The reliability coefficient indicates the extent to which the two forms are related. Alternate forms are used to avoid the practice effects which might occur if exactly the same measure is given twice.

American Psychological Association (APA): The largest organisation in the USA of professional psychologists.

analysis of covariance (ANCOVA): An analysis of variance in which the relation between the dependent variable and one or more other variables is controlled.

analysis of variance (ANOVA): A parametric test which determines whether the variance of an effect differs significantly from the variance expected by chance.

analytic induction: The process of trying to develop working ideas or hypotheses to explain aspects of one's data. It is the opposite of deduction where ideas are developed out of theory.

ANCOVA *see* analysis of covariance.

ANOVA *see* analysis of variance.

APA *see* American Psychological Association.

appendix: The section at the end of a publication or report which contains supplementary further information.

applied research: Research which has as its primary objective the search of solutions to problems.

archive: A collection of documents.

attrition: The loss of research participants during a study such as when they drop out or fail to attend.

basic laboratory experiment: A true or randomised experiment which is conducted in a controlled environment. Random assignment to the groups or conditions is an essential feature.

Behaviourist School of Psychology: An approach which holds that progress in psychology will be advanced by studying the relation between external stimuli and observable behaviour.

between-subjects design: A study in which subjects or participants are randomly assigned to different conditions or groups.

bias: The influence of pre-existing judgements on a research study.

Bonferroni test: The significance level of a test multiplied by the number of comparisons to be made to take account of the fact that a comparison is more likely to be significant the more comparisons that are carried out.

bracketing: The attempt to suspend normal judgements by the researcher/analyst.

British Crime Survey: A regular nationally representative survey of people in Britain looking at their views and experience of crime.

British Psychological Society: The largest organisation of professional psychologists in Britain.

British Social Attitudes Survey: A regular nationally representative survey of adults in Britain about a variety of different social issues.

CAQDAS (Computer-Assisted Qualitative Data Analysis System): Computer software used to help carry out the analysis of qualitative data.

carryover, asymmetrical/differential transfer: The finding of a different effect depending on the order in which conditions are run in a within-subjects design.

CASCOT *see* Computer-Assisted Structured COding Tool.

case: A specific instance of the thing chosen for study – such as a single research participant.

case study: A study based on a single unit of analysis such as a single person or a single factory.

categorisation: The classification of objects of a study into different groups.

category/categorical variables *see* nominal variable.

causal explanation: An explanation in which one or more variables are thought to be determined (affected) by one or more other variables.

causal hypothesis: A hypothesis which states that one or more variables are brought about by one or more other variables in a cause-and-effect sequence.

causality: The idea that one or more variables affect one or more other variables.

cause: A variable that is thought to affect one or more other variables.

chance finding: A result or outcome which generally has a probability of occurring more than five times out of a hundred.

check on experimental manipulation: The process of determining whether a particular intervention has varied what it was supposed to have varied. It assesses whether the experimental manipulation has been effective in creating a particular difference between the experimental and control groups. It cannot be assessed simply by comparing means on the dependent variable.

Chicago School of Sociology: An approach to sociology which emphasised quantification and the study of large groups.

citation: A reference to another source of information such as a research article.

cluster sampling: The selection of spatially separate subgroups which is designed to reduce the time and cost to obtain a sample because members of each cluster are physically close together.

coding data: The process of applying codes or categories to qualitative data (or sometimes quantitative data).

coding frame: The list of codes which may be applied to the data such as in content analysis.

coefficient of determination: The square of a correlation coefficient which gives the proportion of the variance shared between two variables.

comparative method: A comparison of one group of objects of study with one or more other groups to determine how they are similar and different in various respects.

computer grounded-theory analysis: The use of computer software to help carry out a grounded theory analysis.

Computer-Assisted Structured COding Tool (CASCOT): Software for assigning the occupations of participants according to the Standard Occupation Classification 2000.

concept: A general idea which is developed from specific instances.

conclusion: The final section of a publication in which the main arguments are restated.

concurrent validity: The extent to which a measure is related to one or more other measures all assessed at the same time.

condition: A treatment usually in a true experimental design which is part of the independent variable.

confidence interval: The range between a lower and a higher value in which a population estimate may fall within with a certain degree of confidence which is usually 95 per cent or more.

confidentiality: The requirement to protect the anonymity of the data provided by participants in research.

confounding variable: A variable which wholly or partly explains the relation between two or more other variables. It can bring about a misinterpretation of the relationship between the other two variables.

construct validity: The extent to which a measure has been found to be appropriately related to one or more other variables of theoretical relevance to it.

constructivism: The idea that people have a role in creating knowledge and experience.

content analysis: The coding of the content of some data such as television programmes or newspapers.

control condition: A condition which does not contain or has less of the variable whose effect is being determined. It forms a sort of baseline for assessing the effect of the experimental condition.

convenience sample: A group of cases which have been selected because they are relatively easy to obtain.

convergent validity: The extent to which a measure is related to one or more other measures to which it is thought to be related.

conversation analysis: The detailed description of how parts of conversation occur.

correlational research *see* correlational/cross-sectional study.

correlational/cross-sectional study: Research in which all the measures are assessed across the same section or moment of time.

covert observation: Observation of behaviour when those being observed are not aware they are being observed.

crisis: the stage in the development of a discipline when commonly accepted ways of understanding things become untenable thus motivating the search for radically new approaches to the discipline.

critical discourse analysis: A form of discourse analysis which is primarily concerned with observing how power and social inequality are expressed.

critical realism: The idea that there is more than one version of reality.

Cronbach's alpha *see* alpha reliability.

cross-lagged relationship: The association between two different variables which have been measured at different points in time.

cross-sectional design: A design where all variables are measured at the same point in time.

cross-sectional study: Research in which all variables are measured at the same point in time.

data: Information which is used for analysis.

debriefing: Giving participants more information about the study after they have finished participating in it and gathering their experiences as participants.

deception: Deliberately misleading participants or simply not giving them sufficient information to realise that the procedure they are taking part in is not what it appears.

deconstruction: The analysis of textual material in order to expose its underlying contradictions and assumptions.

deduction: The drawing of a conclusion from some theoretical statement.

demand characteristics: Aspects of a study which were assumed not to be critical to it but which may have strongly infiuenced how participants behaved.

dependent variable: A variable which is thought to depend on or be influenced by one or more other variables, usually referred to as independent variables.

design: A general outline of the way in which the main variables are studied.

determinism: The idea that everything is determined by things that went before.

deviant instance: A case or feature which appears to be different from most other cases or features.

Dewey Decimal Classification (DDC) system: A widely used scheme developed by Dewey for classifying publications in libraries which uses numbers to refer to different subjects and their divisions.

dichotomous, binomial/binary variable: A variable which has only two categories such as 'Yes' and 'No'.

dialogical: In the form of a dialogue.

directional hypothesis: A hypothesis in which the direction of the expected results has been stated.

discourse analysis: The detailed description of what seems to be occurring in verbal communication and what language does.

discriminant validity: The extent to which a measure does not relate highly to one or more variables to which it is thought to be unrelated.

discussion: A later section in a publication or report which examines alternate explanations of the main results of the publication and which considers how these are related to the results of previous publications.

disproportionate stratified sampling: The selection of more cases from smaller sized groups or strata than would be expected relative to their size.

Economic and Social Science Research Council (ESRC): A major organisation in Britain which awards grants from government funds for carrying out research and for supporting postgraduate research studentships and fellowships.

electronic database: A source of information which is stored in digital electronic form.

emic: The understanding of a culture through the perspective of members of that culture.

empiricism: The belief that valid knowledge is based on observation.

ESRC *see* Economic and Social Science Research Council.

essentialism: The idea that things have an essential nature which may be identified through research.

ethics: A set of guidelines designed to govern the behaviour of people to act responsibly and in the best interest of others.

ethnography: Research based on the researcher's observations when immersed in a social setting.

etic: The analysis of cultures from perspectives outside of that culture.

evaluation/outcome study: Research which is primarily concerned with the evaluation of some intervention designed to enhance the lives and welfare of others.

experimental condition: A treatment in a true experiment where the variable studied is present or present to a greater extent than in another treatment.

experimental control: A condition in a true experiment which does not contain or has less of the variable whose effect is being determined. It forms a baseline against which the effect of the experimental manipulation is measured.

experimental manipulation: The deliberate varying of the presence of a variable.

experimenter effect: The systematic effect that characteristics of the person collecting the data may have on the outcome of the study.

experimenter expectancy effect: The systematic effect that the results expected by the person collecting the data may have on the outcome of the study.

external validity: The extent to which the results of a study can be generalised to other more realistic settings.

extreme relativism: The assumption that different methods of qualitative research will provide different but valid perspectives of the world.

face validity: The extent to which a measure appears to be measuring what it is supposed to be measuring.

factor analysis: A set of statistical procedures for determining how variables may be grouped together in terms of being more closely related to one another.

factorial design: A design in which there are two or more independent or subject variables.

feasibility study: A pilot study which attempts to assess the viability and practicality of a future major study.

fit: The degree to which the analysis and the data match.

focus group: Usually a small group of individuals who have been brought together to discuss at length a topic or related set of topics.

Grice's maxims of cooperative speech: Principles proposed by Grice which he believed led to effective communication.

grounded theory: A method for developing theory based on the intensive qualitative analysis of qualitative data.

group: A category or condition which is usually one of two or more groups which go to make up a variable.

hierarchical or sequential multiple regression: Entering individual or groups of predictor variables in a particular sequence in a multiple regression analysis.

hypothesis: A statement expressing the expected relation between two or more variables.

hypothetico-deductive method: The idea that hypotheses should be deduced from theory and tested empirically in order to progress scientific knowledge.

idiographic: The intensive study of an individual.

illocution: The effect of saying something.

illocutory act: The function that saying something may have.

independent variable: A variable which is thought and designed to be unrelated to other variables thus allowing its effect to be examined.

in-depth interview: An interview whose aim is to explore a topic or set of topics at length and in detail.

indicator: A measure which is thought to reflect a theoretical concept or construct which may not be directly observed.

induction: The development of theory out of data.

inferring causality: The making of a statement about the causal relation between two or more variables.

informed consent: Agreement to taking part in a study after being informed about the nature of the study and about being able to withdraw from it at any stage.

Institutional Review Board (IRB): A committee or board in universities in the United States which consider the ethics of carrying out research proposals.

interaction: When the relation between the criterion variable and a predictor variable varies according to the values of one or more other predictor variables.

internal reliability: A measure of the extent to which cases respond in a similar or consistent way on all the variables that go to make up a scale.

internal validity: The extent to which the effect of the dependent variable is the result of the independent variable and not some other aspect of the study.

interpretative phenomenological analysis (IPA): A detailed description and interpretation of an account of some phenomenon by one or more individuals.

intervening or mediating variable: A variable which is thought to explain wholly or partly the relation between two variables.

intervention manipulation: An attempt to vary the values of an independent variable.

intervention research: A study which evaluates the effect of a treatment which is thought to enhance well-being.

interview: Orally asking someone questions about some topic or topics.

introduction: The opening section of a research paper which outlines the context and rationale for the study. It is usually not titled as such.

item analysis: An examination of the items of a scale to determine which of them should be included and which can be dispensed with as contributing little of benefit to the measure.

item-whole or item-total approach: The relation between the score of an item and the score for all the items including or excluding that item.

Jefferson transcription: A form of transcription which not only records what is said but also tries to convey some of the ways in which an utterance is made.

known-groups validity: The extent to which a measure varies in the expected way in different groups of cases.

laboratory experiment *see* basic laboratory experiment.

Latin squares: The ordering of conditions in which each condition is run in the same position the same number of times and each condition precedes and follows each other condition once.

levels of treatment: The different conditions in an independent variable.

Library of Congress Classification system: A scheme, developed for the library of the US Congress, that uses letters to classify main subject areas with numbers for their subdivisions.

Likert response scale: A format for answering questions in which three or more points are used to indicate a greater or lesser quantity of response such as the extent to which someone agrees with a statement.

literature review: An account of what the literature search has revealed, which includes the main arguments and findings.

literature search: A careful search for literature which is relevant to the topic being studied.

locution: The act of speaking.

locutory: The adjective for describing an act of speaking.

longitudinal study: Research in which cases are measured at two or more points in time.

MANOVA *see* multivariate analysis of variance.

margin of error *see* sampling error.

matching: The selection of participants who are similar to each other to control for what are seen as being important variables.

materials/apparatus/measures: The subsection in the method section of a research paper or report which gives details of any objects or equipment that are used such as questionnaires or recording devices.

measurement characteristics of variables: A four fold hierarchical distinction proposed for measures comprising nominal, ordinal, equal interval and ratio scales.

mediating variable *see* intervening or mediating variable.

memo-writing: The part of a grounded theory analysis in which a written record is kept of how key concepts may be related to one another.

meta-analytic study: Research which seeks to find all the quantitative studies on a particular topic and to summarise the findings of those studies in terms of an overall effect size.

metaphysics: Philosophical approaches to the study of mind.

method: The section in a research report which gives details of how the study was carried out.

moderating variable: A variable where the relation between two or more other variables seems to vary according to the values of that variable.

multidimensional scale: A measure which assesses several distinct aspects of a variable.

multinomial variables: A variable having more than two qualitative categories.

multiple comparisons: A comparison of the relation or difference between three or more groups two at a time.

multiple dependent variables: More than one dependent variable in the same study.

multiple levels of independent variable: An independent variable having more than two groups or conditions.

multiple regression: A parametric statistical test which assesses the strength and direction of the relation between a criterion variable and two or more predictor variables where the association between the predictor variables is controlled.

multi-stage sampling: A procedure consisting of the initial selection of larger units from which cases are subsequently selected.

multivariate analysis of variance (MANOVA): An analysis of variance which has more than one dependent variable.

national representative survey: A study of cases from a nation state which is designed to reflect all the cases in that state.

national survey: A study of cases which selects cases from various areas of that state.

naturalistic research setting: A situation which has not been designed for a particular study.

Naturalism: The belief that psychology and the social sciences should adopt the methods of the natural sciences such as physics and chemistry.

Neyman–Pearson hypothesis testing model: The formulation of a hypothesis in the two forms of a null hypothesis and an alternate hypothesis.

nominal variable: A variable which has two or more qualitative categories or conditions.

nomothetic: The study of a sufficient number of individuals in an attempt to test psychological principles.

non-causal hypothesis: A statement of the relation between two or more variables in which the causal order of the variables is not specified.

non-directional hypothesis: A statement of the relation between two or more variables in which the direction of the relation is not described.

non-experiment: A study in which variables are not manipulated.

non-manipulation study: A study in which variables are not deliberately varied.

NUD*IST: Computer software designed to aid the qualitative analysis of qualitative data – now known as NVivo.

null hypothesis: A statement which says that two or more variables are not expected to be related.

NVivo: Computer software to aid the qualitative analysis of qualitative data.

objective measure: A test for which trained assessors will agree on what the score should be.

observation: The watching or recording of the behaviour of others.

Occam's razor: The principle that an explanation should consist of the fewest assumptions necessary for explaining a phenomenon.

odd–even reliability: The internal consistency of a test in which the odd numbered variables are summed together and then correlated with the sum of the even-numbered items with a statistical adjustment of the correlation to the full length of the scale.

one-tailed significance level: The significance cut-off point or critical value applied to one end or tail of a probability distribution.

Online Public Access Catalogue (OPAC): A computer software system for recording and showing the location and availability of publications held in a library.

open-ended question: One which does not constrain the responses of the interviewee to a small number of alternatives.

operationalising concepts/variables: The procedure or operation for manipulating or measuring a particular concept or variable.

panel design: A study in which the same participants are assessed at two or more points in time.

paradigm: A paradigm, in Thomas Kuhn's ideas, is a broad way of conceiving or understanding a particular research area which is generally accepted by the scientific/research community.

partial replication: A study which repeats a previous study but extends it to examine the role of other variables.

participant: The recommended term for referring to the people who take part in research.

participant observation: The watching and recording of the behaviour of members of a group of which the observer is part.

passive observational study: Research in which there is no attempt on the part of the researcher to deliberately manipulate any of the variables being studied.

PASW Statistics: The name of SPSS in 2008–9. PASW stands for Predictive Analytic Software.

Pearson correlation coefficient: A measure of the size and direction of the association between two score variables which can vary from –1 to 1.

percentile: The point expressed out of a hundred which describes the percentage of values which fall at and below it.

perlocution: The effect of the speaker's words on a hearer.

phenomenology: The attempt to understand conscious experience as it is experienced.

phi: A measure of association between two binomial or dichotomous variables.

piloting: The checking of the procedures to be used in a study to see that there are no problems.

placebo effect: The effect of receiving a treatment which does not contain the manipulation of the variable whose effect is being investigated.

plagiarism: The use of words of another person without acknowledging them as the source.

point estimate: A particular value for a characteristic of a population inferred from the characteristic in a sample.

point-biserial correlation coefficient: A Pearson correlation between a binomial and a score variable.

pool of items: The statements or questions from which a smaller number are selected to make up a scale.

positivism: A philosophical position on knowledge which emphasises the importance of the empirical study of phenomena.

***post hoc* comparison:** A test to determine whether two or more groups differ significantly from each other which is decided to be made after the data have been collected.

postmodernism: Philosophical positions which are critical of positivism and which concentrate on interpretation.

postpositivism: Philosophical perspectives which are critical of positivism.

pre-coding: The assignment of codes or values to variables before the data have been collected.

predictive validity: A measure of the association between a variable made at one point in time and a variable assessed at a later point in time.

pre-test/post-test sensitisation effects: The effect that familiarity with a measure taken before an intervention may have on a measure taken after the intervention.

probability sampling: The selection of cases in which each case has the same probability of being selected.

procedure: The subsection of the methods section in a research report which describes how the study was carried out.

prospective study: A study in which the same cases are assessed at more than one point in time.
psychological test: A measure which is used to assess a psychological concept or construct.
PsycINFO: An electronic database produced by the American Psychological Association which provides summary details of a wide range of publications in psychology and which for more recent articles includes references.
purposive sampling: Sampling with a particular purpose in mind such as when a particular sort of respondent is sought rather than a representative sample.

qualitative coding: The categorisation of qualitative data.
qualitative data analysis: The analysis of qualitative data which does not involve the use of numbers.
qualitative variable *see* nominal variable.
quantitative data analysis: The analysis of data which at the very least involves counting the frequency of categories in the main variable of interest.
quantitative variable: At its most basic, a variable whose categories can be counted.
quasi-experiment: A study in which cases have not been randomly assigned to treatments or the order in which they are given.
questionnaire item: A statement which is part of a set of statements to measure a particular construct.
quota sample: The selection of cases to represent particular categories or groups of cases.

random assignment: The allocation of cases to conditions in which each case has the same probability of being allocated to any of the conditions.
random sampling *see* stratified random sampling.
randomised experiment: A study in which one or more variables have been manipulated and where cases have been randomly assigned to the conditions reflecting those manipulations or to the order in which the conditions have been run.
realism: A philosophical position which believes that there is an external world which is knowable by humans.
reference: A book or article which is cited in a publication.
register: A list of cases.
Registrar General's Social Class: A measure of the social standing of individuals which is used by the British civil service.
relativism: The philosophical view that there is no fixed reality which can be studied.
reliability: The extent to which a measure or the parts making up that measure give the same or similar classification or score.
replication study: A study which repeats a previous study.
representative sample: A group of cases which are representative of the population of cases from which that group have been drawn.
representativeness of sample: The extent to which a group of cases reflects particular characteristics of the population from which those cases have been selected.
retrospective study: A study in which past details of cases are gathered.
rhetoric: Language designed to impress or persuade others.

sampling: The act of selecting cases.
sampling error (margin of error): The variability of groups of values from the characteristics of the population from which they were selected.
scale: A measuring instrument.
simple random sampling: A method in which each case has the same probability of being chosen.
snowball sampling: The selection of cases who have been proposed by other cases.
socio-demographic characteristic: A variable which describes a basic feature of a person such as their gender, age or educational level.
speech act: The act of making an utterance.
split-half reliability: The association of the two halves of a measure as an index of its internal consistency.
SPSS *see* Statistical Package for the Social Sciences and PASW Statistics.
stability over time: The extent to which a construct or measure is similar at two or more points in time.
stake: The investment that people have in a group.
standard deviation: The square root of the mean or average squared deviation of the scores around the mean. It is a sort of average of the amount that scores differ from the mean.
standard multiple regression: A multiple regression in which all the predictor variables are entered into or analysed in a single step.
Standard Occupational Classification 2000: A system developed in the United Kingdom for categorising occupations.
standardisation of a procedure: Agreement on how a procedure should be carried out.
standardised test: A measure where what it is and how it is to be administered is clear and for which there are normative data from substantial samples of individuals.
statistical hypothesis: A statement which expresses the statistical relation between two or more variables.
Statistical Package for the Social Sciences (SPSS): The name of a widely used computer software for handling and statistically analysing data which was called PASW Statistics in 2008–9.
statistical significance: The adoption of a criterion at or below which a finding is thought to be so infrequent that it is unlikely to be due to chance.
stepwise multiple regression: A multiple regression in which predictor variables are entered or removed one at a time in terms of the size of their statistical significance.
straight replication: The repetition of a previous study.
stratified random sampling: The random selection of cases from particular groups or strata.
structural equation modelling: A statistical model in which there may be more than one criterion or outcome variable and where the specified relations between variables is taken into account.
structured interview: An interview in which at least the exact form of the questions has been specified.
subject variable: A characteristic of the participant which cannot or has not been manipulated.
subjectivism: The philosophical position that there is not a single reality that is knowable.

suppressor variable: A variable which when partialled out of the relation between two other variables substantially increases the size of the relation between those two variables.

synchronous correlation: An association between two variables measured at the same point in time.

systematic sampling: The selection of cases in a systematic way such as selecting every 10th case.

temporal change: The change in a variable over time.

temporal precedence/order of variable: A relation where the association between variable A assessed at one time and variable B assessed later is significantly stronger than the association between variable B measured at the earlier point and variable A at the later point.

test–retest reliability: The correlation between the same or two similar tests over a relatively short period of time such as two weeks.

theism: The belief in gods or a god.

theoretical sampling: A group of values in which each value has the same probability of being selected.

theory: A set of statements which describe and explain some phenomenon or group of phenomena.

third variable issue: The possibility that the relation between two variables may be affected by one or more other variables.

title: A brief statement of about 15 words or less which describe the contents of a publication.

transcription: The process of putting spoken words into a representative written format.

triangulation: The use of three or more methods to measure the same variable or variables.

true or randomised experiment: A study in which the variable thought to affect one or more other variables is manipulated and cases are randomly assigned to conditions that reflect that manipulation or to different orders of those conditions.

***t*-test:** A parametric test which determines whether the means of two groups differ significantly.

two-wave panel design: A panel design in which the same cases are assessed at two points in time or waves.

unidimensional scale: A measure which is thought to assess a single construct or variable.

universalism: The assumption that there are laws or principles which apply to at least all humans.

utterance act: The act of saying something.

validity: An index of the extent to which a measure assesses what it purports to measure.

variable: A characteristic that consists of two or more categories or values.

Web of Science: An electronic database originally developed by the Institute of Information which provides summary details and the references of articles from selected journals in the arts, sciences and social sciences.

web source: The address of information listed on the web.

within-subjects design: A research design in which the same cases participate in all the conditions.

REFERENCES

Allen, R. E. (1992) (ed.). *The Oxford English Dictionary*, 2nd edn, CD-ROM (Oxford: Oxford University Press).

American Psychological Association (2009). 'Updated summary report of journal operations, 2008', *American Psychologist*, *64*, 504–505. http://www.apa.org/pubs/journals/features/2008-operations.pdf

Austin, J. L. (1975). *How To Do Things with Words* (Cambridge, MA: Harvard University Press).

Barber, T. X. (1973). 'Pitfalls in research: Nine investigator and experimenter effects', in R. M. W. Travers (ed.), *Second Handbook of Research on Teaching* (Chicago, IL: Rand McNally), 382–404.

Barber, T. X. (1976). *Pitfalls in Human Research: Ten pivotal points* (New York: Pergamon Press).

Barlow, D. H., and Hersen, M. (1984). *Single Case Experimental Designs: Strategies for studying behavior change*, 2nd edn (New York: Pergamon Press).

Baron, R. M., and Kenny, D. A. (1986). 'The moderator–mediator variable distinction in social psychological research: Conceptual, strategic, and statistical considerations', *Journal of Personality and Social Psychology*, *51*, 1173–82.

Beaugrande, R. D. (1996). 'The story of discourse analysis', in T. van Dijk (ed.), *Introduction to Discourse Analysis* (London: Sage), 35–62.

Beecher, H. K. (1955). 'The powerful placebo', *Journal of the American Medical Association*, *159*, 1602–6.

Benneworth, K. (2004). 'A discursive analysis of police interviews with suspected paedophiles', Doctoral dissertation (Loughborough University, England).

Berenson, M. L., Levine, D. M., and Krehbiel, T. C. (2009). *Basic Business Statistics: Concepts and applications*, 11th edn (Upper Saddle River, NJ: Prentice Hall).

Berkowitz, L. (1962). *Aggression: A social psychological analysis* (New York: McGraw-Hill).

Billig, M., and Cramer, D. (1990). 'Authoritarianism and demographic variables as predictors of racial attitudes in Britain', *New Community: A Journal of Research and Policy on Ethnic Relations*, *16*, 199–211.

Binet, A. and Simon, T. (1904). 'Méthodes nouvelles pour le diagnostic du niveau intellectual des onormaux', *L'Année Psychologique*, *11*, 191–244. Reprinted in H. H. Goddard (ed.) and translated by E. S. Kite (1916) as 'New methods for the diagnosis of the intellectual level of subnormals'. This translation by Elizabeth S. Kite first appeared in 1916 in *The Development of Intelligence in Children* (Baltimorc, MD: Wilkins and Wilkins). http://psychclassics.yorku.ca/Binet/binet1.htm

Bodner, T. E. (2006). 'Designs, participants, and measurement methods in psychological research', *Canadian Psychology*, *47*, 263–72.

Braun, V., and Clarke, V. (2006). 'Using thematic analysis in psychology', *Qualitative Research in Psychology*, *3*, 77–101.

Bridgman, P. W. (1927). *The Logic of Modern Physics* (New York: Macmillan).

Bryman, A. (2008). *Social Research Methods*, 3rd edn (Oxford: Oxford University Press).

Bryman, A., and Bell, E. (2007). *Business Research Methods*, 2nd edn (Oxford: Oxford University Press).

Buss, A. R., and McDermot, J. R. (1976). 'Ratings of psychology journals compared to objective measures of journal impact', *American Psychologist*, *31*, 675–8 (comment).

Campbell, D. T. (1969). 'Prospective: Artifact and control', in R. Rosenthal and R. L. Rosnow (eds), *Artifact in Behavioural Research* (New York: Academic Press), 351–82.

Campbell, D. T., and Fiske, D. W. (1959). 'Convergent and discriminant validation by the multitrait–multimethod matrix', *Psychological Bulletin*, *56*, 81–105.

Campbell, D. T., and Stanley, J. C. (1963). *Experimental and Quasi-experimental Designs for Research* (Boston, MA: Houghton Mifflin).

Campbell, M. L. C., and Morrison, A. P. (2007). 'The subjective experience of paranoia: Comparing the experiences of patients with psychosis and individuals with no psychiatric history', *Clinical Psychology and Psychotherapy*, *14*, 63–77.

Canter, D. (1983). 'The potential of facet theory for applied social psychology', *Quality and Quantity*, *17*, 35–67.

Chan, L. M. (1999). *A Guide to the Library of Congress Classification*, 5th edn (Englewood, CO: Libraries Unlimited).

Chan, L. M., and Mitchell, J. S. (2003). *Dewey Decimal Classification: A practical guide* (Dublin, OH: OCLC).

Charmaz, K. (1995). 'Grounded theory', in J. A. Smith, R. Harre and L. V. Langenhove (eds), *Rethinking Methods in Psychology* (London: Sage), 27–49.

Charmaz, K. (2000). 'Grounded theory: Objectivist and constructivist methods', in N. K. Denzin and

Y. S. E. Lincoln (eds), *Handbook of Qualitative Research*, 2nd edn (Thousand Oaks, CA: Sage), 503–35.

Cicchetti, D. V. (1991). 'The reliability of peer review for manuscript and grant submissions: A cross-disciplinary investigation', *Behavioural and Brain Sciences*, *14*, 119–86.

Clarke, V., Burns, M., and Burgoyne, C. (2006). *'Who would take whose name?': An exploratory study of naming practices in same-sex relationships*. Unpublished manuscript.

Cohen, J. (1988). *Statistical Power Analysis for the Behavioural Sciences*, 2nd edn (Hillsdale, NJ: Lawrence Erlbaum Associates).

Cohen, J., and Cohen, P. (1983). *Applied Multiple Regression/Correlation Analysis for the Behavioral Sciences*, 2nd edn (Hillsdale, NJ: Lawrence Erlbaum Associates).

Cohen, J., Cohen, P., West, S. G., and Aiken, L. S. (2003), *Applied Multiple Regression/Correlation Analysis for the Behavioral Sciences*, 3rd edn (Hillsdale, NJ: Lawrence Erlbaum Associates).

Coleman, L. M., and Cater, S. M. (2005). 'A qualitative study of the relationship between alcohol consumption and risky sex in adolescents', *Archives of Sexual Behavior*, *34*, 649–61.

Cook, T. D., and Campbell, D. T. (1979). *Quasi-experimentation: Design and Analysis Issues for Field Settings* (Chicago, IL: Rand McNally).

Coolican, H. (2009). *Research Methods and Statistics in Psychology*, 5th edn (London: Hodder Education).

Cox, B. D., Blaxter, M., Buckle, A. L. J., Fenner, N. P., Golding, J. F., Gore, M., Huppert, F. A., Nickson, J., Roth, M., Stark, J., Wadsworth, M. E. J., and Whichelow, M. (1987). *The Health and Lifestyle Survey* (London: Health Promotion Research Trust).

Cramer, D. (1991). 'Type A behaviour pattern, extraversion, neuroticism and psychological distress', *British Journal of Medical Psychology*, *64*, 73–83.

Cramer, D. (1994). 'Psychological distress and neuroticism: A two-wave panel study', *British Journal of Medical Psychology*, *67*, 333–42.

Cramer, D. (1995). 'Life and job satisfaction: A two-wave panel study', *Journal of Psychology*, *129*, 261–7.

Cramer, D. (1998). *Fundamental Statistics for Social Research: Step-by-step calculations and computer techniques using SPSS for Windows* (London: Routledge).

Cramer, D. (2003). 'A cautionary tale of two statistics: Partial correlation and standardised partial regression', *Journal of Psychology*, *137*, 507–11.

Cramer, D., and Buckland, N. (1995). 'Effect of rational and irrational statements and demand characteristics on task anxiety', *Journal of Psychology*, *129*, 269–75.

Cramer, D., and Fong, J. (1991). 'Effect of rational and irrational beliefs on intensity and "inappropriateness" of feelings: A test of rational-emotive theory', *Cognitive Therapy and Research*, *15*, 319–29.

Cramer, D., Henderson, S., and Scott, R. (1996). 'Mental health and adequacy of social support: A four-wave panel study'. *British Journal of Social Psychology*, *35*, 285–95.

Cronbach, L. J. (1951). 'Coefficient alpha and the internal structure of tests', *Psychometrika*, *16*, 297–334.

Cronbach, L. J., and Meehl, P. E. (1955). 'Construct validity in psychological tests', *Psychological Bulletin*, *52*, 281–302.

Danziger, K. (1985). 'The origins of the psychological experiment as a social institution', *American Psychologist*, *40* (2), 133–40.

Danziger, K., and Dzinas, K. (1997). 'How psychology got its variables', *Canadian Psychology*, *38*, 43–48.

Denscombe, M. (2002). *Ground Rules for Good Research: A 10 point guide for social researchers* (Buckingham: Open University Press).

Denzin, N. K., and Lincoln, Y. S. E. (2000). 'Introduction: The discipline and practice of qualitative research', in N. K. Denzin and Y. S. E. Lincoln (eds), *Handbook of Qualitative Research*, 2nd edn (Thousand Oaks, CA: Sage), 1–28.

Dereshiwsky, M. (1999). 'The five dimensions of participant observation', http://jan.ucc.nau.edu/~mid/edr725/class/observation/fivedimensions/

Dewey Services (n.d.). http://www.oclc.org/dewey/

Drew, P. (1995). 'Conversation analysis', in J. A. Smith, R. Harre and L. V. Langenhove (eds), *Rethinking Methods in Psychology* (London: Sage), 64–79.

Eatough, V., and Smith, J. A. (2006). 'I feel like a scrambled egg in my head: An idiographic case study of meaning making and anger using interpretative phenomenological analysis', *Psychology and Psychotherapy: Theory, Research and Practice*, *79*, 115–35.

Ebbinghaus, H. (1913). *Memory: A contribution to experimental psychology* (New York: Teacher's College, Columbia University; Reprint edition, New York: Dover, 1964).

Edley, N. (2001). 'Analysing masculinity: Interpretative repertoires, ideological dilemmas and subject positions', in M. Wetherell, S. Taylor and S. J. E. Yates (eds), *Discourse as Data: A guide for analysis* (London: Sage), 189–228.

Edwards, D., and Potter, J. (1993). 'Language and causation: A discursive action model of description and attribution', *Psychological Review*, *100* (1), 23–41.

Engstrom, L., Geijerstam, G., Holmberg, N. G., and Uhrus, K. (1963). 'A prospective study of the relationship between psycho-social factors and course of pregnancy and delivery', *Journal of Psychosomatic Research*, *8*, 151–5.

Epley, N., and Huff, C. (1998). 'Suspicion, affective response, and educational benefit as a result of deception in psychology research', *Personality and Social Psychology Bulletin*, *24*, 759–68.

Eysenck, H. J. (1980). *The Causes and Effects of Smoking* (London: Sage).

Farrington, D. P. (1996). 'Psychosocial influences on the development of antisocial personality', in G. Davies, S. Lloyd-Bostock, M. McMurran and C. Wilson (eds), *Psychology, Law and Criminal Justice: International Developments in Research and Practice* (Berlin: Walter de Gruyter), 424–44.

Ferri, E. (1993) (ed.). *Life at 33: The fifth follow-up of the National Child Development Study* (London: National Children's Bureau).

Fincham, F. D., Beach, S. R. H., Harold, G. T., and Osborne, L. N. (1997). 'Marital satisfaction and depression: Different casual relationships for men and women?', *Psychological Science*, *8*, 351–7.

Forsyth, J. P., Kollins, S., Palav, A., Duff, K., and Maher, S. (1999). 'Has behavior therapy drifted from its experimental roots? A survey of publication trends in mainstream behavioral journals', *Journal of Behavior Therapy and Experimental Psychiatry*, *30*, 205–20.

Garfinkel, H. (1967). *Studies in Ethnomethodology* (Englewood Cliffs, NJ: Prentice-Hall).

Gee, D., Ward, T., and Eccleson, L. (2003). 'The function of sexual fantasies for sexual offenders: a preliminary model', *Behaviour Change*, *20*, 44–60.

Gergen, K.J. (1973). 'Social psychology as history', *Journal of Personality and SocialPsychology*, *26* (2), 309–320.

Gibbs, A. (1997). 'Focus groups', *Social research update*, *19*, http://www.soc.surr.ac.uk/sru19.html

Glaser, B. G., and Strauss, A. L. (1967). *The Discovery of Grounded Theory: Strategies for qualitative research* (New York: Aldine de Gruyter).

Goffman, E. (1959). *The Presentation of Self in Everyday Life* (Garden City, New York: Doubleday).

Goffman, E. (1961). *Asylums: Essays on the social situation of mental patients and other inmates* (Garden City, New York: Anchor).

Goldthorpe, J. H. (1987). *Social Mobility and Class Structure in Modern Britain*, 2nd edn (Oxford: Clarendon Press).

Gottfredson, S. D. (1978). 'Evaluating psychological research reports: Dimensions, reliability, and correlates of quality judgments', *American Psychologist*, *33*, 920–34.

Great Britain Office for National Statistics (2000). *Standard Occupational Classification 2000: Vol. 1, Structure and descriptions of unit groups and Vol. 2, The coding index* (London: Stationery Office). http://www.statistics.gov.uk/nsbase/methods_quality/ns_sec/soc2000.asp

Greene, E. (1990). 'Media effects on jurors', *Law and Human Behavior*, *14* (5), 439–50.

Grice, H. P. (1975). 'Logic and conversation', in P. Cole and J. Morgan (eds), *Syntax and Semantics 3: Speech acts* (New York: Academic Press), 41–58.

Haggbloom, S. J., Warnick, R., Warnick, J. E., Jones, V. K., Yarbrough, G. L., Russell, T. M. *et al.* (2002). The 100 most eminent psychologists of the 20th century. *Review of General Psychology*, 6, 139–52.

Hare, R. D. (1991). *The Hare Psychopathy Checklist – Revised* (Toronto: Multi-Health Systems).

Henriques, J., Hollway, W., Urwin, C., Venn, C., and Walkerdine, V. (1984). *Changing the Subject: Psychology, social regulation and subjectivity* (London: Methuen).

Hepburn, A. (2003). *An Introduction to Critical Social Psychology* (London: Sage).

Hergenhahn, B. R. (2001). *An Introduction to the History of Psychology*, 4th edn (Belmont, CA: Wadsworth Thomson Learning).

Hoinville, G., and Jowell, R. (1978). *Survey Research Practice* (London: Heinemann Educational Books).

Horton-Salway, M. (2001). 'The construction of ME: The discursive action model', in M. Wetherell, S. Taylor and S. J. E. Yates (eds), *Discourse as Data: A Guide for Analysis* (London: Sage), 147–88.

Howell, D. C. (2010). *Statistical Methods for Psychology*, 7th edn (Belmont, CA: Wadsworth).

Howitt, D. (1992a). *Concerning Psychology: Psychology applied to social issues* (Milton Keynes: Open University Press).

Howitt, D. (1992b). *Child Abuse Errors* (London: Harvester Wheatsheaf).

Howitt, D. (1995). *Paedophiles and Sexual Offences Against Children* (Chichester: Wiley).

Howitt, D. (2008). *Introduction to Forensic and Criminal Psychology*, 3rd edn (Harlow: Prentice Hall).

Howitt, D. (2010). *Introduction to Qualitative Research Methods in Psychology* (Harlow: Pearson Education).

Howitt, D., and Cramer, D. (2011a). *Introduction to Statistics in Psychology*, 5th edn (Harlow: Prentice Hall).

Howitt, D., and Cramer, D. (2011b). *Introduction to SPSS Statistics in Psychology*, 5th edn (Harlow: Prentice Hall).

Howitt, D., and Owusu-Bempah, J. (1990). 'Racism in a British journal?', *The Psychologist: Bulletin of the British Psychological Society*, *3* (9), 396–400.

Howitt, D., and Owusu-Bempah, J. (1994). *The Racism of Psychology* (London: Harvester Wheatsheaf).

Husserl, E. (1900/1970). *Logical Investigations*, trans. J. N. Findlay (London: Routledge and Kegan Paul).

Husserl, E. (1913/1962). *Ideas: A General Introduction to Pure Phenomenology*, trans. W. R. Boyce Gibson (London: Collier).

Hutchby, I., and Wooffitt, R. (1998). *Conversation Analysis: Principles, practices and applications* (Cambridge: Polity Press).

Institute for Scientific Information (1994). 'The impact factor', *Current Contents*, 20 June, http://thomsonreuters.com/products_services/science/free/essays/impact_factor/

Jefferson, G. (1984). 'On stepwise transition from talk about a trouble to inappropriately next positioned matters', in J. M. Atkinson and J. Heritage (eds), *Structures of Social Action: Studies in conversation analysis* (Cambridge: Cambridge University Press), 191–222.

Jones, H. (1981). *Bad Blood: The Tuskegee Syphilis Experiment* (New York: Free Press).

Jones, M. C., Bayley, N., MacFarlane, J. W., and Honzik, M. P. (1971). *The Course of Human Development* (Waltham, MA: Xerox Publishing Company).

Keith-Spiegel, P., and Koocher, G. P. (1985). *Ethics in Psychology: Professional standards and cases* (Hillsdale, NJ: Lawrence Erlbaum Associates).

Kelly, G. A. (1955). *The Psychology of Personal Constructs. Volume 1: A theory of personality* (New York: Norton).

Keppel, G., and Wickens, T. D. (2004). *Design and Analysis: A researcher's handbook*, 4th edn (Upper Saddle River, NJ: Pearson).

Kirk, R. C. (1995). *Experimental Design*, 3rd edn (Pacific Grove, CA: Brooks/Cole).

Kitzinger, C., and H. Frith (2001). 'Just say no? The use of conversation analysis in developing a feminist perspective on sexual refusal', in M. Wetherell, S. Taylor and S. J. E. Yates (eds), *Discourse Theory and Practice: A reader* (London, Sage), 167–85.

Korn, J. H. (1997). *Illusions of Reality: A history of deception in social psychology* (New York: State University of New York Press).

Krause, N., Liang, J., and Yatomi, N. (1989). 'Satisfaction with social support and depressive symptoms: A panel analysis', *Psychology and Aging*, *4*, 88–97.

Lana, R. E. (1969). 'Pretest sensitisation', in R. Rosenthal and R. L. Rosnow (eds), *Artifact in Behavioural Research* (New York: Academic Press), 119–41.

Latane, B., and Darley, J. M. (1970). *The Unresponsive Bystander: Why doesn't he help?* (New York: Appleton-Century-Crofts).

Lazarsfeld, P. F. (1948). 'The use of panels in social research', *Proceedings of the American Philosophical Society*, *92*, 405–10.

Leahy, T. H. (2004). *A History of Psychology: Main currents in psychological thought*, 6th edn (Upper Saddle, NJ: Prentice Hall).

Leyens, J. P., Camino, L., Parke, R. D., and Berkowitz, L. (1975). 'The effect of movie violence on aggression in a field setting as a function of group dominance and cohesion', *Journal of Personality and Social Psychology*, *32*, 346–60.

Library of Congress Classification Outline (n.d.). http://www.loc.gov/catdir/cpso/lcco/

Loftus, E. F., and Palmer, J. C. (1974). 'Reconstruction of auto-mobile destruction: An example of the interaction between language and memory', *Journal of Verbal Learning and Verbal Behaviour*, *13*, 585–9.

Lovering, K. M. (1995). 'The bleeding body: Adolescents talk about menstruation', in S. Wilkinson and C. Kitzinger (eds), *Feminism and Discourse: Psychological perspectives* (London: Sage), 10–31.

MacCorquodale, K., and Meehl, P. E. (1948). 'On a distinction between hypothetical variables and intervening variables', *Psychological Review*, *55*, 95–107.

Mace, K. C., and Warner, H. D. (1973). 'Ratings of psychology journals', *American Psychologist*, *28*, 184–6 (Comment).

Mann, E., and Abraham, C. (2006). 'The role of affect in UK commuters' travel mode choices: An interpretative phenomenological analysis', *British Journal of Psychology*, *97*, 155–76.

McArthur, T. (1992). *The Oxford Companion to the English Language* (Oxford: Oxford University Press).

McGuire, W. J. (1997). 'Creative hypothesis generating in psychology: Some useful heuristics', *Annual Review of Psychology*, *48*, 1–30.

McHugh, P. (1968). *Defining the Situation: The Organization of Meaning* (Evanston, IL: Bobbs-Merrill).

Mead, M. (1944). *Coming of Age in Samoa* (Harmondsworth: Pelican).

Medway, C., and Howitt, D. (2003). 'The role of animal cruelty in the prediction of dangerousness', in M. Vanderhallen, G. Vervaeke, P. Van Koppen and J. Goethals (eds), *Much Ado About Crime: Chapters in psychology and law* (Brussels: Politeia), 245–50.

Merton, R., and Kendall, P. (1946). 'The focused interview', *American Journal of Sociology*, *51*, 541–7.

Milgram, S. (1974). *Obedience to Authority* (New York: Harper & Row).

Moser, C. A., and Kalton, G. (1971). *Survey Methods in Social Investigation*, 2nd edn (London: Gower).

Murphy, P. M., Cramer, D., and Lillie, F. J. (1984). 'The relationship between curative factors perceived by patients in their psychotherapy and treatment outcome', *British Journal of Medical Psychology*, *57*, 187–92.

Nunnally, J. (1978). *Psychometric theory*, 2nd edn (New York: McGraw-Hill).

O'Connell, D. C., and Kowal, S. (1995). 'Basic principles of transcription', in J. A. Smith, R. Harré, and L. Van Langenhove (eds), *Rethinking Methods in Psychology* (London: Sage), 93–105.

Ogden, J., Clementi, C., and Aylwin, S. (2006). 'The impact of obesity surgery and the paradox of control: A qualitative study', *Psychology and Health*, *21*, 273–93.

OPCS (1991). *Standard Occupational Classification; Volume 3* (London: HMSO).

Orne, M. T. (1959). 'The nature of hypnosis: Artifact and essence', *Journal of Abnormal and Social Psychology*, *58*, 277–99.

Orne, M. T. (1962). 'On the social psychology of the psychological experiment: With particular reference to demand characteristics and their implications', *American Psychologist*, *17*, 776–83.

Orne, M. T. (1969). 'Demand characteristics and the concept of quasi-controls', in R. Rosenthal and R. L. Rosnow (eds), *Artifact in Behavioural Research* (New York: Academic Press), 143–79.

Orne, M. T., and Scheibe, K. E. (1964). 'The contribution of nondeprivation factors in the production of sensory deprivation effects: The psychology of the "panic button"', *Journal of Abnormal and Social Psychology*, *68*, 3–12.

Owusu-Bempah, J., and Howitt, D. (1995). 'How Eurocentric psychology damages Africa', *The Psychologist: Bulletin of the British Psychological Society*, *8*, 462–5.

Owusu-Bempah, J., and Howitt, D. (2000). *Psychology Beyond Western Perspectives* (Leicester: BPS Books).

Page, M., and Scheidt, R. J. (1971). 'The elusive weapons effect', *Journal of Personality and Social Psychology*, *20*, 304–9.

Park, A., Curtice, J., Thomson, K., Phillips, L., Clery, E., and Butt, S. (2010) (eds). *British Social Attitudes: The 26th report* (London: Sage).

Parker, I. (1989). *The Crisis in Modern Social Psychology* (London: Routledge).

Parker, I. (ed.) (1999). *Deconstructing Psychotherapy* (London: Sage).

Parker, I., Georgaca, E., Harper, D., McLaughlin, T., and Stowell-Smith, M. (1995). *Deconstructing Psychopathology* (London: Sage).

Patton, M. Q. (1986). *How To Use Qualitative Methods in Evaluation* (London: Sage).

Pavlov, I. P. (1927). *Conditioned Reflexes: An investigation of the physiological activity of the cerebral cortex*, trans. G. V. Anrep (London: Oxford University Press).

Pedhazur, E. J., and Schmelkin, L. P. (1991). *Measurement, Design and Analysis: An integrated approach* (Hillsdale, NJ: Lawrence Erlbaum Associates).

Peters, D. P., and Ceci, S. J. (1982). 'Peer-review practices of psychological journals: The fate of published articles, submitted again', *The Behavioral and Brain Sciences*, *5*, 187–255.

Pitcher, J., Campbell, R., Hubbard, P., O'Neill, M., and Scoular, J. (2006). *Living and Working in Areas of Street Sex Work: From Conflict to Co-existence* (Bristol: Policy Press).

Postal Geography (n.d.). http://www.statistics.gov.uk/geography/postal_geog.asp

Potter, J. (1997). 'Discourse analysis as a way of analysing naturally occurring talk', in D. Silverman (ed.), *Qualitative Research: Theory, methods and practice* (London: Sage), 144–60.

Potter, J. (1998). 'Qualitative and discourse analysis', in A. S. Bellack and M. Hersen (eds), *Comprehensive Clinical Psychology*, Vol. 3 (Oxford: Pergamon), 117–44.

Potter, J. (2001). 'Wittgenstein and Austin', in M. Wetherell, S. Taylor and S. J. Yates, *Discourse Theory and Practice: A reader* (London: Sage), 39–56.

Potter, J. (2004). 'Discourse analysis', in M. Hardy and A. Bryman (eds), *Handbook of Data Analysis* (London: Sage), 607–24.

Potter, J., and Wetherell, M. (1987). *Discourse and Social Psychology: Beyond attitudes and behaviour* (London: Sage).

Potter, J., and Wetherell, M. (1995). 'Discourse analysis', in J. A. Smith, R. Harré and L. V. Langenhove (eds), *Rethinking Methods in Psychology* (London: Sage), 80–92.

PsycINFO Database Information (n.d.). http://www.apa.org/pubs/databases/psycinfo/index.aspx

Reis, H. T., and Stiller, J. (1992). 'Publication trends in *JPSP*: A three-decade review', *Personality and Social Psychology Bulletin*, *18*, 465–72.

Rivers, W. H. R. (1908). *The Influence of Alcohol and Other Drugs on Fatigue* (London: Arnold).

Rosenberg, M. (1968). *The Logic of Survey Analysis* (London: Basic Books).

Rosenthal, R. (1963). 'On the social psychology of the psychological experiment: The experimenter's hypothesis as unintended determinant of experimental results', *American Scientist*, *51*, 268–83.

Rosenthal, R. (1969). 'Interpersonal expectations: Effects of the experimenter's hypothesis', in R. Rosenthal and R. L. Rosnow (eds), *Artifact in Behavioural Research* (New York: Academic Press), 181–277.

Rosenthal, R. (1978). 'How often are our numbers wrong?', *American Psychologist*, *33*, 1005–8.

Rosenthal, R. (1991). *Meta-analytic Procedures for Social Research*, rev. edn (Newbury Park, CA: Sage).

Rosenthal, R., and Rosnow, R. L. (1969). 'The volunteer subject', in R. Rosenthal and R. L. Rosnow (eds), *Artifact in Behavioural Research* (New York: Academic Press), 59–118.

Rosenthal, R., and Rubin, D. B. (1978). 'Interpersonal expectancy effects: The first 345 studies', *The Behavioral and Brain Sciences*, *3*, 377–415.

Rosnow, R. L. (2002). 'The nature and role of demand characteristics in scientific enquiry', *Prevention and Treatment*, *5*, no page numbers.

Rushton, J. P., and Roediger, H. L., III. (1978). 'An evaluation of 80 psychology journals based on the *Science Citation Index*', *American Psychologist*, *33*, 520–3 (comment).

Sacks, H. (1992). 'Lecture 1: Rules of conversational sequence', in E. Jefferson (ed.), *H. Y. Sacks Lectures on Conversation*; Vol. 1, 3rd edn (Oxford: Blackwell).

Sacks, O. (1985). *The Man Who Mistook his Wife for a Hat* (London: Picador).

Schlenker, B. R. (1974). 'Social psychology and science', *Journal of Personality and Social Psychology*, *29*, 1–15.

Searle, J. (1969). *Speech Acts: An essay in the philosophy of language* (Cambridge: Cambridge University Press).

Shadish, W. R., and Ragsdale, K. (1996). 'Random versus nonrandom assignment in controlled experiments: Do you get the same answer?', *Journal of Consulting and Clinical Psychology*, *64*, 1290–305.

Shadish, W. R., Cook, T. D., and Campbell, D. T. (2002). *Experimental and Quasi-experimental Designs for Generalised Causal Inference* (New York: Houghton Mifflin).

Sheldon, K., and Howitt, D. (2007). *Sex Offenders and the Internet* (Chichester: Wiley).

Sheldrake, R. (1998). 'Experimenter effects in scientific research: How widely are they neglected?', *Journal of Scientific Exploration*, *12*, 1–6.

Sherman, R. C., Buddie, A. M., Dragan, K. L., End, C. M., and Finney, L. J. (1999). 'Twenty years of *PSPB*: Trends in content, design, and analysis', *Personality and Social Psychology Bulletin*, *25*, 177–87.

Shye, S., and Elizur, D. (1994). *Introduction to Facet Theory: Content Design and Intrinsic Data Analysis in Behavioural Research* (Thousand Oaks, CA: Sage).

Silverman, D. (1997). 'The logics of qualitative research', in G. Miller and R. Dingwall (eds), *Context and Method in Qualitative Research* (London: Sage), 12–25.

Skinner, B. F. (1938). *The Behavior of Organisms* (New York: Appleton-Century-Crofts).

Smith, J. A. (1996). 'Beyond the divide between cognition and discourse: Using interpretative phenomenological analysis in health psychology', *Psychology and Health*, *11*, 261–71.

Smith, J. A., and Eatough, V. (2006). 'Interpretative phenomenological analysis', in G. M. Breakwell, S. Hammond, C. Fife-Schaw and J. A. Smith (eds), *Research Methods in Psychology*, 3rd edn (London: Sage), 322–41.

Smith, J. A., and Osborn, M. (2003). 'Interpretative phenomenological analysis', in J. A. Smith (ed.), *Qualitative Psychology: A practical guide to research methods* (London: Sage), 51–80.

Smith, J. A., and Osborn, M. (2007). 'Pain as an assault on the Self: An interpretative phenomenological analysis of the psychological impact of chronic benign low back pain', *Psychology and Health*, *22*, 517–34.

Smith, S. S., and Richardson, D. (1983). 'Amelioration of deception and harm in psychological research: The important role of debriefing', *Journal of Personality and Social Psychology*, *44*, 1075–82.

Solomon, R. L. (1949). 'An extension of control group design', *Psychological Bulletin*, *46*, 137–50.

Steffens, L. (1931). *Autobiography of Lincoln Steffens* (New York: Harper & Row).

Stevens, S. S. (1946). 'On the theory of scales of measurement', *Science*, *103*, 677–80.

Strauss, A., and Corbin, J. (1999). 'Grounded theory methodology: An overview', in A. Bryman and R. G. Burgess (eds), *Qualitative Research*; Vol. 3 (Thousand Oaks, CA: Sage), 73–93.

Stubbs, M. (1983). *Discourse Analysis* (Oxford: Blackwell).

Tannen, D. (2007). 'Discourse analysis', Linguistic Society of America, http://www.lsadc.org/info/ling-fields-discourse.cfm

Taylor, S. (2001). 'Locating and conducting discourse analytic research', in M. Wetherell, S. Taylor and S. J. E. Yates (eds), *Discourse as Data: A Guide for Analysis* (London: Sage), 5–48.

ten Have, P. (2007). 'Methodological issues in conversation analysis', http://www2.fmg.uva.nl/emca/mica.htm

Thomson Reuters (n.d.). 'The Thomson Reuters journal selection process', http://isiwebofknowledge.com/benefits/essays/journalselection/

Tomer, C. (1986). 'A statistical assessment of two measures of citation: The impact factor and the immediacy index', *Information Processing and Management*, *22*, 251–8.

Trochim, W. M. K. (2006). 'Positivism and post-positivism', http://www.socialresearchmethods.net/kb/positvsm.htm

van Dijk, T. (2001). 'Principles of critical discourse analysis', in M. Wetherell, S. Taylor and S. J. E. Yates (eds), *Discourse Theory and Practice: A reader* (London: Sage), 300–17.

Velten, E., Jr (1968). 'A laboratory task for induction of mood states', *Behaviour Research and Therapy*, *6*, 473–82.

Watson, J. B., and Rayner, R. (1920). 'Conditioned emotional reactions', *Journal of Experimental Psychology*, *3*, 1–14.

Westermann, R., Spies, K., Stahl, G., and Hesse, F. W. (1996). 'Relative effectiveness and validity of mood induction procedures: A meta-analysis', *European Journal of Social Psychology*, *26*, 557–80.

Wetherell, M. S., and Taylor, S. (2001) (eds). *Discourse as Data: A guide for analysis* (London: Sage).

Wilson, V. L., and Putnam, R. R. (1982). 'A meta-analysis of pretest sensitization effects in experimental design', *American Educational Research Journal*, *19*, 249–58.

Woodworth, R. S. (1938). *Experimental Psychology*, New York: Holt.

Wooffitt, R. (2001). 'Researching psychic practitioners: Conversation analysis', in M. Wetherell, S. Taylor and S. J. E. Yates (eds), *Discourse as Data: A Guide for Analysis* (London: Sage), 49–92.

INDEX

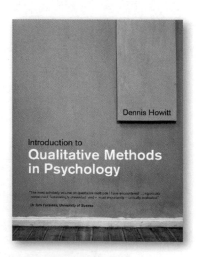